D0780822

LINEAR SYSTEM FUNDAMENTALS

Continuous and Discrete, Classic and Modern

McGraw-Hill Series in Electrical Engineering

Consulting Editor
Stephen W. Director, Carnegie-Mellon University

Networks and Systems
Communications and Information Theory
Control Theory
Electronics and Electronic Circuits
Power and Energy
Electromagnetics
Computer Engineering and Switching Theory
Introductory and Survey
Radio, Television, Radar, and Antennas

Previous Consulting Editors

Ronald M. Bracewell, Colin Cherry, James F. Gibbons, Willis W. Harman, Hubert Heffner, Edward W. Herold, John G. Linvill, Simon Ramo, Ronald A. Rohrer, Anthony E. Siegman, Charles Susskind, Frederick E. Terman, John G. Truxal, Ernst Weber, and John R. Whinnery

Networks and Systems

Consulting Editor
Stephen W. Director, Carnegie-Mellon University

Antoniou: *Digital Filters: Analysis and Design*
Belove and Drossman: *Systems and Circuits for Electrical Engineering Technology*
Belove, Schachter, and Schilling: *Digital and Analog Systems, Circuits, and Devices*
Bracewell: *The Fourier Transform and Its Applications*
Cannon: *Dynamics of Physical Systems*
Chirlian: *Basic Network Theory*
Desoer and Kuh: *Basic Circuit Theory*
Fitzgerald, Higginbotham, and Grabel: *Basic Electrical Engineering*
Hammond and Gehmlich: *Electrical Engineering*
Hayt and Kemmerly: *Engineering Circuit Analysis*
Hilburn and Johnson: *Manual of Active Filter Design*
Jong: *Methods of Discrete Signal and System Analysis*
Liu and Liu: *Linear Systems Analysis*
Papoulis: *The Fourier Integral and Its Application*
Peatman: *Design of Digital Systems*
Reid: *Linear System Fundamentals: Continuous and Discrete, Classic and Modern*
Sage: *Methodology for Large Scale Systems*
Schwartz and Friedland: *Linear Systems*
Temes and LaPatra: *Introduction to Circuit Synthesis*
Truxal: *Introductory System Engineering*

LINEAR SYSTEM FUNDAMENTALS

Continuous and Discrete, Classic and Modern

J. Gary Reid, Ph.D.

Lear Siegler, Inc.
Astronics Division
Adjunct Professor
Wright State University

McGraw-Hill Book Company

New York St. Louis San Francisco Auckland Bogotá Hamburg
Johannesburg London Madrid Mexico Montreal New Delhi
Panama Paris São Paulo Singapore Sydney Tokyo Toronto

This book was set in Times Roman by York Graphic Services, Inc.
The editors were T. Michael Slaughter and J. W. Maisel;
the production supervisor was Leroy A. Young.
The drawings were done by Burmar.
The cover was designed by Miriam Reccio.
Halliday Lithograph Corporation was printer and binder.

LINEAR SYSTEM FUNDAMENTALS
Continuous and Discrete, Classic and Modern

1234567890 HALHAL 898765432

ISBN 0-07-051808-4

Library of Congress Cataloging in Publication Data

Reid, J. Gary.
Linear system fundamentals.

(McGraw-Hill series in electrical engineering.
Networks and systems)
Includes bibliographical references and index.
1. System analysis. I. Title. II. Series.
QA402.R39 1983 621.319′2 82-9977
ISBN 0-07-051808-4 AACR2

FOR
Wendee

CONTENTS

*An asterisk indicates an advanced topic area that may be omitted from the course of study with no loss in continuity to later material.

PREFACE

This text gives a thorough presentation of the foundations of linear time-invariant dynamic systems theory. It goes from classic analysis in the time and frequency domains to the modern state-space techniques, while interweaving both continuous-time analysis and treatment of discrete-time and digital computation methods.

With respect to the organization of the text, Chapters 1 to 5 present classic theory, both continuous and discrete. Chapter 1 begins on an input-output basis, motivated by discussions of an example of a physical system. The meaning of properties such as linearity, time-invariance, causality, and stability are explored on a physical input-output basis. The discrete pulse response model is introduced, and the principle of linear superposition is used to derive the discrete convolution representation of the linear time-invariant system. This concept is extended to derive the continuous impulse response model and the continuous convolution representation of the linear time-invariant system. Ideas of the impulse response model and convolution integral (or discrete pulse response model and convolution summation) serve as unifying threads to the linear systems theory; and the concepts and methods discussed in the later chapters consistently close the loop back to these basics of Chapter 1.

Chapter 2 gives a more mathematical treatment of the linear time-invariant system by considering, in depth, the classic solution of linear time-invariant systems. Input-output stability concepts of Chapter 1 are related to the system poles and the homogeneous response. A systematic technique for determination of the particular solution is given in terms of linear operations on the system-transfer function. This technique for finding the particular solution has considerable system-theoretic properties that are used in later chapters.

Chapter 3 extends treatment of the particular solution concepts of Chapter 2 and the convolution integral concepts of Chapter 1 to "derive" the Fourier transform relation between the system-transfer function and the impulse response model. The chapter further develops basic frequency-response concepts and useful properties of the Fourier transform.

Discussion of the Fourier transform concept is extended to include that of the Laplace transform in Chapter 4. Laplace transform properties allow algebraic simplification of a complex interconnection of linear systems and provide an alternative to the classic solution approach of Chapter 2.

Chapter 5 covers the discrete finite-difference equation model starting from principles introduced in Chapter 1 regarding the discrete pulse response model. A technique is thereby derived to calculate the finite-difference equation model for a given linear time-invariant system. This procedure may be used on an experimental input-output basis, and techniques for doing so are discussed. The chapter then proceeds to run parallel with the classic homogeneous and particular response analysis and the transform analysis—now using the z transform.

Chapters 6 to 12 present the "modern" state-variable analysis and simulation of the linear time-invariant system. Chapter 6 examines the derivation of several alternative state-variable canonical forms from the transfer function; introduces physical variable state-space models; and introduces the derivation of the multi-input, multi-output (MIMO) state-space model.

Chapter 7 derives the continuous-time solution of the state-space model starting from Laplace transform concepts. The chapter is complete in its coverage, relating state-space ideas back to classical concepts, but it does not use extensive matrix theory concepts of eigenvalues and eigenvectors, which are so important to the modern state-space methods.

Chapter 8 is a direct extension of ideas from Chapter 7, but now for the discrete state model analysis. A squaring algorithm is presented for discretizing the continuous-state model at any discrete sample interval. Relations are referred back to the Chapter 5 classic discrete time system analysis. And finally, the notions of discrete system controllability and observability are introduced.

Although Chapters 6, 7, and 8 give an introductory summary of state variable methods, it is presented with a small degree of mathematical sophistication. To progress further into the modern theory, the student must have a greater command of linear vector space concepts (Chapter 9) and matrix theory concepts, including that of the central role of the system eigenvalues and eigenvectors (Chapter 10). Chapter 11 thus furnishes a more extensive discussion of controllability and observability built around the topics presented in Chapters 9 and 10.

Finally, Chapter 12 introduces the linear time-varying state model, the time-varying state transition matrix, numerical integration, and the procedure for linearizing a nonlinear model about a nominal solution path. The concept of nonlinear system equilibrium points and their stability consideration vis-à-vis stability of the linearized model at the equilibrium points are briefly discussed. Although Chapter 12 is the last chapter, Chapters 1 and 7 are the only prerequisite for Chapter 12. The more mathematical treatment of state-space concepts provided by Chapters 9, 10, and 11 need not be covered prior to Chapter 12. Hence, given insufficient time in a given course, the instructor may wish to cover Chapter 12 in lieu of Chapters 9 to 11.

Considerable care is given to discussion of practical digital computation while treating theoretical concepts. Sometimes practical digital algorithms are developed in the text itself: for example, the squaring algorithm for discretizing a linear state variable model as presented in Chapter 8 fits this category. More often, however, it is beyond the scope here to present the details of modern digital methods. Instead, simple techniques are presented for understanding and illustrating concepts in computing simple hand examples, and then references and discussion are given to direct the reader to the more practical digital methods. An example of this is the treatment of the eigenvalue and eigenvector problem in Chapter 10. An easy-to-understand elementary row operation technique is presented for finding the system eigenvectors. This technique is used in hand examples to provide a better understanding of the concept of eigenvectors and, more importantly, to illustrate how and why the eigenvalues and eigenvectors apply to the linear dynamic system. At the conclusion, then, of Chapter 10 there is reference and discussion of the more practical digital methods of solving the eigenvalue/eigenvector problem (i.e., the available software of EISPACK). The same holds true in discussing the concepts of rank, range space, and null space of a matrix. Elementary row operations provide concept understanding in Chapter 10, and Appendix E gives a brief introduction to the use of the singular value decomposition for the practical digital computation of these quantities. Reference is given to readily available software packages which accomplish the modern digital techniques; and Appendix F discusses applicable software packages which make the modern digital techniques available to the nonspecialist.

Prerequisites for the text are relatively modest. Some prior introduction to differential equations and exposure to elementary matrix operations is assumed, but even these skills are developed completely within the text. Some prior experience in deriving differential equation models for simple linear networks (electrical and/or mechanical) is also required background to completely comprehend the examples of Sections 2.1 and 6.4. Such background is probably available to students in engineering or science by the junior level. Yet the timeliness of topics, particularly in regards to state variable techniques and discrete system/digital methods, ensures that it will be a text of interest to higher-level students and to practicing engineers and analysts.

ACKNOWLEDGMENTS

In a project such as this one, a great many people contribute to the final product, and the author owes appreciation to each and every one of them. Perhaps the first to mention should be the many students at both the Air Force Institute of Technology and Wright State University who worked with earlier versions of this material. Without their patience, constructive criticism, and valuable comments this textbook would not have been possible.

Thanks go to Professor Jack D'Azzo for using and evaluating earlier versions of this material in his courses. Thanks also go to Professors Robert Fontana, Dino Houpis, and Peter Maybeck for their continual encouragement and support throughout the project.

A special word of thanks must go to Professor Max Mintz for his thorough review of the final manuscript and his very helpful suggestions. So, too, is the author indebted to Professors Stephen Director and Ronald Schafer, and Dr. Narendra Gupta for their very helpful suggestions on the organization and their comments on the text.

Words cannot express the author's appreciation to Beverly Spalding for her absolutely superb job in typing the manuscript, or to Diane Williams for persevering through an almost endless string of changes.

Thanks go to Dr. Larry Gearhart and Mr. Michael Bise for their expertise and assistance in the preparation of so many of the figures, and to Mr. Robert Schneble for his expert proofreading.

Finally, the author must express his deep gratitude to Wendee, Jennifer, and Gregory for their continuous support and encouragement.

J. Gary Reid

LINEAR SYSTEM FUNDAMENTALS

Continuous and Discrete, Classic and Modern

PART ONE

CLASSIC ANALYSIS

CHAPTER

ONE

INPUT-OUTPUT CHARACTERISTICS OF A LINEAR DYNAMIC SYSTEM: TIME DOMAIN

Linear dynamic-systems theory is an exciting area of study because there are so many real-world applications in which a linear-system mathematical model can be used to represent or model a given physical system, and the linear-system model will very closely *approximate* the actual behavior of the physical system. The diverse mathematical tools underlying linear systems theory may then be used to analyze, simulate, and control the particular system at hand. For instance, analysis can determine the basic time and frequency-response characteristics of a given system. Simulation can provide a prediction of the time-evolution characteristics of the system for a selected input and an initial energy condition. Then, if the time and frequency-response characteristics of the system prove to be unsatisfactory, feedback-control-theory techniques can be applied to design a compensator system that will make the system respond more favorably.

It is gratifying when mathematics can provide such analysis and design capabilities, and the mathematics of linear dynamic-systems theory can be applied to practical physical systems in a great many instances. However, the word *approximate* is emphasized above because, as we will discuss in more detail, physical processes are never strictly linear. *System linearity* has a precise mathematical definition, which is presented in the first section of this chapter. No actual physical system could ever exactly conform to the requirements of this definition.

There are very few physical systems, however, that do not meet the requirements of system linearity if the operating region is sufficiently restricted. Conversely, it may be stated that almost all physical systems may be described by linear dynamic-system models if the region of applicability is sufficiently restricted. For example, common electrical elements such as resistors, capacitors, and inductors, as well as mechanical elements such as springs, masses, and dampers, all obey linear relationships until they are forced outside of their normal range of operation. A mechanical spring is a good illustration of this: by Hooke's law, the displacement of the spring is linearly proportional to the force applied; this is true until the spring is either compressed to its limit or stretched so far that the coils unwind. At either extreme, the displacement-force relationship becomes nonlinear and the normal linear relation no longer remains valid.

This chapter explores the requirements of system linearity and some useful consequences that result when a system is linear. This discussion is made on an input-output physical basis, with the hope that the reader will gain an understanding of the physical motivation for linearity and its importance by this discussion. This is a foundation for the more mathematical presentations in later chapters.

1-1 BASIC PROPERTIES OF A LINEAR DYNAMIC-SYSTEM TIME RESPONSE

This section attempts to create a mental image of a *dynamic system*. The particular system chosen for illustration is a hypothetical rocket sled as shown in Fig. 1-1. The forcing or input function is the rocket motor. This function is given the symbol $u(t)$, which throughout the text will be used to denote system input. The output, or measured, variable is the velocity of the *payload mass,* m_2; throughout the text the output variable will be given the symbol $y(t)$. Since the sled in Fig. 1-1 is constrained to travel a perfectly straight and horizontal track, the force input and the velocity output are both scalar variables with positive orientation to the right. Hence this is a single-input, single-output (SISO) system. If the sled were not constrained to linear motion, then the vector force input and the vector velocity output would represent a three-input, three-output, or multi-input, multi-output (MIMO) system. Initially attention is limited to the SISO system, but later chapters deal with the more complicated MIMO situation.

In addition to the input variables and the physically measured output variables, the dynamic system generally has additional internal variables and elements that are *not* directly observed (if they were directly measured or observed they would be, by definition, additional output variables). Internal elements of the

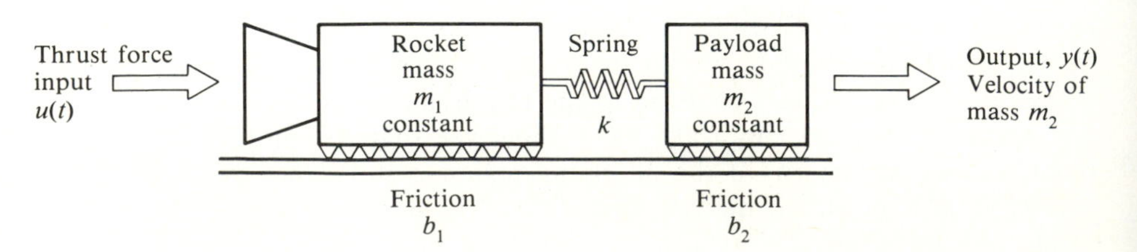

Figure 1-1 Rocket-sled system.

rocket-sled system are masses m_1 and m_2, the spring k, and the friction elements b_1 and b_2. For simplicity, it is assumed that each of the elements is constant. This is not entirely realistic in this particular case, because the rocket mass m_1 would not remain constant as the rocket consumes the propellant. Such a simplifying assumption yields a *linear, time-invariant dynamic system*—a term that will be given precise meaning shortly. If the mass m_1 were dependent on the rocket thrust, then the rocket would be a nonlinear and time-varying system. Why this is so will also be examined later.

The internal elements of a dynamic system interact according to certain physical laws. For the simple mechanical system illustrated here the appropriate laws of interaction are Newton's laws. The interaction of these internal elements produces a given output function of time, $y(t)$ (in this case the velocity of mass m_2 as a function of time), for a given input function of time, $u(t)$ (the thrust curve of the rocket motor).

These internal elements and the physical laws that relate these elements can be analyzed and used to derive a differential-equation *internal* model of the system. Such an internal model is examined in the next chapter and in most of the remainder of the text; however, this chapter considers only those aspects of a dynamic system which are physically and directly measured—the system inputs and outputs. Therefore, no differential-equation model of the rocket-sled dynamic system in our example will be attempted at this time.

What are defined as the system input and output variables are, of course, system- and problem-dependent. The only reason for picking the rocket sled is for the relative ease in visualization that it provides. Other examples of inputs and outputs are:

System	Possible input variables	Possible output variables
Electric	Battery or voltage source; current source	Voltage across a resistor or a capacitor; current through an inductor or a resistor
Mechanical	External forces or torques; external velocity field	Velocity or position of masses; force through springs; rotational rate or angle of spinning masses
Fluid	Reservoirs or external pressure	Internal pressures; flow rates
Thermal	External heat source	Temperatures
Biological	Environmental factors; food supplies and intake	Physical activities and function; population size
Socioeconomic	Price controls, interest rates, wages, etc.	Cash flow, capital goods, gross national product, etc.

The key requirement for the input variable is that it be an externally generated *forcing function* on the system. There are often *internal forces* created when the elements of a system interact, but these are not considered *inputs* because they are not created external to the defined system. For example, when m_1 moves in response to the rocket-thrust input, it compresses the spring k, which in turn transmits a force to mass m_2. As m_1 and m_2 both move, they create internal friction forces as they slide down the track. But only the rocket-thrust force is considered to be the input function, because it alone is *external* to the rocket-sled and track system.

Also, in this text it is always assumed that the input function, $u(t)$, is known. In contrast, inputs do not always have to be known in a random or probabilistic system. Quite frequently random "noise" inputs or disturbances can corrupt an otherwise known, or *deterministic,* input. For instance, random forces might be generated as the sled moves down the track, and the actual force of the rocket motor might differ somewhat from that which might be modeled or predicted. Although the linear system theory to be developed in this text can quite readily handle such random inputs, a purely mathematical treatment of these random inputs would have to be based on statistics or on probability theory. Such topics are beyond the scope of this text.

The key requirement on the output variable, $y(t)$, is that it be *physically measured* as a function of time. Quite frequently the output observations are only made and/or recorded at discrete time points. Such a situation would result in a *sample-data,* or simply *sampled,* output system. This is depicted in Fig. 1-2 for a system that is sampled every T units of time.

With this rocket-sled system in mind, the time characteristics of the system output—the velocity of mass m_2—are examined for a given rocket-thrust input, $u(t)$. It is assumed that the various internal parameters—m_1, m_2, k, b_1, and b_2—are predetermined; but from an input-output viewpoint it is really not necessary to know what these specific values are. For an internal, differential-equation model, these specific values would have to be known; however, for an external, input-output viewpoint, one only needs to determine, experimentally if you will, a relationship between the output function of time for a given input function of time, and a set of given initial conditions, or the *state.*

A new term has been introduced, that of *state.* Clearly, if the system were initially at rest—at the zero initial state—then the velocity of mass m_2 would remain zero until the input rocket motor were ignited. The system response to the

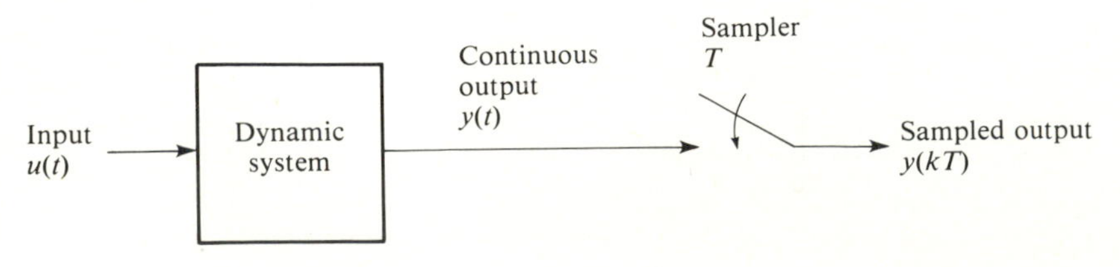

Figure 1-2 Sampled output response.

input, starting from the zero initial state of rest, is termed the system *zero-state response* and is given the symbol $y_{zs}(t)$.

The zero-state response, $y_{zs}(t)$, for the input thrust depicted in Fig. 1-3 is shown in Fig. 1-4. Notice that the output velocity remains zero until time $t = t_0 = 0$, when the input begins taking a nonzero value. The symbol t_0 is used to denote the initial time, unless, as is frequently the case, $t_0 = 0$.

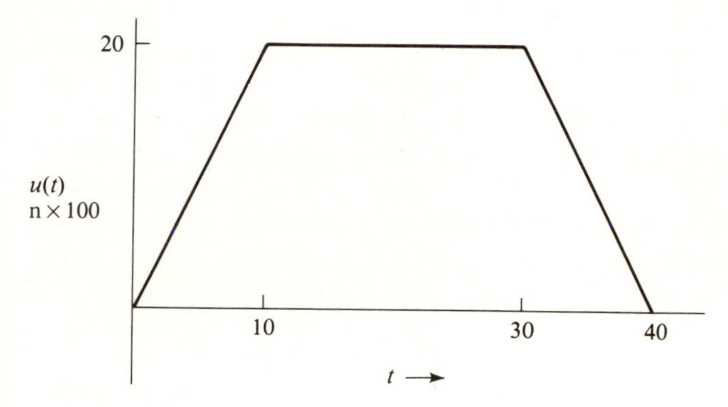

Figure 1-3 Thrust input curve of the rocket-sled system.

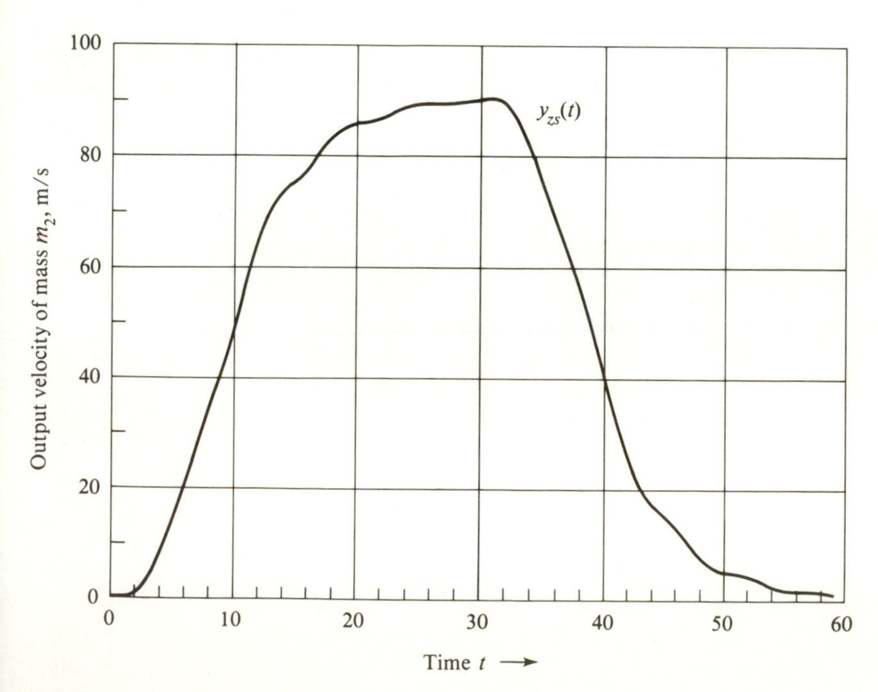

Figure 1-4 Rocket-sled zero-state response to input, Fig. 1-3; zero initial state, $t_0 = 0$.

Causality

Because the system response does not begin until the thrust itself begins at time t_0, this system is said to be *causal*. All systems dealt with in this text are causal—the present output is not influenced by future inputs. Systems *not* having this property are said to be *anticipatory* and are not discussed in this text.

The Zero-Input Response

The zero state is a fairly simple concept, because it merely corresponds to an initial rest condition. Instead, suppose that the system were not initially at rest, but that it had some initial stored energy so that the initial state is not zero. For example, masses m_1 and m_2 might both have some initial velocity at the initial time t_0. Additionally, the spring k might be in an initial condition of compression, and hence it would contribute to the motion of masses m_1 and m_2 when released. If one considers the system output due only to the initial energy or state of the system, with zero input, one obtains what is defined as the *zero-input response* of the system, which is given the symbol $y_{zi}(t)$.

The assumed initial state for the rocket-sled system is

$$x(t_0) = \begin{bmatrix} x_1(t_0) \\ x_2(t_0) \\ x_3(t_0) \end{bmatrix} = \begin{bmatrix} 10 \\ 50 \\ 1 \end{bmatrix} = \begin{bmatrix} \text{Initial velocity } m_2\text{, m/s} \\ \text{Initial velocity } m_1\text{, m/s} \\ \text{Initial compression of spring } k, \end{bmatrix} \quad (1\text{-}1)$$

The units are meters per second for velocities and meters for displacement. The state vector is always given the symbol x. This initial state or energy level is arbitrary, and the system zero-input response depends on the particular value of the initial state chosen.

The particular values of the initial state also depend on the coordinate system selected for state. For example, if a different set of units were picked—say, kilometers per second for velocity and centimeters for position—then the initial state vector would be

$$x'(t_0) = \begin{bmatrix} 0.01 \\ 0.5 \\ 100 \end{bmatrix} = \begin{bmatrix} \text{Initial velocity } m_2\text{, km/s} \\ \text{Initial velocity } m_1\text{, km/s} \\ \text{Initial compression of spring } k\text{, cm} \end{bmatrix} \quad (1\text{-}2)$$

The $x(t_0)$ and $x'(t_0)$ represent exactly the same initial energy level, and they produce exactly the same system zero-input response. The only difference is a different *state coordinate system*. As will be shown in later chapters, the state coordinate system is really quite arbitrary. Whatever state coordinates one picks for a system's internal differential-equation model, the basic, physical, input-output characteristics of the system do not change.

The zero-input response for the initial state value of Eq. (1-1) [or Eq. (1-2), since they are equivalent energy levels] is shown in Fig. 1-5. The response is what

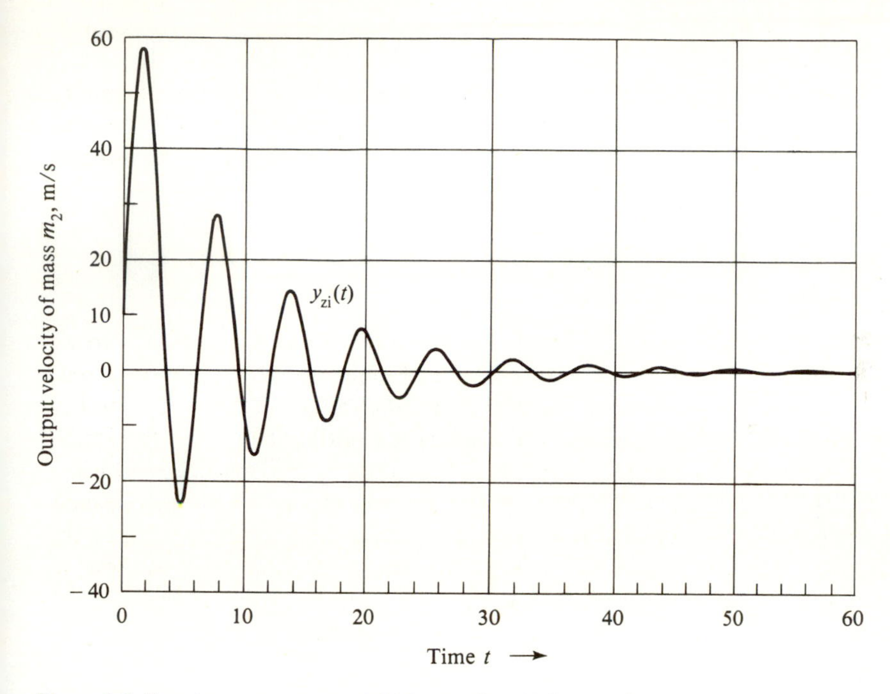

Figure 1-5 Zero input response to initial state, Eq. (1-1), $t_0 = 0$.

one might expect. The output, $y(t) \equiv$ velocity of m_2, starts at $y(t_0) = 10$ m/s. Then it increases because of the initial compression of the spring and the increased velocity of m_1. The output velocity reaches a peak of 58 m/s at $t = t_0 +$ 2 s and then decays to zero as frictions b_1 and b_2 gradually absorb all the energy. This brings the system to rest, or to the zero state. This occurs at approximately time $t_0 + 50$ s.

The Complete Response

Now suppose that the system is at the initial state $x(t_0)$, as in Eq. (1-1), and that the input $u(t)$ shown in Fig. 1-3 is then applied. The complete system response, $y(t)$, to both the initial state and to the input is shown in Fig. 1-6.

The question now arises—do the zero-input response, $y_{zi}(t)$, and the zero-state response, $y_{zs}(t)$, tell us anything about what should be expected from the total response $y(t)$? If the system is *linear*, then it is always true that

$$y(t) = y_{zi}(t) + y_{zs}(t) \tag{1-3}$$

The rocket-sled system depicted here does happen to be linear, and so Eq. (1-3) is true, as is illustrated in Fig. 1-7. A nonlinear system would not necessarily meet this requirement.

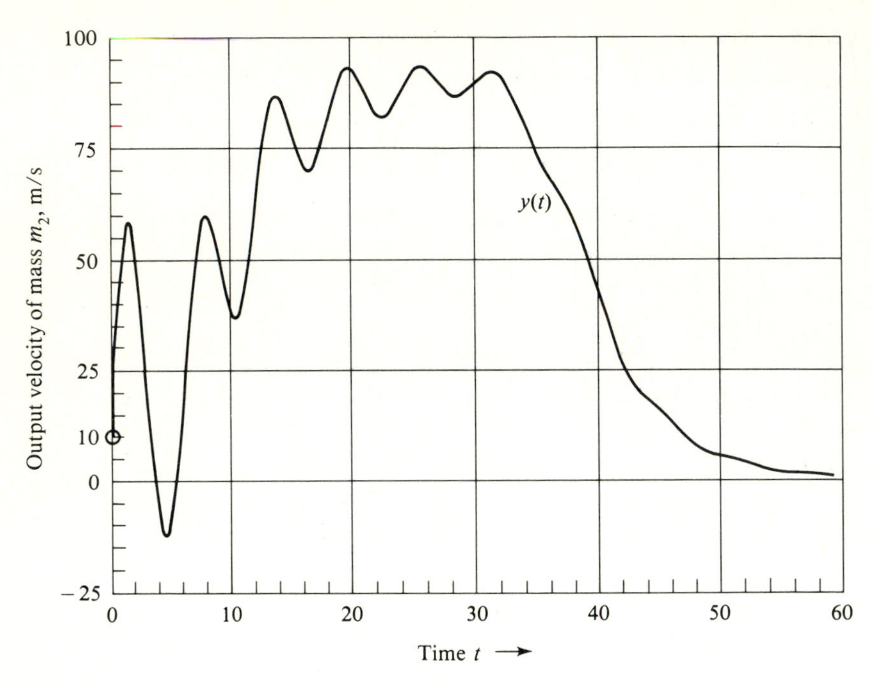

Figure 1-6 Complete response to input, Fig. 1-3, and initial state, Eq. (1-1).

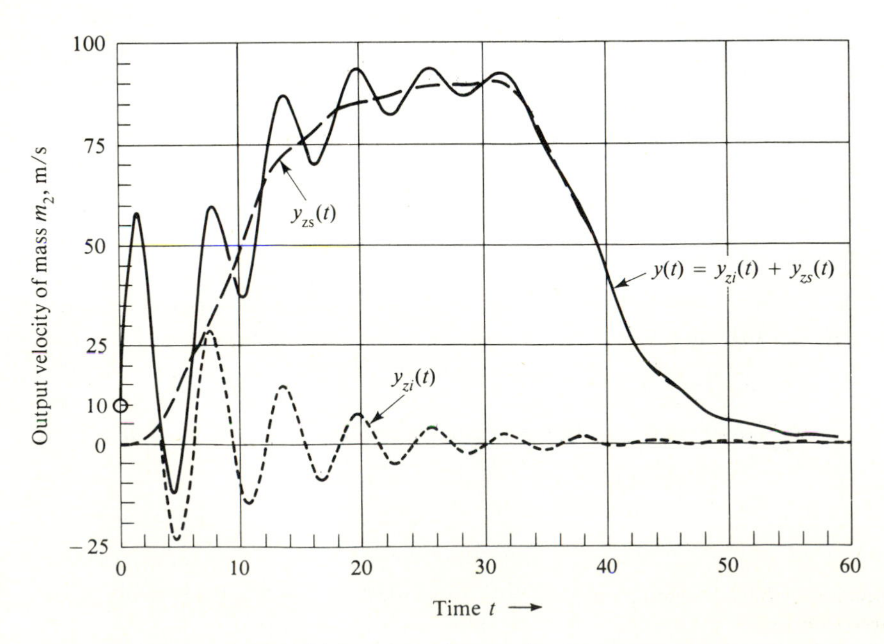

Figure 1-7 Illustration of additivity of zero-input and zero-state response.

The Three Requirements for System Linearity

The additivity of the zero-input and zero-state response relations is only the first of three separate requirements that must be met before any given system may be called linear. A system that does not satisfy *all three* separate requirements is by definition nonlinear.

For example, if a hypothetical system had the input-output-initial-state relation

$$y(t) = \sin(t)x^2(t_0) + \int_{t_0}^{t} e^{-2(t-\tau)}u^2(\tau)\,d\tau \tag{1-4}$$

it would have the additive decomposition of the zero-input and zero-state response, since

$$y_{zi}(t) = \sin(t)x^2(t_0) \tag{1-5}$$

$$y_{zs}(t) = \int_{t_0}^{t} e^{-2(t-\tau)}u^2(\tau)\,d\tau \tag{1-6}$$

$$y(t) = y_{zi}(t) + y_{zs}(t) \tag{1-7}$$

However, the system would still be nonlinear because it does not satisfy either the second or third requirements for a linear system: linearity with respect to the initial state and linearity with respect to the input.

Again a new term has been introduced—*linearity*. Linearity of a given function, or a *linear function,* is defined as:

Definition The function $f(x)$ is a linear function of the independent variable x if and only if it satisfies two properties:
(*a*) Additivity: $f(x_1 + x_2) = f(x_1) + f(x_2)$ *for all* x_1 and x_2 in the domain of f.
(*b*) Homegeneity: $f(\alpha x) = \alpha f(x)$ *for all* x in the domain of f and all scalars* α.

Considering the input-output-initial-state relation Eq. (1-4), it is seen that Eq. (1-4) does not satisfy linearity with respect to the initial state, since

$$\sin(t)[x_1(t_0) + x_2(t_0)]^2 = \sin(t)[x_1^2(t_0) + 2x_1(t_0)x_2(t_0) + x_2^2(t_0)]$$
$$\neq \sin(t)x_1^2(t_0) + \sin(t)\cdot x_2^2(t_0)$$

or

$$\sin(t)[\alpha x(t_0)]^2 = \sin(t)\cdot\alpha^2x^2(t_0)$$
$$\neq \alpha\sin(t)x^2(t_0)$$

*Depending on the *field* of scalars over which the function $f(\cdot)$ is defined, the scalar α may range over all real-valued scalars or over all complex-valued scalars. In this text, linearity is always derived over the field of real-valued scalars.

Thus $y_{zi}(t)$ does not satisfy either the additivity or homogeneity conditions with respect to the initial state $x(t_0)$. To be a linear function of the initial state, $y_{zi}(t)$ must satisfy both properties.

Notice that sin (t) does not enter into the consideration of linearity of $y_{zi}(t)$ with respect to the initial state. The sin (t) factor is merely a time-varying multiplier, and it does not influence linearity with respect to $x(t_0)$. Indeed,

$$y_{zi}(t) = \sin(t)x(t_0) \tag{1-8}$$

would be linear with respect to the initial state.

In a similar fashion, Eq. (1-4) does not satisfy linearity with respect to the input $u(t)$ since

$$\int_{t_0}^{t} e^{-2(t-\tau)}[u_1(\tau) + u_2(\tau)]^2 \, d\tau = \int_{t_0}^{t} e^{-2(t-\tau)}[u_1^2(\tau) + 2u_1(\tau)u_2(\tau) + u_2^2(\tau)] \, d\tau$$

$$\neq \int_{t_0}^{t} e^{-2(t-\tau)}u_1^2(\tau) \, d\tau + \int_{t_0}^{t} e^{-2(t-\tau)}u_2^2(\tau) \, d\tau$$

or

$$\int_{t_0}^{t} e^{-2(t-\tau)}[\alpha u(\tau)]^2 \, d\tau = \int_{t_0}^{t} e^{-2(t-\tau)}\alpha^2 u^2(\tau) \, d\tau$$

$$\neq \alpha \int_{t_0}^{t} e^{-2(t-\tau)}u^2(\tau) \, d\tau$$

Thus the zero-state response satisfies neither additivity nor homogeneity with respect to the input function. To be linear, it must satisfy both.

Again, neither the time-varying factor $e^{-2(t-t_0)}$ nor the integral influence the linearity of Eq. (1-4) with respect to the input function. Indeed, the zero-state response

$$y_{zs}(t) = \int_{t_0}^{t} e^{-2(t-\tau)}u(\tau) \, d\tau \tag{1-9}$$

would be linear, since it is linear in the input function $u(t)$. This may be checked by observing that

$$\int_{t_0}^{t} e^{-2(t-\tau)}[u_1(\tau) + u_2(\tau)] \, d\tau = \int_{t_0}^{t} e^{-2(t-\tau)}u_1(\tau) \, d\tau + \int_{t_0}^{t} e^{-2(t-\tau)}u_2(\tau) \, d\tau$$

and

$$\int_{t_0}^{t} e^{-2(t-\tau)}[\alpha u(\tau)] \, d\tau = \alpha \int_{t_0}^{t} e^{-2(t-\tau)}u(\tau) \, d\tau$$

If one is given a mathematical input-output-initial-state relation such as Eq. (1-4), it may be checked, mathematically, whether or not that relation satisfies the three conditions for a *linear system:*

1. Additivity of the zero-input and zero-state response, $y(t) = y_{zi}(t) + y_{zs}(t)$
2. Linearity with respect to the initial state
3. Linearity with respect to the input function

But what if one has only a physical system to work with, and no mathematical model? One may then simply check these three conditions *experimentally*.

Figure 1-7 illustrates that the rocket sled satisfies the condition of zero-input and zero-state additivity. Figure 1-8 illustrates the linearity of the zero-input response with respect to additivity of the initial-state vector. Figure 1-9 illustrates the linearity of the zero-state response with respect to homogeneity of the rocket-force input.

From these experiments it cannot yet be concluded for sure that the rocket sled is a linear system. If one goes back to analyze the strict mathematical requirement for system linearity, one sees that the three requirements must be true *for all* initial states, system inputs, scalars α, etc. Thus an infinite number of experiments would have to be performed, and all physical systems would fail the test for linearity under an infinite range of operation. Experimental verification of linearity is therefore not really practical. Fortunately, nearly all physical dynamic systems will behave as linear dynamic systems if the operating region is restricted suitably; therefore, our main concern is to understand what the mathematical consequences of system linearity happen to be. These consequences are the focus of this text, and their consideration begins in the next section.

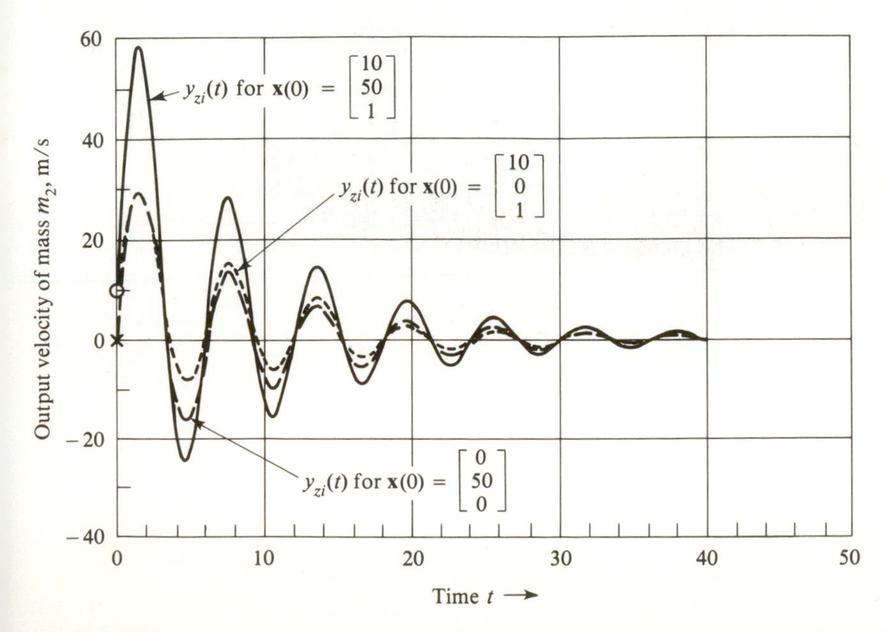

Figure 1-8 Illustration of additivity of zero-input response for linear rocket-sled system.

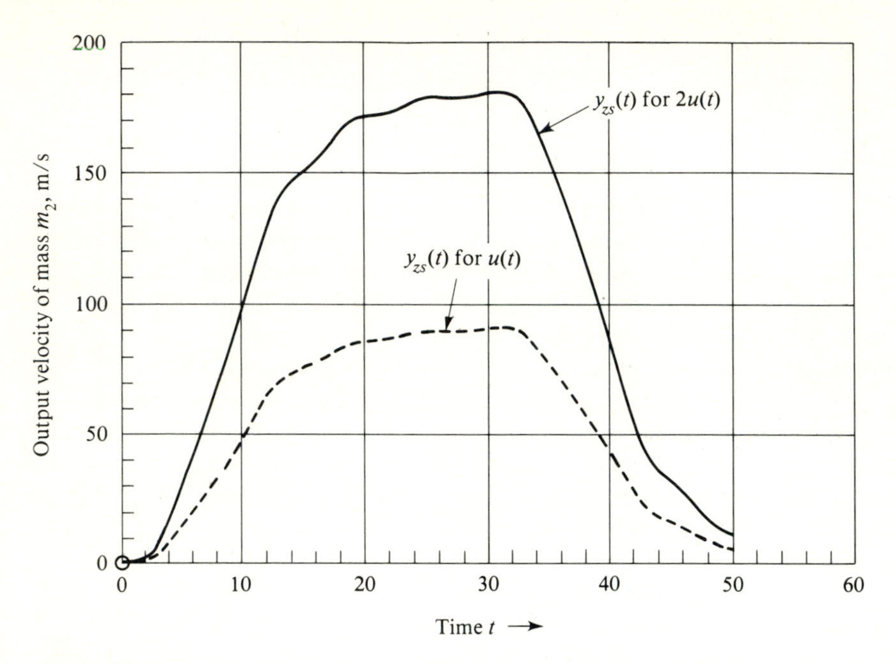

Figure 1-9 Illustration of linear homogeneity of rocket-sled zero-state response for $u(t)$ = Fig. 1-3 and $2u(t)$.

1-2 THE PRINCIPLE OF LINEAR SUPERPOSITION

It is important to note that a linear system response can be separated by Eq. (1-3),

$$y(t) = y_{\text{zi}}(t) + y_{\text{zs}}(t) \tag{1-3}$$

where $y_{\text{zi}}(t)$ is the response to only the initial stored energy and $y_{\text{zs}}(t)$ is the response to only the input if the initial energy were zero. This separation will be the basis of many later discussions. Likewise, the fact that $y_{\text{zi}}(t)$ is a linear function of the initial-state vector is also important, but its significance cannot really be dealt with mathematically until after Chap. 7, in which state-variable solution methods are first introduced. Therefore, the discussion in this section focuses on the linearity of the zero-state response and on the consequence of this key fact.

The significance of zero-state linearity lies in the fundamental truism for a linear system: If a linear system starts from rest, and if the subsequent input $u(t)$ may be expressed as the linear combination of other known inputs, $u_i(t)$, as

$$u(t) = \sum_{i=1}^{N} \alpha_i u_i(t) \qquad t \geq t_0 \tag{1-10}$$

then the response to $u(t)$ satisfies

$$y_{zs}(t) = \sum_{i=1}^{N} \alpha_i y_{zs_i}(t) \tag{1-11}$$

where $y_{zs_i}(t)$, $i = 1, 2, \ldots, N$, are the zero-state response to each of the $u_i(t)$ inputs separately. This fact is known as the *principle of linear superposition.* It is redundant to say that the system starts from the zero initial state and to also use the notation $y_{zs}(t)$, but it is important to emphasize that the principle of linear superposition applies only to the zero-state portion of the response.

Figure 1-10 illustrates this principle of linear superposition for the linear rocket sled described in the previous section. The system is assumed to start initially from rest (from the zero initial state). The $u_i(t)$, $i = 1, 2, 3, 4$, are a series of constant pulse inputs each held for a duration $T = 10$ s. This input is a coarse approximation to the continuous input of Fig. 1-3. If we halve the pulse width to $T = 5$ s, we get a better approximation to the continuous-curve input of Fig. 1-3. The principle of linear superposition still applies to finding the output for the sequence of pulse inputs. This is illustrated in Fig. 1-11. If one decreases the pulse width T to zero, one obtains in the limit the continuous-curve input of Fig. 1-3 with the response of Fig. 1-4. Figures 1-12 and 1-13 compare the rocket-sled response to the continuous-curve input, Fig. 1-3, with the four-pulse approximation and the eight-pulse approximation, respectively.

The principle of superposition applies in the limit for the continuous-curve input, and one thereby obtains the fundamental superposition integral relationship for the zero-state response of a linear dynamic system. But before going into this integral representation, it is instructive to systematize the discussion here with the *discrete-pulse* inputs as illustrated in Fig. 1-10 and 1-11. Indeed, this discussion not only provides insights to the limiting continuous-time case, but it is important in its own right. With the widespread use of the digital computer in either the simulation of mathematical dynamic-system models or actual on-line control, the *digital-pulse* input is fundamental. A great deal of this text explicitly discusses how such digital-pulse inputs may be handled mathematically. It is only fitting that the discussion begin right here on a physical input-output basis.

Time Invariance

To considerably simplify the notation and discussion, it is helpful to assume that the dynamic system dealt with is both linear and time-invariant. When a system has the properties of being both linear and time-invariant, it is an especially nice situation and some powerful analysis methods then accrue.

But what is meant by the system property of *time invariance?* If one is given a differential-equation model of a dynamic system, it is fairly easy to determine whether that mathematical model is time-varying or time-invariant: if any of the coefficients in the differential-equation model are time-varying (explicit functions of time), then the system mathematical model is time-varying. If all of the coeffi-

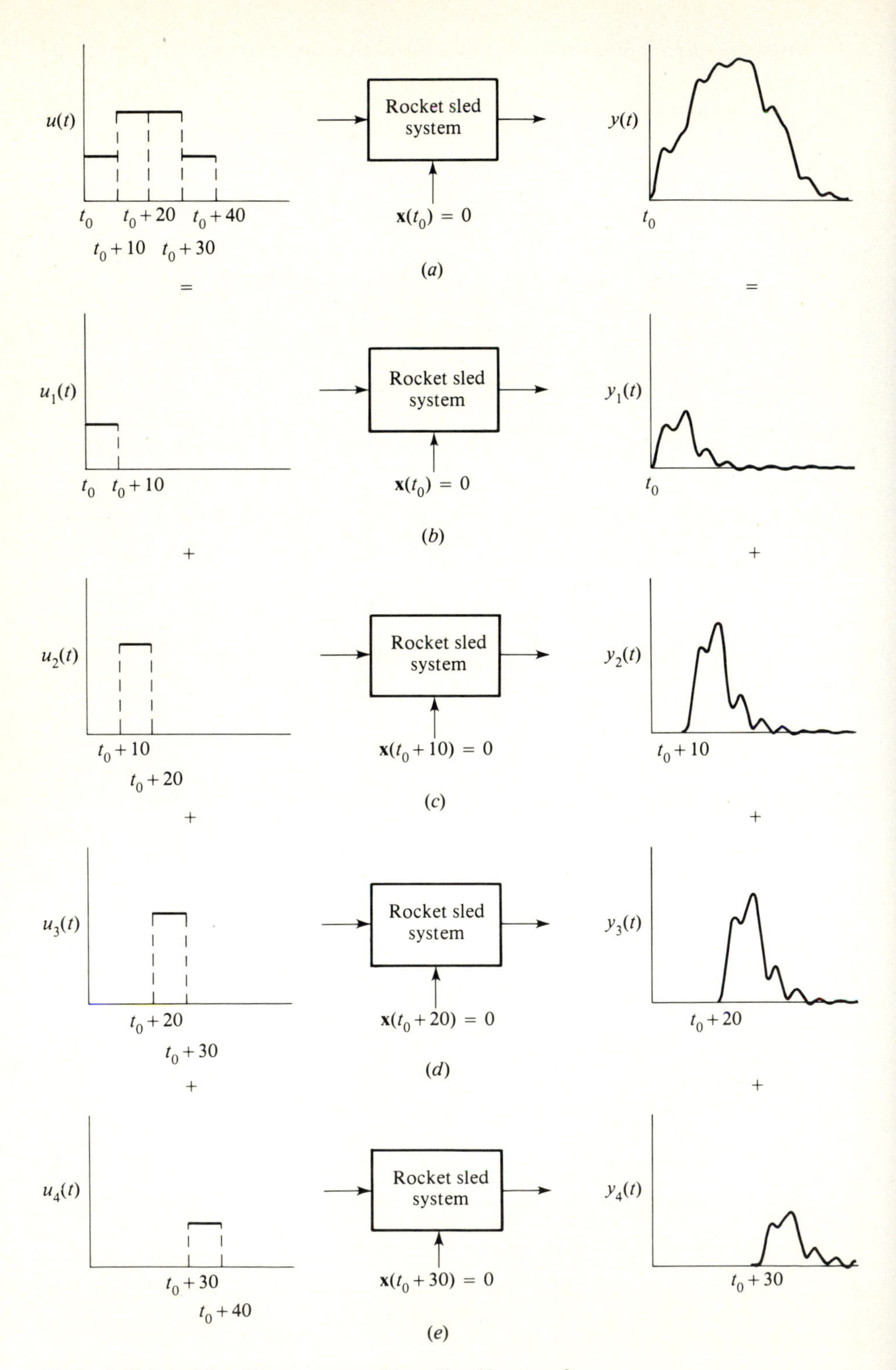

Figure 1-10 Principle of linear superposition; $T = 10$ s, $t_0 = 0$.

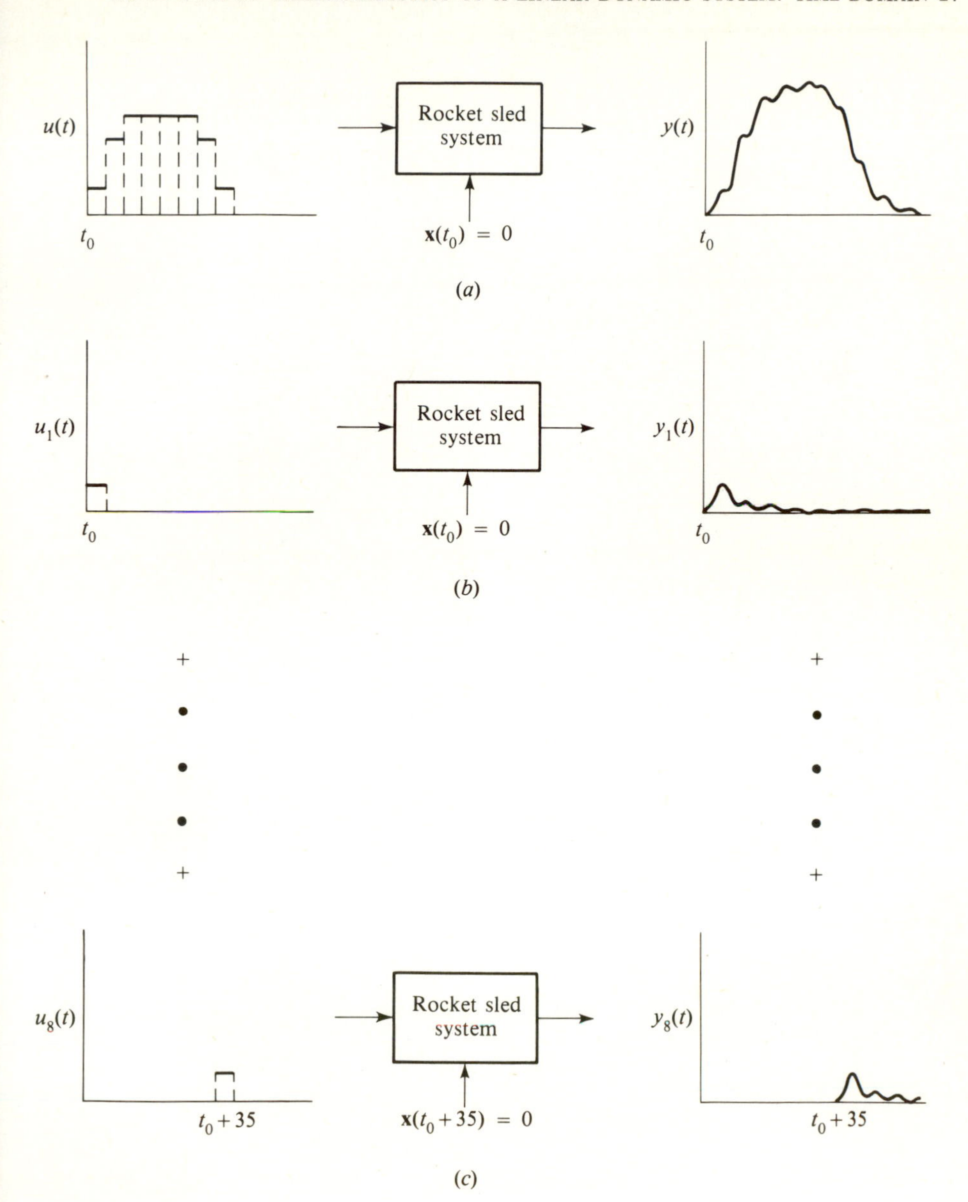

Figure 1-11 Principle of linear superposition; $T = 5$ s, $t_0 = 0$.

cients are constant, then it is a time-invariant model. For example, the thrust-velocity relation of mass m_1 in the rocket sled is given by*

$$m_1 \frac{d}{dt} v_1(t) = u(t) \tag{1-12}$$

*This is merely an application of Newton's law $f(t) = \frac{d}{dt} p(t)$, where the force is $u(t)$ and the momentum $p(t)$ is $m_1(t)v_1(t)$.

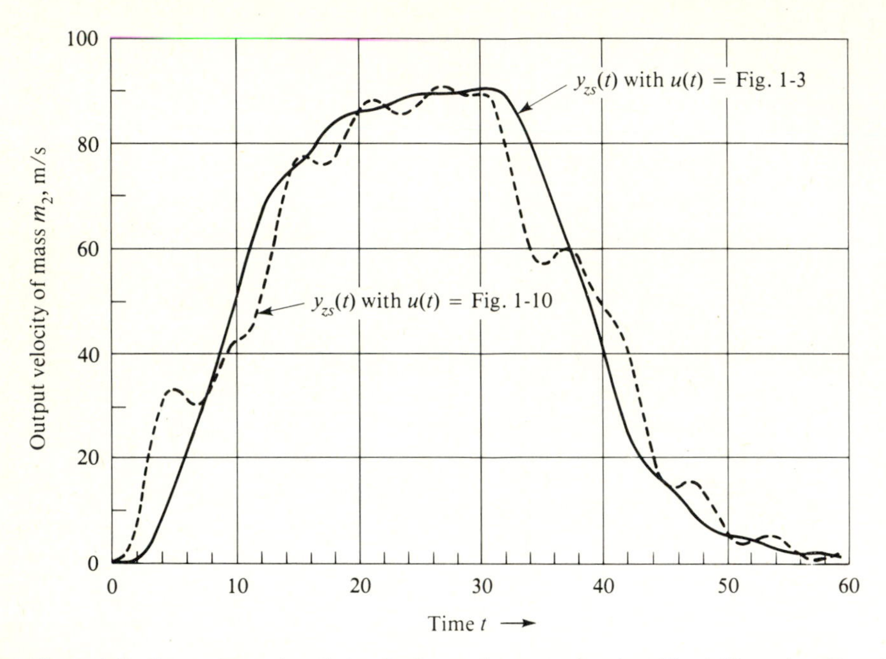

Figure 1-12 Comparison of $y_{zs}(t)$ for the four-pulse approximation, Fig. 1-10, to the $u(t)$ = Fig. 1-3.

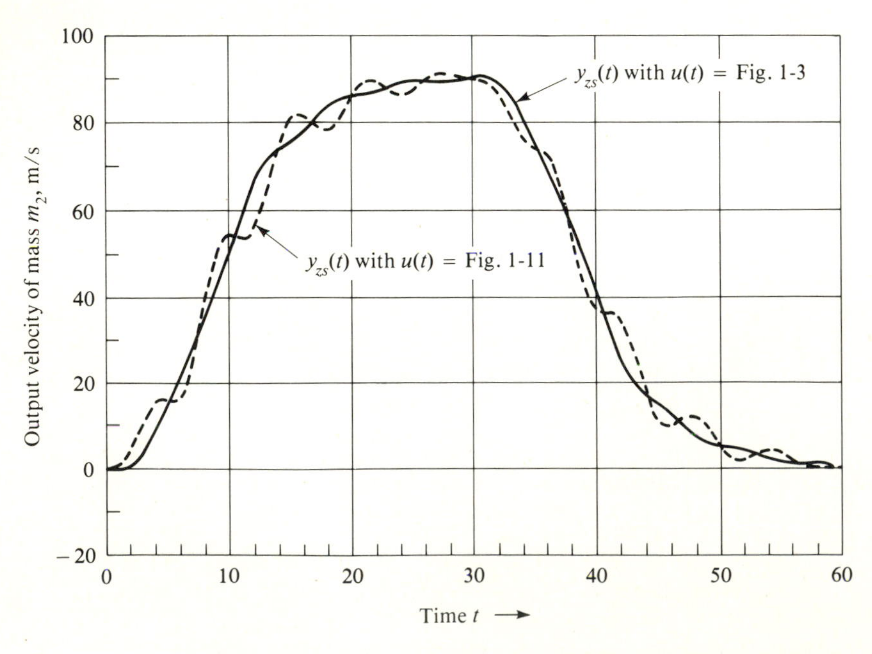

Figure 1-13 Comparison of $y_{zs}(t)$ for the eight-pulse approximation, Fig. 1-11, to the $u(t)$ = Fig. 1-3.

where m_1 is constant. Since m_1 is assumed constant, this is a time-invariant model. However, in a more realistic model the mass of m_1 would be changing as propellant is consumed, and so the differential-equation internal relation would be

$$m_1(t)\frac{d}{dt}v_1(t) + v_1(t)\frac{d}{dt}m_1(t) = u(t) \tag{1-13}$$

This is a time-varying model, in which the coefficients in the internal relation are now functions of time.

If an input-output-initial-state mathematical relation is given (the "solution" of the differential-equation model), the determination of time invariance is not quite so transparent. The output response for a time-invariant system may always be written as a function of the time difference $(t - t_0)$ (the difference between the current time, t, and the initial time, t_0). For example, the zero-input response relation

$$y_{zi}(t) = \sin(t)x^2(t_0) \tag{1-14}$$

is time-varying because of the time-varying $\sin(t)$ factor. However, the zero-input response relation

$$y_{zi}(t) = \sin(t - t_0)x^2(t_0)$$

would represent a time-invariant response because the $\sin(t - t_0)$ factor is dependent only on the time difference between t and t_0; the response is dependent on the time difference between the current time, t, and the initial time, t_0, and not on the absolute t or t_0.

Based solely on experimental input-output observations, the property of time invariance is difficult to determine. One usually must know something about the internal characteristics or properties of a system to determine that the input-output-initial-state relations are not changing based on the absolute initial time, t_0, or the absolute current time, t. Like linearity, time invariance must be demonstrated for *all* initial times, t_0, for *all* initial states, $x(t_0)$, and for *all* inputs, $u(t)$. All physical systems would be subject to some aging and changing of system characteristics over an infinite interval of time. But, again like linearity, nearly all physical systems may be quite nearly approximated by time-invariant models if the interval of time is sufficiently restricted. Linearity and time invariance are distinct system properties (a nonlinear system may be time-invariant and a linear system may be time-varying); but if a system possesses both linearity and time invariance, then some especially powerful analysis methods are available.

The Discrete Pulse-Response and Convolution Summation

If a system is linear and time-invariant, then its response to a unit pulse input of duration T is always the same shape but shifted in time regardless of the absolute initial time of pulse application. This is not true of a time-varying linear system. This is illustrated in Fig. 1-14. In the time-varying system, the pulse response is dependent on the absolute time of application of the pulse.

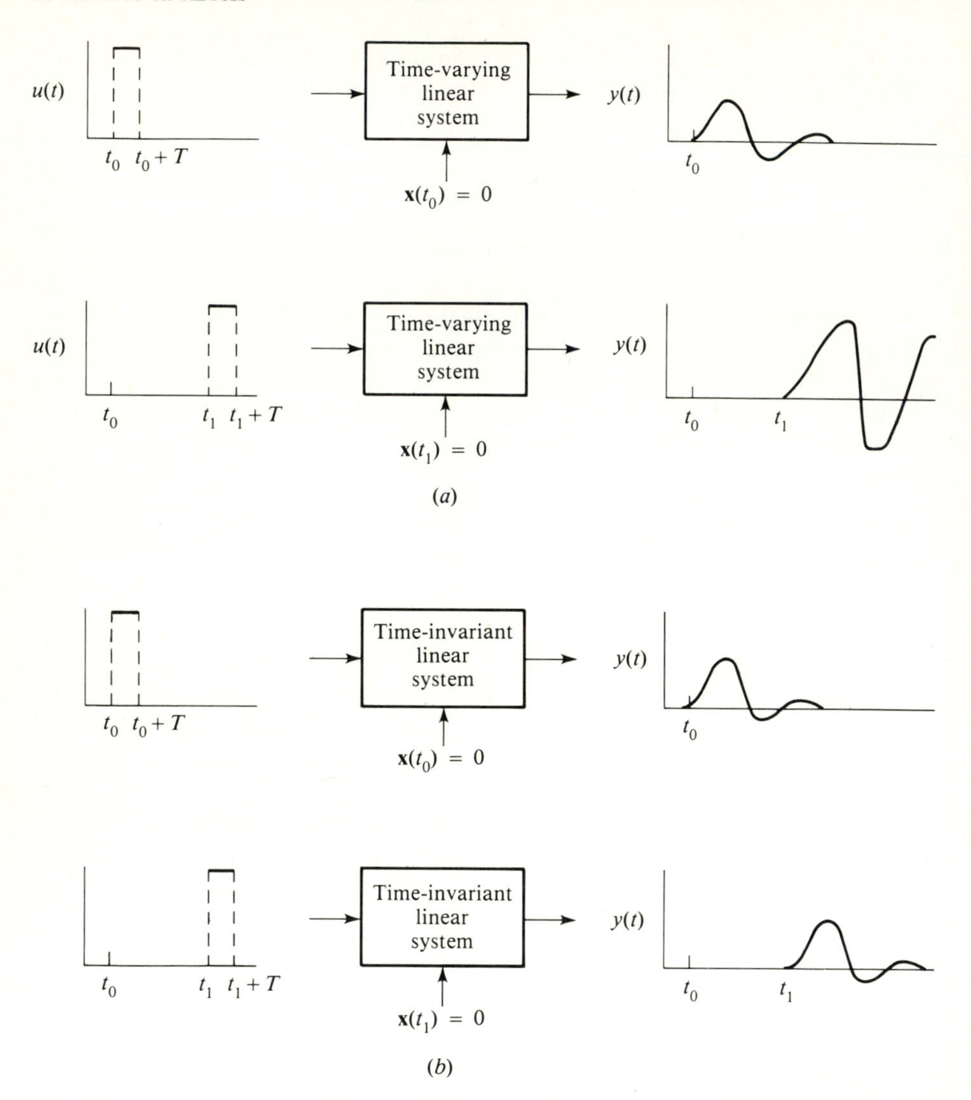

Figure 1-14 (*a*) Pulse response, time-varying linear system. (*b*) Pulse response, time-invariant linear system.

Apply the unit-amplitude pulse of duration T to a linear time-invariant system (the rocket-sled dynamics are used once again) and sample the output at every T units of time. Since the system is time-invariant, the initial time, t_0, is arbitrarily set to zero. Denote the sampled output to the unit-amplitude pulse input as

$$h_T(k) \equiv y_{zs}(kT) \qquad k = 1, 2, 3, \ldots \tag{1-15}$$

The sequence $h_T(k)$ is termed the *discrete pulse response* of the system. This is depicted in Fig. 1-15. Choosing $T = 1$ s, a unit thrust force of 100 N, and output

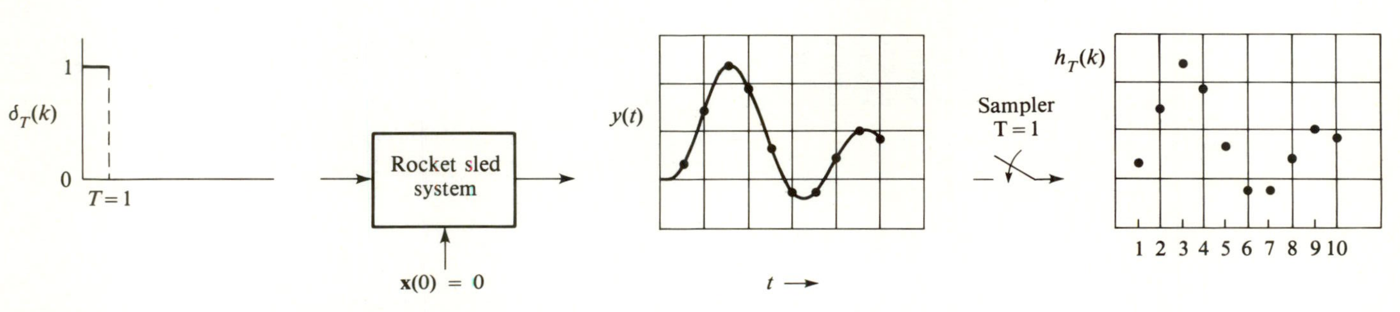

Figure 1-15 Generation of the discrete pulse response, $h_T(k)$, $k = 1, 2, 3, \ldots$.

units of m/s, the numerical values of $h_T(k)$ for the rocket-sled system are shown in Table 1-1.

Notice that the units chosen for the input and output are arbitrary so long as one keeps things consistent and two other practical requirements are met. First, the unit-amplitude pulse input over a time duration T must be large enough to produce a measurable response. For instance, a thrust input of just 1 N for 1 s

Table 1-1 Discrete pulse-response model, rocket-sled system

$h_T(k)$		$h_T(k)$	
(1) =	1.4287E − 01	(41) =	9.0919E − 04
(2) =	7.3603E − 01	(42) =	−1.1369E − 02
(3) =	1.1658E + 00	(43) =	−1.0882E − 02
(4) =	9.6007E − 01	(44) =	−4.2399E − 04
(5) =	3.2998E − 01	(45) =	8.6613E − 03
(6) =	−1.4931E − 01	(46) =	8.3401E − 03
(7) =	−1.3050E − 01	(47) =	5.8889E − 04
(8) =	2.2909E − 01	(48) =	−6.2168E − 03
(9) =	4.9995E − 01	(49) =	−6.0762E − 03
(10) =	4.1849E − 01	(50) =	−3.8750E − 04
(11) =	9.5216E − 02	(51) =	4.6638E − 03
(12) =	−1.5465E − 01	(52) =	4.5944E − 03
(13) =	−1.3567E − 01	(53) =	3.8938E − 04
(14) =	7.1121E − 02	(54) =	−3.3845E − 03
(15) =	2.3141E − 01	(55) =	−3.3793E − 03
(16) =	1.9785E − 01	(56) =	−2.8824E − 04
(17) =	2.7825E − 02	(57) =	2.5170E − 03
(18) =	−1.0614E − 01	(58) =	2.5362E − 03
(19) =	−9.4356E − 02	(59) =	2.5505E − 04
(20) =	2.1521E − 02	(60) =	−1.8376E − 03
(21) =	1.1359E − 01	(61) =	−1.8750E − 03
(22) =	9.9437E − 02	(62) =	−1.9681E − 04
(23) =	8.7017E − 03	(63) =	1.3599E − 03
(24) =	−6.4301E − 02	(64) =	1.4014E − 03
(25) =	−5.8097E − 02	(65) =	1.6416E − 04
(26) =	5.9520E − 03	(66) =	−9.9595E − 04
(27) =	5.8027E − 02	(67) =	−1.0388E − 03
(28) =	5.2051E − 02	(68) =	−1.2832E − 04
(29) =	3.2299E − 03	(69) =	7.3491E − 04
(30) =	−3.6908E − 02	(70) =	7.7460E − 04
(31) =	−3.3956E − 02	(71) =	1.0380E − 04
(32) =	1.1812E − 03	(72) =	−5.3913E − 04
(33) =	3.0395E − 02	(73) =	−5.7498E − 04
(34) =	2.7931E − 02	(74) =	−8.1288E − 05
(35) =	1.5445E − 03	(75) =	3.9714E − 04
(36) =	−2.0628E − 02	(76) =	4.2815E − 04
(37) =	−1.9349E − 02	(77) =	6.4607E − 05
(38) =	−1.5567E − 04	(78) =	−2.9157E − 04
(39) =	1.6158E − 02	(79) =	−3.1801E − 04
(40) =	1.5203E − 02	(80) =	−5.0483E − 05

would produce no measured response. And second, for a physical system, the amplitude of the unit pulse must not be so large that it violates the linearity region of the physical system. If these practical considerations are met, the discrete pulse-response sequence can be determined experimentally.

Next, suppose that the sequence of pulse inputs as depicted in Fig. 1-16 are input to the linear rocket-sled system, which is initially at rest. Denote the pulse values of the input by

$$u_T(k) \equiv u(t),\ t\,\varepsilon\,(kT,\,(k+1)T] \qquad k = 0, 1, 2, \ldots \tag{1-16}$$

Now, since the output response to a single-unit-amplitude pulse is known, the principle of linear superposition may be applied to determine the output response to the entire sequence of pulse inputs. In particular, the output at time $t = T = 1$ is only the result of the first pulse. Thus,

$$\begin{aligned} y_{zs}(T) &= h_T(1) \cdot u_T(0) \\ &= 0.14287 \cdot 1 \\ &= 0.14287 \end{aligned} \tag{1-17}$$

However, the output at time $t = 2T = 2$ is the result of both the first pulse and the second pulse. Thus,

$$\begin{aligned} y_{zs}(2T) &= h_T(2) \cdot u_T(0) + h_T(1) \cdot u_T(1) \\ &= 0.73603 \cdot 1 + 0.14287 \cdot 2 \\ &= 1.02177 \end{aligned} \tag{1-18}$$

The first term in Eq. (1-18) is the output, at time $t = 2T$, caused by the first pulse, and the second term is the output, at time $t = 2T$, caused by the second pulse. By the principle of linear superposition, the total response is just the sum of the responses caused by both pulses.

In like fashion, the output at time $t = 3T = 3$ is the sum of three terms:

$$\begin{aligned} y_{zs}(3T) &= [h_T(3) \cdot u_T(0)] + [h_T(2) \cdot u_T(1)] + [h_T(1) \cdot u_T(2)] \\ &= [1.1658 \cdot 1] + [0.73603 \cdot 2] + [0.14228 \cdot 3] \\ &= 3.0147 \end{aligned} \tag{1-19}$$

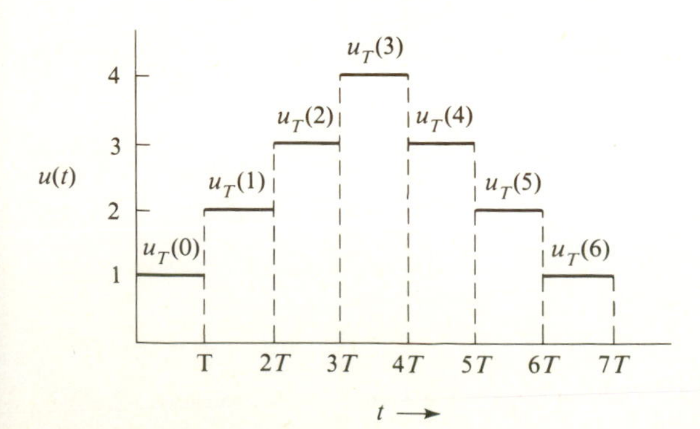

Figure 1-16 Discrete pulse input sequence, $u_T(k)$, $k = 1, 2, 3, 4, 5, 6$.

The first term is the output, at time $t = 3T$, caused by the first pulse; the second term is that caused by the second pulse; and the third term is that caused by the third pulse. Again, the principle of linear superposition applies, so that the complete output at time $t = 3T$ is just the sum of the output caused by all previous pulses.

This may be continued. The output sampled at $t = kT$ is then the sum of the responses caused by all previous inputs:

$$y_{zs}(kT) = [h_T(k) \cdot u_T(0)] + [h_T(k-1) \cdot u_T(1)] + \cdots + [h_T(2) \cdot u_T(k-2)] + [h_T(1) \cdot u_T(k-1)] \quad (1\text{-}20)$$

Notice that, by the notation chosen for the discrete pulse-response sequence $h_T(k)$ and the discrete pulse input $u_T(k)$, the arguments of every term on the right side of Eq. (1-20) add up to k. This is the desired sample time, $t = kT$. The notation is chosen this way as a matter of convenience.

The actual output response to the sequence of discrete pulse inputs shown in Fig. 1-16 is shown in Fig. 1-17. The output at each sample time, $t = kT$, is shown by the dots on the smooth-curve output response.

Equation (1-20) may be written in compact notation using either of two equivalent *convolution summations:*

$$\boxed{y_{zs}(kT) = \sum_{i=0}^{k-1} h_T(k-i) \cdot u_T(i)} \quad (1\text{-}20a)$$

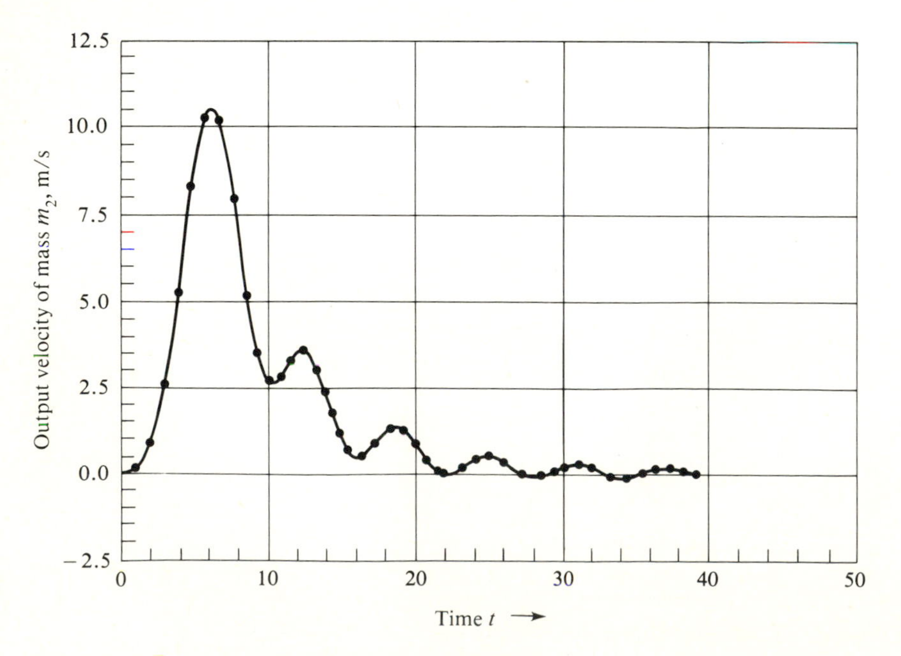

Figure 1-17 Zero-state response to input, Fig. 1-16.

or

$$y_{zs}(kT) = \sum_{j=1}^{k} h_T(j) \cdot u_T(k-j) \tag{1-20b}$$

The reader is invited to check that both of these two convolution summations, Eqs. (1-20*a*) and (1-20*b*), yield exactly the same set of terms as Eq. (1-20). The order of the terms in an expansion of Eq. (1-20*b*) is reversed, but the sum of the terms is exactly the same. Notice that the combined arguments of h_T and u_T always add up to k, the current sample time.

The Zero-Input Response

It should be stressed that the convolution summation Eq. (1-20) is written only for the zero-state response; it is assumed that the initial state at $t_0 = 0$ is zero. If the initial state were not initially zero, then

$$y(kT) = y_{zi}(kT) + y_{zs}(kT) \tag{1-21}$$

Here $y_{zi}(kT)$ is the sampled zero-input response caused by the initial state, $x(0)$, at time zero. The $y_{zs}(kT)$ is the zero-state portion of the response given by Eq. (1-20).

But the initial time $t_0 = 0$ is arbitrary. Therefore, if the entire past history of the input were known, assuming that the past input had always been a sequence of pulse inputs, an expression like Eq. (1-20*a*) could be extended backwards in time all the way to $t_0 = -\infty T$ to yield

$$y(kT) = \sum_{i=-\infty}^{k-1} h_T(k-i) \cdot u_T(i) \tag{1-21a}$$

Splitting the summation over two intervals yields

$$y(kT) = \sum_{i=-\infty}^{-1} h_T(k-i) \cdot u_T(i) + \sum_{i=0}^{k-1} h_T(k-i) \cdot u_T(i) \tag{1-21b}$$

The second summation in Eq. (1-21*b*) is seen to simply be $y_{zs}(kT)$. From Eq. (1-21) it is apparent that the first summation in Eq. (1-21*b*) must be $y_{zi}(kT)$:

$$\begin{aligned} y_{zi}(kT) &= \sum_{i=-\infty}^{-1} h_T(k-i) \cdot u_T(i) \\ &\equiv \text{sampled zero-input response at time } t = kT \text{ caused by the initial state, } x(0) \end{aligned} \tag{1-22}$$

This provides another interesting interpretation of the concept of state: the state at any time t is a summary of all past information about the system, which in the absence of any further inputs would allow one to predict the future response of the system. Knowledge of the initial state is thus equivalent to knowing all previous input information.

This is quite a compression of information, and a discussion of state is obviously a very important topic. The detailed discussion of state-space methods begins in Chap. 6 and continues until the end of the text. A state-space interpretation of all the results we have seen in this section is given in these later chapters.

1-3 THE CONVOLUTION INTEGRAL OF A LINEAR TIME-INVARIANT SYSTEM

In the previous section the principle of linear superposition was used to derive the convolution summation relation, Eq. (1-21*a*), for the sampled output response, $y(kT)$, of a linear time-invariant system to the sequence of past digital pulse inputs, $u_T(i)$, $i = k, k-1, k-2, \ldots, -\infty$. If the sample interval is made smaller and smaller, then one approaches the continuous time response as $T \to 0$. But as $T \to 0$, one must also increase the magnitude of the input pulse so that the infinitesimal-duration pulse still produces a measurable change on the system output. For example, if one holds the area under the curve of the input at unity, then in the limit as $T \to 0$, one obtains an input pulse of infinite amplitude and zero time duration. This is depicted in Fig. 1-18.

This infinite pulse, of zero time duration, with "unit area under its curve," is called the *unit impulse function,* or *delta function,* and given the symbol $\delta(t)$. It has the property that if one integrates across the impulse [when the argument of $\delta(t)$ goes to zero—i.e., at $t = 0$] then the integral value is unity:

$$\int_{-\infty}^{\infty} \delta(\tau)\, d\tau = \int_{0^-}^{0^+} \delta(\tau)\, d\tau = 1 \tag{1-23}$$

This integral relation generalizes to provide a fundamental property of the unit-impulse function known as the *sampling property of the unit inpulse:*

$$\boxed{\int_{-\infty}^{\infty} f(\tau)\, \delta(\tau - a)\, d\tau = f(a)} \tag{1-24}$$

for any function f which is continuous at a. This sampling property is key in a number of later applications.

The impulse function is not a function in the ordinary sense, since its value is undefined when the argument goes to zero. Its definition is really Eq. (1-24), which defines its property under the integral; the unit-impulse function really has no meaning except under the integral.

Fortunately, the use required of the impulse function is always within the integral. Indeed, the generalization of the convolution summation, Eq. (1-21*a*), to

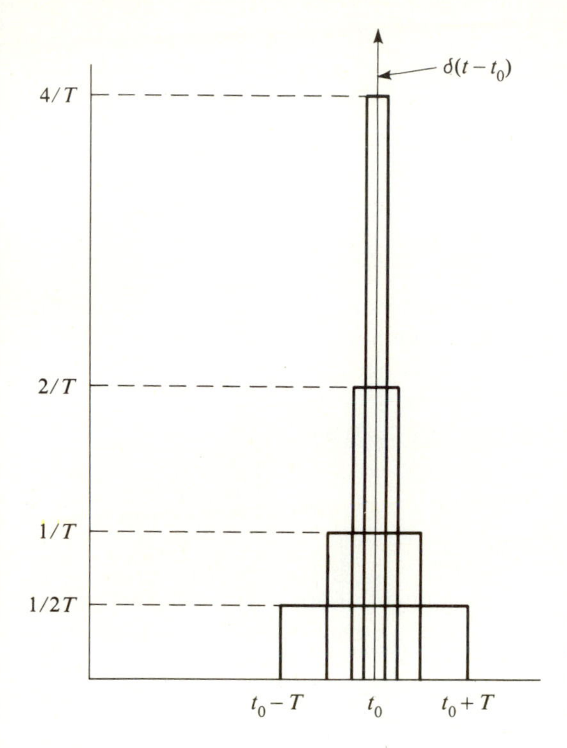

Figure 1-18 Illustration of the unit impulse function as the limit of the discrete pulse input of unit area as the pulse duration goes to zero.

the continuous-time, zero-T case, is the *convolution integral:*

$$y(t) = \int_{-\infty}^{t} h(t - \tau)u(\tau)\, d\tau \tag{1-25a}$$

where $h(t - \tau)$ is the continuous-impulse-response function to an ideal impulse occurring at time $t = \tau$. The $u(\tau)$, $-\infty < \tau \leq t$, is the past input. Notice that the convolution integral has the same form as the convolution summation in that the arguments for integration, $t - \tau$ and τ, add up to the current time t. In essence, the impulse response $h(t - \tau)$ is a weighting factor which describes how much the past input, $u(\tau)$, $-\infty < \tau \leq t$, influences the present output, $y(t)$.

The impulse-response function, $h(t)$, is analogous to the discrete pulse-response sequence, $h_T(k)$, in that $h(t)$ is the zero-state response to an ideal unit-impulse input occurring at time zero. Using the unit impulse as the input in Eq. (1-25a) gives

$$y(t) = \int_{-\infty}^{t} h(t - \tau)\, \delta(\tau)\, d\tau$$

When the argument of $\delta(\tau)$ goes to zero (i.e., when $\tau = 0$), the impulse occurs. The output until then is identically zero, since the system is at rest prior to time zero.

Then, as a consequence of Eq. (1-24),

$$y(t) = \int_{-\infty}^{t} h(t - \tau)\, \delta(\tau)\, d\tau$$

$$= h(t - \tau)|_{\tau=0} \qquad \text{the value of } \tau \text{ at which the impulse occurs}$$

$$= h(t) \qquad \text{the impulse response function}$$

If the impulse is made to occur at some other time, say, at time $t = 7$, then the impulse would be written as $\delta(t - 7)$ (the argument is zero when $t = 7$). The impulse response with impulse at $t = 7$ would then be given by

$$y(t) = \int_{-\infty}^{t} h(t - \tau)\, \delta(\tau - 7)\, d\tau$$

$$= h(t - \tau)|_{\tau=7}$$

$$= h(t - 7)$$

Notice that the value of time, t, when the argument of $h(t - 7)$ goes to zero is the time at which the impulse occurs. Hence, in the convolution representation

$$y(t) = \int_{-\infty}^{t} h(t - \tau)u(\tau)\, d\tau \tag{1-25a}$$

past inputs are being weighted by a whole continuum of impulse response functions, each with impulse occurring at time τ, $-\infty < \tau \leq t$.

Equivalent Forms of the Convolution Integral

Similar to the two forms of the convolution summation as represented by Eqs. (1-20*a*) and (1-20*b*), a change of variable of integration $\xi \equiv t - \tau$, $d\xi = -d\tau$, may be made in Eq. (1-25*a*) to yield the equivalent form of the convolution integral:

$$\boxed{y(t) = \int_{0}^{\infty} h(\xi)u(t - \xi)\, d\xi} \tag{1-25b}$$

Next, since $h(t - \tau) = 0$ for $\tau > t$, or equivalently since $h(\xi) \equiv 0$ for $\xi < 0$, Eqs. (1-25*a*) and (1-25*b*) may be written as

$$\boxed{y(t) = \int_{-\infty}^{\infty} h(t - \tau)u(\tau)\, d\tau} \tag{1-25c}$$

$$\boxed{y(t) = \int_{-\infty}^{\infty} h(\xi)u(t - \xi)\, d\xi} \tag{1-25d}$$

Finally, it is instructive to split Eq. (1-25*a*) (or any of the previous equivalent forms) into the zero-input and zero-state portions of the response:

$$
\boxed{\begin{aligned} y(t) &= \int_{-\infty}^{0^-} h(t-\tau)u(\tau)\,d\tau + \int_{0^-}^{t} h(t-\tau)u(\tau)\,d\tau \\ &= y_{zi}(t) + y_{zs}(t) \end{aligned}} \tag{1-25e}
$$

where zero is arbitrarily taken as the initial time. This is the equivalent of Eq. (1-21*b*) in the discrete time domain. The time 0^- is used to indicate that there might be a discontinuity of the input, $u(t)$, occurring at the initial time zero.

All of these forms are equivalent, and each is utilized in later applications and analysis.

Example 1-1 To illustrate the process of convolution, consider the $h(t)$ impulse response function shown in Fig. 1-19*a* convolved with the input $u(t)$ shown in Fig. 1-19*b*. The response to this input is given by the convolution integral

$$y(t) = \int_{-\infty}^{t} h(t-\tau)u(\tau)\,d\tau$$

The impulse response function is given by the expression

$$h(t) = \begin{cases} 0 & t < 0,\ t > 8 \\ 4 - \dfrac{t}{2} & 0 < t < 8 \end{cases}$$

Therefore, it is true that

$$h(-\tau) = \begin{cases} 0 & \tau > 0,\ \tau < -8 \\ 4 + \dfrac{\tau}{2} & -8 < \tau < 0 \end{cases}$$

This essentially flips the $h(\tau)$ curve about the y axis. Then $h(t-\tau)$ shifts the $h(-\tau)$ curve by t units, giving

$$h(t-\tau) = \begin{cases} 0 & \tau > t,\ \tau < t-8 \\ 4 - \dfrac{t}{2} + \dfrac{\tau}{2} & t-8 < \tau < t \end{cases}$$

This is seen in Fig. 1-19*c* through 1-19*g*.

The complicated part of computing the convolution of h and u is determination of the limits of integration. This is assisted by Fig. 1-19*c* through 1-19*g*. From Fig. 1-19*c* it is seen that $y(t) = 0$ for $t < -2$; the curve of $h(t-\tau)$ and the curve of $u(\tau)$ do not overlap for $t < -2$, and so the integral of their

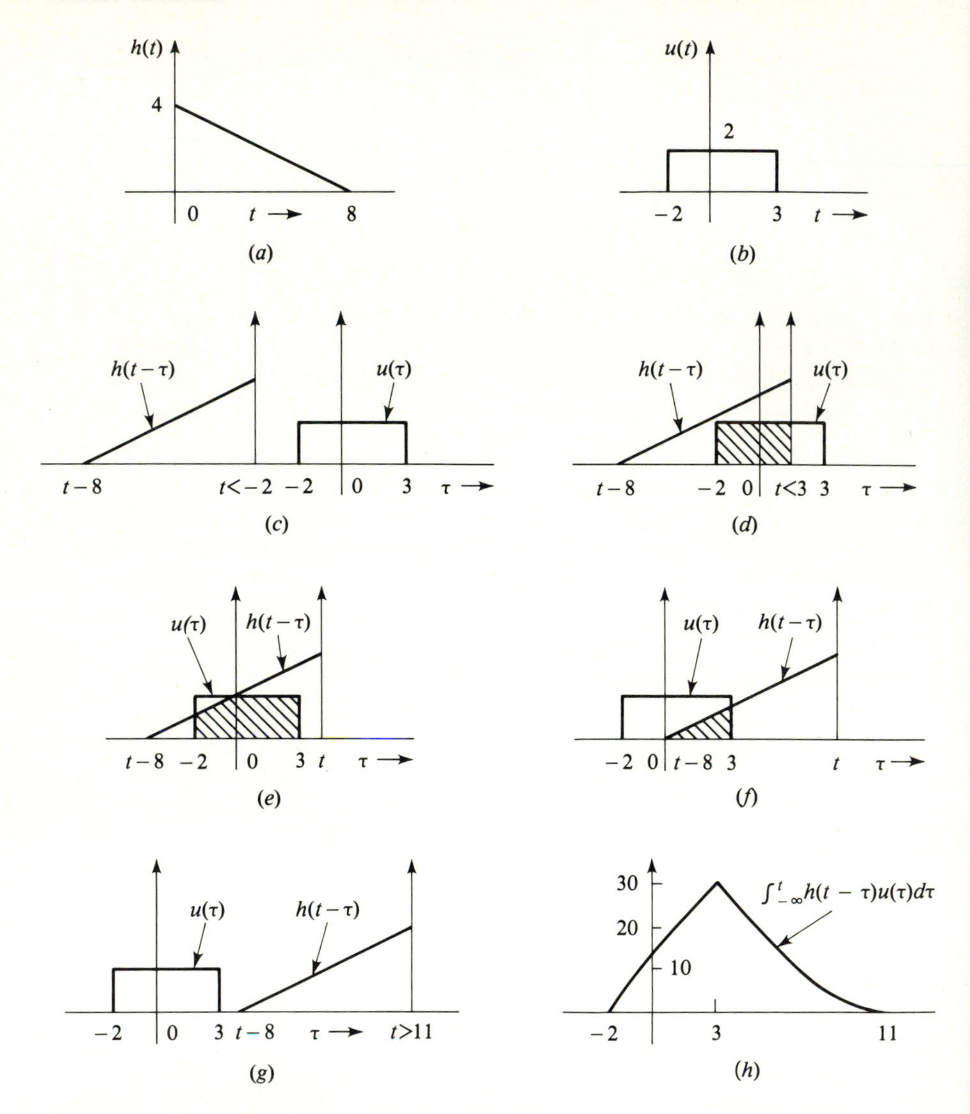

Figure 1-19 Graphical illustration of the process of convolution.

product is zero. From Fig. 1-19*d* it is seen that

$$y(t) = \int_{-\infty}^{t} h(t - \tau)u(\tau)\, d\tau$$

$$= \int_{-2}^{t} \left(4 - \frac{t}{2} + \frac{\tau}{2}\right)(2)\, d\tau$$

$$= 14 + 6t - \frac{t^2}{2} \qquad \text{for } -2 < t < 3$$

Next, from Fig. 1-19*e* it is seen that

$$y(t) = \int_{-2}^{3} \left(4 - \frac{t}{2} + \frac{\tau}{2}\right)(2)\, d\tau$$

$$= 85/2 - 5t \qquad \text{for } 3 < t < 6$$

Then, from Fig. 1-19*f* it is true that

$$y(t) = \int_{t-8}^{3} \left(4 - \frac{t}{2} + \frac{\tau}{2}\right)(2)\, d\tau$$

$$= 60\frac{1}{2} - 11t + \frac{t^2}{2} \qquad \text{for } 6 < t < 11$$

Finally, from Fig. 1-19*g* it follows that $y(t) = 0$ for $t > 11$, since the curves no longer overlap. The overall convolution response is shown in Fig. 1-19*h*.

As seen in this example, the process of convolution can be rather complex when $h(t)$ and $u(t)$ are both of finite extent. Fortunately, the impulse-response function for a linear time-invariant system modeled by ordinary differential equations always has an exponential form that allows one to convert the convolution integral into equivalent ordinary integrals.

Example 1-2 The continuous impulse response function may be determined mathematically from the internal model. By such a procedure the impulse response function of the rocket-sled system is found to be

$$h(t) = 0.90909e^{-0.2t} - 0.90909e^{-0.1t}\cos(1.044t) + 0.087078e^{-0.1t}\sin(1.044t) \tag{1-26}$$

Methods for such a determination are presented in Chap. 2. A plot of this continuous impulse response function is shown in Fig. 1-20.

Once determined, the continuous impulse response function may be used in the convolution integral to calculate the system output response to a known input function:

$$y(t) = \int_{-\infty}^{t} \{0.90909e^{-0.2(t-\tau)} - 0.90909e^{-0.1(t-\tau)}\cos[1.044(t-\tau)] + 0.087078e^{-0.1(t-\tau)}\sin[1.044(t-\tau)]\}u(\tau)\, d\tau \tag{1-27}$$

Fortunately, the impulse response function for a linear time-invariant system modeled by ordinary differential equations* always has this convenient exponential form. Therefore, the somewhat difficult convolution integral

* Linear time-invariant systems may also require use of time-delay or partial-differential-equation models. The convolution-integral representation of such systems is still applicable, but the impulse response function for such "infinite-dimension" systems can be considerably more complicated. See Sec. 1.8.

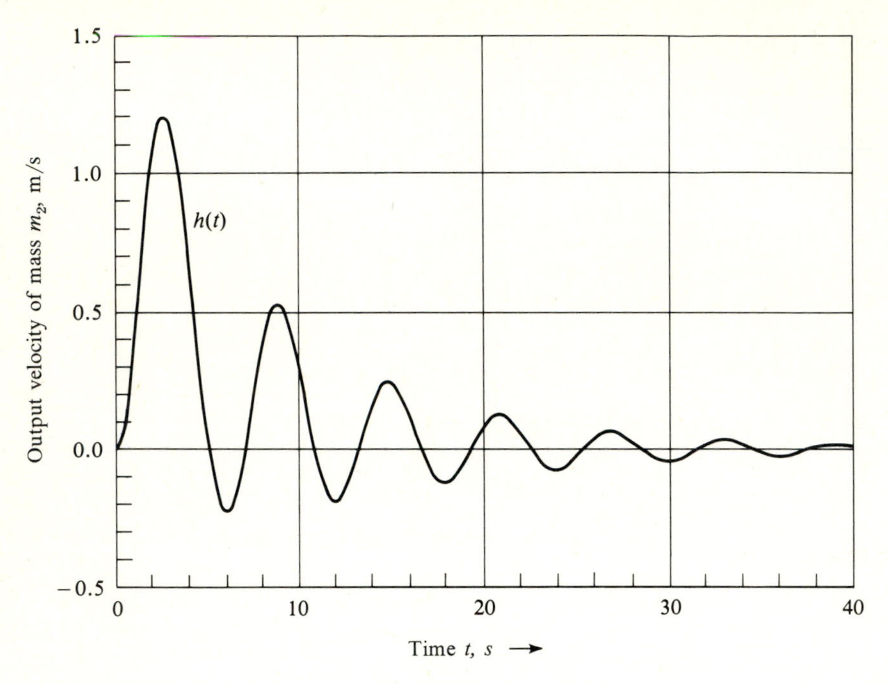

Figure 1-20 Impulse responses of rocket-sled system.

for such linear time-invariant systems may always be converted to ordinary integrals as in

$$\begin{aligned} y(t) &= 0.90909e^{-0.2t} \int_0^t e^{0.2\tau} u(\tau)\, d\tau \\ &\quad - 0.90909e^{-0.1t}[\cos(1.044t) \int_0^t e^{0.1\tau} \cos(1.044\tau) u(\tau)\, d\tau \\ &\quad + \sin(1.044t) \int_0^t e^{0.1\tau} \sin(1.044\tau) u(\tau)\, d\tau] \\ &\quad + 0.087078e^{-0.1t}[\sin(1.044t) \int_0^t e^{0.1\tau} \cos(1.044\tau) u(\tau)\, d\tau \\ &\quad - \cos(1.044t) \int_0^t e^{0.1\tau} \sin(1.044\tau) u(\tau)\, d\tau] \end{aligned} \tag{1-28}$$

Once the actual input, $u(t)$, is known, the response can be calculated from Eq. (1-28). For example, suppose the input is the unit ramp input:

$$u(t) = \begin{cases} 0 & t < 0 \\ t & t > 0 \end{cases} \tag{1-29}$$

After taking the integral of Eq. (1-28) with the unit ramp input, and performing considerable algebra, the unit ramp response is found to be

$$y(t) = 22.7272e^{-0.2t} + 0.8265e^{-0.1t}\cos(1.044t) + 0.0792e^{-0.1t}\sin(1.044t) - 23.5537 + 4.5455t \qquad (1\text{-}30)$$

If one actually attempts to carry out the integration of Eq. (1-28) using the unit ramp input Eq. (1-29) to obtain Eq. (1-30), one will quickly hope for "a better way." The algebra is quite tedious, even for this relatively simple input function. Fortunately, there are much easier techniques for finding the output solution given a differential-equation internal model of the system. These solution techniques are fully developed in later chapters.

1-4 RELATION BETWEEN THE DISCRETE PULSE-RESPONSE SEQUENCE AND THE CONTINUOUS IMPULSE RESPONSE

The discrete pulse-response sequence, $h_T(k)$, $k = 1, 2, 3, \ldots$, is defined as the sampled response of the linear time-invariant system forced by the unit pulse input:

$$\delta_T(t) \equiv \begin{cases} 0 & t < 0 \\ 1 & 0 < t < T \\ 0 & t > T \end{cases} \qquad (1\text{-}31)$$

But from the convolution integral in Eq. (1-25a), the output response for all time t may be computed. Namely, for $t \geq T$ the input, Eq. (1-31), may be substituted in Eq. (1-25a) to yield

$$y(t) = \int_{-\infty}^{t} h(t-\tau)\,\delta_T(\tau)\,d\tau$$

$$= \int_{0}^{T} h(t-\tau)\,d\tau \qquad (1\text{-}32a)$$

Alternatively, the second form of the convolution relation Eq. (1-25b) yields for $t \geq T$

$$y(t) = \int_{0}^{\infty} h(\xi)\delta_T(t-\xi)\,d\xi$$

$$= \int_{t-T}^{t} h(\xi)\,d\xi \qquad (1\text{-}32b)$$

Since $h_T(k)$ is the sampled response to the unit pulse input, Eqs. (1-32a) or (1-32b) may be used to relate $h_T(k)$ to $h(t)$. Namely, from Eq. (1-32a),

$$\boxed{h_T(k) = \int_{0}^{T} h(kT-\tau)\,d\tau} \qquad (1\text{-}33a)$$

or, from Eq. (1-32*b*),

$$h_T(k) = \int_{(k-1)T}^{kT} h(\xi)\, d\xi \tag{1-33b}$$

Thus $h_T(k)$ is the integral or "average" of $h(t)$ over the interval $(k-1)T < t < kT$. Indeed, from the mean value theorem of integrals it is true that $h_T(k) = T \cdot h(t)$ for some $t \in [(k-1)T, kT]$.

Example 1-3 The impulse response function for the rocket-sled system is given by Eq. (1-26) in Example 1-2. Applying Eq. (1-33), the $h_T(k)$ sequence is thus found to equal

$$h_T(k) = -4.54545(1 - e^{0.2T})e^{-0.2kT} - 0.87078\{e^{-0.1kT} \sin(1.044kT) - e^{-0.1(k-1)T} \sin[1.044(k-1)T]\}$$

Figure 1-21 plots the continuous $h(t)$ versus $h_T(k)/T$ for various values of T, including $T = 0.1$, $T = 1$, and $T = \pi/1.044$ ($\cong 3$). For the very short $T = 0.1$, the discrete $h_T(k)/T$ curve is seen to track the continuous $h(t)$ quite closely. However, for $T = \pi/1.044$ there is no resemblence to the continuous $h(t)$. For this T it is seen that

$$h_T(k) = -4.54545(1 - e^{0.2T})e^{-0.2kT} = 3.75214e^{-0.60184k}$$

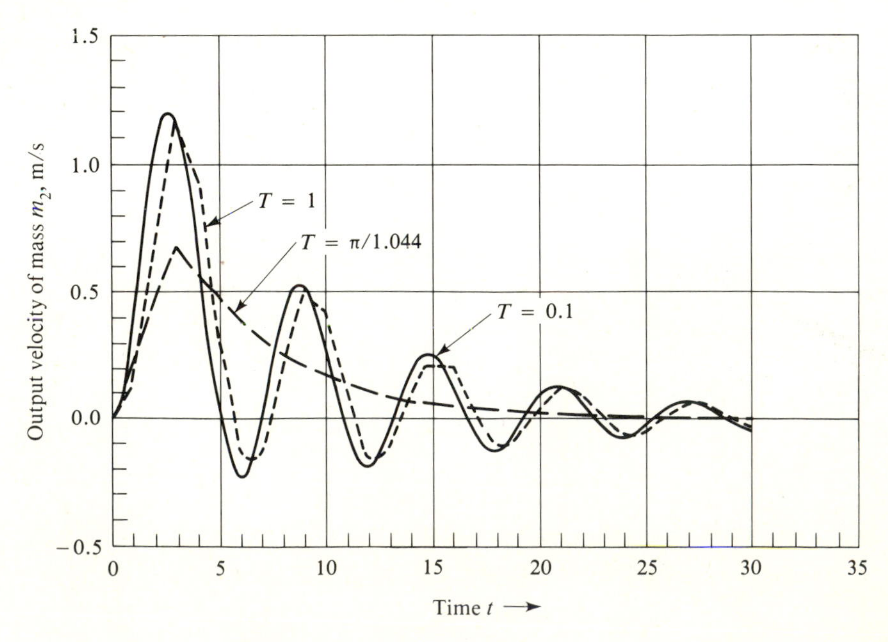

Figure 1-21 Plot of $h_T(k)/T$ for $T = 0.1, 1, \pi/1.044$s.

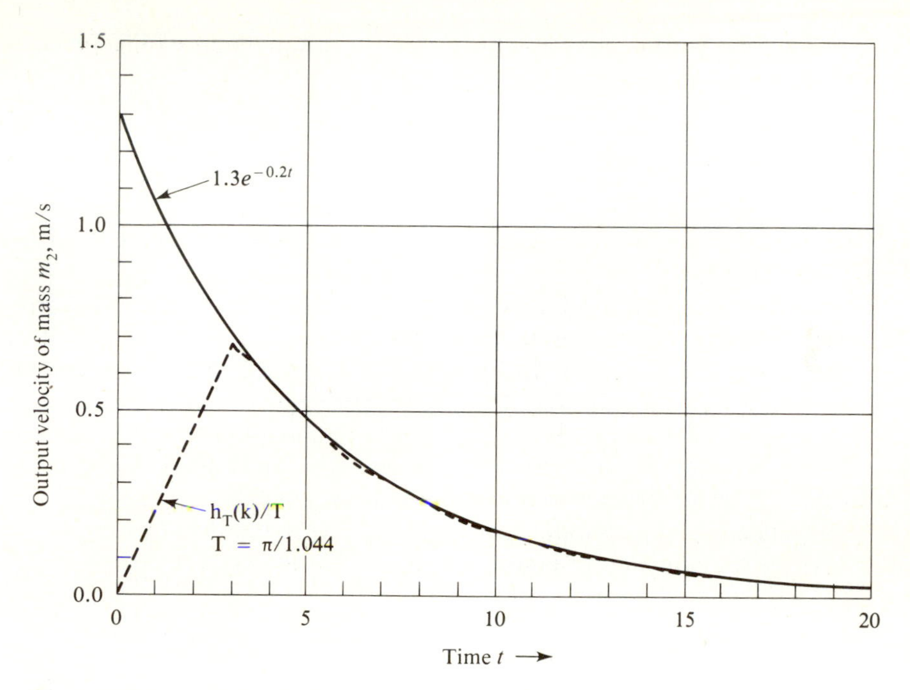

Figure 1-22 Plot of $h_T(k)/T$ for $T = \pi/1.044$ versus $1.3e^{-0.2t}$.

The $h_T(k)/T$ curve now approximates the continuous impulse response $h(t) = 1.3e^{-0.2t}$, as shown in Fig. 1-22.

This example illustrates some important aspects of the discrete time representation of a continuous system. Picking the wrong sampling interval, T, can obscure internal dynamics. This topic is explored further in later chapters.

1-5 SYSTEM STEP RESPONSE

There are certain system inputs that are basic to the analysis of linear dynamic-system response because they reveal important system behavior. By the principle of linear superposition, this behavior may be generalized to other arbitrary inputs. One of these has already been encountered—the impulse input $\delta(t)$, yielding the very important impulse response function $h(t)$. Another one of these special inputs is the *unit step function:*

$$u_{-1}(t) \equiv \begin{cases} 0 & t < 0 \\ 1 & t > 0 \end{cases} \tag{1-34}$$

Like the unit impulse function, the discontinuity of the unit step function occurs at the time when the argument goes to zero. For example,

$$u_{-1}(t-7) = \begin{cases} 0 & t < 7 \\ 1 & t > 7 \end{cases} \tag{1-35}$$

The unit step function may also be related to the unit impulse function via the integral relationship

$$u_{-1}(t-7) = \int_{-\infty}^{t} \delta(\tau - 7)\, d\tau \tag{1-36}$$

The integral of $\delta(\tau - 7)$ is zero up until $\tau = 7$, at which point it samples the integrand (which happens to be a unity multiplier). Thus the integral of $\delta(\tau - 7)$ is unity for all $t > 7$.

Because of the integral relation Eq. (1-36), which is mathematically correct, the unit impulse is often referred to as the *derivative of the unit step;* if one differentiates the definite integral with upper limit of integration t, one normally gets the integrand as the answer. This is mathematically incorrect in this case, since a function like the unit step, which is discontinuous, is certainly not differentiable at the point of discontinuity. Nevertheless, it is sometimes a useful *concept* with which to relate the unit step and the unit impulse.

Applying the unit step input with step at $t = 0$ to the linear time-invariant system yields the unit step response, which is given the symbol $r(t)$:

$$r(t) = \int_{-\infty}^{t} h(t-\tau) u_{-1}(\tau)\, d\tau$$

Since $u_{-1}(t)$ is zero for $t < 0$ and unity for $t > 0$, this may be written as

$$\boxed{r(t) = \int_{0}^{t} h(t-\tau)\, d\tau \qquad t > 0} \tag{1-37a}$$

The equivalent convolution relation Eq. (1-25*b*) may be used to give an alternative relationship:

$$r(t) = \int_{0}^{\infty} h(\xi) u_{-1}(t-\xi)\, d\xi$$

or, again, since $u_{-1}(t-\xi) = 0$ for $\xi > t$ and $u_{-1}(t-\xi) = 1$ for $\xi < t$,

$$\boxed{r(t) = \int_{0}^{t} h(\xi)\, d\xi \qquad t > 0} \tag{1-37b}$$

The Sampled Unit Step Response

By sampling the unit step response every T time units, a relation may also be found to the discrete pulse-response sequence, $h_T(k)$, $k = 1, 2, 3, \ldots$. Namely, using Eq. (1-37*b*) and sampling at $t = kT$ gives

$$r(kT) = \int_{0}^{kT} h(\xi)\, d\xi \qquad k = 1, 2, 3, \ldots \tag{1-38}$$

The integral from zero to kT may then be rewritten as the sum of integrals

$$r(kT) = \sum_{j=1}^{k} \int_{(j-1)T}^{jT} h(\xi)\, d\xi \tag{1-39}$$

But from Eq. (1-33) it is recognized that

$$h_T(j) = \int_{(j-1)T}^{jT} h(\xi)\, d\xi$$

Substitution into Eq. (1-39) then gives the relation

$$\boxed{r(kT) = \sum_{j=1}^{k} h_T(j) \qquad k = 1, 2, 3, \ldots} \tag{1-40}$$

The usefulness of the relation is perhaps somewhat surprising. Namely, one may use Eq. (1-40) to calculate $h_T(k)$ itself from an experimentally sampled step response, $r(kT)$. Although an ideal impulse input, $\delta(t)$, is impossible to physically produce, even the digital pulse input, Eq. (1-31), is difficult to produce, particularly if T is very short. On the other hand, the unit step input is typically an input that can be produced with reasonably high fidelity. For instance, in an electrical system, producing a step input may be as easy as closing the switch to a battery input source.

To understand better how Eq. (1-40) may be used to find $h_T(k)$, rewrite Eq. (1-40) as

$$r(kT) = \sum_{j=1}^{k-1} h_T(j) + h_T(k)$$

But the summation from $j = 1$ to $k - 1$ is $r(kT - T)$, and so

$$r(kT) = r(kT - T) + h_T(k)$$

This may be reformed as

$$\boxed{h_T(k) = r(kT) - r(kT - T) \qquad k = 1, 2, 3 \ldots} \tag{1-41}$$

This says that the discrete pulse-response model, $h_T(k)$, is simply the difference between the sampled step response at time $t = kT$ and time $t = (k - 1)T$.

Example 1-4 The unit step response for the rocket-sled linear time-invariant system is shown in Fig. 1-23. Both the continuous response and the sampled points with a $T = 1$ s are shown. The continuous curve is computed with the help of Eq. (1-37b):

$$r(t) = \int_0^t [0.90909e^{-0.2\xi} - 0.90909e^{-0.1\xi} \cos (1.044\xi) + 0.087078e^{-0.1\xi} \sin (1.044\xi)]\, d\xi$$

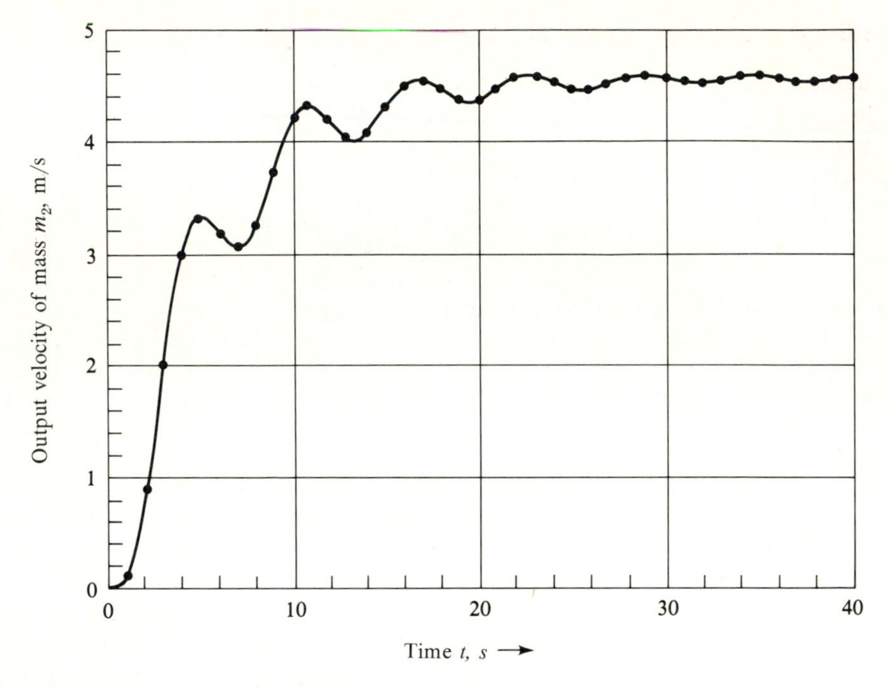

Figure 1-23 Sampled step-response rocket-sled system, $T = 1$ s.

or $$r(t) = -4.54545e^{-0.2t} - 0.87078e^{-0.1t} \sin(1.044t) + 4.54545$$

The sampled values of $r(kT)$ for $T = 1$ are shown in Table 1-2. Also shown are the values of $h_T(k)$, the discrete pulse-response sequence. These may be computed by repeated application of Eq. (1-41).

Table 1-2 Producing the discrete pulse response from the sampled step response

Sampled step response, $r(kT)$; $T = 1$	Discrete pulse response; $h_T(k) = r(kT) - r(kT - T)$
$r(0) = 0.0000E + 00$	$h_T(1) = 1.4287E - 01$
$r(1) = 1.4287E - 01$	$h_T(2) = 7.3603E - 01$
$r(2) = 8.7890E - 01$	$h_T(3) = 1.1658E + 00$
$r(3) = 2.0447E + 00$	$h_T(4) = 9.6007E - 01$
$r(4) = 3.0048E + 00$	$h_T(5) = 3.2998E - 01$
$r(5) = 3.3348E + 00$	$h_T(6) = -1.4931E - 01$
$r(6) = 3.1855E + 00$	$h_T(7) = -1.3050E - 01$
$r(7) = 3.0550E + 00$	$h_T(8) = 2.2909E - 01$
$r(8) = 3.2841E + 00$	$h_T(9) = 4.9995E - 01$
$r(9) = 3.7840E + 00$	$h_T(10) = 4.1849E - 01$
$r(10) = 4.2025E + 00$	

Finding the Impulse Response from the Step Response

Equation (1-41) gives a relation by which to determine the discrete pulse-response sequence, $h_T(k)$, from the sampled step response, $r(kT)$. The continuous impulse response, $h(t)$, may also be found from the continuous step response, $r(t)$. Namely, referring back to Eq. (1-37*b*), it is true that

$$r(t) = \int_{-\infty}^{t} h(\xi)\, d\xi \qquad t > 0$$

or, in other words, the step response is the definite integral of the impulse response. Now, provided that the step response is differentiable for all $t > 0$,* both sides of Eq. (1-37*b*) may be differentiated to give

$$h(t) = \frac{d}{dt}[r(t)] \qquad t > 0 \tag{1-42}$$

The question now arises as to whether Eq. (1-42) does or does not provide an experimental method to determine the continuous impulse response function. Namely, the system step response may be determined experimentally by forcing the system with a unit step input—an ideal input, but one which can still be produced with reasonably high fidelity. The answer would be yes, the equation does provide such a model, if the measurements of the step response could be made with no error and if a perfect differentiator could be produced. However, differentiation is typically a procedure that is very sensitive to noise corruption, and so this still cannot be considered a practical experimental method. The best experimental method to work with is the discrete pulse-response model, as previously described.

1-6 PERFORMANCE CHARACTERISTICS OF THE STEP RESPONSE

The time-response characteristics of the rocket-sled step response shown in Fig. 1-23 are typical of the step response for many other linear time-invariant systems. From an input-output viewpoint it doesn't matter whether the system is electrical, mechanical, fluidic, ecological, or whatever; these same basic characteristics will repeat themselves. Because of their universality, and because the step-response characteristics reveal many intrinsic performance characteristics of the system, the parameters of the step response have received special names

*The point $t = 0$ may be a point at which the step response is not differentiable (its "slope" is discontinuous at $t = 0$), or it may even be discontinuous at $t = 0$, jumping from an identically zero value for $t < 0$ to a nonzero value for $t > 0$. This is examined more closely in the next chapter.

and are used as performance indicators to compare the response of one system versus another.

The most common step-response performance parameters are *rise time, duplication time, peak time, settling time, peak value,* and *steady-state final value*. These are defined as:

Rise time. The duration of time for the output response to go from 10 to 90 percent of its final value.
Duplication time. The time at which the output response first crosses the final value.
Peak time. The time at which the response reaches its peak.
Settling time. The time at which the output reaches a certain percentage of its final value (usually 2 or 5 percent) and stays within that percentage.
Peak value. The response value at its peak.
Final value. The steady-state final value.

For the rocket-sled system these parameters are

Rise time:	$TR = 8.1$
Duplication time:	$TD = 16.7$
Peak time:	$TP = 22.0$
Settling time:	$TS = 25.0$
Peak value:	$MP = 4.56$
Final value:	$FV = 4.545$

Analytic expressions can be found for these parameters, but usually these parameters are determined graphically as shown for the rocket sled in Fig. 1-24.

Figure 1-25 shows the typical step response for some other hypothetical linear time-invariant systems. Curve *a* is characteristic of a system that is *overdamped*—the response is purely exponential, with no overshoot of the final value. Curve *b* is characteristic of a system that is said to be *critically damped*. This is usually a desirable type of response, in that the rise time is comparatively short and is accompanied by a minimal overshoot and a short settling time. Curve *c* is characteristic of a system that is *lightly damped*. It has a large overshoot of its final value, many oscillations, and a long settling time. This would usually be an undesirable system behavior and would have to be corrected by a control system.

Figure 1-26 shows the impulse response for these same three systems. The information regarding overshoot, damping, and settling time is much the same. However, a key piece of information required in system design and analysis is the steady-state value as compared to a constant input value. This is revealed by the unit step response but not by the impulse response. Therefore, it is typical to use the step response as a performance indicator of a linear system rather than the impulse response.

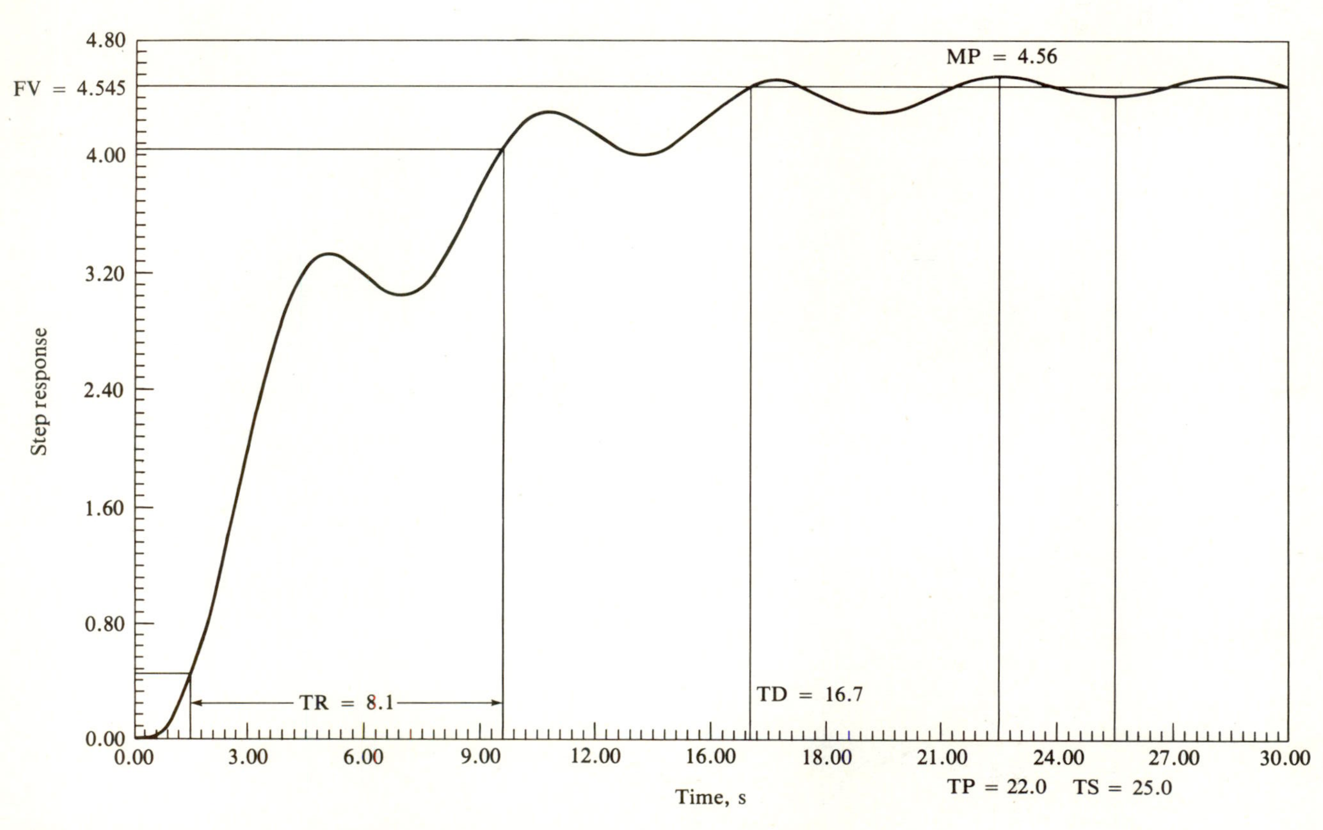

Figure 1-24 Illustration of performance parameters, rocket-sled step response.

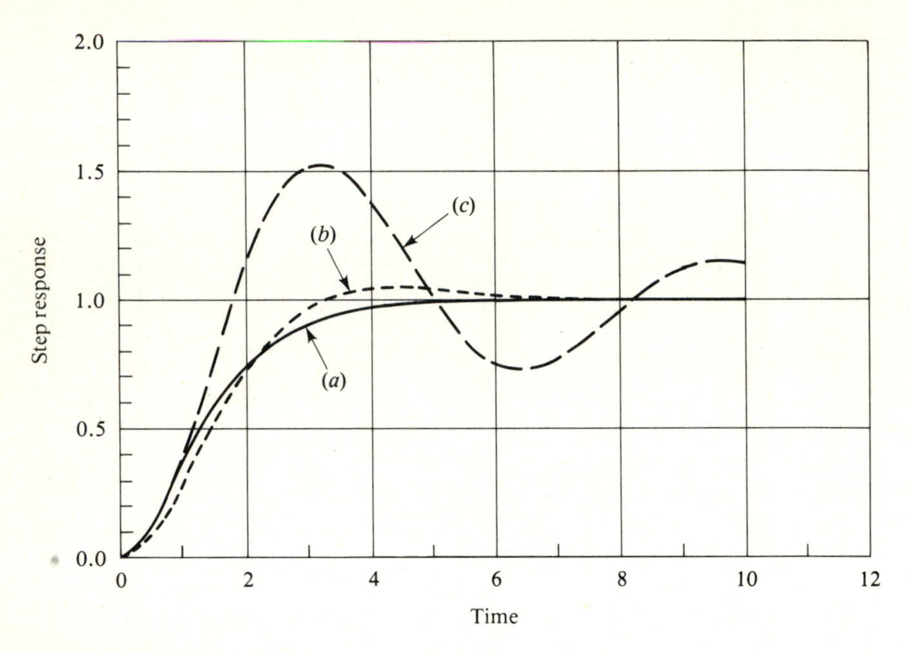

Figure 1-25 Typical step responses for stable linear time-invariant systems.

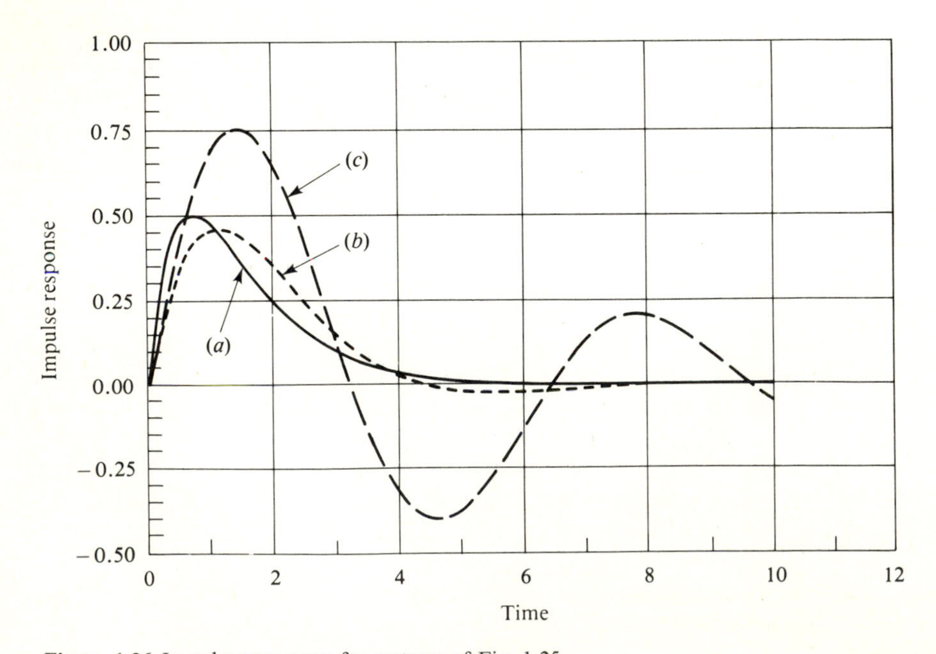

Figure 1-26 Impulse responses for systems of Fig. 1-25.

1-7 STABILITY

All three step responses in Fig. 1-25 represent the response of a stable linear time-invariant system, because all three eventually come to a finite steady-state value. Figure 1-27 depicts the step responses for three hypothetical ideal linear systems which are either unstable or marginally stable.

Figure 1-28 shows the corresponding impulse responses. Curve *a* is an unstable system with a pure exponential growth to infinity. Curve *b* is an unstable system with an exponentially growing sinusoidal response. Finally, curve *c* is a marginally stable system in which the sinusoidal oscillations do not decay to zero, nor do they increase without bound. This would be the response of a perfect harmonic oscillator—say, an ideal pendulum—in which there is zero damping and the harmonic motion continues indefinitely.

Clearly, the unstable behavior of systems *a* and *b* is undesirable, because the growing response will cause the system to destroy itself. The marginally stable system may or may not be desirable, depending on the situation. Regardless, though, stability is a very important system property to determine.

In later chapters, when dealing with an internal differential-equation model of the linear time-invariant system, some precise mathematical conditions can be made for system stability. Yet even at this point, in treating the system from an input-output viewpoint, some mathematical conditions can be placed on system stability. In particular, the step response is given by

$$r(t) = \int_0^t h(\xi)\, d\xi \qquad t > 0 \tag{1-37b}$$

Unstable behavior is exhibited by the unbounded growth of $r(t)$ as $t \to \infty$. Stable behavior occurs when $r(t)$ reaches a constant steady-state value as $t \to \infty$.

Precise mathematical requirements for linear system stability can thus be derived from Eq. (1-37*b*), and from the concepts above. In particular, a linear time-invariant system is *stable** if, and only if,

$$\int_0^\infty |h(\xi)|\, d\xi < \infty \tag{1-43}$$

From a practical viewpoint, Eq. (1-43) means that the impulse-response function for a stable linear time-invariant system decays exponentially to zero. Also, given any $\varepsilon > 0$, no matter how small, a time $T_m > 0$ can be found such that $|h(t)| < \varepsilon$ for all $t \geq T_m$.

For the discrete-system representation, it is also true that there exists some N such that $|h_T(k)| < \varepsilon$ for all $k > N$. Thus if ε is picked small enough to be insignifi-

*This is also known as *bounded-input, bounded-output* (BIBO) stability because bounded inputs are guaranteed of producing bounded outputs. For the linear time-invariant system all "forms" of stability are equivalent.

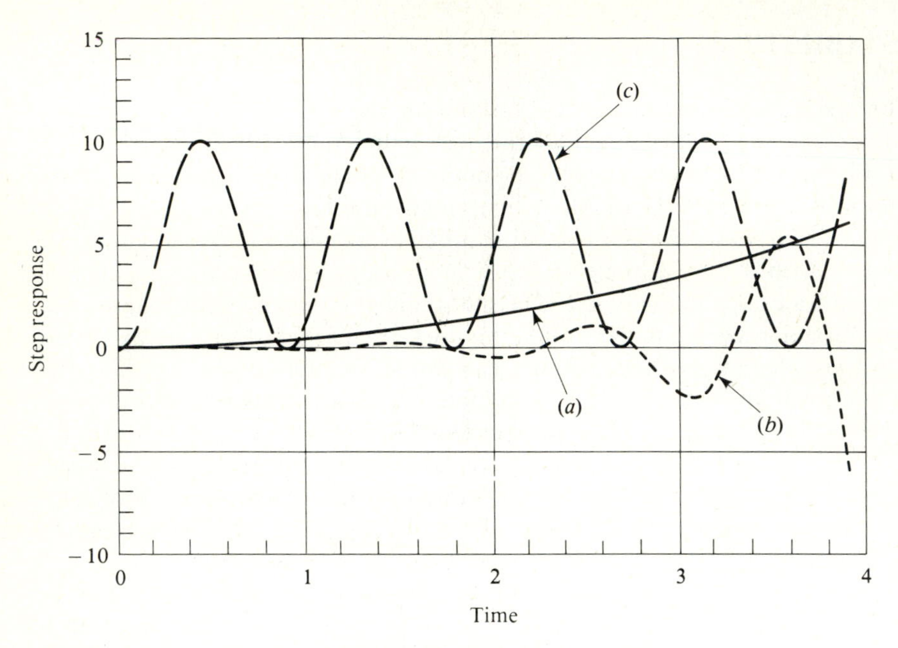

Figure 1-27 Step responses for three unstable linear time-invariant systems.

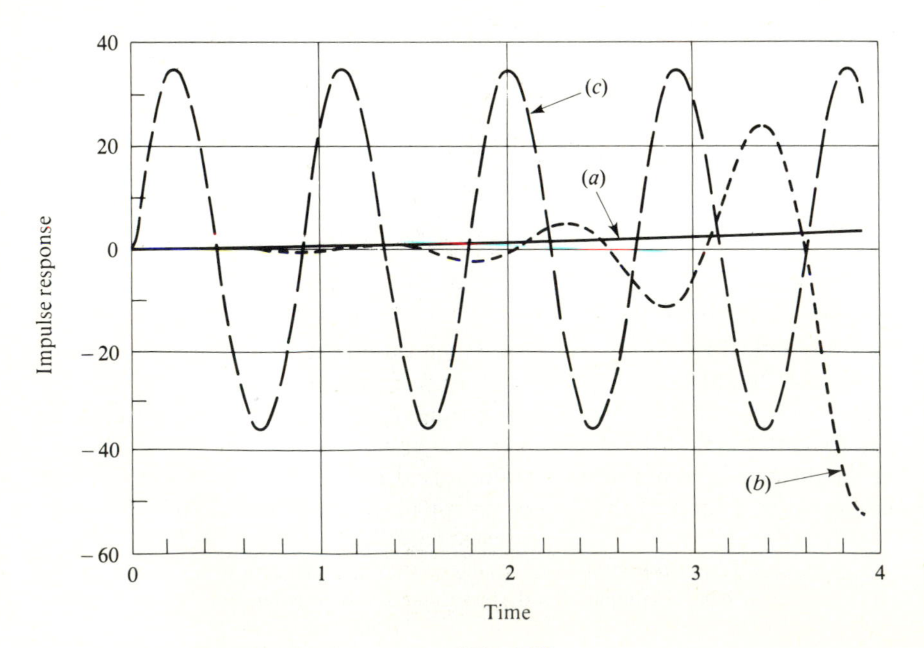

Figure 1-28 Impulse responses for systems of Fig. 1-27.

cant, the infinite discrete convolution

$$y(kT) = \sum_{i=-\infty}^{k-1} h_T(k-i) \cdot u_T(i) \qquad (1\text{-}21a)$$

may be well approximated by the finite convolution summation

$$y(kT) \cong \sum_{i=-N+k}^{k-1} h_T(k-i) u_T(i) \qquad (1\text{-}44)$$

The same holds for the alternative way of writing the discrete convolution summation as exhibited by Eq. (1-20*b*).

The duration of time $N \cdot T = T_m$ thus has an interesting physical interpretation. From Eq. (1-44) it is clear that past inputs as far back in time as $N \cdot T$ time units will continue to have a noticeable effect on the present output. Thus $N \cdot T$ may be viewed as the *effective memory length* of the dynamic system. As the system becomes more and more lightly damped, $N \cdot T$ gets larger and larger. For an unstable system, $N \cdot T = \infty$, and it would require infinite past knowledge to use the input-output convolution representation of the linear dynamic system. Fortunately there are other, more efficient system representations that avoid these problems.

Example 1-5 Consider the pulse-response sequence of the rocket-sled system as presented in Table 1-1. If one uses the criterion $\varepsilon = 0.001$ as being insignificant, one sees from Table 1-1 that

$$|h_T(k)| < \varepsilon$$

for all $k \geq 68$. Since $T = 1$, the effective memory length for this rocket-sled system is $T_m = 68$ s. Thus, past inputs as far back in time as 68 s are influencing the present output with a weighting factor greater than 0.001.

*1-8 TIME-DELAY AND DISTRIBUTED-PARAMETER SYSTEMS

Almost exclusively, this text considers linear time-invariant systems that can be modeled by ordinary differential equations. As will be seen later, such systems admit a finite-dimension-state model in which the state derivative at any time t is only dependent on the state and input at the current time t. Oftentimes, though, one encounters systems with significant delays. Biological systems are prime examples. There is often a significant delay between an input and a reaction to that input. A human operator typically exhibits about a 0.2-s delay before reacting to a stimulus. Economic systems are also good examples; for example, there might be a substantial delay before the effect of a change in monetary policy takes place.

In electrical systems, there is delay in the transmission of signals over long lines or distances. In chemical systems there may be delays before a reaction takes place.

The examples could go on and on. Many such time-delay systems are linear and time-invariant. They may be modeled by what are called linear delay-differential equations. For example, consider the three differential-equation input-outut models:

$$\dot{y}(t) + y(t) = u(t) \tag{1-45}$$

$$\dot{y}(t) + y(t) = u(t - 0.3) \tag{1-46}$$

$$\dot{y}(t) + y(t - 0.3) = u(t) \tag{1-47}$$

$$\dot{y}(t) + 5y(t - 0.3) = u(t) \tag{1-48}$$

Equation (1-45) is an ordinary differential-equation model. Equation (1-46) is a differential-equation model with a simple delay on the input of 0.3; the output derivative, $\dot{y}(t)$, reacts to the current output $y(t)$ and the past input $u(t - 0.3)$. Equations (1-47) and (1-48) are delay-differential equations in which $\dot{y}(t)$ is reacting to the current $u(t)$ input and to the past output value $y(t - 0.3)$. All four models are linear and time-invariant; however, Eqs. (1-46) to (1-48) are time-delay-system models.

Time-delay-systems models are much more difficult to deal with mathematically. Generally, a state model for time-delay systems must be dimensionally infinite. Reasons for this are discussed briefly in Chap. 7. However, if the systems are linear and time-invariant, all of the results regarding the impulse-response model presented in this chapter will still apply. The impulse-response model can be significantly different, though, than the linear time-invariant system without delay. Figures 1-29 and 1-30 show the impulse response for the four systems of Eqs. (1-45) to (1-48). The impulse response of Eq. (1-46) is the same as Eq. (1-45), only it is delayed by 0.3 time units. The impulse responses of Eqs. (1-47) and (1-48) look nothing like that of Eq. (1-45). Obviously, the delay can cause significantly different system response characteristics.

Little treatment can be given in this text to such infinite-dimensional delay differential systems. Mathematically, they are quite complicated. But they do arise frequently in various applications, and so it is important to have some awareness of them.

Another important class of systems referred to as infinite-dimensional are distributed-parameter systems, or systems modeled by partial differential equations. An example is modeling the time evolution of temperature distribution in a rod of finite length; the output variable of interest is the entire distribution of temperature in the rod—an infinite number of data points along the continuous rod. Systems modeled by ordinary differential equations are often referred to as *lumped-parameter* models because one "lumps" a distributed effect into a finite number of nodes. The continuous temperature distribution in the rod might be approximated by the temperatures measured at a finite number of node points.

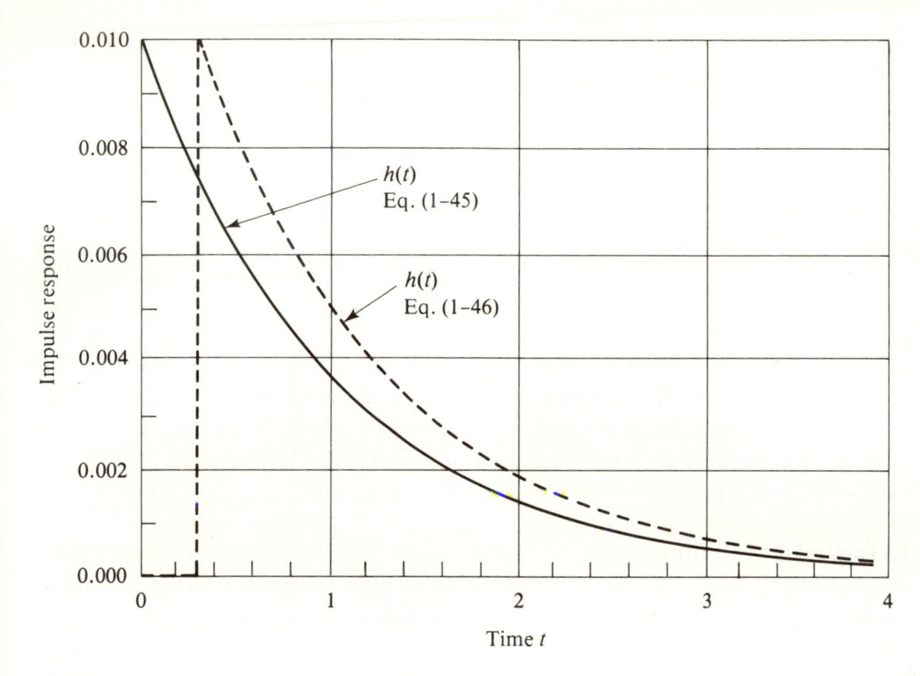

Figure 1-29 Impulse responses for systems Eq. (1-45) and Eq. (1-46).

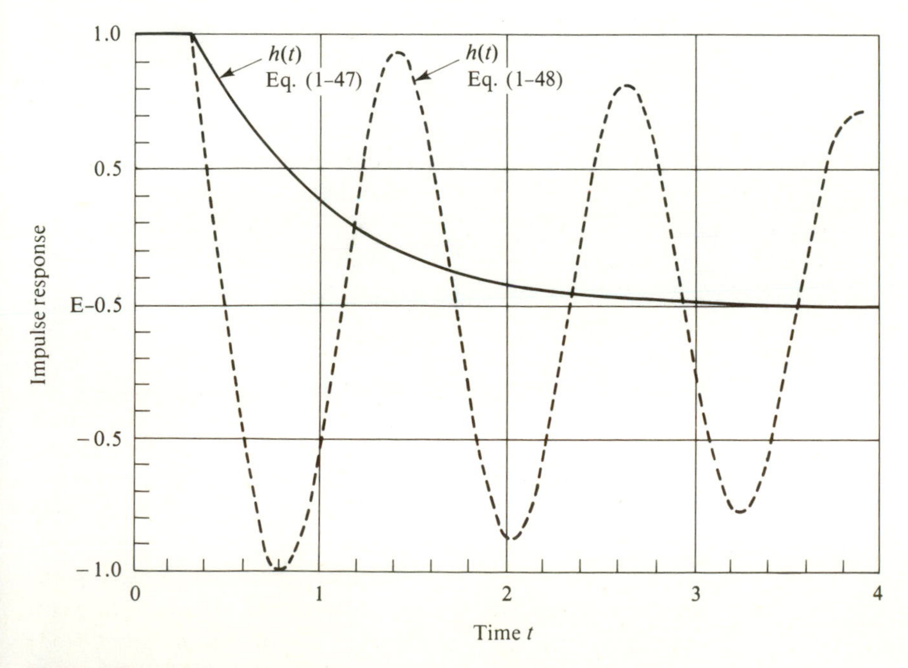

Figure 1-30 Impulse responses for systems Eq. (1-47) and Eq. (1-48).

The distributed-parameter system may also be linear and time-invariant, and thus may have an impulse-response model description. The treatment of such systems, though, is beyond the scope of this book.

1-9 LINEAR TIME-VARYING SYSTEMS

Treatment is given to the time-varying linear system in Chap. 12 with state-variable analysis.

But even on an input-output basis, we may understand basic characteristics of a time-varying linear system and what distinguishes it from a time-invariant linear system. One primary difference is that the impulse response of a time-varying system changes in time as a function of the time at which the impulse is applied to the system. The impulse-response function of a time-invariant system is invariant with the absolute initial time of the impulse.

Example 1-6 The time-invariant linear rocket sled has a constant rocket mass $m_1 = 100$ kg. Figure 1-31 compares the impulse-response function of the rocket-sled system for the system when $m_1 = 100$ kg (curve *a*) to the impulse response when $m_1 = 50$ kg (curve *b*). For the constant-mass system with $m_1 = 100$ kg, the impulse response would be the same for the whole duration of the thrust-force input. For the time-varying mass system the impulse re-

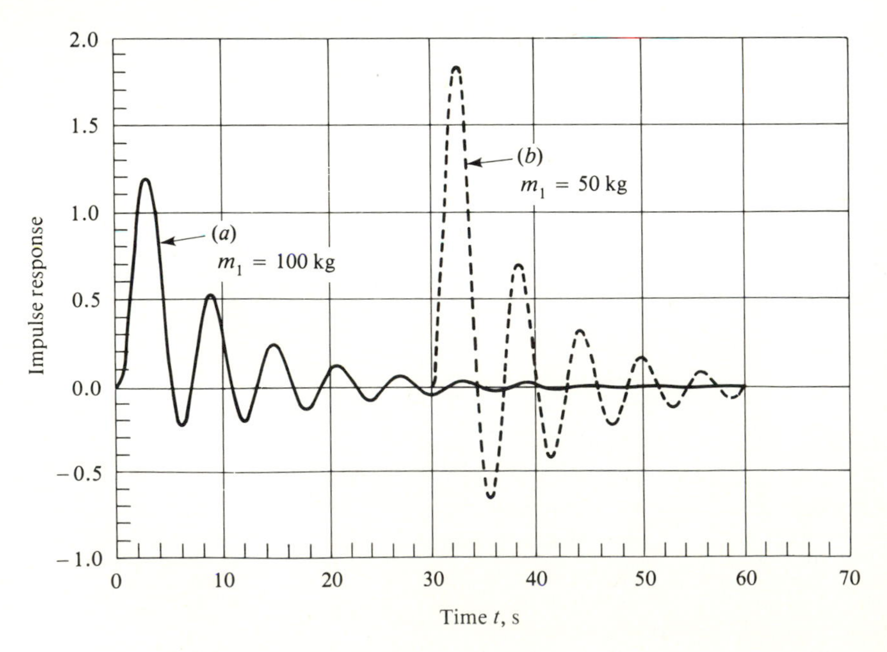

Figure 1-31 Impulse responses rocket-sled system with $m_1 = 100$ kg versus $m_1 = 50$ kg.

sponse function would be continuously changing, depending on the instantaneous value of $m_1(t)$.

To denote the dependence on the time of the impulse, the impulse response function of a time-varying system is denoted by $h(t, \tau)$. The first argument, t, is the current time, and the second argument, τ, is the time that the impulse is applied.

The time-varying linear system is still linear, and the principle of linear superposition still applies. Thus the system response to arbitrary inputs may still be written as an integral expression involving the impulse response function:

$$\begin{aligned} y(t) &= \int_{-\infty}^{t} h(t, \tau)u(\tau)\, d\tau \\ &= \int_{-\infty}^{t_0} h(t, \tau)u(\tau)\, d\tau + \int_{t_0}^{t} h(t, \tau)u(\tau)\, d\tau \\ &= y_{zi}(t) + y_{zs}(t) \end{aligned} \tag{1-49}$$

The integral relation Eq. (1-49) is now called a *superposition integral* rather than the convolution integral associated with the linear time-invariant system. Such an integral is more difficult to deal with. No more is done with the time-varying linear system until the appropriate state-variable methods are discussed in Chap. 12.

1-10 SUMMARY

This chapter introduces the linear time-invariant dynamic system on a physical input-output basis. The three basic properties required of a linear dynamic system are explored:

1. additive decomposition of the zero-input response and the zero-state response;
2. linearity of the zero-input response with respect to the initial state;
3. linearity of the zero-state response with respect to the input

Then the concept of linear superposition is analyzed and used to derive the fundamental convolution representation of the linear time-invariant system. The discrete pulse-response sequence can be found on an experimental basis, and this can, in turn, be used to predict the sampled output for any arbitrary sequence of pulse inputs. Performance characteristics revealed by the system step response are examined along with an input-output concept of system stability. Finally, the impulse response function of a time-varying linear system is introduced.

The next chapter departs from this physical input-output discussion by introducing the differential-equation model and its solution via classic methods. Many of the topics discussed in this present chapter will reappear. Their interpretations are made in terms of characteristics of the differential-equation model.

PROBLEMS

1-1 Demonstrate whether or not the following input-output-initial-state relations satisfy the three requirements of a linear system: (1) additive decomposition between the zero-input response and the zero-state response; (2) linearity of the zero-input response; and (3) linearity of the zero-state response.

(*a*) $y(t) = x(t_0)u(t) + tx(t_0) + \int_{t_0}^{t} e^{\tau}u(\tau)\, d\tau$

(*b*) $y(t) = \sin[x(t_0)t] + \int_{t_0}^{t} \tau u(\tau)\, d\tau$

(*c*) $y(t) = tx(t_0) + \int_{t_0}^{t} \sin(t)u(\tau)\, d\tau$

1-2 Repeat Prob. 1-1 for the following:

(*a*) $y(t) = [x(t_0) + u(t)]^2$

(*b*) $y(t) = tx(t_0)u^2(t) + e^{-t}x(t_0) + \int_{t_0}^{t} \tau^2 u(\tau)\, d\tau$

(*c*) $y(t) = t^2x(t_0) + \int_{t_0}^{t} (e^{-\tau})^2 u(\tau)\, d\tau$

1-3 Suppose the zero-state response of a linear system to the input $u(t) = (\sin 2t)u_{-1}(t)$ is $y_{zs}(t) = (e^{-2t} + 3\sin 2t + \cos 2t)u_{-1}(t)$. Find the system zero-state response to the input $u(t) = (10\sin 2t)u_{-1}(t)$.

1-4 Suppose the zero-state response of a linear system to the input $u_1(t) = u_{-1}(t)$ is $y_{zs_1}(t) = (2e^{-5t} + 3)u_{-1}(t)$ and that the zero-state response to the input $u_2(t) = tu_{-1}(t)$ is $y_{zs_2}(t) = (0.5e^{-5t} - 0.5 + 2t)u_{-1}(t)$. Find the system zero-state response to the input $u(t) = (2 + t/2)u_{-1}(t)$.

1-5 Suppose the zero-state response of a linear time-invariant system to the input $u_1(t) = tu_{-1}(t)$ is $y_{zs_1}(t) = (e^{-t} - 1 + t)u_{-1}(t)$ and to the input $u_2(t) = u_{-1}(t)$ is $y_{zs_2}(t) = (10 - 10e^{-t})u_{-1}(t)$. Find the system zero-state response to the input $u(t) = 2(t - 5)u_{-1}(t - 5)$.

1-6 Given the same system as Prob. 1-5, find the system zero-state response to the input $u(t) = 2tu_{-1}(t - 5)$. (*Hint:* draw a sketch of what the input in Prob. 1.5 looks like versus what this input looks like.)

1-7 Determine if the following differential-equation system models are time-varying or time-invariant:

(*a*) $\dot{y}(t) + 5y(t) = 7u(t)$

(*b*) $\dot{y}(t) + 5(t - 2)y(t) = 7u(t)$

(*c*) $\dot{y}(t) = u^2(t)$

(*d*) $\dot{y}(t) + 2y(t)u(t) = u(t)$

1-8 Determine if the following differential-equation system models are time-varying or time-invariant:

(*a*) $\ddot{y}(t) + y(t) = \dot{u}(t) + 2u(t)$

(*b*) $\ddot{y}(t) + y^2(t) = \dot{y}(t)u(t)$

(*c*) $\dot{y}(t) + \sin(t - 2)y(t) = u(t)$

(*d*) $\dot{y}(t) + \ln(t)y(t) = \dot{u}(t)$

1-9 A linear time-invariant system has a pulse-response model $h_T(1) = 2.2$, $h_T(2) = 1.0$, $h_T(3) = 0.4$, $h_T(4) = 0.1$, $h_T(5) = 0$, $h_T(6) = 0$, $\cdots$. Find $y_T(1)$, $y_T(2)$, $y_T(3)$ for the following input sequences:

(*a*) $u_T(k) = 0,\ k < 0;\ u_T(k) = 1,\ k \geq 0$. This is the unit step input.

(*b*) $u_T(k) = 0,\ k < 0;\ u_T(k) = k + 1,\ k \geq 0$. This is a stair-step input.

(*c*) $u_T(k) = 0,\ k < 0;\ u_T(k) = (-1)^k,\ k \geq 0$. This is a square-wave input.

(*d*) $u_T(-3) = 2,\ u_T(-2) = 1,\ u_T(-1) = -1;\ u_T(k) = 0,\ k \geq 0$. This produces the zero-input response.

(*e*) $u_T(k) = 0, k < 0;\ u_T(0) = 10,\ u_T(1) = -2,\ u_T(2) = 5$. This produces the zero-state response.

(*f*) Input (*d*) and (*e*) together.

1-10 Repeat the inputs for Prob. 1-11 for the linear time-invariant system with the pulse-response model $h_T(1) = 1.0$, $h_T(2) = -0.4$, $h_T(3) = 0.2$, $h_T(4) = -0.1$, $h_T(k) = 0,\ k \geq 5$

1-11 Suppose $h(t)$ has the shape shown in Fig. 1-32*a*. For the input shown in Fig. 1-32*b* find

(*a*) $y_{zi}(t)$, $t \geq 0$
(*b*) $y_{zs}(t)$, $t \geq 0$
(*c*) $y(t)$, $t \geq 0$

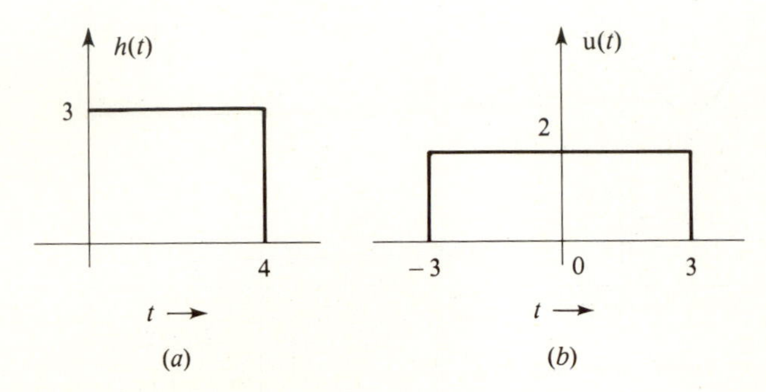

Figure 1-32 System and input for Prob. 1-11.

1-12 For the $h(t)$ shown in Fig. 1-33*a* with the input shown in Fig. 1-33*b* find

(*a*) $y_{zi}(t)$, $t \geq 0$
(*b*) $y_{zs}(t)$, $t \geq 0$
(*c*) $y(t)$, $t \geq 0$

1-13 A linear time-invariant system has an impulse response $h(t) = 2e^{-t}$. The system is at rest at $t = 0$. Find:

(*a*) $y_{zs}(t)$ for the input $u(t) = u_{-1}(t)$, the unit step input.
(*b*) $y(t)$ for the input $u(t) = u_{-1}(t)$.
(*c*) $y(t)$ for the input $u(t) = tu_{-1}(t)$, the unit ramp input.

1-14 A linear time-invariant system has the impulse response $h(t) = 2 \sin t$. The system is at rest at $t = 0$. Find:

(*a*) $y(t)$ for the input $u(t) = u_{-1}(t)$, the unit step input.
(*b*) $y(t)$ for the input $u(t) = e^{-t}u_{-1}(t)$.

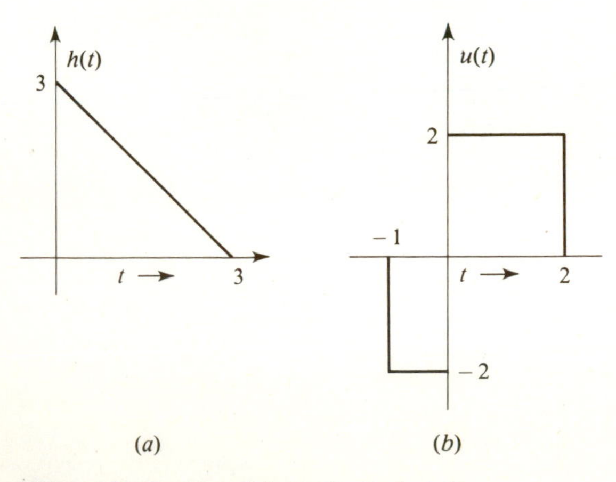

Figure 1-33 System and input for Prob. 1-12.

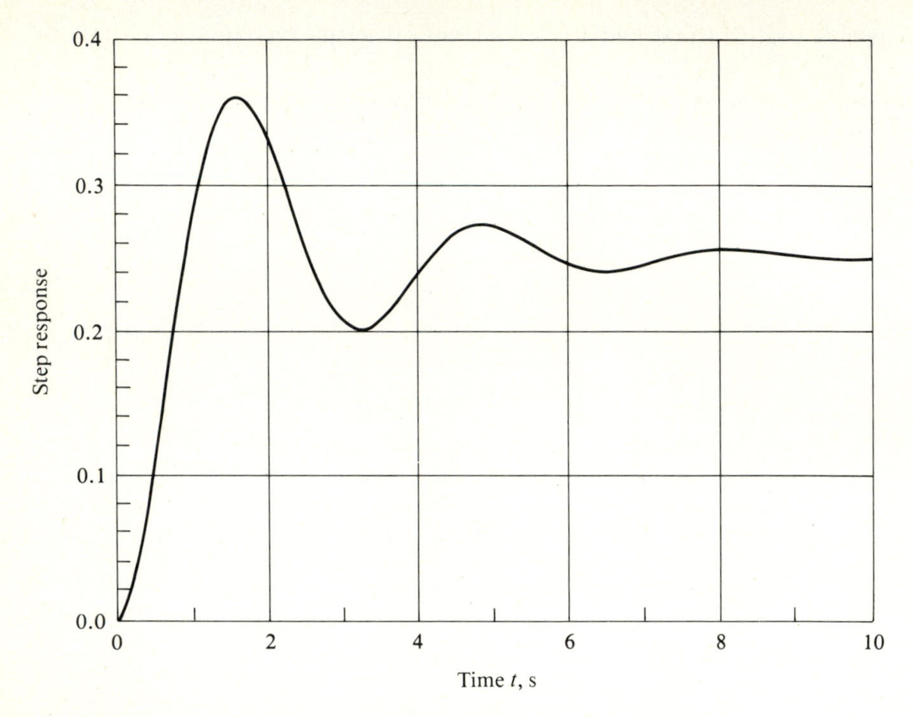

Figure 1-34 System unit step response, Prob. 1-23.

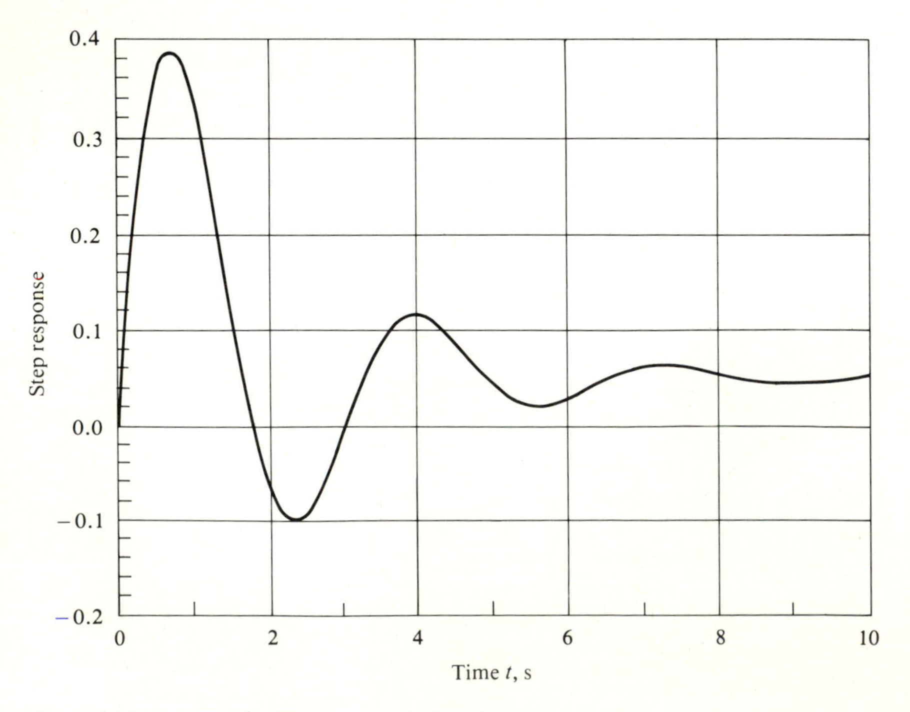

Figure 1-35 System unit step response, Prob. 1-24.

1-15 Evaluate the integral

$$\int_{-\infty}^{\infty} \sin [2(t - \tau)]\, \delta(3 - \tau)\, d\tau$$

1-16 Evaluate the integral

$$\int_{-\infty}^{\infty} e^{-2t+2\tau}\, \delta(\tau + 3)\, d\tau$$

1-17 Suppose the continuous impulse response for a given system is $h(t) = e^{-t}$. Find $h_T(k)$, $k = 1, 2, 3, 4$, with $T = 0.5$.

1-18 Suppose the continuous impulse response for a given system is $h(t) = e^{-t} - e^{-2t}$. Find $h_T(k)$, $k = 1, 2, 3, 4$, with $T = 0.1$.

1-19 Suppose a linear time-invariant system has a sampled unit step response $r(T) = 1.0$, $r(2T) = 1.8$, $r(3T) = 1.4$, $r(4T) = 2.2$. Find $h_T(1)$, $h_T(2)$, $h_T(3)$, $h_T(4)$.

1-20 Suppose a linear time-invariant system has a sampled unit step response $r(T) = -1.2$, $r(2T) = 2.4$, $r(3T) = -2.8$, $r(4T) = 4.2$. Find $h_T(1)$, $h_T(2)$, $h_T(3)$, $h_T(4)$.

1-21 Given the following unit step response, determine the unit impulse response: $r(t) = (-e^{-4t} + e^{-2t} \sin t + 1)u_{-1}(t)$.

1-22 Given the following unit step response, determine the unit impulse response: $r(t) = (-e^{-t} \cos 2t + 1)u_{-1}(t)$.

1-23 For the system step response shown in Fig. 1-34, determine the performance parameters of rise time, duplication time, peak time, settling time, peak value, and final value.

1-24 Repeat Prob. 1-25 for the system step response shown in Fig. 1-35.

1-25 Determine if the following systems are stable or unstable:

(*a*) $h(t) = te^{-t} + e^{-2t} \sin (3t)$

(*b*) $h(t) = e^{-t} - e^{0.01t}$

(*c*) $h(t) = e^{-t} \sin (2t) - t$

1-26 Determine if the following systems are stable or unstable:

(a) $h(t) = e^{t} \sin (2t)$

(*b*) $h(t) = e^{-2t} + e^{-t}$

(*c*) $h(t) = t^4 e^{-0.01t}$

NOTES AND REFERENCES

References 1 and 2, for example, give further background on the problem of estimation of the discrete pulse-response model from experimental observations of the system input and output. Usually measurement noise is a primary concern in an actual experimental procedure. Reference 3 provides some advanced theoretical background on why determination of the continuous impulse response function is such a difficult experimental problem. Reference 4, for example, gives further background on time-delay systems. Reference 5 treats the time-varying linear system.

1. Sage, A. P., and J. L. Melsa: *System Identification,* Academic, New York, 1971, pp. 6–12.
2. Graupe, D.: *Identification of Systems,* Van Nostrand, New York, 1972, pp. 66–71.
3. Audley, D. R., and D. A. Lee: "Ill-Posed and Well-Posed Problems in System Identification", *IEEE Trans. Auto Control,* AC-19, no. 6, December 1974, pp. 738–747.
4. Bellman, R., and K. L. Cooke: *Differential Difference Equations,* Academic, New York, 1963.
5. D'Angelo, H.: *Linear Time-Varying Systems: Analysis and Synthesis,* Allyn and Bacon, Boston, 1970.

CHAPTER

TWO

CLASSIC SOLUTION TO THE LINEAR TIME-INVARIANT ORDINARY DIFFERENTIAL EQUATION MODEL

This chapter reviews the procedure for finding a differential equation model to relate the input forcing function to the output dynamics of the linear time-invariant, single-input, single-output (SISO) system. Classic methods of solution are then derived to solve this output differential equation for selected classes of input signals.

Deriving the differential equation model is a vast topic area, and Sec. 2-1 presents only the briefest review of this subject. For more extensive treatments on modeling, the interested reader is referenced to such texts as Shearer, Murphy, and Richardson [1] and Ogata [2]. Also, Appendix A presents a method to find the transfer function of a given circuit using the theory of linear graphs.

The real heart of the material in this chapter is the solution method that is used once a differential equation model has been derived. The method presented is the so-called *classic method* of solution, in which the total solution is found as the sum of the homogeneous and particular solution parts. This turns out to be an especially easy method of solution for the SISO problem. Furthermore, the homogeneous and particular solution portions the way they are solved for here have special physical significance. For a stable system, the homogeneous part of the solution corresponds to the *transient* portion of the response in that the homogeneous solution decays to zero with increasing time. The particular solution, on the other hand, corresponds to the portion of the response that tracks the input. It will persist for as long as the input persists. For an input that remains for an extended period of time beyond the point at which the initial transients of the homogeneous response have decayed to zero, the particular solution then corresponds to the *steady-state* response of the system to that persistent input.

2-1 DERIVING THE DIFFERENTIAL EQUATION MODEL

The starting point for nearly all of the topics in the remainder of the text is a differential equation model of the given linear time-invariant dynamic system of interest. The output differential equation sought has the general form

$$y^{(n)}(t) + a_{n-1}y^{(n-1)}(t) + \cdots + a_1 y^{(1)}(t) + a_0 y(t)$$
$$= b_m u^{(m)}(t) + \cdots + b_0 u(t) \quad (2\text{-}1)$$

where $y(t)$ is the output variable, $u(t)$ is the input variable, the superscript in parentheses represents differentiation to the appropriate order, $m \leq n$, and all the coefficients (the a_i and b_i) are constant. This first section reviews one method for deriving such a differential equation model.

As discussed in the previous chapter, a physical dynamic system is made up of certain internal elements. For instance, the ideal rocket-sled system (see Fig. 2-1) is made up of the rocket mass m_1 (assumed to be constant); the payload mass m_2; the spring k connecting masses m_1 and m_2; and friction or damper elements b_1 and b_2, which oppose the velocity of masses m_1 and m_2, respectively. These elements react to the external force input, and they interact according to certain laws of physics.

For this mechanical system, the appropriate laws of physics are Newton's laws of motion and Hooke's law governing the force of the spring. Newton's law of interaction states that every action (force) is opposed by an equal and opposite reaction (force in the opposite direction). If one draws a free-body diagram of each mass in the rocket-sled system, the algebraic sum of all forces on the mass (including the reactionary force of the mass itself, $F = ma$) is zero (see Fig. 2-2). Hooke's law states that the force exerted by an ideal linear spring is proportional to the displacement of the spring from its relaxed position.

If the system of interest were, say, an electrical network, then the basic elements might be voltage or current input sources, plus passive elements such as inductors, capacitors, and resistors. The appropriate laws of interaction are Kirchhoff's current and voltage laws. No matter what type of system, though, the procedure is basically similar to the rocket-sled example. Other systems are considered in the homework exercises.

Considering a free-body diagram of the rocket-sled system (see Fig. 2-2), the reaction force of mass m_2, plus the friction force* $F_{b_2} = b_2 v_2$, balance the spring force, F_k:

$$m_2 \frac{d}{dt} v_2(t) + b_2 v_2(t) = F_k(t) \quad (2\text{-}2)$$

where $v_2(t)$ is the velocity of mass m_2. This velocity $v_2(t)$ is the measured output variable, $y(t)$:

$$y(t) = v_2(t) \quad (2\text{-}3)$$

*The friction force is proportional to the velocity, $F_{b_2}(t) = b_2 v_2(t)$, and has a direction opposite to that of v_2.

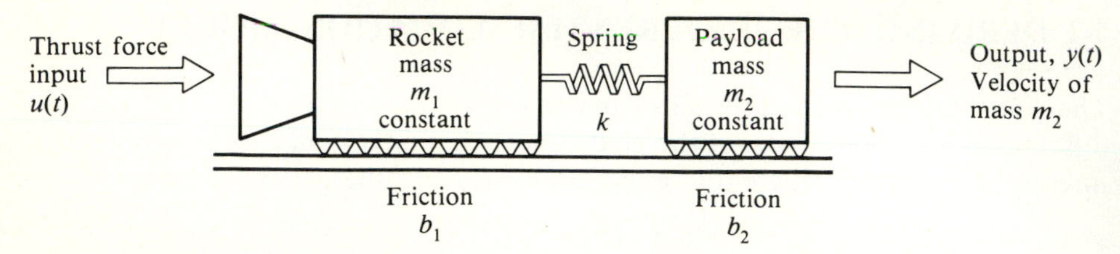

Figure 2-1 Rocket-sled system.

This output relation links the system output to the internal variables. The spring force is given by Hooke's law as

$$F_k(t) = k[x_1(t) - x_2(t)] \tag{2-4}$$

where $x_2(t)$ and $x_1(t)$ are reference positions of the two ends of the spring. Finally, the force balance equation applied to mass m_1 yields

$$m_1 \frac{d}{dt} v_1(t) + b_1 v_1(t) + F_k(t) = u(t) \tag{2-5}$$

where $u(t)$ is the force input of the rocket motor.

Equations (2-2) through (2-5) give all the required equations that relate the input $u(t)$ to the output $y(t)$. But these are not yet in the desired SISO form of Eq. (2-1).

To achieve this objective, the equations must be combined and manipulated. As a first step, Eq. (2-4) is differentiated one time and the substitution of Eq. (2-3) is made, yielding

$$\begin{aligned}\frac{d}{dt} F_k(t) &= k\left[\frac{d}{dt} x_1(t) - \frac{d}{dt} x_2(t)\right] \\ &= k[v_1(t) - v_2(t)] \\ &= k v_1(t) - k y(t)\end{aligned} \tag{2-6}$$

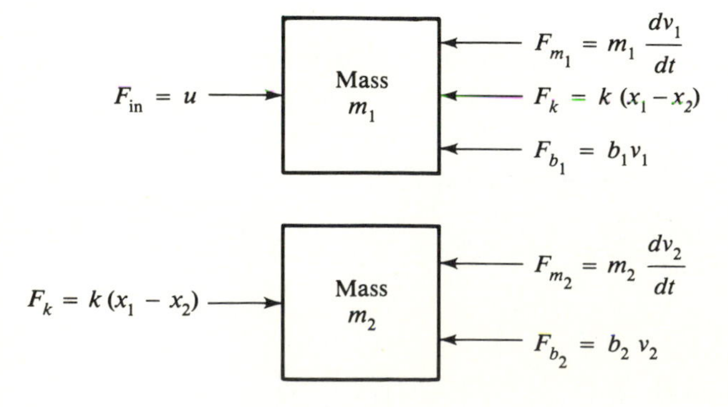

Figure 2-2 Free-body diagram of rocket-sled system.

Next, Eq. (2-2) is differentiated once, and substitutions from Eqs. (2-3) and (2-6) are made to yield

$$m_2 y^{(2)}(t) + b_2 y^{(1)}(t) + ky(t) = kv_1(t) \tag{2-7}$$

The same is done to Eq. (2-5) to give

$$m_1 v_1^{(2)}(t) + b_1 v_1^{(1)}(t) + kv_1(t) - ky(t) = u^{(1)}(t) \tag{2-8}$$

where the superscript notation for differentiation has now been adopted. Equation (2-7) may be solved for $v_1(t)$ in terms of $y(t)$, and the result substituted into Eq. (2-8). This yields

$$\begin{aligned} m_1\left[\frac{m_2}{k}y^{(4)}(t) + \frac{b_2}{k}y^{(3)}(t) + y^{(2)}(t)\right] & \\ + b_1\left[\frac{m_2}{k}y^{(3)}(t) + \frac{b_2}{k}y^{(2)}(t) + y^{(1)}(t)\right] & \\ + k\left[\frac{m_2}{k}y^{(2)}(t) + \frac{b_2}{k}y^{(1)}(t) + y(t)\right] - ky(t) &= u^{(1)}(t) \end{aligned} \tag{2-9}$$

Combining terms yields

$$\begin{aligned} \frac{m_1 m_2}{k}y^{(4)}(t) + \left[\frac{m_1 b_2}{k} + \frac{b_1 m_2}{k}\right]y^{(3)}(t) & \\ + \left[m_1 + \frac{b_1 b_2}{k} + m_2\right]y^{(2)}(t) + (b_1 + b_2)y^{(1)}(t) &= u^{(1)}(t) \end{aligned} \tag{2-10}$$

Both sides of this expression may be integrated once, and both sides may then be divided by $k/m_1 m_2$ to yield the final, standard form of the output differential equation for this system:

$$\boxed{\begin{aligned} y^{(3)}(t) + \left[\frac{m_1 b_2 + m_2 b_1}{m_1 m_2}\right]y^{(2)}(t) + \left[\frac{(m_1 + m_2)k + b_1 b_2}{m_1 m_2}\right]y^{(1)}(t) & \\ + \frac{(b_1 + b_2)k}{m_1 m_2}y(t) &= \frac{k}{m_1 m_2}u(t) \end{aligned}} \tag{2-11}$$

Specific values of the elements m_1, m_2, b_1, b_2, and k may be substituted into Eq. (2-11) to yield the final numerical form of the output differential equation. In this particular case, the following values are utilized:

$$m_1 = 100 \text{ kg}$$

$$m_2 = 10 \text{ kg}$$

$$b_1 = 20 \text{ N} \cdot \text{s/m}$$

$$b_2 = \quad 2\ \text{N} \cdot \text{s/m}$$

$$k = \quad 10\ \text{N/m}$$

Velocities are measured in meters per second and forces are measured in newtons.

With these values, the output differential becomes

$$y^{(3)}(t) + 0.4y^{(2)}(t) + 1.14y^{(1)}(t) + 0.22y(t) = u(t) \qquad (2\text{-}12)$$

The thrust input force is given an additional gain of 100, so that one unit of the thrust force input is actually 100 N.

These are the same rocket-sled dynamics as are utilized in the previous chapter, in which analysis was done on an "experimental" input-output basis. By the methods of this chapter, the output solution is found mathematically for the various input functions that had been considered experimentally in the previous chapter.

2-2 THE FORM OF THE COMPLETE SOLUTION

Once having obtained an output differential equation model in the standard form

$$y^{(n)}(t) + a_{n-1}y^{(n-1)}(t) + \cdots + a_0y(t)$$
$$= b_mu^{(m)}(t) + b_{m-1}u^{(m-1)}(t) \cdots + b_0u(t) \qquad (2\text{-}1)$$

the next step is to obtain a solution of this differential equation for some desired input forcing function; that is, for a given input function $u(t)$, $t \geq t_0$, one wishes to find the output solution, $y(t)$, $t \geq t_0$. This chapter develops one very useful method for accomplishing this—the *classic solution method.*

The classic solution method considers the total solution as the sum of two solutions:

$$y(t) = y_h(t) + y_p(t) \qquad (2\text{-}13)$$

The $y_h(t)$ is termed the *homogeneous* or *complementary solution;* the $y_p(t)$ is termed the *particular solution.*

Although it looks similar, this decomposition of the total solution into the homogeneous and particular solutions is not exactly the same as the zero-input and zero-state decomposition as introduced in the previous chapter:

$$y(t) = y_{zi}(t) + y_{zs}(t) \qquad (2\text{-}14)$$

The total solution is unique, and Eqs. (2-13) and (2-14) are equal; however, in general, $y_h(t) \neq y_{zi}(t)$ and $y_p(t) \neq y_{zs}(t)$. Only their sums, $y(t)$, are equal.

Later chapters will fully develop methods that yield the zero-input and zero-state portions of the solution. Both solution methods are important, and both decompositions—the homogeneous/particular and the zero-input/zero-state—have special physical significance. The physical significance of the zero-input and zero-state portions of the response was explored in the previous chapter. The homogeneous solution for stable systems corresponds to the *natural,* or transient, part of the system response—that portion of the response which is only temporary and which decays to zero with increasing time. The particular solution corresponds to the part of the response that is tracking the input. The difference between $y_{zi}(t)/y_{zs}(t)$ and $y_h(t)/y_p(t)$ is shown explicitly in Chap. 4 when the solution for $y_{zi}(t)$ and $y_{zs}(t)$ is derived.

The Characteristic Polynomial, the Zero Polynomial, and the Transfer Function

Basic to the classic solution are several key polynomials associated with the output differential equation, Eq. (2-1). With $m \leq n$, these polynomials are:

1. The characteristic polynomial:

$$\begin{aligned}\Delta(s) &= s^n + a_{n-1}s^{n-1} + \cdots + a_1 s + a_0 \\ &= (s - p_1)(s - p_2) \cdots (s - p_n)\end{aligned} \tag{2-15}$$

2. The zero polynomial:

$$\begin{aligned}N(s) &= b_m s^m + b_{m-1}s^{m-1} + \cdots + b_1 s + b_0 \\ &= b_m(s - z_1)(s - z_2) \cdots (s - z_m)\end{aligned} \tag{2-16}$$

3. The system transfer function:

$$\begin{aligned}H(s) &= \frac{N(s)}{\Delta(s)} \\ &= \frac{b_m s^m + b_{m-1}s^{m-1} + \cdots + b_1 s + b_0}{s^n + a_{n-1}s^{n-1} + \cdots + a_1 s + a_0} \\ &= \frac{b_m(s - z_1)(s - z_2) \cdots (s - z_m)}{(s - p_1)(s - p_2) \cdots (s - p_n)}\end{aligned} \tag{2-17}$$

The characteristic and zero polynomials are obtained from the left and right sides of the differential equation, Eq. (2-1), by merely replacing the operator variable s for each derivative. The roots of the characteristic polynomial, the p_i, are termed the *system poles.* They play an essential role in the transient time-response characteristics of the system. They are also key to the system property of stability. The roots of the zero polynomial, the z_i, are called the *system zeros.* They also play an essential part in the time-response characteristics of the system. However, the exact nature of their role is not quite so transparent. It will take considerable study to fully appreciate the very important part that the system zeros have in the system behavior.

Example 2-1 The rocket-sled output differential equation is

$$y^{(3)}(t) + 0.4y^{(2)}(t) + 1.14y^{(1)}(t) + 0.22y(t) = u(t)$$

The characteristic and zero polynomials are

$$\Delta(s) = s^3 + 0.4s^2 + 1.14s + 0.22$$
$$= (s + 0.2)(s + 0.1 - 1.044j)(s + 0.1 + 1.044j)$$
$$N(s) = 1$$

where j denotes the complex number $\sqrt{-1}$.

The rocket sled has three system poles and no zero. Notice that the two complex poles p_2 and p_3 occur in complex conjugate pairs. With real-valued polynomial coefficients this will always be true for either complex poles or complex zeros whenever they occur. The transfer function is

$$H(s) = \frac{1}{s^3 + 0.4s^2 + 1.14s + 0.22}$$

$$= \frac{1}{(s + 0.2)(s + 0.1 - 1.044j)(s + 0.1 + 1.044j)}$$

Plotting the Poles and Zeros

After finding the transfer function for the output differential equation, Eq. (2-1), in its factored form, Eq. (2-17),

$$H(s) = \frac{b_m(s - z_1)(s - z_2) \cdots (s - z_m)}{(s - p_1)(s - p_2) \cdots (s - p_n)}$$

it is often helpful to plot the location of the poles and the zeros on the complex plane. The location of the poles are shown by X's and the location of the zeros by 0's.

Example 2-2 For the rocket-sled system there are no zeros, but the poles may be plotted as shown in Fig. 2-3.

Example 2-3 Consider the system

$$y^{(3)}(t) + 4y^{(2)}(t) + 9y^{(1)}(t) + 10y(t) = 5u^{(2)}(t) + 20u^{(1)}(t) + 25u(t)$$

The transfer function is

$$H(s) = \frac{5(s^2 + 4s + 5)}{s^3 + 4s^2 + 9s + 10}$$

$$= \frac{5(s + 2 - j)(s + 2 + j)}{(s + 2)(s + 1 - 2j)(s + 1 + 2j)}$$

A plot of the poles and zeros on the complex plane is shown in Fig. 2-4.

The preceding examples have all poles and zeros in the left half-plane; that is, all the poles and zeros have negative real parts and so they are to the left of the

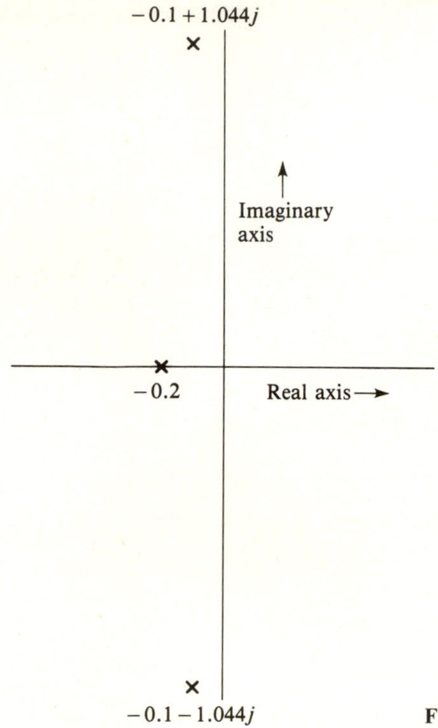

Figure 2-3 Plot of system poles for rocket-sled system.

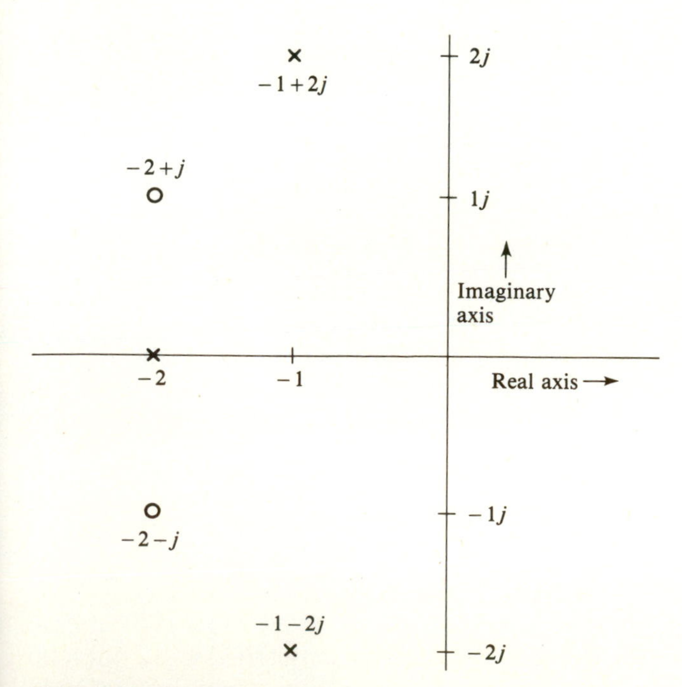

Figure 2-4 Pole-zero plot of the system of Example 2-3.

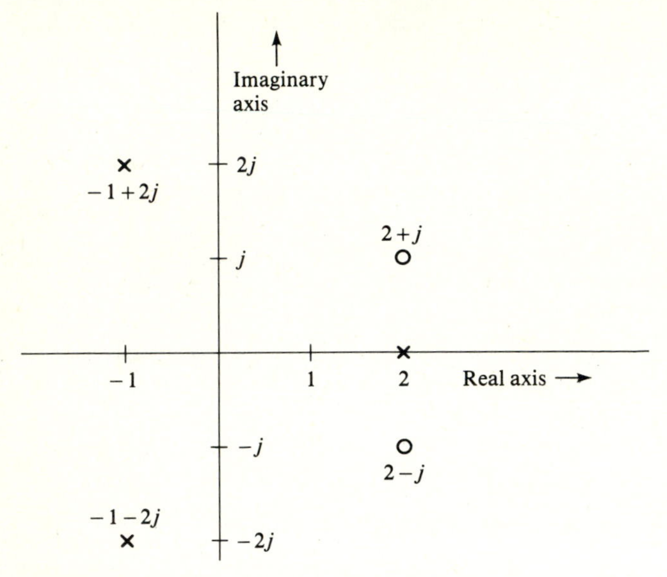

Figure 2-5 Pole-zero plot of the system of Example 2-4.

imaginary $j\omega$ axis. The next example has some poles and zeros in the right half-plane.

Example 2-4

$$y^{(3)}(t) + y^{(1)}(t) - 10y(t) = 5u^{(2)}(t) - 20u^{(1)}(t) + 25u(t)$$

$$H(s) = \frac{5(s^2 - 4s + 5)}{s^3 + s - 10}$$

$$= \frac{5(s - 2 + j)(s - 2 - j)}{(s - 2)(s + 1 - 2j)(s + 1 + 2j)}$$

A pole-and-zero plot of this transfer function is shown in Fig. 2-5.

The significance of the poles being in the right and left half-planes will be seen shortly. The significance of the zero locations is somewhat more difficult to appreciate and will be treated later in the chapter.

2-3 THE HOMOGENEOUS SOLUTION

The homogeneous solution, $y_h(t)$, is required to satisfy the *homogeneous differential equation*

$$y_h^{(n)}(t) + a_{n-1}y_h^{(n-1)}(t) + \cdots + a_1 y_h^{(1)}(t) + a_0 y_h(t) = 0 \tag{2-18}$$

In words, Eq. (2-18) says that the derivative relations of the homogeneous solution are null or add nothing to the derivative relations of $y(t)$, which must add up to the right-hand, forcing function side of Eq. (2-1).

The homogeneous solution is relatively easy to find. Let $p_1, p_2, \ldots, p_n$ be the n roots or poles of the characteristic polynomial, Eq. (2-15). Assume initially that these roots are all distinct; i.e., $p_i \neq p_j$ for $i \neq j$. Then the homogeneous solution is

$$y_h(t) = c_1 e^{p_1 t} + c_2 e^{p_2 t} + \cdots + c_n e^{p_n t} \tag{2-19}$$

The $c_1, c_2, \ldots, c_n$ are arbitrary constants which are determined based upon the boundary conditions on the problem. These are determined *after* forming the complete solution from the sum of the homogeneous and particular solutions, Eq. (2-13). The procedure for doing this will be discussed in Sec. 2.6.

To see that Eq. (2-19) is the homogeneous solution, it is necessary merely to substitute Eq. (2-19) into the homogeneous differential equation, Eq. (2-18), and ensure that equality is satisfied. Doing this yields

$$\begin{aligned} c_1(p_1^n + a_{n-1}p_1^{n-1} + \cdots + a_1 p_1 + a_0)e^{p_1 t} \\ + c_2(p_2^n + a_{n-1}p_2^{n-1} + \cdots + a_1 p_2 + a_0)e^{p_2 t} \\ + \cdots + c_n(p_n^n + a_{n-1}p_n^{n-1} + \cdots + a_1 p_n + a_0)e^{p_n t} = 0 \end{aligned} \tag{2-20}$$

Each term within parentheses is the characteristic polynomial evaluated at one of the poles, p_i:

$$\begin{aligned} \Delta(s)|_{s=p_i} &= (s^n + a_{n-1}s^{n-1} + \cdots + a_1 s + a_0)|_{s=p_i} \\ &= (s - p_1)(s - p_2) \cdots (s - p_n)|_{s=p_i} = 0 \end{aligned} \tag{2-21}$$

Thus each term within parentheses is identically zero, and there is equality for any value of the c_i whatsoever. This makes the homogeneous solution an infinite family of solutions which satisfy the homogeneous differential equation.

Example 2-5 The homogeneous differential equation for the rocket-sled example is

$$y_h^{(3)}(t) + 0.4y_h^{(2)}(t) + 1.14y_h^{(1)}(t) + 0.22y_h(t) = 0 \tag{2-22}$$

From Example 2-1 the system poles are

$$p_1 = -0.2 \qquad p_2 = -0.1 + 1.044j \qquad p_3 = -1 - 1.044j$$

The homogeneous solution is

$$y_h(t) = c_1 e^{-0.2t} + c_2 e^{(-0.1+1.044j)t} + c_3 e^{(-0.1-1.044j)t} \tag{2-23a}$$

The reader is invited to substitute Eq. (2-23*a*) into Eq. (2-22) and see that there is indeed equality. This ensures that Eq. (2-23*a*) is the homogeneous solution for *any* c_1, c_2, and c_3 that are found later.

Because of the complex conjugate poles, p_2 and p_3, there are also several other *equivalent* ways to write the homogeneous solution that avoid the use of the complex exponentials. Namely, because $y_h(t)$ must always be real-valued, it is always true that c_2 and c_3 are complex conjugates of each other. If $c_2 = a_2 + b_2 j$, then $c_3 = a_2 - b_2 j$. Using Euler's identities,* the last two terms in Eq. (2-23*a*) may be rewritten as

$$c_2 e^{(-0.1+1.044j)t} + c_3 e^{(-0.1-1.044j)t}$$

$$= e^{-0.1t}[a_2(e^{1.044jt} + e^{-1.044jt}) + b_2(je^{1.044jt} - je^{-1.044jt})]$$

$$= e^{-0.1t}[(2a_2)\cos(1.044t) + (-2b_2)\sin(1.044t)]$$

Since $c_2 = a_2 + b_2 j$ and $c_3 = a_2 - b_2 j$ are arbitrary constants, one may assign two new arbitrary constants $c_2' \equiv 2a_2$ and $c_3' \equiv -2b_2$. The equivalent homogeneous solution is

$$y_h(t) = c_1 e^{-0.2t} + c_2' e^{-0.1t}\cos(1.044t) + c_3' e^{-0.1t}\sin(1.044t) \tag{2-23b}$$

The arbitrary constants c_2' and c_3' are both real-valued. It is really not necessary to find the complex-valued c_2 and c_3 at all.

Still a third way to write the homogeneous solution is found via the trig identity†

$$c_2'\cos(1.044t) + c_3'\sin(1.044t) = c_2''\sin(1.044t + c_3'')$$

where

$$c_2'' = \sqrt{c_2'^2 + c_3'^2}$$

$$c_3'' = \sin^{-1}(c_2'/c_2'')$$

The equivalent homogeneous solution is

$$y_h(t) = c_1 e^{-0.2t} + c_2'' e^{-0.1t}\sin(1.044t + c_3'') \tag{2-23c}$$

The arbitrary constants are now c_1, c_2'', and c_3''. Again, it is unnecessary to find c_2' and c_3' or c_2 and c_3.

Which form to use for the complex-pole case—form (2-23*a*), (2-23*b*), or (2-23*c*)—depends on the particular application. They are all equivalent, and we have occasion to use each of them.

Repeated Roots

The case in which the characteristic polynomial has roots or poles that are repeated is only slightly more complicated. The rule is simply this: for each repeti-

*$\cos a = (e^{ja} + e^{-ja})/2$
$\sin a = (e^{ja} - e^{-ja})/2$
$e^{ja} = \cos a + j\sin a$

†This may be derived from the relation $\sin(a + b) = \sin b\cos a + \cos b\sin a$.

tion of a pole, p_i, there is an additional $e^{p_i t}$ term in the homogeneous solution with a successively higher factor of t multiplying it. For instance, consider a characteristic polynomial

$$\Delta(s) = s^3(s + 5)^2(s + 7) \tag{2-24}$$

There are three poles, $p_1 = 0$, $p_2 = -5$, and $p_3 = -7$. Pole p_1 is repeated three times and pole p_2 is repeated twice. The homogeneous solution is

$$y_h(t) = c_1 + c_2 t + c_3 t^2 + c_4 e^{-5t} + c_5 t e^{-5t} + c_6 e^{-7t} \tag{2-25}$$

Complex repeated roots work the same way. Consider

$$\Delta(s) = (s + 1 + 2j)^2(s + 1 - 2j)^2 \tag{2-26}$$

The homogeneous solution is

$$y_h(t) = c_1 e^{(-1-2j)t} + c_2 e^{(-1+2j)t} + c_3 t e^{(-1-2j)t} + c_4 t e^{(-1+2j)t} \tag{2-27a}$$

or

$$y_h(t) = c_1' e^{-t} \cos 2t + c_2' e^{-t} \sin 2t + c_3' t e^{-t} \cos 2t + c_4' t e^{-t} \sin 2t \tag{2-27b}$$

or

$$y_h(t) = c_1'' e^{-t} \sin (2t + c_2'') + c_3'' t e^{-t} \sin (2t + c_4'') \tag{2-27c}$$

These forms are all equivalent. See Example 2-5 for discussion.

2-4 SYSTEM STABILITY

The n roots or poles of the characteristic polynomial dictate the form of the homogeneous response regardless of the input applied. Based on this, the linear time-invariant system response is termed *stable* if and only if all system poles are in the negative (left) half-plane. The homogeneous response then decays exponentially to zero. The system is *unstable* if any one or more poles are in the positive (right) half-plane. The homogeneous response then increases exponentially and is unbounded.

If the system has a distinct pole at the origin ($p_i = 0$), then the corresponding term in the homogeneous solution is $c_i e^{0t} = c_i$, a constant. This term neither decays nor increases. The system is said to be *marginally stable.*

If the system has two pure complex conjugate poles with zero real parts, $p_i = j\omega_i$ and $p_{i+1} = -j\omega_i$, the corresponding terms in the homogeneous response are

$$\begin{aligned} c_i e^{j\omega_i t} + c_{i+1} e^{-j\omega_i t} &= c_i' \cos \omega_i t + c_{i+1}' \sin \omega_i t \\ &= c_i'' \sin (\omega_i t + c_{i+1}'') \end{aligned} \tag{2-28}$$

These terms represent a constant-amplitude sinusoid. This is the pure harmonic oscillator. Again the system is said to be marginally stable.

If the system has repeated poles at the origin or on the imaginary axis, then the corresponding terms in the homogeneous response have a t multiplier. For instance, the system with characteristic polynomial $\Delta(s) = s^2(s + 1)$ has homogeneous response

$$y_h(t) = c_1 + c_2 t + c_3 e^{-t} \tag{2-29}$$

If the constant $c_2 \neq 0$, then this represents a growing response, and it is unstable.

These stability conditions for decay or growth of the homogeneous response are reminiscent of the Chap. 1 stability conditions on the exponential decay of the impulse response function.

Later in this chapter it will be shown that the impulse response and the homogeneous response have exactly the same exponential terms, $e^{p_i t}$. Therefore, the Chap. 1 condition for stability of the exponential decay of the impulse response corresponds exactly to the condition here that the system poles all lie in the left half-plane. We may now see, mathematically, why this condition for stability is so.

2-5 THE PARTICULAR SOLUTION FOR EXPONENTIAL INPUTS

The second part of the total solution to Eq. (2-1) is the particular solution, $y_p(t)$. The particular solution is required to satisfy the differential equation, Eq. (2-1) itself with the particular input, $u(t)$, whatever it may be. Namely, $y_p(t)$ is *any* solution satisfying

$$y_p^{(n)}(t) + a_{n-1} y_p^{(n-1)}(t) + \cdots + a_0 y_p(t) = b_m u^{(m)}(t) + \cdots + b_0 u(t) \tag{2-30}$$

There are quite a few different methods for finding a particular solution, but only one is developed here. This method provides an essentially unique form for the particular solution; it is especially simple to utilize; and it has a great deal of physical significance. It equals the steady-state response for cases of a persistent input, and it may be used to define the frequency-response behavior of the system.*

Exponential-Class Inputs

The particular solution technique developed here is limited to input signals that may be written in the form of some linear operation and/or a linear combination of exponential inputs. This is not a particularly severe limitation; Table 2-1 shows that most common closed-form input signals may be written as a linear operation on an exponential input.

*It is also remarked that this same technique may be used to find the particular solution for more general classes of linear time-invariant systems than simply ordinary differential equation systems as considered here. For example, with proper definition of $H(s)$ and $y_p(t)$, this technique may be applied to time-delay and distributed parameter systems as introduced briefly in Sec. 1.8.

Table 2-1 Common inputs written as linear operation on e^{st}

Input $u(t)$	$u(t) = L\{e^{st}\}$
1. ke^{qt}	$(ke^{st})_{s=q} = ke^{qt}$
2. k	$(ke^{st})_{s=0} = ke^{0t} = k$
3. kt	$\frac{d}{ds}(ke^{st})\vert_{s=0} = (kte^{st})\vert_{s=0} = kt$
4. kte^{qt}	$\frac{d}{ds}(ke^{st})\vert_{s=q} = (kte^{st})\vert_{s=q} = kte^{qt}$
5. $k \sin(\omega t + \theta)$	$\text{Im}\,\{ke^{j\theta}e^{st}\}_{s=j\omega}$ $= \text{Im}\,\{k[\cos(\omega t + \theta) + j\sin(\omega t + \theta)]\}$ $\equiv k\sin(\omega t + \theta)$
6. $k \cos(\omega t + \theta)$	$\text{Re}\,\{ke^{j\theta}e^{st}\}_{s=j\omega}$ $= \text{Re}\,\{k[\cos(\omega t + \theta) + j\sin(\omega t + \theta)]\}$ $\equiv k\cos(\omega t + \theta)$

To demonstrate what is meant by a linear operation on e^{st}, consider first the t input, item 3 of Table 2-1. It should be clear that

$$\frac{d}{ds}(e^{st})\bigg|_{s=0} = (te^{st})\bigg|_{s=0} = t \tag{2-31}$$

What we claim is that

$$L\{\cdot\} \equiv \frac{d}{ds}\{\cdot\}\bigg|_{s=0} \tag{2-32}$$

is a linear operation on its argument—any differentiable function of s. Showing that Eq. (2-31) is a linear operation is conceptually identical to showing that a given function $f(x)$ is a linear function of its argument: $f(x)$ must satisfy the two conditions of linear additivity and linear homogeneity (see Sec. 1.1). To this end, let $g_1(s)$ and $g_2(s)$ be two arbitrary differentiable functions of s. Then

$$\begin{aligned} L\{g_1(s) + g_2(s)\} &\equiv \frac{d}{ds}[g_1(s) + g_2(s)]\bigg|_{s=0} \\ &= \frac{d}{ds}[g_1(s)]\bigg|_{s=0} + \frac{d}{ds}[g_2(s)]\bigg|_{s=0} \\ &= L\{g_1(s)\} + L\{g_2(s)\} \end{aligned} \tag{2-33}$$

and so $L\{\cdot\}$ satisfies linear additivity. Second,

$$\begin{aligned} L\{kg_1(s)\} &\equiv \frac{d}{ds}[kg_1(s)]\bigg|_{s=0} \\ &= k\frac{d}{ds}[g_1(s)]\bigg|_{s=0} \\ &= kL\{g_1(s)\} \end{aligned} \tag{2-34}$$

and so $L\{\cdot\}$ satisfies linear homogeneity. Since both linear additivity and linear homogeneity are satisfied, it follows that $L\{\cdot\} \equiv (d/ds)\{\cdot\}|_{s=0}$ is a linear operation.

Consider next the sine and the cosine inputs, $u(t) = \sin(\omega t)$ and $u(t) = \cos(\omega t)$. From Euler's identity it is known that

$$e^{j\omega t} = \cos \omega t + j \sin \omega t \tag{2-35}$$

Therefore, defining the linear operations Re $\{\cdot\} \equiv$ the real part of complex argument and Im $\{\cdot\} \equiv$ the imaginary part of the complex argument, it is seen that

$$\cos(\omega t) = \text{Re}\,\{e^{j\omega t}\} = \text{Re}\,\{e^{st}\}|_{s=j\omega} \tag{2-36}$$

$$\sin(\omega t) = \text{Im}\,\{e^{j\omega t}\} = \text{Im}\,\{e^{st}\}|_{s=j\omega} \tag{2-37}$$

Now it only remains to be shown that Re $\{\cdot\}$ and Im $\{\cdot\}$ are linear operations. To show this, let

$$c_1 = a_1 + b_1 j \qquad c_2 = a_2 + b_2 j$$

be two arbitrary complex variables. Then

$$\begin{aligned} \text{Re}\,\{c_1 + c_2\} &= \text{Re}\,\{a_1 + b_1 j + a_2 + b_2 j\} \\ &= a_1 + a_2 \\ &= \text{Re}\,\{c_1\} + \text{Re}\,\{c_2\} \end{aligned} \tag{2-38}$$

and

$$\begin{aligned} \text{Re}\,\{kc_1\} &= \text{Re}\,\{ka_1 + kb_1 j\} \\ &= ka_1 \\ &= k\,\text{Re}\,\{c_1\} \end{aligned} \tag{2-39}$$

for the real constant k. Thus Re $\{\cdot\}$ satisfies both linear additivity and linear homogeneity. Thus it is a linear operation. The linearity of Im $\{\cdot\}$ may be proven in exactly the same way.

In a similar manner all of the inputs in Table 2-1 may be shown to be linear operations of e^{st} of the form $u(t) = L\{e^{st}\}$, where the linear operator $L\{\cdot\}$ is defined appropriately. Problems 2-23 and 2-24 give the reader some additional practice in demonstrating that the entries in Table 2-1 are linear, and they also show how the list can be expanded to other common input functions.

Solution Technique: Simplest Case

The technique for finding the particular solution is first shown for the simplest case: a real exponential input signal

$$u(t) = ke^{qt} \tag{2-40}$$

where it is assumed that k and q are both real and q is not equal to any of the system poles.* To illustrate, consider the examples:

*We note that in this technique for finding the particular solution it is immaterial as to just when the initial time happens to be. It doesn't matter when the input *starts* having the form Eq. (2-40), but only that it does have this form over the time period of interest.

Example 2-6 Switching to the notation $\dot{y}(t) \equiv (d/dt)y(t)$, $\ddot{y}(t) = (d^2/dt^2)y(t)$, etc., consider

$$\dot{y}(t) + 5y(t) = u(t)$$

with input

$$u(t) = 3e^{2t} \qquad t > 0$$

The system transfer function is

$$H(s) = \frac{1}{s+5}$$

It is claimed that a particular solution is

$$\begin{aligned} y_p(t) &= [H(s)3e^{st}]|_{s=2} \\ &= 3H(2)e^{2t} \\ &= \frac{3}{2+5}e^{2t} \\ &= \frac{3}{7}e^{2t} \end{aligned}$$

To be a particular solution, $y_p(t) = 3/7e^{2t}$ must satisfy the original differential equation

$$\frac{d}{dt}\left(\frac{3}{7}e^{2t}\right) + 5\left(\frac{3}{7}e^{2t}\right) = 3e^{2t}$$

Carrying out the differentiation on the left and combining terms, it is seen that there is indeed equality. This proves that the claimed solution is a particular solution.

Note that the particular solution takes the form of the input with the weighting factor $H(s)|_{s=2}$, where $s = 2$ is the exponent of the exponential input.

Example 2-7

$$\ddot{y}(t) + 3\dot{y}(t) + 2y(t) = \dot{u}(t) + 3u(t)$$

Now let $u(t)$ have the general form

$$u(t) = ke^{qt}$$

The system transfer function is

$$H(s) = \frac{s+3}{s^2+3s+2} = \frac{s+3}{(s+1)(s+2)}$$

and so the system has a zero at -3, a pole at -1, and a pole at -2. The particular solution is

$$\begin{aligned} y_p(t) &= [H(s)ke^{st}]|_{s=q} = H(q)ke^{qt} \\ &= k\frac{q+e}{(q+1)(q+2)}e^{qt} \end{aligned}$$

provided that $q \neq -1$ or -2. This is treated as a special case later. The claimed particular solution may be checked by verifying that

$$\frac{d^2}{dt^2}[kH(q)e^{qt}] + 3\frac{d}{dt}[kH(q)e^{qt}] + 2[kH(q)e^{qt}] = \frac{d}{dt}(ke^{qt}) + 3(ke^{qt})$$

The derivatives on both sides of the equation may be taken to yield

$$kH(q)q^2e^{qt} + 3kH(q)qe^{qt} + 2kH(q)e^{qt} = kqe^{qt} + 3ke^{qt}$$

The common factor ke^{qt} may be divided from every term, and $H(q)$ factored from every term on the left, to give

$$H(q)(q^2 + 3q + 2) = q + 3$$

Dividing both sides by $(q^2 + 3q + 2)$ yields

$$H(q) = \frac{q+3}{q^2 + 3q + 2}$$

which indeed is an equality. Thus $y_p(t)$ has been shown to be a particular solution for any value of q except $q = -1$ or $q = -2$.

Exponential input corresponding to a system zero An interesting case occurs when

$$u(t) = ke^{-3t} \tag{2-41}$$

with $q = -3$ coinciding with the system zero at -3 in Example 2-7. Then $H(-3) = 0$, and

$$y_p(t) = H(-3)ke^{-3t} = 0 \tag{2-42}$$

Hence there is zero particular solution; the system response is simply the natural or homogeneous response alone (which is, in general, not zero). No component of the input waveform e^{-3t} appears in the total response, as it would under circumstances when the input exponent did not coincide with the system zero; the location of the system zero simply zeroes out the exponential input of that value. Again, this does not mean that the total system response is zero, as there is still the natural or homogeneous part of it—just the duplication of the input waveform is missing. This is one "physical" property of the system zero—a property that could actually be demonstrated experimentally. Other interesting properties of the system zeros are discussed as we proceed.

Constant input A second interesting case for the general form

$$u(t) = ke^{qt} \tag{2-40}$$

is when $q \equiv 0$. Then $u(t) = k$ is the constant input, and

$$\begin{aligned} y_p(t) &= H(0)k \\ &= \frac{b_0}{a_0}k \end{aligned} \tag{2-43}$$

For this case, $y_p(t)$ is the steady-state, or final, value of the system response to the constant input.* The particular solution, the way it is derived here, is later seen to be the steady-state response in a number of other significant cases.

Summary Summarizing the results of these two examples, it is true that:

Given

1. An input-output, linear time-invariant system with transfer function $H(s)$.
2. An input of the form

$$u(t) = \sum_{i=1}^{N} k_i e^{q_i t} \tag{2-44}$$

where none of the q_i input exponents exactly equals a pole of the system transfer function $H(s)$, and t ranges over any given time span of interest.

Then

1. The system response has the particular solution

$$y_p(t) = \sum_{i=1}^{N} k_i H(q_i) e^{q_i t} \tag{2-45}$$

2. If any of the q_i corresponds to a system zero, z_i, then $H(q_i) = 0$, and that particular exponent value does not appear in the particular solution response and hence does not appear in the total system response $y(t)$.

Note that the transfer function, $H(s)$, does not have to be in factored form to calculate $H(q_i)$. This is fortunate, since factoring large-degree polynomials is often a difficult proposition, especially by hand.

General Case, Exponential-Class Inputs

Table 2-1 shows how quite a number of common input signals [e.g., $u(t) = t$, $u(t) = \sin \omega t$, etc.] may be written as a "linear operation" on an exponential input of the form e^{st}, for the complex variable s. In the general form this is written as

$$u(t) = L\{e^{st}\} \tag{2-46}$$

Table 2-2 then shows that the particular solution for such an input has the form

$$y_p(t) = L\{H(s)e^{st}\} \tag{2-47}$$

*This assumes, of course, that the system is stable so that the homogeneous portion of the response decays to zero with increasing time.

Table 2-2 Particular solution for exponential-class inputs

$u(t)$	$u(t) = L\{e^{st}\}$	$y_p(t) = L\{H(s)e^{st}\}$
1. ke^{qt}	ke^{qt}	$y_p(t) = kH(q)e^{qt}$
2. k	ke^{0t}	$y_p(t) = kH(0) = k\dfrac{b_0}{a_0}$
3. kt	$k\dfrac{d}{ds}(e^{st})_{s=0}$	$y_p(t) = k\dfrac{d}{ds}[H(s)e^{st}]_{s=0}$ $= k\dfrac{d}{ds}[H(s)]_{s=0} + kH(0)t$ $= k\left(\dfrac{b_1}{a_0} - \dfrac{b_0a_1}{a_0^2}\right) + k\dfrac{b_0}{a_0}t$
4. kte^{qt}	$k\dfrac{d}{ds}(e^{st})_{s=q}$	$y_p(t) = k\dfrac{d}{ds}[H(s)e^{st}]_{s=q}$ $= k\dfrac{d}{ds}[H(s)]_{s=q}e^{qt} + kH(q)te^{qt}$
5. $k\sin(\omega t + \theta)$	$k\,\text{Im}\,\{e^{j\theta}e^{st}\}_{s=j\omega}$	$y_p(t) = k\,\text{Im}\,\{H(j\omega)e^{j(\omega t+\theta)}\}$ $= k\|H(j\omega)\|\sin(\omega t + \theta + \phi)$
6. $k\cos(\omega t + \theta)$	$k\,\text{Re}\,\{e^{j\theta}e^{st}\}_{s=j\omega}$	$y_p(t) = k\,\text{Re}\,\{H(j\omega)e^{j(\omega t+\theta)}\}$ $= k\|H(j\omega)\|\cos(\omega t + \theta + \phi)$

where $H(j\omega) = |H(j\omega)|e^{j\phi}$

This is nothing more than a generalization of the simplest case.

Having already established that inputs such as t, $\sin\omega t$, or $\cos\omega t$, can be written as linear operations on e^{st}, it is now necessary to show that the particular solution for such an input has the form of Eq. (2-47). To do this for the general system is notationally cumbersome; therefore, it is done for only one example system:

$$\ddot{y}(t) + 3\dot{y}(t) + 2y(t) = \dot{u}(t) + 3u(t) \tag{2-48}$$

If Eq. (2-47) is to be a particular solution for the input $u(t) = L\{e^{st}\}$, it must satisfy the differential equation:

$$\frac{d^2}{dt^2}[L\{H(s)e^{st}\}] + 3\frac{d}{dt}[L\{H(s)e^{st}\}] + 2L\{H(s)e^{st}\} = \frac{d}{dt}[L\{e^{st}\}] + 3L\{e^{st}\} \tag{2-49}$$

Now, since $L\{\cdot\}$ is a linear operation on a function of s, the derivatives with respect to time t may be brought inside the linear operation to give

$$L\{H(s)s^2e^{st}\} + 3L\{H(s)se^{st}\} + 2L\{H(s)e^{st}\} = L\{se^{st}\} + 3L\{e^{st}\}$$

Again, the fact that $L\{\cdot\}$ is a linear operation can be used to combine terms inside $L\{\cdot\}$, giving

$$L\{H(s)(s^2 + 3s + 2)e^{st}\} = L\{(s + 3)e^{st}\}$$

Finally, the fact that

$$H(s)(s^2 + 3s + 2) = \frac{s + 3}{s^2 + 3s + 2}(s^2 + 3s + 2)$$
$$= (s + 3)$$

may be used to give

$$L\{(s + 3)e^{st}\} = L\{(s + 3)e^{st}\}$$

Thus $L\{H(s)e^{st}\}$ is indeed a particular solution for the example system. This same procedure may be generalized to any $H(s)$ rational transfer function.

Now to carry through with this example system, Eq. (2-48), and using the t input

$$u(t) = t = \frac{d}{ds}(e^{st})\bigg|_{s=0} \tag{2-50}$$

it is seen that

$$y_p(t) = \frac{d}{ds}(H(s)e^{st})\bigg|_{s=0}$$
$$= \frac{d}{ds}\left(\frac{s + 3}{s^2 + 3s + 2}e^{st}\right)\bigg|_{s=0}$$
$$= \left(\frac{1}{2} - \frac{3 \cdot 3}{4} + \frac{3}{2}t\right)$$

or finally

$$y_p(t) = -\frac{7}{4} + \frac{3}{2}t = \left(\frac{b_1}{a_0} - \frac{b_0 a_1}{a_0^2}\right) + \frac{b_0}{a_0}t \tag{2-51}$$

To check that this is a particular solution, one substitutes $y_p(t)$ into the differential equation, Eq. (2-51), and checks for equality:

$$\frac{d^2}{dt^2}\left(\frac{-7}{4} + \frac{3}{2}t\right) + 3\frac{d}{dt}\left(\frac{-7}{4} + \frac{3}{2}t\right) + 2\left(\frac{-7}{4} + \frac{3}{2}t\right) = \frac{d}{dt}(t) + 3(t) \tag{2-52}$$

After taking the derivatives on both sides this becomes

$$1 + 3t = 1 + 3t$$

which checks.

By the same procedure, the particular solution for the input

$$u(t) = \sin\left(2t + \frac{\pi}{4}\right) = \text{Im}\,\{e^{st}e^{j\pi/4}\}\bigg|_{s=2j} \tag{2-53}$$

is found to be

$$y_p(t) = \text{Im}\,\{H(s)e^{st}e^{j\pi/4}\}\,|_{s=2j}$$

$$= \text{Im}\left\{\frac{s+3}{s^2+3s+2}e^{st}e^{j\pi/4}\right\}\Bigg|_{s=2j}$$

$$= \text{Im}\left\{\frac{2j+3}{(2j)^2+3(2j)+2}e^{j(2t+\pi/4)}\right\}$$

$$= \text{Im}\left\{\frac{3+2j}{-2+6j}\left[\cos\left(2t+\frac{\pi}{4}\right)+j\sin\left(2t+\frac{\pi}{4}\right)\right]\right\}$$

$$= \text{Im}\left\{\frac{6-22j}{40}\left[\cos\left(2t+\frac{\pi}{4}\right)+j\sin\left(2t+\frac{\pi}{4}\right)\right]\right\}$$

or finally

$$y_p(t) = \frac{-11}{20}\cos\left(2t+\frac{\pi}{4}\right)+\frac{3}{20}\sin\left(2t+\frac{\pi}{4}\right) \tag{2-54a}$$

The reader is invited to check that Eq. (2-54) is the particular solution by substituting it into Eq. (2-48) with input Eq. (2-53) and checking for equality.

This example brings up another interesting point. Just like the case of the complex terms in the homogeneous response, Euler's identities may be used to write the particular solution, Eq. (2-54), in several different forms. An especially revealing form is obtained as follows:

$$y_p(t) = \text{Im}\,\{H(s)e^{st}e^{j\pi/4}\}\,|_{s=2j}$$

$$= \text{Im}\left\{\frac{3+2j}{-2+6j}e^{(2t+\pi/4)j}\right\}$$

$$= \text{Im}\left\{\frac{3.6\angle 33.7^\circ}{6.32\angle 108.4^\circ}e^{(2t+\pi/4)j}\right\}$$

$$= \text{Im}\,\{0.57\angle -74.7^\circ e^{(2t+\pi/4)j}\}$$

or finally

$$y_p(t) = 0.57\sin(2t-0.518) \tag{2-54b}$$

Notice that

$$H(2j) = |H(j\omega)|\angle\phi$$
$$= 0.57\angle -74.7^\circ \tag{2-55}$$

The factor 0.57 is the magnitude of $H(j\omega)$ at $\omega = 2$, the frequency of the input, and -74.7 degrees is the phase angle of $H(j\omega)$ at $\omega = 2$. This concept forms the foundation of the discussion of the next chapter dealing with the frequency analysis of a linear dynamic system.

All of the inputs considered in Table 2-2, or any linear combination of these inputs, may be considered in a similar fashion. The exercises at the end of the chapter provide some more practice in doing this. The next step in solving the general linear differential equation, Eq. (2-1), is to add together the homogeneous and particular solutions to form the complete response. This is considered in the next section; however, before doing so we may now tidy up one remaining loose end.

A Special Case

By applying the same sort of linear operator wizardry as in the previous subsection, it is now possible to consider the special case when

$$u(t) = ke^{qt} \tag{2-40}$$

and $q = p_i$, one of the system poles. As stated earlier, it is impossible to directly use the form

$$y_p(t) = kH(q)e^{qt}$$

because $H(q) = H(p_i)$ does not exist (it becomes infinite). However, by considerable foresight it is seen that

$$u(t) = ke^{p_i t} = k\left[\frac{d}{ds}(s - p_i)e^{st}\right]_{s=p_i} \tag{2-56}$$

Therefore, by the previous discussion it follows that

$$y_p(t) = k\frac{d}{ds}[H(s)(s - p_i)e^{st}]_{s=p_i} \tag{2-57}$$

Now, since $(s - p_i)$ is a factor multiplying $H(s)$, the troublesome pole at p_i will divide out, leaving the result safe to be evaluated at $s = p_i$.

If there were repeated roots at p_i, then higher factors of $(s - p_i)$ and higher derivatives of s would have to be utilized.

Example 2-8

Given

$$H(s) = \frac{1}{(s + 1)(s + 5)} \qquad \text{and} \qquad u(t) = 2e^{-t}$$

Then

$$y_p(t) = 2\frac{d}{ds}\left[\frac{1}{(s + 1)(s + 5)}(s + 1)e^{st}\right]_{s=-1}$$

$$= 2\frac{d}{ds}\left[\frac{1}{(s + 5)}e^{st}\right]_{s=-1}$$

$$y_p(t) = 2\left(\frac{-1}{16}\right)e^{-t} + 2\left(\frac{1}{4}\right)te^{-t}$$

The reader may check that this satisfies

$$\ddot{y}_p(t) + 6\dot{y}_p(t) + 5y_p(t) = 2e^{-t}$$

2-6 THE COMPLETE SOLUTION

Having found $y_h(t)$ and $y_p(t)$, the complete solution to the general system Eq. (2-1) is given by

$$y(t) = y_h(t) + y_p(t)$$

$$= \sum_{i=1}^{n} c_i e^{p_i t} + y_p(t) \tag{2-58}$$

where c_i, $i = 1, 2, \ldots, n$, are arbitrary constants yet to be determined and the p_i, $i = 1, 2, \ldots, n$, are the system poles of the transfer function.

The problem is how to determine the n arbitrary constants $c_1, c_2, \ldots, c_n$. The situation is analogous to the previous chapter, in which it was necessary to know the system initial state at the initial time t_0. Here, in the classic solution approach, it is assumed that the output $y(t_0^+)$ and $n - 1$ of its higher derivatives, $y^{(1)}(t_0^+)$, $y^{(2)}(t_0^+), \ldots, y^{(n-1)}(t_0^+)$, are all known at the initial time t_0^+. These n parameters are equivalent to knowing the initial state information, and they provide an easy way to match boundary conditions on the $y(t)$ solution. This creates n simultaneous linear equations to find the n unknown c's.

The notation t_0^+ is used above rather than merely t_0 to indicate that the boundary conditions are matched *after* any discontinuities in the input have occurred. Take, for example, a unit step input with the step from zero to unity occurring at time t_0. The discontinuity in the input from zero at t_0^- to unity at t_0^+ frequently causes a discontinuity in the output boundary conditions from t_0^- to t_0^+. This is often a point of confusion in applying the classic solution method. Section 2-8 addresses this topic. For now it is assumed that the boundary conditions are known at t_0^+.

Example 2-9

$$\dot{y}(t) + 5y(t) = u(t)$$

with $u(t) = 2e^{-3t}$, $t_0^+ = 0.1^+$, and $y(0.1^+) = 4$. Then

$$y_h(t) = c_1 e^{-5t}$$

$$y_p(t) = 2H(-3)e^{-3t} = \frac{2}{2}e^{-3t} = e^{-3t}$$

$$y(t) = c_1 e^{-5t} + e^{-3t}$$

Evaluating at $y(0.1^+) = 4$ gives

$$y(0.1^+) = 4 = c_1 e^{-0.5} + e^{-0.3}$$

$$= c_1(0.6065) + 0.7408$$

Solving for c_1 gives

$$c_1 = 5.3735$$

Thus

$$y(t) = 5.3735e^{-5t} + e^{-3t} \qquad t \geq 0.1$$

is the unique complete solution. It satisfies both the differential equation *and* the boundary condition. (Note that the particular solution also satisfies the differential equation, but it does not, in general, satisfy the boundary conditions.)

Example 2-10

$$\ddot{y}(t) + \dot{y}(t) + y(t) = \dot{u}(t) + 2u(t)$$

with input $u(t) = 3 + 5t$, and boundary conditions at $t_0^+ = \pi/4^+$ of

$$y\left(\frac{\pi}{4^+}\right) = 0 \qquad \dot{y}\left(\frac{\pi}{4^+}\right) = 2$$

The system transfer function is

$$H(s) = \frac{s+2}{s^2+s+1} = \frac{s+2}{\left(s + 0.5 - \frac{\sqrt{3}}{2}j\right)\left(s + 0.5 + \frac{\sqrt{3}}{2}j\right)}$$

The homogeneous solution is

$$y_h(t) = c_1 e^{-0.5t} \cos\left(\frac{\sqrt{3}}{2}t\right) + c_2 e^{-0.5t} \sin\left(\frac{\sqrt{3}}{2}t\right)$$

Using Table 2-2, the particular solution is

$$y_p(t) = 3H(0) + 5[(1-2) + 2t]$$

or

$$y_p(t) = 1 + 10t$$

Combining the homogeneous and particular solutions yields

$$y(t) = c_1 e^{-0.5t} \cos\left(\frac{\sqrt{3}}{2}t\right) + c_2 e^{-0.5t} \sin\left(\frac{\sqrt{3}}{2}t\right) + 1 + 10t \quad (2\text{-}59)$$

Evaluating this expression at $t^+ = \pi/4^+$ and using the first boundary condition gives

$$y\left(\frac{\pi}{4^+}\right) = 0 = 0.6752c_1 + 0.01187c_2 + 8.54 \quad (2\text{-}60)$$

Differentiating the expression for $y(t)$ gives

$$\dot{y}(t) = -0.5\left[c_1 e^{-0.5t} \cos\left(\frac{\sqrt{3}}{2}t\right) + c_2 e^{-0.5t} \sin\left(\frac{\sqrt{3}}{2}t\right)\right] + \frac{\sqrt{3}}{2}\left[c_2 e^{-0.5t} \cos\left(\frac{\sqrt{3}}{2}t\right) - c_1 e^{-0.5t} \sin\left(\frac{\sqrt{3}}{2}t\right)\right] + 10$$

Using the boundary condition $\dot{y}(\pi/4^+) = 2$ yields

$$\dot{y}\left(\frac{\pi}{4}\right) = 2 = -0.3479c_1 + 0.5788c_2 + 10 \tag{2-61}$$

Equations (2-60) and (2-61) are two simultaneous linear equations for the unknowns c_1 and c_2

$$0.6752c_1 + 0.01187c_2 = -8.54 \tag{2-62}$$

$$-0.3479c_1 + 0.5788c_2 = -8 \tag{2-63}$$

These may be solved as

$$c_1 = -12.26 \qquad \text{and} \qquad c_2 = -21.19$$

These values of c_1 and c_2 may be substituted into Eq. (2-59) to give the unique solution, $y(t)$, for $t > \pi/4$. A systematic technique for solving the simultaneous linear equations, Eqs. (2-62) and (2-63), is presented in the next section.

Example 2-11

$$\dddot{y}(t) + 2\ddot{y}(t) - \dot{y}(t) - 2y(t) = 2\dot{u}(t) + 8u(t)$$

with input

$$u(t) = 3e^{-4t}$$

and known boundary conditions at $t_0^+ = 0^+$

$$y(0^+) = 1 \qquad \dot{y}(0^+) = -2 \qquad \ddot{y}(0^+) = 3$$

The system transfer function is

$$H(s) = \frac{2(s+4)}{s^3 + 2s^2 - s - 2} = \frac{2(s+4)}{(s+1)(s-1)(s+2)}$$

Notice that the system is unstable due to the pole at $p_2 = 1$. The complete solution has the form

$$y(t) = c_1e^{-t} + c_2e^{t} + c_3e^{-2t} + 3H(-4)e^{-4t}$$

Because of the system zero at $z_1 = -4$, $H(-4) = 0$, and the particular solution response is identically zero. However, this does not imply that $y(t)$ is identically zero, since $y(t) = y_h(t)$, and the arbitrary constants c_1, c_2, and c_3 must still be evaluated. This problem leads to one of three equations in three unknowns from the three boundary-condition relations

$$y(0^+) = 1 = c_1 + c_2 + c_3$$

$$\dot{y}(0^+) = -2 = -c_1 + c_2 - 2c_3$$

$$\ddot{y}(0^+) = 3 = c_1 + c_2 + 4c_3$$

These may be reformulated in matrix vector form as

$$\begin{bmatrix} 1 & 1 & 1 \\ -1 & 1 & -2 \\ 1 & 1 & 4 \end{bmatrix} \begin{bmatrix} c_1 \\ c_2 \\ c_3 \end{bmatrix} = \begin{bmatrix} 1 \\ -2 \\ 3 \end{bmatrix} \tag{2-64}$$

Equation (2-64) is in the form of the classic linear equation problem. A general solution procedure to solve such linear equation problems is treated in the next section.

2-7 ELEMENTARY ROW OPERATIONS AND THE LINEAR EQUATION PROBLEM

One of the basic problems encountered over and over in the text is the solution of a matrix vector linear equation problem of the form

$$M\mathbf{x} = \mathbf{d} \tag{2-65}$$

where M is an $n \times n$ coefficient matrix, $\mathbf{d}$ is an $n \times 1$ known vector of data, and $\mathbf{x}$ is an $n \times 1$ vector of unknowns whose solution is sought. There are many techniques for solving this linear equation problem, but one of the easiest to learn and the easiest to use for hand computations is a method known as *Gauss-Jordan elimination.* Indeed, not only is this method one of the easiest for calculation by hand, but it (or its variations, such as straight gaussian elimination or gaussian elimination with pivoting) is still one of the most effective methods for large, computer-based algorithms.*

The strategy of the solution method is really very simple: one simply appends the data vector to the original coefficient matrix to obtain an $n \times (n+1)$ dimension matrix, $[M, \mathbf{d}]$. Then elementary row operations are performed on the augmented matrix until it has the form $[I, \mathbf{d}']$, where I is the $n \times n$ identity matrix. The solution to the problem $M\mathbf{x} = \mathbf{d}$ is then $\mathbf{x} = \mathbf{d}'$.

Elementary row operations are any combination of three basic operations performed on the rows of a given matrix:

1. Any two rows may be interchanged.
2. Any row may be multiplied (or divided) by any nonzero scalar.
3. Any nonzero multiple of one entire row may be added to (or subtracted from) any other row.

Example 2-11 (continued) Consider the linear equation problem from the previous section:

$$\begin{bmatrix} 1 & 1 & 1 \\ -1 & 1 & -2 \\ 1 & 1 & 4 \end{bmatrix} \begin{bmatrix} c_1 \\ c_2 \\ c_3 \end{bmatrix} = \begin{bmatrix} 1 \\ -2 \\ 3 \end{bmatrix} \tag{2-64}$$

*The interested reader is referenced to Stewart [8] or LINPACK [9], for example, for reference material on computer algorithms for solving the linear equation problem.

Forming the augmented matrix, and then performing elementary row operations to add the first row to the second and subtract the first row from the third, yields:

$$\begin{bmatrix} 1 & 1 & 1 & 1 \\ -1 & 1 & -2 & -2 \\ 1 & 1 & 4 & 3 \end{bmatrix} \sim \begin{bmatrix} 1 & 1 & 1 & 1 \\ 0 & 2 & -1 & -1 \\ 0 & 0 & 3 & 2 \end{bmatrix}$$

The $\sim$ is used to denote that an elementary row operation has been performed. Next, the second row is divided by 2, and the third row is divided by 3, to give

$$\sim \begin{bmatrix} 1 & 1 & 1 & 1 \\ 0 & 1 & -1/2 & -1/2 \\ 0 & 0 & 1 & 2/3 \end{bmatrix}$$

In general, this is the desired strategy: use elementary row operations until the matrix is upper triangular (has all zero elements below the main diagonal) and there are all unity elements on the diagonal. In the method of gaussian elimination, back-substitution would now be used to solve the upper triangular problem that exists at this stage of the elementary row operations; however, it is convenient for later applications to use Gauss-Jordan elimination, in which the elementary row operations are continued all the way to producing an identity matrix on the left.

By adding 1/2 times the third row to the second row, and subtracting the third row from the first row, we get

$$\sim \begin{bmatrix} 1 & 1 & 0 & 1/3 \\ 0 & 1 & 0 & -1/6 \\ 0 & 0 & 1 & 2/3 \end{bmatrix}$$

Finally, the second row may be subtracted from the first row to yield the desired form

$$\sim \begin{bmatrix} 1 & 0 & 0 & 1/2 \\ 0 & 1 & 0 & -1/6 \\ 0 & 0 & 1 & 2/3 \end{bmatrix}$$

The answer to the original linear equation problem, Eq. (2-64), is

$$\begin{bmatrix} c_1 \\ c_2 \\ c_3 \end{bmatrix} = \begin{bmatrix} 1/2 \\ -1/6 \\ 2/3 \end{bmatrix}$$

The reader may verify that this is correct.

Using this solution for the arbitrary constants, the total solution to the example problem becomes the sum of the homogeneous and particular solutions [recall $y_p(t) \equiv 0$]:

$$y(t) = \frac{1}{2}e^{-t} - \frac{1}{6}e^{t} + \frac{2}{3}e^{-2t}$$

Example 2-12 Consider the linear equation problem Eqs. (2-63) and (2-64) of Example 2-10. The augmented matrix is formed, and elementary row operations proceed as

$$\begin{bmatrix} 0.6752 & 0.01187 & -8.54 \\ -0.3479 & 0.5788 & -8 \end{bmatrix} \sim \begin{bmatrix} 1 & 0.0178 & -12.6305 \\ -0.3479 & 0.5788 & -8 \end{bmatrix}$$

$$\sim \begin{bmatrix} 1 & 0.0178 & -12.6305 \\ 0 & 0.5849 & -12.3942 \end{bmatrix} \sim \begin{bmatrix} 1 & 0.0178 & -12.6305 \\ 0 & 1 & -21.19 \end{bmatrix}$$

$$\sim \begin{bmatrix} 1 & 0 & -12.26 \\ 0 & 1 & -21.19 \end{bmatrix}$$

The answer is $c_1 = -12.26$ and $c_2 = -21.19$, which agrees with the answer given in Example 2-10.

An important point about the procedure is that the steps taken to arrive at the final form are not unique, but the answer obtained is always unique no matter what path is followed in getting there.

It is also probably not clear just *why* the method works. Basically, performing elementary row operations on the augmented matrix $[M, \mathbf{d}]$ is equivalent to premultiplying both sides of

$$M\mathbf{x} = \mathbf{d} \tag{2-65}$$

by a nonsingular matrix, E, to give

$$EM\mathbf{x} = E\mathbf{d} \tag{2-66}$$

If E is a nonsingular (invertible) matrix, then the solution to Eq. (2-66) is exactly the same as the solution to Eq. (2-65). If $EM = I$, the identity matrix, then the solution to

$$EM\mathbf{x} = I\mathbf{x} = E\mathbf{d}$$

is particularly simple; $\mathbf{x} = E\mathbf{d}$ is the solution.

This is the strategy. It only remains to demonstrate that each of the three elementary row operations is equivalent to premultiplication by some nonsingular E matrix. Consider the linear equation problem

$$\begin{bmatrix} 0 & 1 & 0 \\ 2 & 0 & 0 \\ 3 & 0 & 1 \end{bmatrix} \begin{bmatrix} x_1 \\ x_2 \\ x_3 \end{bmatrix} = \begin{bmatrix} 4 \\ -2 \\ 6 \end{bmatrix} \tag{2-67}$$

Forming the augmented matrix $[M, \mathbf{d}]$ and premultiplying by

$$\begin{bmatrix} 0 & 1 & 0 \\ 1 & 0 & 0 \\ 0 & 0 & 1 \end{bmatrix} \begin{bmatrix} 0 & 1 & 0 & 4 \\ 2 & 0 & 0 & -2 \\ 3 & 0 & 1 & 6 \end{bmatrix} = \begin{bmatrix} 2 & 0 & 0 & -2 \\ 0 & 1 & 0 & 4 \\ 3 & 0 & 1 & 6 \end{bmatrix}$$

interchanges rows 1 and 2. Next.

$$\begin{bmatrix} 1/2 & 0 & 0 \\ 0 & 1 & 0 \\ 0 & 0 & 1 \end{bmatrix} \begin{bmatrix} 2 & 0 & 0 & -2 \\ 0 & 1 & 0 & 4 \\ 3 & 0 & 1 & 6 \end{bmatrix} = \begin{bmatrix} 1 & 0 & 0 & -1 \\ 0 & 1 & 0 & 4 \\ 3 & 0 & 1 & 6 \end{bmatrix}$$

multiplies the first row by 1/2. Finally,

$$\begin{bmatrix} 1 & 0 & 0 & 1 \\ 0 & 1 & 0 & 0 \\ -3 & 0 & 1 & 3 \end{bmatrix} \begin{bmatrix} 0 & 0 & -1 \\ 1 & 0 & 4 \\ 0 & 1 & 6 \end{bmatrix} = \begin{bmatrix} 1 & 0 & 0 & -1 \\ 0 & 1 & 0 & 4 \\ 0 & 0 & 1 & 9 \end{bmatrix}$$

subtracts 3 times the first row from the third. The final solution is

$$\begin{bmatrix} x_1 \\ x_2 \\ x_3 \end{bmatrix} = \begin{bmatrix} -1 \\ 4 \\ 9 \end{bmatrix} \tag{2-68}$$

The nonsingular E matrix such that EM is the identity matrix is the product of the three transformations

$$E = \begin{bmatrix} 1 & 0 & 0 \\ 0 & 1 & 0 \\ -3 & 0 & 1 \end{bmatrix} \begin{bmatrix} 1/2 & 0 & 0 \\ 0 & 1 & 0 \\ 0 & 0 & 1 \end{bmatrix} \begin{bmatrix} 0 & 1 & 0 \\ 1 & 0 & 0 \\ 0 & 0 & 1 \end{bmatrix}$$

$$= \begin{bmatrix} 0 & 1/2 & 0 \\ 1 & 0 & 0 \\ 0 & -3/2 & 1 \end{bmatrix} \tag{2-69}$$

The reader is invited to check that $EM = I$ and that $E\mathbf{d}$ is the solution above. Notice that ordinarily it is completely unnecessary to find the transformation E, as the row operations are carried out without regard to what the transformation involved happens to be.

Matrix Inversion

This same procedure of elementary row operations may be used to calculate the inverse of a given matrix, M. The technique is simply to form the augmented matrix $[M, I]$, and then perform elementary row operations on this $n \times 2n$ dimension matrix until it is in the form $[I, E]$. Then $M^{-1} = E$.

Example 2-13 Take

$$M = \begin{bmatrix} 0 & 1 & 0 \\ 2 & 0 & 0 \\ 3 & 0 & 1 \end{bmatrix}$$

as considered in the previous example. Forming $[M, I]$ and performing elementary row operations yields

$$\begin{bmatrix} 0 & 1 & 0 & 1 & 0 & 0 \\ 2 & 0 & 0 & 0 & 1 & 0 \\ 3 & 0 & 1 & 0 & 0 & 1 \end{bmatrix} \sim \begin{bmatrix} 2 & 0 & 0 & 0 & 1 & 0 \\ 0 & 1 & 0 & 1 & 0 & 0 \\ 3 & 0 & 1 & 0 & 0 & 1 \end{bmatrix}$$

$$\sim \begin{bmatrix} 1 & 0 & 0 & 0 & 1/2 & 0 \\ 0 & 1 & 0 & 1 & 0 & 0 \\ 3 & 0 & 1 & 0 & 0 & 1 \end{bmatrix} \sim \begin{bmatrix} 1 & 0 & 0 & 0 & 1/2 & 0 \\ 0 & 1 & 0 & 1 & 0 & 0 \\ 0 & 0 & 1 & 0 & -3/2 & 1 \end{bmatrix}$$

Then

$$M^{-1} = \begin{bmatrix} 0 & 1/2 & 0 \\ 1 & 0 & 0 \\ 0 & -3/2 & 1 \end{bmatrix}$$

which agrees with the result from Eq. (2-69).

This is a fairly simple example, but it illustrates another important point. Namely, it illustrates why it is computationally so much easier to find the solution to $M\mathbf{x} = \mathbf{d}$ via direct elementary row operations on $[M, \mathbf{d}]$ than it is to first find M^{-1} and then find the solution $\mathbf{x} = M^{-1}\mathbf{d}$. The latter computation requires elementary row operations on the $n \times 2n$-dimensioned $[M, I]$ rather than row operations on the $n \times (n + 1) =$ dimensioned $[M, \mathbf{d}]$. Still this method or variations of this method are the most efficient way to find M^{-1}. About the worst method to use, computationally, for any matrix above a 3×3 is the oft-used Cramer's rule (see Appendix B); to invert a 10×10 matrix using Cramer's rule would take approximately 36 million multiplications and a like number of additions. Using the above elementary row operation method to find M^{-1} would take approximately 1925 multiplies and 1770 adds. To solve $M\mathbf{x} = \mathbf{d}$ via Gauss-Jordan elimination for a 10×10 M would take approximately 600 multiplications and a like number of additions. The best advice for any large numerical problem, though, is to use expertly written software packages like LINPACK [9] or IMSL [3]. These packages are readily available at nearly all major computer centers, and they incorporate such techniques as pivoting in order to help make them computationally stable. For more background on the theory and computational aspects of solving the linear equation problem, the interested reader is referred to the text by Stewart [8].

2-8 A VERY SUBTLE POINT: THE DIFFERENCE BETWEEN t_0^- AND t_0^+

In the foregoing sections, care is always taken to specify that the boundary conditions are at t_0^+, *after* any discontinuities in the input might have occurred. For example, if the input is the unit step input

$$u(t) = u_{-1}(t - t_0) \equiv \begin{cases} 0 & t < t_0 \\ 1 & t > t_0 \end{cases} \tag{2-70}$$

a discontinuity in the input occurs at the initial time t_0. It is therefore important to use the boundary conditions on y and its derivatives *after* this discontinuity in the input has occurred.

If the system has no zeros, then there is no difficulty, because the boundary conditions are always continuous even for discontinuous (but bounded nonimpulsive) inputs. However, whenever there are system zeros, the highest derivatives on y—equal in number to the number of zeros—are discontinuous for discontinuous inputs.

To see how this comes about, consider a general second-order system with a single system zero at $-b_0/b_1$:

$$\ddot{y}(t) + a_1\dot{y}(t) + a_0 y(t) = b_1\dot{u}(t) + b_0 u(t) \tag{2-71}$$

Let it be assumed that the boundary conditions $y(t_0^-)$ and $\dot{y}(t_0^-)$ are both known, and so it is desired to find the boundary conditions at t_0^+.

To find the boundary condition on $y(t_0^+)$ the differential equation is integrated twice from t_0^- to t_0^+ to give*

$$\iint_{t_0^-}^{t_0^+} (\ddot{y} + a_1\dot{y} + a_0 y) = \iint_{t_0^-}^{t_0^+} (b_1\dot{u} + b_0 u) \tag{2-72}$$

or $y(t_0^+) - y(t_0^-) = 0$. The double integral of $\ddot{y}$ gives y evaluated at the upper and lower limits of integration. The double integral of all other terms is zero, since the integral is over a zero time interval. Thus $y(t_0^+) = y(t_0^-)$.

To find $\dot{y}(t_0^+)$, a first integral of the differential equation is made, giving

$$\int_{t_0^-}^{t_0^+} (\ddot{y} + a_1\dot{y} + a_0 y) = \int_{t_0^-}^{t_0^+} (b_1\dot{u} + b_0 u)$$

or

$$\dot{y}(t_0^+) - \dot{y}(t_0^-) + a_1[\overbrace{y(t_0^+) - y(t_0^-)}^{0}] = b_1[u(t_0^+) - u(t_0^-)]$$

*Use is being made of the facts that

$$\int_{t_0^-}^{t_0^+} \ddot{y} = \dot{y}(t_0^+) - \dot{y}(t_0^-), \quad \iint_{t_0^-}^{t_0^+} \ddot{y} = y(t_0^+) - y(t_0^-),$$

$$\int_{t_0^-}^{t_0^+} \dot{y} = y(t_0^+) - y(t_0^-), \quad \iint_{t_0^-}^{t_0^+} \dot{y} = \int_{t_0^-}^{t_0^+} y, \text{ etc.}$$

Again, all other terms are zero, because the integration is over a zero time interval. Since, from above, $y(t_0^+) - y(t_0^-) = 0$, the result for $\dot{y}(t_0^+)$ is

$$\dot{y}(t_0^+) = \dot{y}(t_0^-) + b_1[u(t_0^+) - u(t_0^-)]$$

Thus if u has a jump discontinuity at time t_0, $u(t_0^+) \neq u(t_0^-)$, the boundary condition on $\dot{y}(t_0^+)$ does not equal $\dot{y}(t_0^-)$.

Example 2-14 Three second-order systems are considered:

(*a*) $\ddot{y}(t) + 7\dot{y}(t) + 10y(t) = u(t)$

(*b*) $\ddot{y}(t) + 7\dot{y}(t) + 10y(t) = \dot{u}(t) + 3u(t)$

(*c*) $\ddot{y}(t) + 7\dot{y}(t) + 10y(t) = \ddot{u}(t) + 2\dot{u}(t) - 3u(t)$

Each has input

$$u(t) = 2\cos(4t)u_{-1}(t) = \begin{cases} 0 & t < 0 \\ 2\cos(4t) & t > 0 \end{cases}$$

and boundary conditions at time 0^- are zero [that is, $y(0^-) = \dot{y}(0^-) = 0$].

The characteristic polynomial in each case is

$$\Delta(s) = s^2 + 7s + 10 = (s + 2)(s + 5)$$

and so the homogeneous response for all three systems is the same:

$$y_h(t) = c_1 e^{-2t} + c_2 e^{-7t}$$

However, the transfer function is different in each case, and so the particular solution

$$y_p(t) = \text{Re}\,\{H(4j)2e^{4jt}\}$$

is different for each system. The particular solutions are:

(*a*) $$y_p(t) = \text{Re}\left\{\frac{2}{(4j)^2 + 28j + 10}(\cos 4t + j\sin 4t)\right\}$$

or $$y_p(t) = -\frac{3}{196}\cos 4t - \frac{14}{196}\sin 4t$$

(*b*) $$y_p(t) = \text{Re}\left\{\frac{2(4j) + 6}{(4j)^2 + 28j + 10}(\cos 4t + j\sin 4t)\right\}$$

or $$y_p(t) = \frac{47}{196}\cos 4t + \frac{54}{196}\sin 4t$$

(*c*) $$y_p(t) = \text{Re}\left\{\frac{2(4j)^2 + 4(4j) - 2}{(4j)^2 + 28j + 10}(\cos 4t + j\sin 4t)\right\}$$

or $$y_p(t) = \frac{163}{196}\cos 4t - \frac{214}{196}\sin 4t$$

The complete solution is the sum of the homogeneous and particular solutions. The next step is to evaluate the arbitrary constants c_1 and c_2 in the homogeneous part of the response. This in turn requires finding $y(0^+)$ and $\dot{y}(0^+)$ from the known $y(0^-) = \dot{y}(0^-) = 0$. These boundary conditions are different for the three different systems because of the zeros in (*b*) and (*c*) and because the input is discontinuous at time zero. Namely, $u(0^-) = \dot{u}(0^-) = 0$, but $u(0^+) = 2$ and $\dot{u}(0^+) = 0$. Thus the input is discontinuous at 0, but its first derivative is continuous. The result is

(*a*) Since there are no system zeros, the first and second integrals of the differential equation from 0^- to 0^+ reveals that $y(0^+) = y(0^-) = 0$ and $\dot{y}(0^+) = \dot{y}(0^-) = 0$.

(*b*) Since there is one system zero, performing two integrals of the differential equation from 0^- to 0^+ reveals that $y(0^+) = y(0^-) = 0$, while a single integral gives

$$\dot{y}(0^+) - \dot{y}(0^-) + 7[y(0^+) - y(0^-)] = u(0^+) - u(0^-)$$

or

$$\dot{y}(0^+) = 2.$$

(*c*) Since there are two system zeros, two integrals of the differential equation gives

$$y(0^+) - y(0^-) = u(0^+) - u(0^-)$$

or

$$y(0^+) = 2.$$

A single integral gives

$$\dot{y}(0^+) - \dot{y}(0^-) + 7[y(0^+) - y(0^-)] = \dot{u}(0^+) - \dot{u}(0^-) + 2[u(0^+) - u(0^-)]$$

or

$$\dot{y}(0^+) = -10$$

It is interesting to note in this third case that the initial slope of the output is negative, even though the initial value of the input is a positive $+2$, $u(0^+) = 2$. This is caused by the positive system zero at $z_2 = +1$, a zero in the right half-plane. More will be discussed about this phenomenon and the influence of system zeros in the next section.

The boundary conditions at time 0^+ have now been found, and so the final step is to set up linear equation problems to find the unknown constants c_1 and c_2. In each case the complete solution has the form

$$y(t) = c_1 e^{-2t} + c_2 e^{-5t} + a \cos 4t + b \sin 4t$$

where a and b are different for each of the three systems. Using the boundary conditions on $y(0^+)$ and $\dot{y}(0^+)$ gives two equations in two unknowns for c_1 and c_2:

$$y(0^+) = c_1 + c_2 + a$$

$$\dot{y}(0^+) = -2c_1 - 5c_2 + 4b$$

The resulting linear equations are

(*a*) $$\begin{bmatrix} 1 & 1 \\ -2 & -5 \end{bmatrix} \begin{bmatrix} c_1 \\ c_2 \end{bmatrix} = \begin{bmatrix} 3/196 \\ 56/196 \end{bmatrix}$$

(*b*) $$\begin{bmatrix} 1 & 1 \\ -2 & -5 \end{bmatrix} \begin{bmatrix} c_1 \\ c_2 \end{bmatrix} = \begin{bmatrix} -47/196 \\ 2 - 216/196 \end{bmatrix}$$

(*c*) $$\begin{bmatrix} 1 & 1 \\ -2 & -5 \end{bmatrix} \begin{bmatrix} c_1 \\ c_2 \end{bmatrix} = \begin{bmatrix} 2 - 163/196 \\ -10 + 856/196 \end{bmatrix}$$

Solving these for c_1 and c_2, the complete solutions become

(*a*) $$y(t) = 0.12075e^{-2t} - 0.10544e^{-5t} - \frac{3}{196}\cos 4t - \frac{14}{196}\sin 4t$$

(*b*) $$y(t) = -0.10034e^{-2t} - 0.13946e^{-5t} + \frac{47}{196}\cos 4t + \frac{54}{196}\sin 4t$$

(*c*) $$y(t) = 0.06973e^{-2t} + 1.09864e^{-5t} + \frac{163}{196}\cos 4t - \frac{214}{196}\sin 4t$$

Figure 2-6 shows a plot of the three time responses from Example 2-14. It is interesting to recall that all three systems were at rest at 0^-, had the same input $u(t) = 2\cos(4t) \cdot u_{-1}(t)$, and had the same characteristic polynomial, $\Delta(s) = (s + 2)(s + 5)$. The only difference was the different system zeros. The next section explores in greater detail the influence of the poles and zeros upon the system step response.

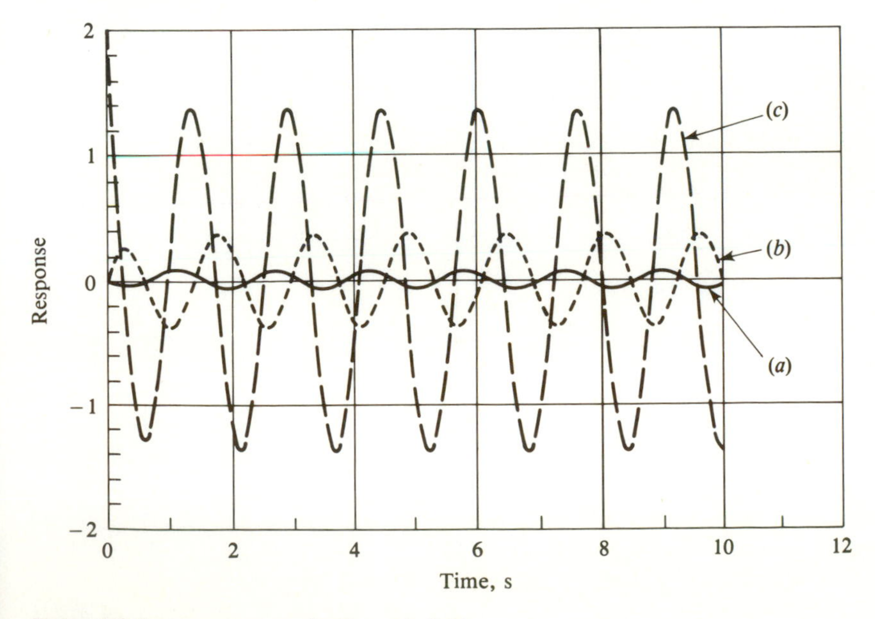

Figure 2-6 System responses for Example 2-14.

2-9 THE SYSTEM STEP RESPONSE

In the previous chapter the system response to the unit step input (the step response) was analyzed on an input-output basis. Now, with a differential equation model of a given system in hand, the mathematics of the step response may be analyzed.

Step Response for the First-Order System

Figure 2-7 shows the step response for the first-order system

$$\dot{y}(t) + a_0 y(t) = a_0 u(t) \tag{2-73}$$

as a_0 is varied from 1 to 10.

The steady-state final value is unity, because

$$y_p(t) = H(0) = \frac{a_0}{a_0} = 1 \tag{2-74}$$

Notice that as the system pole moves farther into the left half-plane, the transient time response becomes faster and faster. For positive pole values (when $a_0 < 0$), the system step response is unbounded and the system is unstable.

The step-response solution for this system has the form

$$y(t) = c_1 e^{-a_0 t} + 1 \tag{2-75}$$

Since the system has no zero, the boundary conditions are continuous, and so

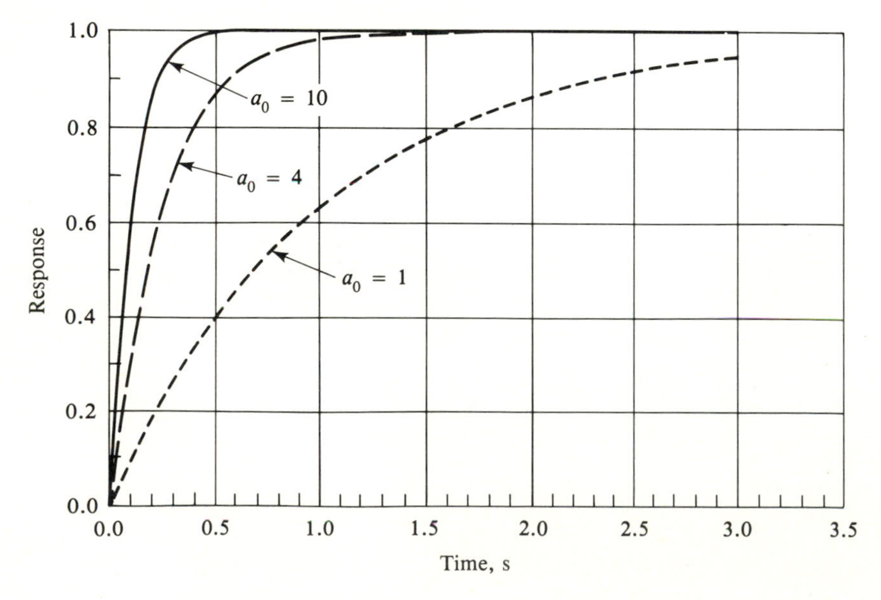

Figure 2-7 Step responses for $\dot{y} + a_0 y = a_0 u$.

$y(0^-) = y(0^+) = 0$. This gives $c_1 = -1$, and the final complete solution is

$$y(t) = -e^{-a_0 t} + 1 \tag{2-76}$$

If the system has a zero, the form of the solution is the same, but the boundary conditions change. In particular, consider the system

$$\dot{y}(t) + y(t) = \frac{\dot{u}(t)}{-z_1} + u(t) \tag{2-77}$$

The transfer function is

$$H(s) = \frac{-s/z_1 + 1}{s + 1} \tag{2-78}$$

Still, $y_p(t) = H(0) = 1$ is the steady-state final value. However, because of the system zero at z_1, the boundary conditions are discontinuous for the discontinuous step input. A first integral of the differential equation from 0^- to 0^+ yields

$$y(0^+) - y(0^-) = -\frac{1}{z_1}[u(0^+) - u(0^-)]$$

or $y(0^+) = -1/z_1$. The complete solution is

$$y(t) = \left(-1 - \frac{1}{z_1}\right)e^{-t} + 1 \qquad t > 0 \tag{2-79}$$

This is shown plotted in Fig. 2-8 for different zero locations ranging from $z_1 = +2$ to $z_1 = -2$. Notice that in all cases the system is stable regardless of the zero

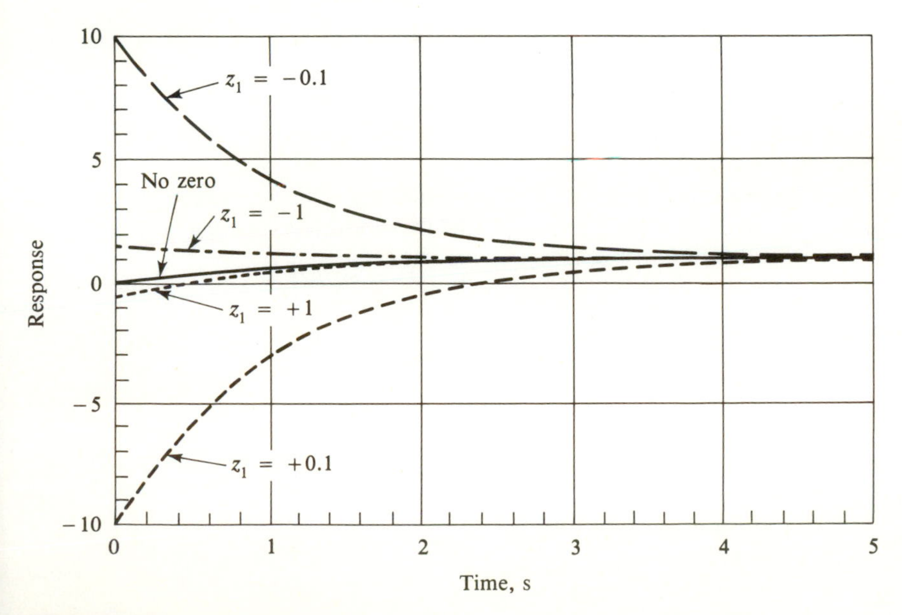

Figure 2-8 Step response for $\dot{y} + y = \dot{u}_1/z_1 + u$, $z_1 = \pm 1$, $z = \pm 0.1$, no zero.

location in the left or right half-plane; stability is only influenced by the location of the pole. Notice, however, that the zero has a dramatic influence on the nature of the time response. The most pronounced effect is when the zero approaches the imaginary axis from either the left or the right. Also notice that whenever the zero is in the right half-plane the *initial* response of the system is *negative,* even though the steady-state value is $+1$. Such a system, with a zero in the right half-plane, is termed *nonminimum phase* for reasons to be explained in the next chapter. This type of initial response opposite to the final value is typical of the nonminimum-phase system, and it presents a very difficult control situation. Imagine trying to drive a car in which every time you turned the wheel to the left, the initial response was to the right before it eventually came back to the left. This is the type of situation one must contend with in a nonminimum-phase system with a zero in the right half-plane.

Also notice that the output for discontinuous inputs is also discontinuous. A change in the input immediately changes the output with no lag. The system is thus said to have a *feedforward* element to it; there is an immediate feedforward of the input to the output. Such a system is characterized by an *improper* transfer function in which the order of the zero polynomial is equal to the order of the characteristic polynomial—the system has as many zeros as it does poles. A *proper* transfer function has a zero polynomial that is of lower order than the characteristic polynomial. Improper-transfer-function systems are dealt with further in later chapters.

Step Response for the Second-Order System

Consider the general second-order system with no zeros

$$\ddot{y}(t) + a_1\dot{y}(t) + a_0 y(t) = b_0 u(t) \tag{2-80a}$$

Again it is assumed that $a_0 = b_0$ so that the steady-state final value to the step input is $y_p(t) = H(0) = 1$.

It is often convenient to write the general second-order system in terms of parameters called the *damping ratio,* ζ, and the *natural frequency,* ω_n, as

$$\ddot{y}(t) + 2\zeta\omega_n\dot{y}(t) + \omega_n^2 y(t) = \omega_n^2 u(t) \tag{2-80b}$$

where $2\zeta\omega_n = a_1$, $\omega_n^2 = a_0$.

The pole locations are given by

$$p_{1,2} = -\zeta\omega_n \pm \sqrt{1-\zeta^2}\,\omega_n j \tag{2-81}$$

For all values of the damping ratio, $0 < \zeta < 1$, the system is stable and has complex conjugate poles. The system step response is given by

$$y(t) = c_1 e^{-\zeta\omega_n t}\cos\left(\sqrt{1-\zeta^2}\,\omega_n t\right) + c_2 e^{-\zeta\omega_n t}\sin\left(\sqrt{1-\zeta^2}\,\omega_n t\right) + 1 \tag{2-82}$$

Since the system has no zeros, the boundary conditions are all continuous and zero. The zero boundary conditions give the arbitrary constants c_1 and c_2 as

$$c_1 = -1 \qquad c_2 = -\zeta/\sqrt{1-\zeta^2} \tag{2-83}$$

Figure 2-9 plots this step response for values of the damping ratio, ζ, as it varies between 0 and 1. The natural frequency remains constant at $\omega_n = 1$.

Influence of a System Zero

Figure 2-10 shows the step response for a second-order system with $\zeta = 0.5$ and $\omega_n = 1$, and a system zero at location z_1. The zero location z_1 is varied between the bounds $-1 < z_1 < 1$. The system differential equation is

$$\ddot{y}(t) + \dot{y}(t) + y(t) = \frac{-\dot{u}(t)}{z_1} + u(t) \tag{2-84}$$

Notice that in each case the steady-state final value is $y_p(t) = H(0) = 1$. Also notice that all of the systems are stable no matter where the zero is located; the poles are at $p_{1,2} = -0.5 \pm \sqrt{3}/2j$, which are in the left half-plane and stable. However, as with the first-order system, the zero has a dramatic effect on the system response. The effect is more pronounced as the zero gets closer to the imaginary axis. This is to be expected; when the zero is at the origin, this corresponds to a pure differentiator of the input. Differentiation of the unit step input would be the unit impulse.

Also notice the nonminimum phase behavior when the zero is in the right half-plane. The initial response is opposite in direction to the final steady-state value.

Finally, note that the initial slope of the response is not zero as it is with a second-order system without a zero. Taking a first integral of Eq. (2-84) it is seen that $\dot{y}(0^+) = -1/z_1$. Thus if the zero, z_1, is positive, the initial slope is negative.

2-10 THE IMPULSE RESPONSE FUNCTION

In Chap. 1 it was shown that the system impulse response function is the time derivative of the system step response. Now, given a differential equation model of a linear dynamic system, the step response may be determined and the derivative taken to yield a mathematical expression for the impulse response.

For example, the output differential equation for the rocket-sled system derived in Sec. 2-1 is

$$\dddot{y}(t) + 0.4\ddot{y}(t) + 1.14\dot{y}(t) + 0.22y(t) = u(t)$$

Since the system has no zeros, the boundary conditions are all continuous and hence are zero. The system poles are $p_1 = -0.2$; $p_2 = -0.1 + 1.044j$; and $p_3 = -0.1 - 1.044j$. The steady-state final value for the unit step input is $y_p(t) = H(0) = 4.54545$. Thus the step-response solution has the form

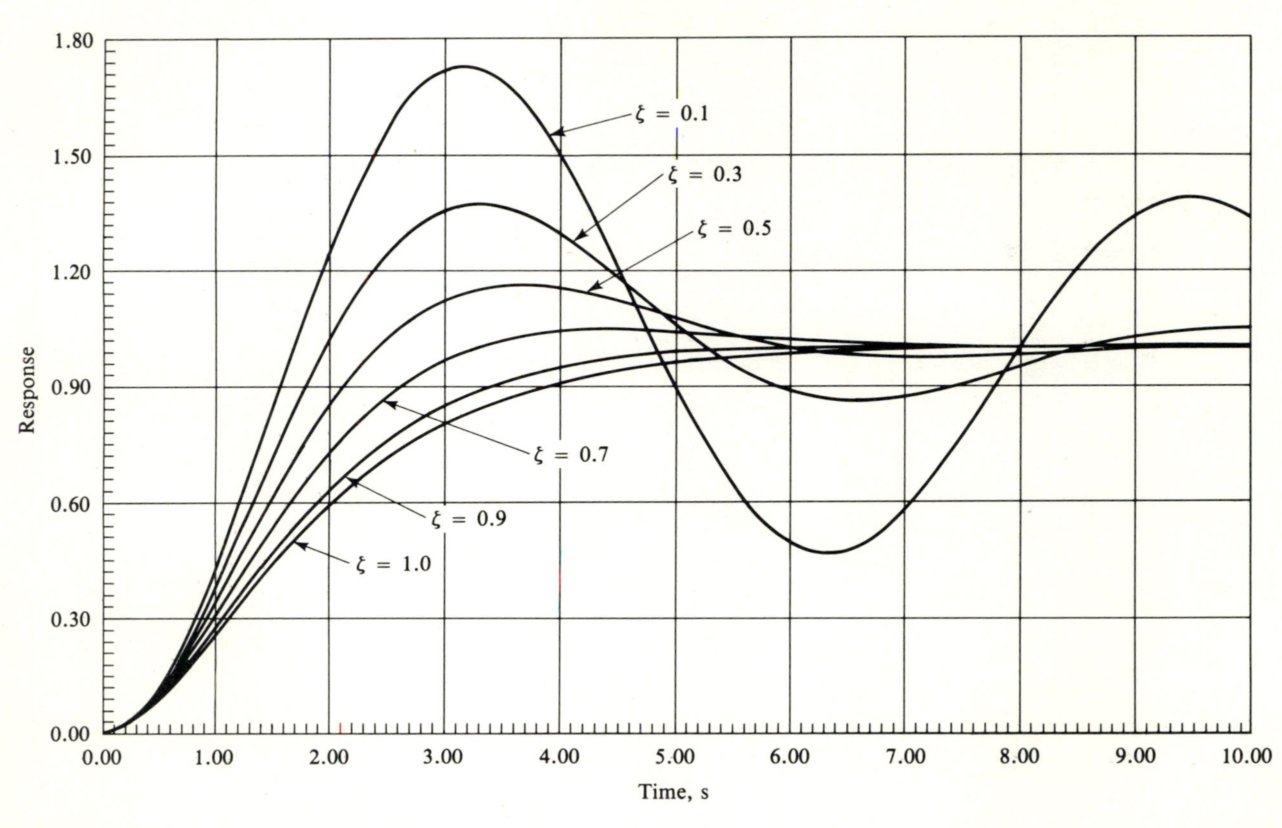

Figure 2-9 Step-response general second-order systems, $\omega_n = 1$.

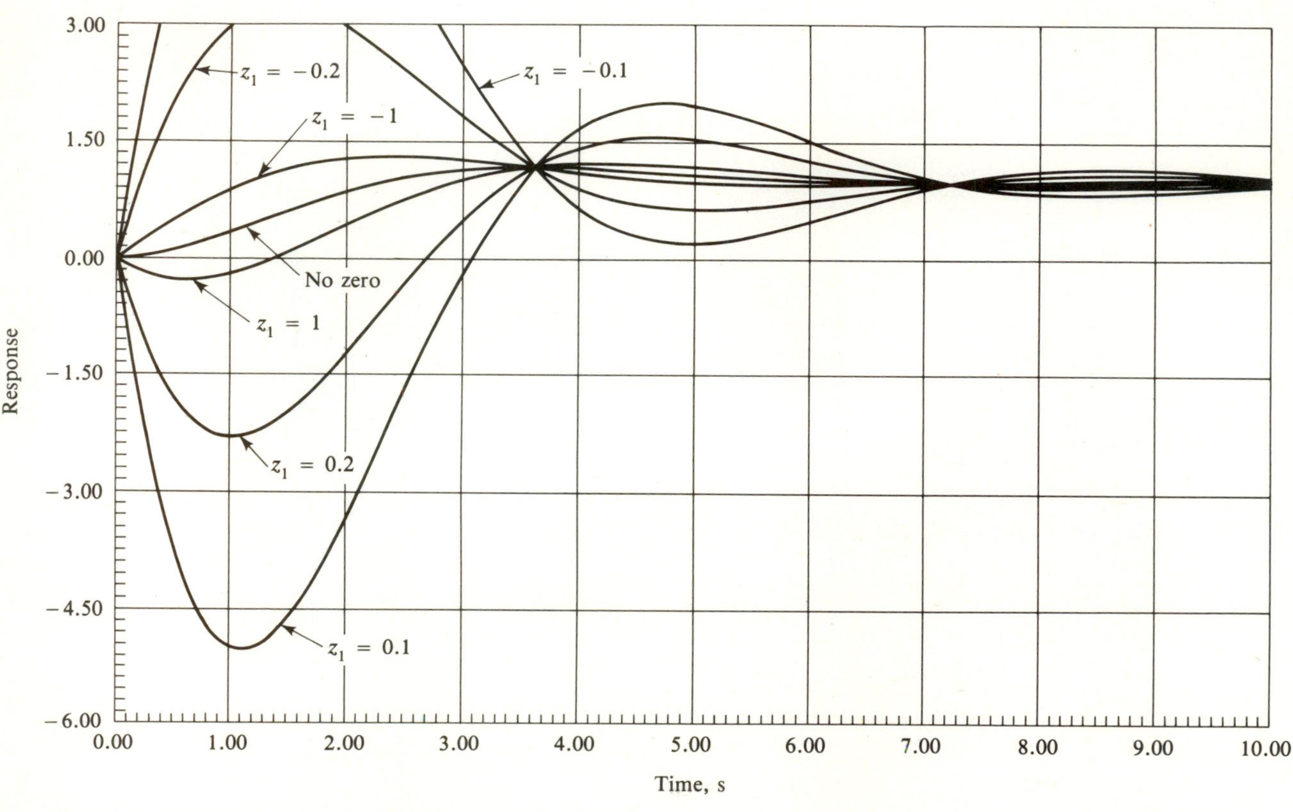

Figure 2-10 The influence of the system zero; $\zeta = 0.5$, $\omega_n = 1$.

$$r(t) = [c_1 e^{-0.2t} + c_2 e^{-0.1t}\cos(1.044t) + c_3 e^{-0.1t}\sin(1.044t) + 4.54545]u_{-1}(t) \tag{2-85}$$

where the arbitrary constants are found from

$$\begin{bmatrix} 1 & 1 & 0 \\ -0.2 & -0.1 & 1.044 \\ 0.04 & -1.07994 & -0.2088 \end{bmatrix} \begin{bmatrix} c_1 \\ c_2 \\ c_3 \end{bmatrix} = \begin{bmatrix} -4.54545 \\ 0 \\ 0 \end{bmatrix} \tag{2-86}$$

Solving Eq. (2-86), the step-response solution is

$$r(t) = [-4.54545e^{-0.2t} - 0.87078e^{-0.1t}\sin(1.044t) + 4.54545]u_{-1}(t) \tag{2-87}$$

The time derivative of the step response yields the impulse response:

$$\begin{aligned} h(t) &= \frac{d}{dt}[r(t)] \\ &= 0.90909e^{-0.2t} - 0.90909e^{-0.1t}\cos(1.044t) \\ &\quad + 0.087078e^{-0.1t}\sin(1.044t) \end{aligned} \tag{2-88}$$

Notice that the impulse response has exactly the same exponential terms as the general homogeneous response.

Therefore, the Chap. 1 stability requirement for the exponential decay of the impulse response corresponds exactly with this chapter's requirement for all poles lying in the left half-plane.

Also notice that the derivative of the unit step $u_{-1}(t)$ yields, symbolically, the unit impulse, $\delta(t)$. But since $r(0^+) = 0$, there is no impulse appearing in the impulse response. This is true in all cases except the improper, feedforward system, in which the step input causes an instantaneous nonzero value of $r(0^+)$.

Feedforward-System Impulse Response

Consider the first-order feedforward system

$$\dot{y}(t) + 2y(t) = 3\dot{u}(t) + 12u(t) \tag{2-89}$$

with improper transfer function

$$H(s) = \frac{3s + 12}{s + 2} \tag{2-90}$$

The step response has the form

$$r(t) = [c_1 e^{-2t} + H(0)]u_{-1}(t) \tag{2-91}$$

where $H(0) = 6$ and c_1 is found from the boundary condition

$$r(0^+) = c_1 + 6$$

To find the boundary condition at 0^+, the differential equation, Eq. (2-89), is integrated from 0^- to 0^+ to yield

$$y(0^+) - \overset{0}{\cancel{y(0^-)}} = 3[\overset{1}{\cancel{u_{-1}(0^+)}} - \overset{0}{\cancel{u_{-1}(0^-)}}]$$

Thus $r(0^+) = y(0^+) = 3$ and $c_1 = -3$. The step response is

$$r(t) = (-3e^{-2t} + 6)u_{-1}(t) \tag{2-92}$$

Now for $t > 0$ the time derivative of the unit step response is simply

$$h(t) = 6e^{-2t} \qquad t > 0 \tag{2-93}$$

But at $t = 0$, the time derivative of $u_{-1}(t)$ yields the unit impulse, $\delta(t)$, which is now weighted by 3 since $r(0^+) = -3 + 6 = 3$. Therefore, the final representation of the impulse response function for this improper system is

$$h(t) = 3\delta(t) + 6e^{-2t}u_{-1}(t) \tag{2-94}$$

To see that the convolution representation of the system yields the same step response, consider:

$$\begin{aligned} r(t) &\equiv \int_0^t h(t-\tau)u_{-1}(\tau)\,d\tau \\ &= \int_0^t [3\delta(t-\tau) + 6e^{-2(t-\tau)}u_{-1}(t-\tau)]u_{-1}(\tau)\,d\tau \end{aligned} \tag{2-95}$$

By the sampling property of the delta function, the integral of the first term is always 3 for any $t > 0$. For the second term, $u_{-1}(t - \tau) = 1$ for all $t > \tau$. Integrating the exponential and evaluating at the upper and lower limits of integration then gives

$$r(t) = 3 + 6e^{-2t}\left(\frac{e^{2\tau}}{2}\right)\Bigg|_0^t$$

or

$$r(t) = 6 - 3e^{-2t} \qquad t > 0 \tag{2-96}$$

which agrees with Eq. (2-92) found via the classic solution method.

2-11 SUMMARY

This chapter has first briefly reviewed the process for deriving a linear time-invariant ordinary differential equation model for a given dynamic system of interest. The classic solution method is then fully developed. The homogeneous and particular solutions are seen to have important physical significance, but they are different than the zero-input and zero-state responses discussed in Chap. 1. For a stable system, the homogeneous response corresponds to the transient part of the solution, while the particular solution corresponds to the steady-state part for a persistent input. Conditions for stability are given in terms of all system poles lying in the left half-plane. This is directly related to the Chap. 1 stability condition on the exponential decay of the impulse response function. This analysis thus brings us full circle with the topics discussed in Chap. 1.

The next chapter focuses on the steady-state response to one very important class of inputs—the sinusoidal input of varying frequencies. Once again this may be considered as an input-output analysis of the linear time-invariant system, but now in the frequency domain.

PROBLEMS

2-1 Find the output differential equation model in standard form for the electrical network shown in Fig. 2-11 with:

(*a*) $u(t) \equiv V_{in}(t)$, $y(t) = I(t)$, the current in the loop.
(*b*) $u(t) \equiv V_{in}(t)$, $y(t) = V_2(t)$, the voltage drop across the capacitor.

2-2 Find the output differential equation in standard form for the electrical network shown in Fig. 2-12 with:

(*a*) $u(t) \equiv V_{in}(t) \quad y(t) \equiv I_2(t)$
(*b*) $u(t) \equiv V_{in}(t) \quad y(t) \equiv I_1(t)$
(*c*) $u(t) \equiv V_{in}(t) \quad y(t) \equiv V_1(t)$

2-3 Find the output differential equation in standard form for the mechanical network shown in Fig. 2-13 with:

(*a*) $u(t) \equiv F_{in}(t)$, $y(t) \equiv V_1(t)$, the velocity of the mass, m.
(*b*) $u(t) \equiv F_{in}(t)$, $y(t) \equiv F_k(t)$, the force measured through the spring, k.

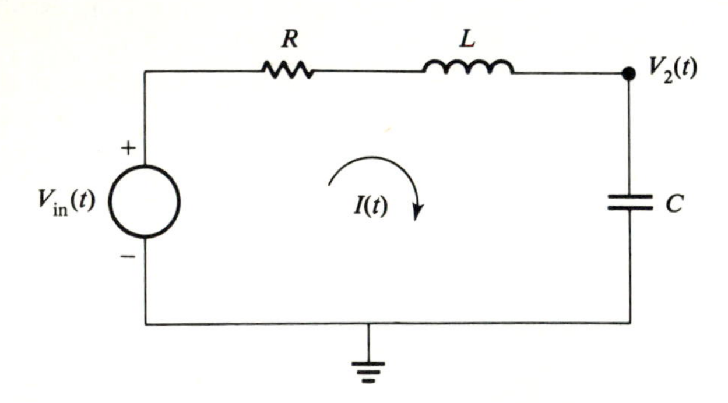

Figure 2-11 Electric circuit for Prob. 2-1.

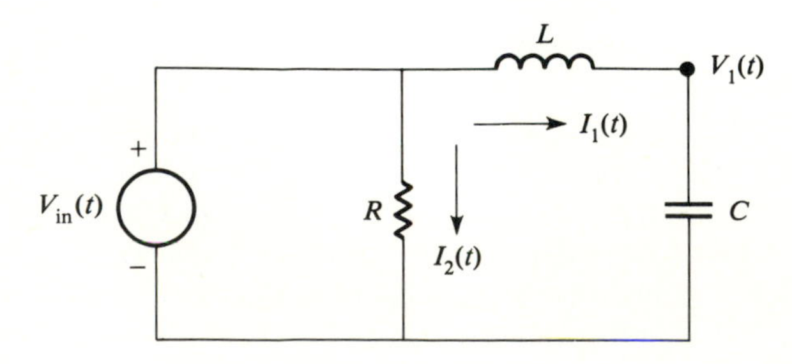

Figure 2-12 Electric circuit for Prob. 2-2.

2-4 Find the output differential equation in standard form for the mechanical network shown in Fig. 2-14 with:

(*a*) $u(t) \equiv F_{in}(t)$, $y(t) \equiv V_1(t)$, the velocity of the mass.
(*b*) $u(t) \equiv F_{in}(t)$, $y(t) \equiv F_k(t)$, the force measured through the spring, k.

2-5 Find the output differential equation in standard form for the mechanical network of Fig. 2-15. Take $y(t) \equiv v_2(t)$, the velocity of mass m_2; and take $u(t) = F_{in}(t)$, the input force.

2-6 Based on the results of Probs. 2-1 through 2-4, comment on the statement: "A system characteristic polynomial is dependent on the basic elements of a system and their interconnection; the system zero polynomial is dependent on the selected output variable."

2-7 Find the transfer function for each of the following systems:

(*a*) $\dot{y} + 2y = 3u$ (*b*) $\dot{y} + 2y = 3\dot{u} + 2u$
(*c*) $\ddot{y} + 4y = \dot{u}$ (*d*) $\ddot{y} = \ddot{u} + 2\dot{u} + 10u$

2-8 Find the transfer function for each of the following systems:

(*a*) $\dot{y} = 2\dot{u} + 4u$ (*b*) $\ddot{y} - 7u = \ddot{u}$
(*c*) $\dot{y} + 5y = 4u$ (*d*) $y^{(3)} - 3y^{(2)} + 2y^{(1)} = u^{(2)} + 10u^{(1)}$

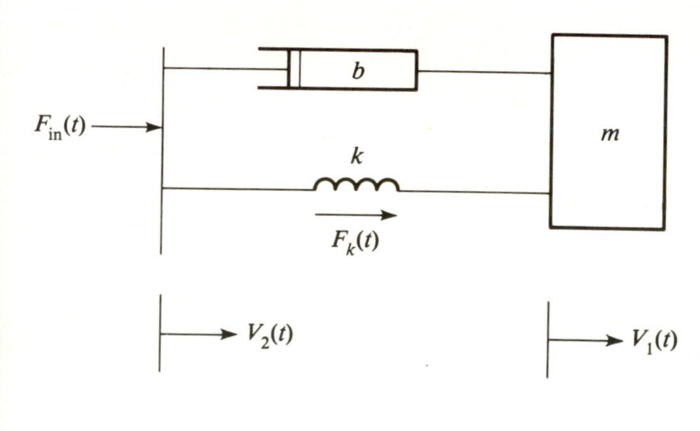

Figure 2-13 Mechanical network for Prob. 2-3.

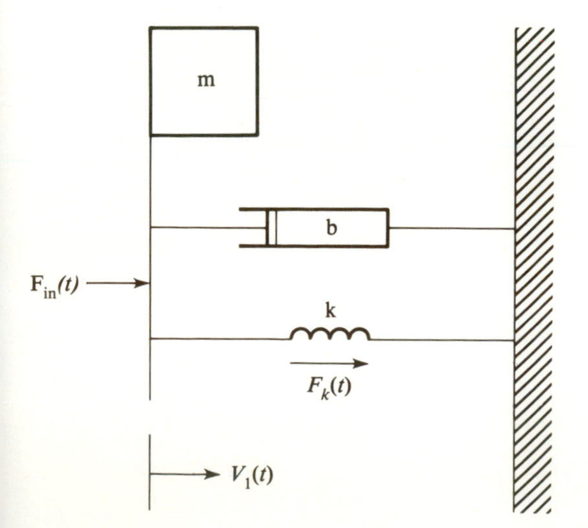

Figure 2-14 Mechanical network for Prob. 2-4.

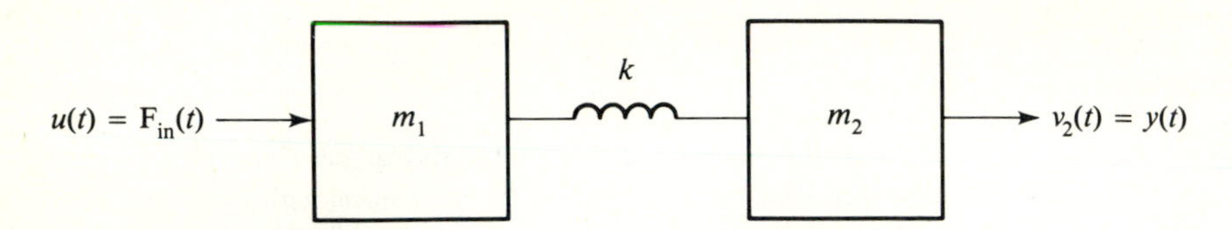

Figure 2-15 Mechanical network for Prob. 2-5.

2-9 Find the form of the homogeneous response for the following transfer functions. For complex conjugate poles write the form in three separate but equivalent ways.

(*a*) $H(s) = \dfrac{1}{(s+7)(s+8)}$ (*b*) $H(s) = \dfrac{(s-1)}{(s+7)(s+8)}$

(*c*) $H(s) = \dfrac{1}{s(s^2+0.4s+4.04)}$ (*d*) $H(s) = \dfrac{1}{(s+1)^2}$

2-10 Find the form of the homogeneous response for the following transfer functions. For complex conjugate poles write the form in three separate but equivalent ways.

(*a*) $H(s) = \dfrac{(s-2)}{s(s+5)}$ (*b*) $H(s) = \dfrac{(s+4)}{(s+1)(s+2)(s-3)}$

(*c*) $H(s) = \dfrac{1}{(s+1)^3(s+2)}$ (*d*) $H(s) = \dfrac{1}{(s^2+4)}$

2-11 Determine if each of the systems in Prob. 2-7 is stable, unstable, or marginally stable. Plot the poles and zeros on the complex plane.

2-12 Determine if each of the systems in Prob. 2-8 is stable, unstable, or marginally stable. Plot the poles and zeros on the complex plane.

2-13 Find the particular solution for the system $H(s) = (s+1)/(s^2+10s+1)$ to the input $u(t) = 2 + 3t - e^t$.

2-14 Find the particular solution for the system $H(s) = (s-1)/(s^2+2s+1)$ to the input $u(t) = -1 + 2t + 3e^{-2t}$.

2-15 With the input $u(t) = \sin(2t)$, find the particular solution for the following systems. Find the particular solution in two equivalent forms.

(*a*) $H(s) = \dfrac{s+1}{s-1}$

(*b*) $H(s) = \dfrac{s-1}{s+1}$

2-16 With the input $u(t) = \cos(t - \pi/4)$, find the particular solution for the following systems. Find the particular solution in two equivalent forms.

(*a*) $H(s) = \dfrac{1}{s^2+2s+2}$

(*b*) $H(s) = \dfrac{1}{s^2+0.2s+1.01}$

2-17 Find the particular solution for the input $u(t) = 2e^{-3t}u_{-1}(t)$ to the system

$$H(s) = \frac{(s+1)}{(s+2)(s+3)}$$

2-18 Demonstrate that the particular solution to Prob. 2-17 is in fact a particular solution.

2-19 Find $y_p(t)$ for the input $u(t) = (3te^{-t})u_{-1}(t)$ for the system $H(s) = 4/(s+2)$.

2-20 Demonstrate that the particular solution to Prob. 2-21 is in fact a particular solution.

2-21 (*a*) Show that the input $u(t) = e^{\sigma t}\cos(\omega t)$ may be written as

$$u(t) = \text{Re}\,\{e^{st}\}\big|_{s=\sigma+j\omega}$$

(*b*) What would be the particular solution for the input $u(t) = e^{\sigma t}\cos(\omega t)$ for the general system, $H(s)$?

(*c*) Find the particular solution for the input $u(t) = e^{-2t}\cos(4t)$ for the system

$$H(s) = \frac{3}{s+1}$$

2-22 Repeat Prob. 2-22 for the general input $u(t) = e^{\sigma t}\sin\omega t$. For part (*c*) use $u(t) = e^{-2t}\sin(4t)$ for the same system.

2-23 Demonstrate that $L\{\cdot\} \equiv \text{Im}\,\{\cdot\}$ is a linear operation.

2-24 Demonstrate that $L\{\cdot\} \equiv \frac{d}{ds}(\cdot)\Big|_{s=q}$ corresponding to the input $u(t) = kte^{qt}$ (entry 4 in Table 2-1) is a linear operation.

2-25 Find the complete solutions for $t > 0$ for the following systems:

(*a*) $\dot{y} + 7y = 2u$
$y(t) = u_{-1}(t)$
$y(0^+) = 0$

(*b*) $\dot{y} + 2y = \dot{u} + 3u$
$u(t) = 4\cos(5t)u_{-1}(t)$
$y(0^+) = -1$

(*c*) $\ddot{y} + 7\dot{y} + 12y = 2\dot{u} + 2u$
$u(t) = 5e^{-t}u_{-1}(t)$
$y(0^+) = 2 \qquad \dot{y}(0^+) = -1 \qquad u(0^+) = 5$

2-26 Find the complete solutions for $t > 0$ for the following systems:

(*a*) $\dot{y} + 5y = u$
$u(t) = (3 + t)u_{-1}(t)$
$y(0^+) = 1$

(*b*) $\dot{y} - y = 2\dot{u} + 4u$
$u(t) = u_{-1}(t)$
$y(0^+) = 1 \qquad u(0^+) = 1$

(*c*) $\ddot{y} + \dot{y} - 12y = \dot{u} + u$
$u(t) = (2e^{-t})u_{-1}(t)$
$y(0^+) = 0 \qquad \dot{y}(0^+) = 0 \qquad u(0^+) = 2$

2-27 Solve the following linear equation problems via Gauss-Jordan elimination:

(*a*) $$\begin{bmatrix} 2 & -4 & 6 \\ 1 & 0 & 2 \\ 0 & 8 & -5 \end{bmatrix} \begin{bmatrix} x_1 \\ x_2 \\ x_3 \end{bmatrix} = \begin{bmatrix} 10 \\ 6 \\ -2 \end{bmatrix}$$

(*b*) $$\begin{bmatrix} 4 & 3 & -1 & 3 \\ 8 & 7 & -2 & 6 \\ -12 & -9 & 4 & 0 \\ 0 & 0 & 1 & 10 \end{bmatrix} \begin{bmatrix} x_1 \\ x_2 \\ x_3 \\ x_4 \end{bmatrix} = \begin{bmatrix} 1 \\ 0 \\ -3 \\ 0 \end{bmatrix}$$

2-28 Solve the following linear equation problems via Gauss-Jordan elimination:

(*a*) $$\begin{bmatrix} 2 & 10 & -2 \\ 3 & 15 & 4 \\ 4 & 0 & 2 \end{bmatrix} \begin{bmatrix} x_1 \\ x_2 \\ x_3 \end{bmatrix} = \begin{bmatrix} 2 \\ 0 \\ 4 \end{bmatrix}$$

(*b*) $$\begin{bmatrix} 3 & 0 & -6 & 0 \\ 2 & 1 & -4 & 0 \\ 0 & 2 & -1 & 1 \\ 0 & 3 & 0 & -1 \end{bmatrix} \begin{bmatrix} x_1 \\ x_2 \\ x_3 \\ x_4 \end{bmatrix} = \begin{bmatrix} 0 \\ 1 \\ 2 \\ 3 \end{bmatrix}$$

2-29 Find the unit step response for the following systems, which are initially at rest [$y(0^-) = \dot{y}(0^-) = 0$]:

(*a*) $H(s) = 1/(s^2 + 4s + 5)$

(*b*) $H(s) = (s + 1)/(s^2 + 4s + 5)$

(*c*) $H(s) = (s - 1)/(s^2 + 4s + 5)$

(*d*) $H(s) = (s^2 + 3s + 2)/(s^2 + 4s + 5)$

(*e*) $H(s) = (s^2 + s - 2)/(s^2 + 4s + 5)$

2-30 Find the unit step response for the following systems, which are initially at rest [$y(0^-) = \dot{y}(0^-) = 0$]:

(*a*) $H(s) = 1/(s^2 + 3s + 2)$

(*b*) $H(s) = (10s + 1)/(s^2 + 3s + 2)$

(*c*) $H(s) = (10s - 1)/(s^2 + 3s + 2)$

(*d*) $H(s) = (10s^2 + 31s + 3)/(s^2 + 3s + 2)$

(*e*) $H(s) = (10s^2 + 29s - 3)/(s^2 + 3s + 2)$

2-31 Find the slope of the initial response [$\dot{y}(0^+)$] and the steady-state final value for the systems in Prob. 2-29.

2-32 Find the slope of the initial response [$\dot{y}(0^+)$] and the steady-state final value for the systems in Prob. 2-30.

2-33 Find the impulse response function, $h(t)$, for the systems in Prob. 2-29 by differentiating the unit step response.

2-34 Find the impulse response function, $h(t)$, for the systems in Prob. 2-30 by differentiating the unit step response.

2-35 How would you characterize, in general, $h(0^+)$ for a system with and without a zero? Use your results from either Prob. 2-33 or Prob. 2-34 as your example.

2-36 What is a characteristic of $h(t)$ for a system that is improper (feedforward) which is not characteristic of a proper system?

NOTES AND REFERENCES

References 1 and 2, for example, are good references for further background on the process of system modeling. In regard to the method of using the system transfer function for finding the system particular solution, it is well known (e.g., Ref. 6) that the transfer function can be used in this way for finding the particular solution for exponential inputs, including the sine and cosine inputs. However, generalizing this to the case of any linear operation on an exponential input is felt to be original here. See, for example, Ref. 7 for discussion of other more classic methods of finding the particular solution. In regard to solving the basic linear equation problem, the reader is advised to always make use of expertly written software whenever possible (e.g., Refs. 3 and 9). For further discussion on the general theory of solving the linear equation problem see, for example, Ref. 8.

1. Shearer, J. L., A. T. Murphy, and H. H. Richardson: *Introduction to System Dynamics,* Addison-Wesley, Reading, Mass., 1971.
2. Ogata, K.: *Modern Control Engineering,* Prentice-Hall, Englewood Cliffs, N.J., 1979.
3. *IMSL Reference Manual,* IMSL Inc., Houston, Tex., 1973.
4. D'Azzo, J. J., and C. H. Houpis: *Linear Control System Analysis and Design, Conventional and Modern,* McGraw-Hill, New York, 1981, pp. 188–189.
5. Fortmann, T. E., and K. L. Hitz: *An Introduction to Linear Control Systems,* Marcel Dekker, New York, 1977, pp. 166–167.
6. DeRusso, P. M., R. J. Roy, and C. M. Close: *State Variables for Engineers,* Wiley, New York, 1965.
7. Coddington, E., and N. Levison: *Theory of Ordinary Differential Equations,* McGraw-Hill, New York, 1965.
8. Stewart, G. W.: *Introduction to Matrix Computations,* Academic, New York, 1973.
9. Dongarra, J. J., et al.: *LINPACK User's Guide,* SIAM, Philadelphia, 1979.

CHAPTER

THREE

INPUT-OUTPUT CHARACTERISTICS OF A LINEAR DYNAMIC SYSTEM: FREQUENCY DOMAIN

This chapter once again considers the input-output characteristics of a linear time-invariant system. In particular, the steady-state output response is examined for the special input signal—a sinusoid input of varying frequency, ω. By consideration of this special sinusoid input, some helpful relationships are derived between the system impulse-response function introduced in Chap. 1 and the system transfer function introduced in Chap. 2. They are seen to be related by the very useful Fourier transformation. This also leads to a frequency-domain analysis of the linear time-invariant system. These results are some of the most important in linear systems theory, and they follow in a natural way from the material of the previous two chapters.

3-1 STEADY-STATE RESPONSE TO SINUSOID INPUTS

This section examines the steady-state response of a stable, linear, time-invariant, ordinary differential-equation system to a sinusoid input signal, $u(t) = \sin(\omega t)$, where ω is a given input frequency measured in radians per time unit. To help make the discussion specific, the dynamics of the rocket sled are once again utilized:

$$\dddot{y}(t) + 0.4\ddot{y}(t) + 1.14\dot{y}(t) + 0.22y(t) = u(t) \tag{3-1}$$

The transfer function of this system is

$$H(s) = \frac{1}{s^3 + 0.4s^2 + 1.14s + 0.22} \tag{3-2}$$

The steady-state response (the response after all of the transients associated with the homogeneous response of the stable system have decayed to zero) is the particular solution

$$y_{ss}(t) = y_p(t) = \text{Im}\,\{H(j\omega)e^{j\omega t}\}$$
$$= \text{Im}\left\{\frac{1}{(j\omega)^3 + 0.4(j\omega)^2 + 1.14(j\omega) + 0.22}e^{j\omega t}\right\} \tag{3-3}$$

There are several ways to evaluate the above expression, but the one used here is to write the complex $H(j\omega)$ in terms of its magnitude, $|H(j\omega)|$, and phase angle, $\phi(\omega)$:

$$H(j\omega) = |H(j\omega)|e^{j\phi(\omega)}$$

For the rocket-sled system,

$$|H(j\omega)| = \left[\sqrt{(0.22 - 0.4\omega^2)^2 + (1.14\omega - \omega^3)^2}\right]^{-1} \tag{3-4}$$

and

$$\phi(\omega) = -\arctan\left(\frac{1.14\omega - \omega^3}{0.22 - 0.4\omega^2}\right) \tag{3-5}$$

In general, the steady-state sinusoidal response is

$$y_{ss}(t) = |H(j\omega)| \sin[\omega t + \phi(\omega)] \tag{3-6}$$

For the rocket-sled system, the steady-state sinusoidal response has this form where $|H(j\omega)|$ and $\phi(\omega)$ are given by Eqs. (3-4) and (3-5), respectively.

Figure 3-1 shows the total response for the rocket-sled system for several input sinusoids of differing frequency, ω. Notice that the response immediately has the sinusoidal appearance even before the transients have died. This is be-

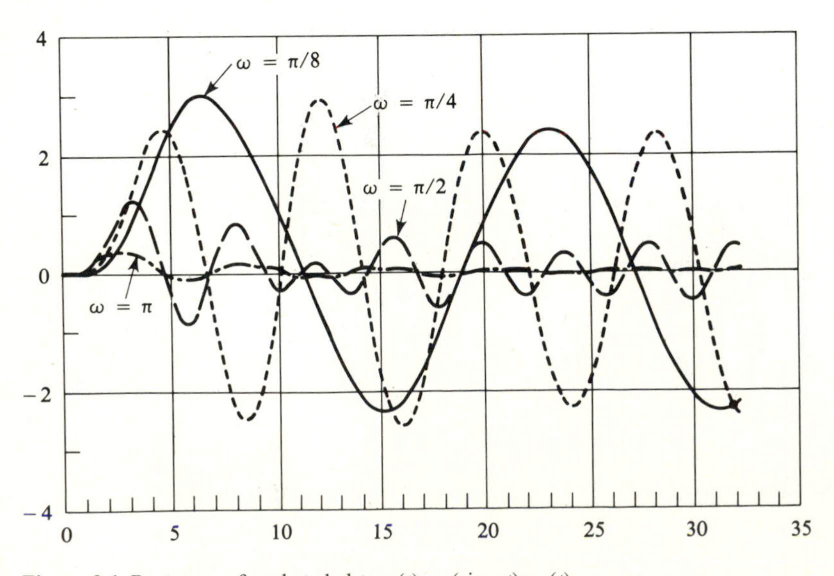

Figure 3-1 Response of rocket sled to $u(t) = (\sin \omega t)u_{-1}(t)$.

cause the particular solution or forced part of the response is present for all $t > 0$ (time 0 is when the sinusoidal input is initiated; before that time the system is assumed to be at rest):

$$\begin{aligned} y(t) &= y_h(t) + y_p(t) \\ &= y_{\text{trans}}(t) + y_{\text{ss}}(t) \\ &= y_{\text{trans}}(t) + |H(j\omega)| \sin(\omega t + \phi(\omega)) \end{aligned} \tag{3-7}$$

The $y_{\text{trans}}(t)$ decays to zero with increasing time, but the $y_p(t) = y_{\text{ss}}(t)$ is present for the entire time that the forcing input is present. For this reason the $y_p(t) = y_{\text{ss}}(t)$ is often termed the *forced response* of the system.

Also notice that as ω gets very large, the system response becomes very small. This is because $|H(j\omega)| \to 0$ as $\omega \to \infty$. The system does not allow the high-frequency signal inputs to pass. The system acts like what might be called a low-pass filter; high-frequency inputs are diminished.

The phase angle may also be determined from Fig. 3-1, but this is difficult to do unless we plot the output and the input on the same graph. This is shown in Fig. 3-2 for the example $\omega = \pi/4$.

The phase angle is $\phi(\pi/4) = -93.7°$ and the magnitude is $|H(j\pi/4)| = 2.43$.

Experimental Determination of the Magnitude and Phase

Since the steady-state response to the sinusoidal input may be measured by simply finding the steady-state output when forcing the system with the input $\sin \omega t$, ω being fixed, the magnitude $|H(j\omega)|$ and phase angle $\phi(\omega)$ may actually be deter-

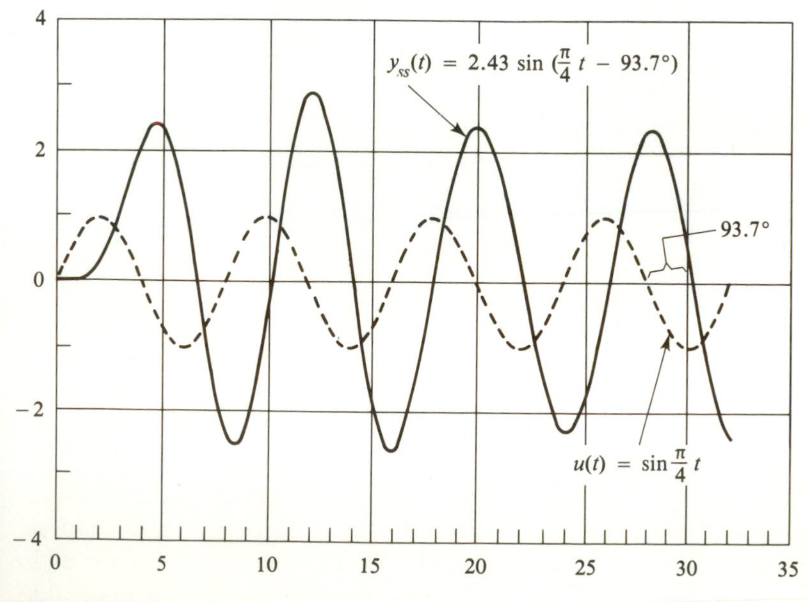

Figure 3-2 Rocket-sled response to $u(t) = \sin \pi/4t$.

mined on an experimental basis. By repeating the measurement over some frequency band $\omega_1 < \omega < \omega_2$, an entire plot of $|H(j\omega)|$ and $\phi(\omega)$ versus ω may be determined. This is such useful data in the analysis of a given system that there are actually commercial hardware devices, known as frequency or spectrum analyzers, that automatically perform this operation (or an analogous operation). Once the magnitude and phase angle have been determined, the analyzer plots the data in a desired format.

There are several ways of plotting $|H(j\omega)|$ and $\phi(\omega)$ versus ω, but one of the most useful techniques was developed by Bode [1]. It is termed the *system Bode plot.* Plots of $|H(j\omega)|$ and $\phi(\omega)$ are each made separately.

3-2 THE BODE MAGNITUDE PLOT

In plotting the Bode magnitude plot, the logarithm of $|H(j\omega)|$, in decibels, is plotted versus the logarithm (to the base 10) of the frequency ω. This choice of coordinate axes greatly simplifies the computations involved in computing the magnitude versus frequency plot. In fact, it enables one to sketch a close approximation to the actual Bode magnitude plot almost by inspection.

To see how this comes about, consider a general fourth-order transfer function

$$\begin{aligned} H(s) &= \frac{b_1 s + b_0}{s^4 + a_3 s^3 + a_2 s^2 + a_1 s} \\ &= \frac{b_1(s + z_1)}{s(s + p_1)(s^2 + 2\zeta\omega_n s + \omega_n^2)} \\ &= \frac{K_m(s/z_1 + 1)}{s(s/p_1 + 1)[s^2/\omega_n^2 + (2\zeta/\omega_n)s + 1]} \end{aligned} \tag{3-8}$$

where $K_m = (b_1 z_1)/(p_1 \omega_n^2) = b_0/a_1$. Then

$$H(j\omega) = \frac{K_m(j\omega/z_1 + 1)}{(j\omega)(j\omega/p_1 + 1)[(j\omega)^2/\omega_n^2 + (2\zeta/\omega_n)j\omega + 1]} \tag{3-9}$$

The logmagnitude of $H(j\omega)$ is given by

$$\begin{aligned} \mathrm{Lm}[H(j\omega)] &\equiv 20 \log |H(j\omega)| \\ &= 20[\log |K_m| + \log |j\omega/z_1 + 1| \\ &\quad - \log |j\omega| - \log |j\omega/p_1 + 1| \\ &\quad - \log |(j\omega)^2/\omega_n^2 + (2\zeta/\omega_n)j\omega + 1|] \end{aligned} \tag{3-10}$$

Thus Lm $[H(j\omega)]$ is the algebraic sum of all the individual factors of $H(j\omega)$. Factors in the numerator of $H(j\omega)$ add, while factors in the denominator subtract. Therefore, if the plot of each individual factor is determined, the plot of the entire transfer function is determined simply by their algebraic sum.

Lm $|K_m|$

Lm $|K_m|$ is a constant, and its graph is a straight horizontal line for all ω. The gain K_m thus shifts the Bode plot up or down by an amount Lm $|K_m| = 20 \log |K_m|$.

Lm $|j\omega/z_1 + 1|$

For $\omega \ll |z_1|$, Lm $|j\omega/z_1 + 1| \simeq$ Lm $|1| = 0$. For $\omega = |z_1|$, Lm $|j\omega/z_1 + 1| = 20 \log |j + 1| = 20 \log \sqrt{2} \simeq 3.01$. For $\omega \gg |z_1|$, Lm $|j\omega/z_1 + 1| \simeq$ Lm $|j\omega/z_1| = 20 \log \omega - 20 \log |z_1|$. Taking the partial of this last expression with respect to $\log \omega$ (which is the scale of the frequency axis), it is seen that the plot has a $+20$ dB/decade of frequency change.

The actual Bode plot for this factor is shown in Fig. 3-3. A straight-line approximation to the Bode plot for this factor is also shown. The break point for the straight-line approximation is at $\omega = |z_1|$. This is called the *corner frequency*. Notice that the actual plot is $+3.01$ dB above the straight-line approximation at the corner frequency, but the actual plot approaches the straight-line approximation asymptotically for $\omega \ll |z_1|$ and $\omega \gg |z_1|$.

$-$Lm $|j\omega|$

For the pole at the origin, $-\text{Lm}\,|j\omega| = -\text{Lm}\,\omega = -20 \log \omega$. This has a value of zero at $\omega = 1$, and a negative slope of 20 dB/decade of frequency change. A plot is shown in Fig. 3-3.

$-$Lm $|j\omega/p_1 + 1|$

The analysis for a distinct real pole at p_1 is just the opposite of the distinct real zero at z_1. Thus

$$-\text{Lm}\,|j\omega/p_1 + 1| = \begin{cases} 0 & \omega \ll |p_1| \\ -3.01 & \omega = |p_1| \\ \text{Lm}\,\omega - \text{Lm}\,|p_1| & \omega \gg |p_1| \end{cases}$$

A plot along with the straight-line approximation is shown in Fig. 3-3.

$-$Lm $|-\omega^2/\omega_n^2 + (2\zeta/\omega_n)j\omega + 1|$

The complex conjugate pole (with $0 < \zeta < 1$) is more complicated than the previous cases. For $\omega \ll \omega_n$, the term is approximately zero. For $\omega \gg \omega_n$, the term is approximately equal to

$$-\text{Lm}\,\omega^2 + \text{Lm}\,\omega_n^2 = -40 \log \omega + 40 \log \omega_n$$

This produces a straight asymptote with zero crossing at $\omega = \omega_n$ and a negative slope of -40 dB/decade of frequency change. At $\omega = \omega_n$ the term equals $-\text{Lm}\,|2\zeta| = -20 \log |2\zeta|$. Hence the curve behavior in the vicinity of $\omega = \omega_n$ is

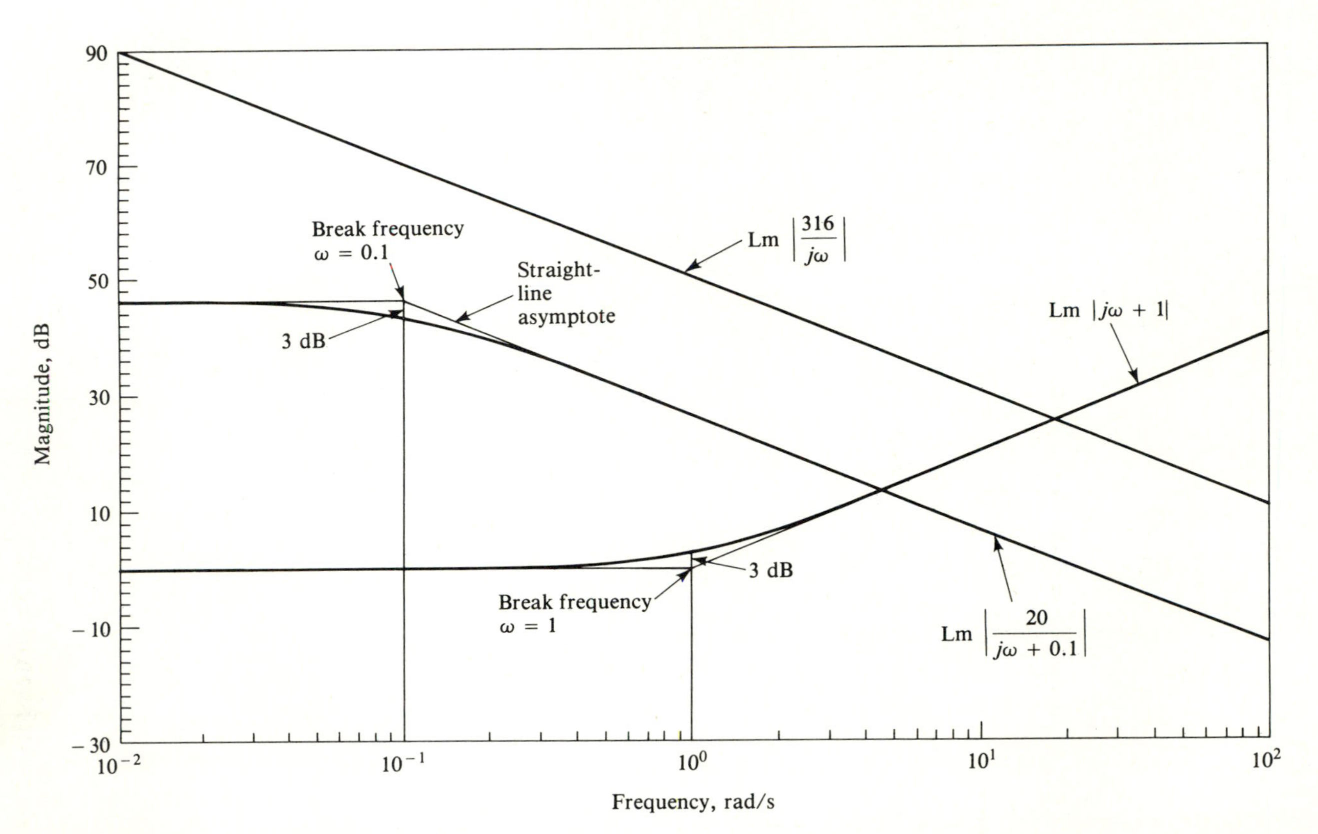

Figure 3-3 Sketching the bode magnitude plot, distinct poles, and zeroes, system Eq. (3-11).

highly dependent on the damping ratio, ζ. Typical curves for various values of ζ are shown in Fig. 3-4.

It is interesting to reflect on these curves. Recall that a lightly damped system ($\zeta < 0.5$) has a step response with a large overshoot, large transient oscillations, and a long settling time before reaching a steady-state constant value. Now for the sinusoidal inputs we see another undesirable behavior; but now it is in the *steady state.* Namely, $y_{ss}(t) = |H(j\omega)| \sin[\omega t + \phi(\omega)]$, and if $\omega = \omega_n$ and $\zeta < 0.5$ (the system is lightly damped), the factor $|H(j\omega)| = 1/2\zeta > 1$. Hence the steady-state response to the sinusoidal input of frequency $\omega = \omega_n$ is increased rather than decreased! The frequency ω_n is termed the *resonant* or *natural frequency* of the system. Forcing a system at its resonant frequency is responsible for such disasters as the destruction of suspension bridges when the wind buffets at just the wrong frequency or when troops march across at just the wrong cadence. The factor $|H(j\omega_n)| = 1/2\zeta$ becomes a large multiplier at the resonant frequency for a lightly damped system.

Drawing the Complete Bode Magnitude Plot

The last step is to algebraically add together the Bode plots for each individual factor. This may be done graphically, but with a little practice a good approximation may be drawn based purely on the gain and the pole and zero locations. The first step is to locate the poles and zeros (the corner frequencies) on the frequency axis. The low-frequency curve is dictated by the gain. Then asymptotes are drawn at each break frequency. The slope changes by +20 or −20, depending on whether the break point is a zero or a pole. Repeated poles (or zeros) or complex conjugate poles (or zeros) have a slope of −40 (+40). This procedure is shown in Fig. 3-5 for the transfer function

$$H(s) = \frac{10(s/0.1 + 1)}{s(s/1 + 1)[s^2/100 + (0.4/10)s + 1]} \tag{3-11}$$

Poles or Zeros in the Right Half-Plane

In drawing the Bode magnitude plot it makes no difference whether the poles or zeros are in the right or left half-plane; the plot is the same regardless. For instance,

$$\left|\frac{j\omega}{j\omega + 1}\right| = \left|\frac{j\omega}{(-j\omega) + 1}\right|$$

for all ω. The only way to determine the pole or zero locations relative to the right or left half-planes is by also considering the Bode phase-angle plot as discussed in the next section.

3-3 THE BODE PHASE-ANGLE PLOT

The same technique of individually determining the contribution of each separate pole or zero may also be applied to find the Bode phase-angle plot. Again con-

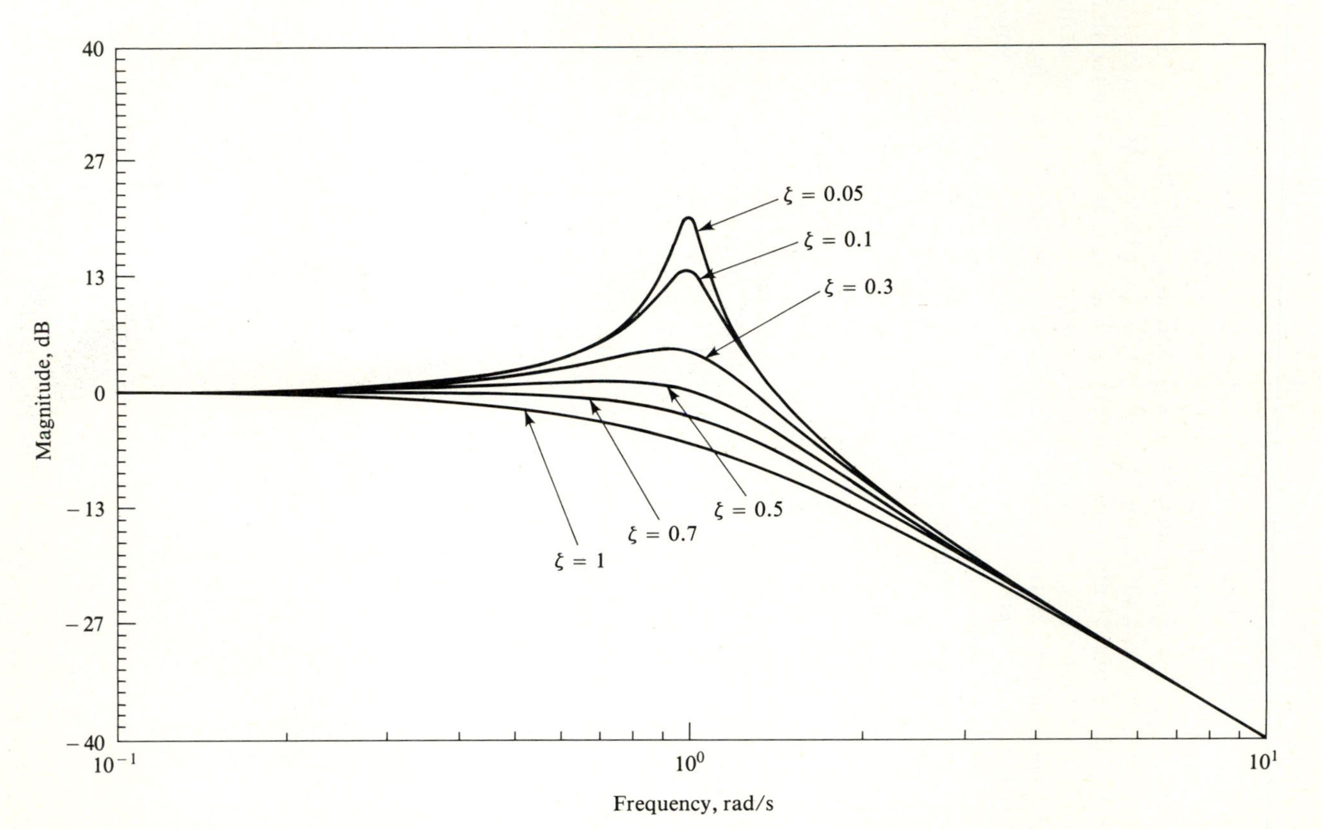

Figure 3-4 Bode magnitude plot: $H(s) = 1/(s^2 + 2\zeta s + 1)$.

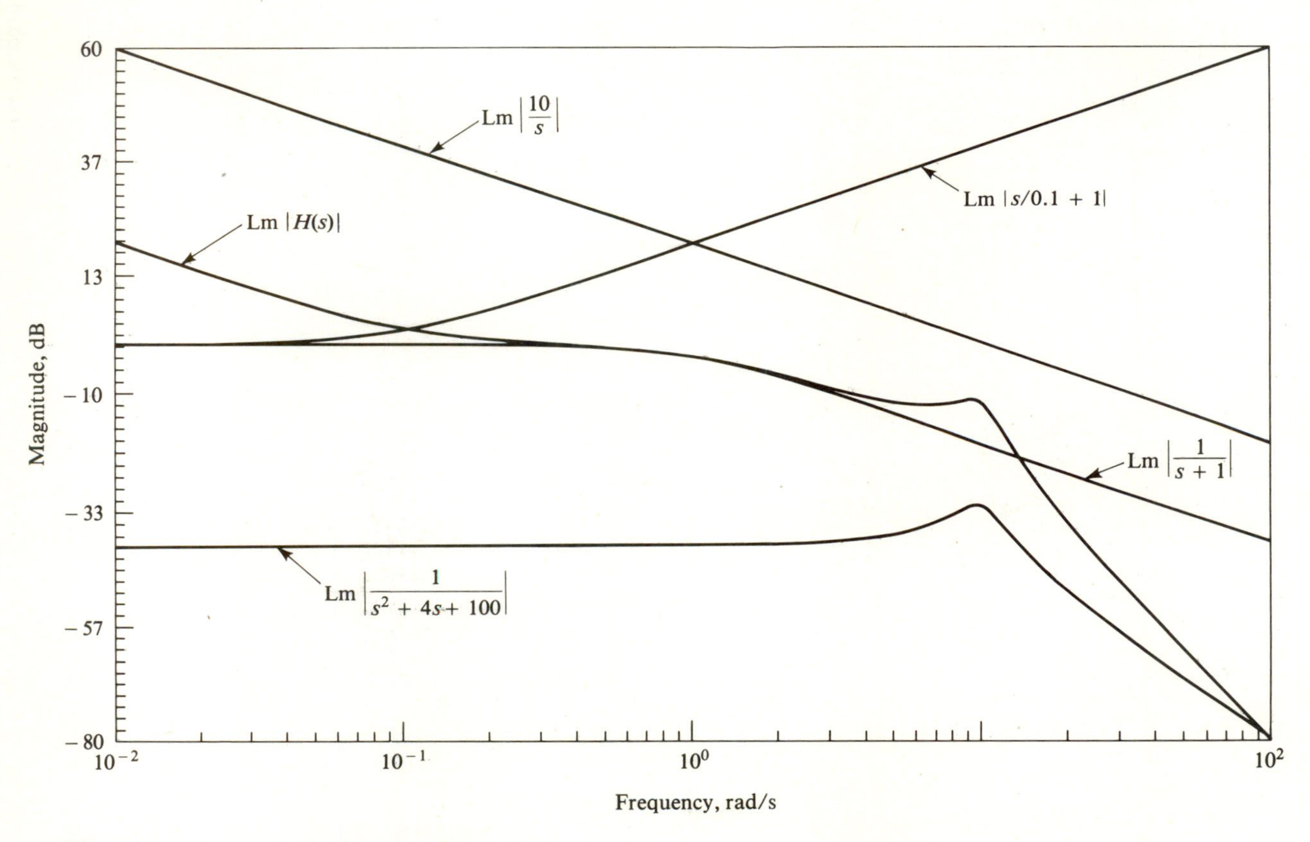

Figure 3-5 Bode magnitude plot: $H(s)$, Eq. (3-11).

sider the general fourth-order system

$$H(j\omega) = \frac{K_m(j\omega/z_1 + 1)}{(j\omega)(j\omega/p_1 + 1)[(j\omega)^2/\omega_n^2 + (2\zeta/\omega_n)(j\omega) + 1]} \qquad (3\text{-}9)$$

The phase angle of the complex number $H(j\omega)$ is then given by

$$\phi(\omega) = \phi_{z_1}(\omega) - \phi_{j\omega}(\omega) - \phi_{p_1}(\omega) - \phi_{p_2}(\omega)$$

where $\phi_{z_1}(\omega) = \tan^{-1}(\omega/z_1) \qquad \phi_{j\omega}(\omega) = 90°$

$$\phi_{p_1}(\omega) = \tan^{-1}(\omega/p_1) \qquad \phi_{p_2}(\omega) = \tan^{-1}\left(\frac{2\zeta/\omega_n}{1 - \omega^2/\omega_n^2}\right)$$

These are plotted individually and added together in Figs. 3-6 and 3-7 for the example system

$$H(s) = \frac{10(s/0.1 + 1)}{s(s/1 + 1)[s^2/100 + (0.4/10)s + 1]} \qquad (3\text{-}11)$$

Notice that for each distinct pole in the left half-plane the phase-angle contribution is $-45°$ at $\omega = p_i$, and it is $-90°$ at $\omega = \infty$. For each zero in the left half-plane the contribution is $+45°$ at $\omega = z_1$ and $+90°$ at $\omega = \infty$. Thus poles tend to give the system *lag*, while zeros tend to give the system *lead*. A *lead-lag* circuit has both a zero and a pole.

Poles and Zeros in the Right Half-Plane

If one considers the Bode magnitude and phase-angle plots by themselves, there is an ambiguity which arises with regard to poles and zeros in the right and left half-planes. However, taken together there is no ambiguity. For example, the systems $H_1(s) = (s + 1)/(s + 10)$, $H_2(s) = (s + 1)/(s - 10)$, $H_3(s) = (s - 1)/(s + 10)$, and $H_4(s) = (s - 1)/(s - 10)$ all have the same magnitude plot, but they have different phase-angle plots (see Figs. 3-8 and 3-9). Systems $H_2(s)$, $H_3(s)$, and $H_4(s)$ with poles and/or zeros in the right half-plane are termed *nonminimum-phase.*

There is also an ambiguity if one considers the phase-angle plot alone. For instance, $H_1(s)$ and $H_5(s) = (s - 10)/(s - 1)$ have the same phase-angle plot but different magnitude plots (see Figs. 3-10 and 3-11). Hence, reconstruction of the system transfer function from the Bode plot requires both the magnitude and phase information. If the system is known to be minimum-phase, the magnitude plot alone will suffice.

Stability

The Bode magnitude plot and phase-angle plot taken together provide valuable information about closed-loop, feedback system stability (see Fig. 3-12). The phase angle of $\phi = -180°$ is an important value in this regard. The output is then

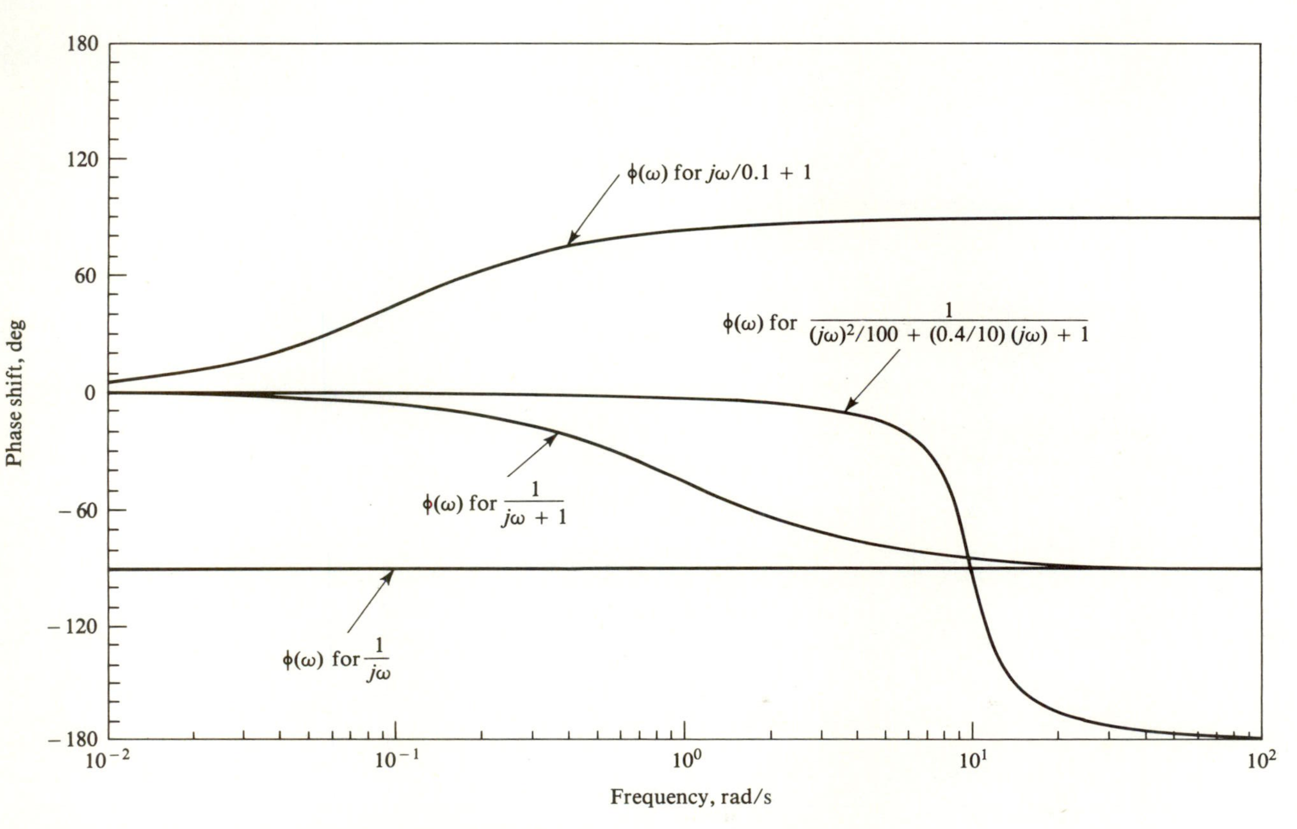

Figure 3-6 Bode phase-angle plot for elements of $H(s)$, Eq. (3-11).

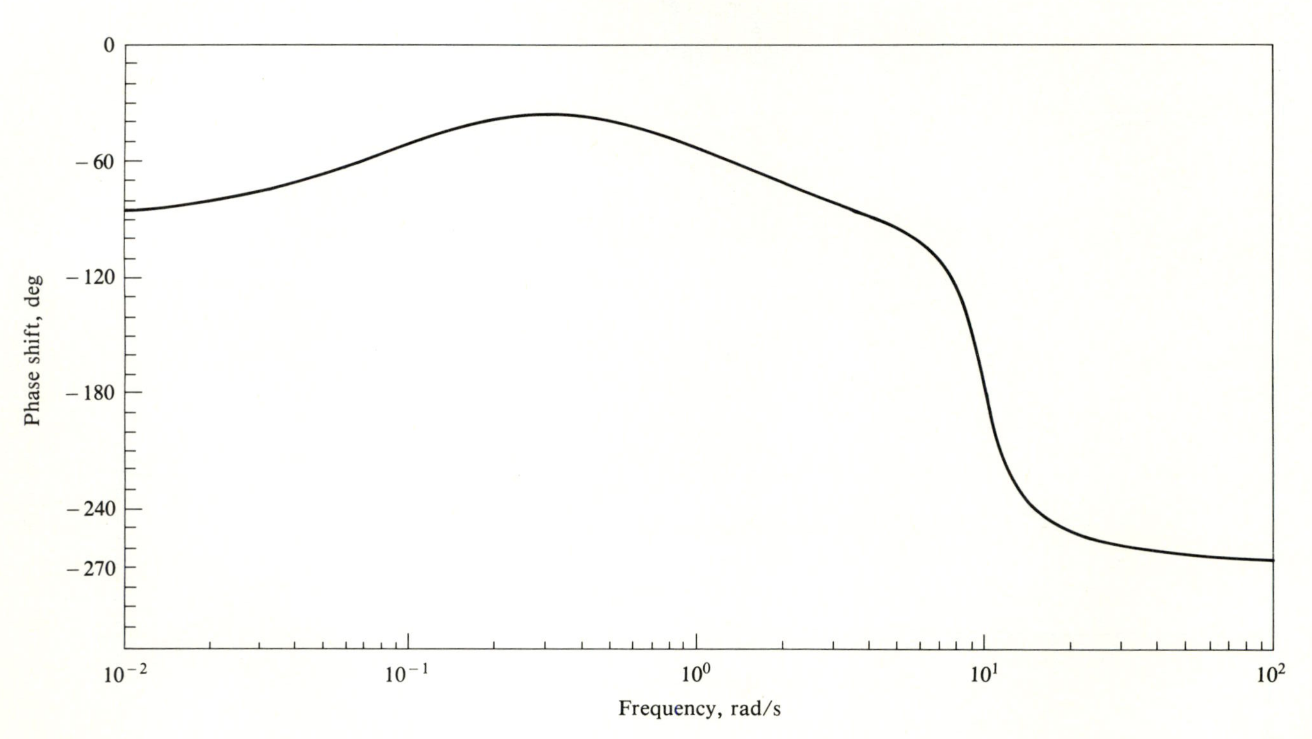

Figure 3-7 Bode phase-angle plot for $H(s)$, Eq. (3-11), and the sum of elements, Fig. 3-6.

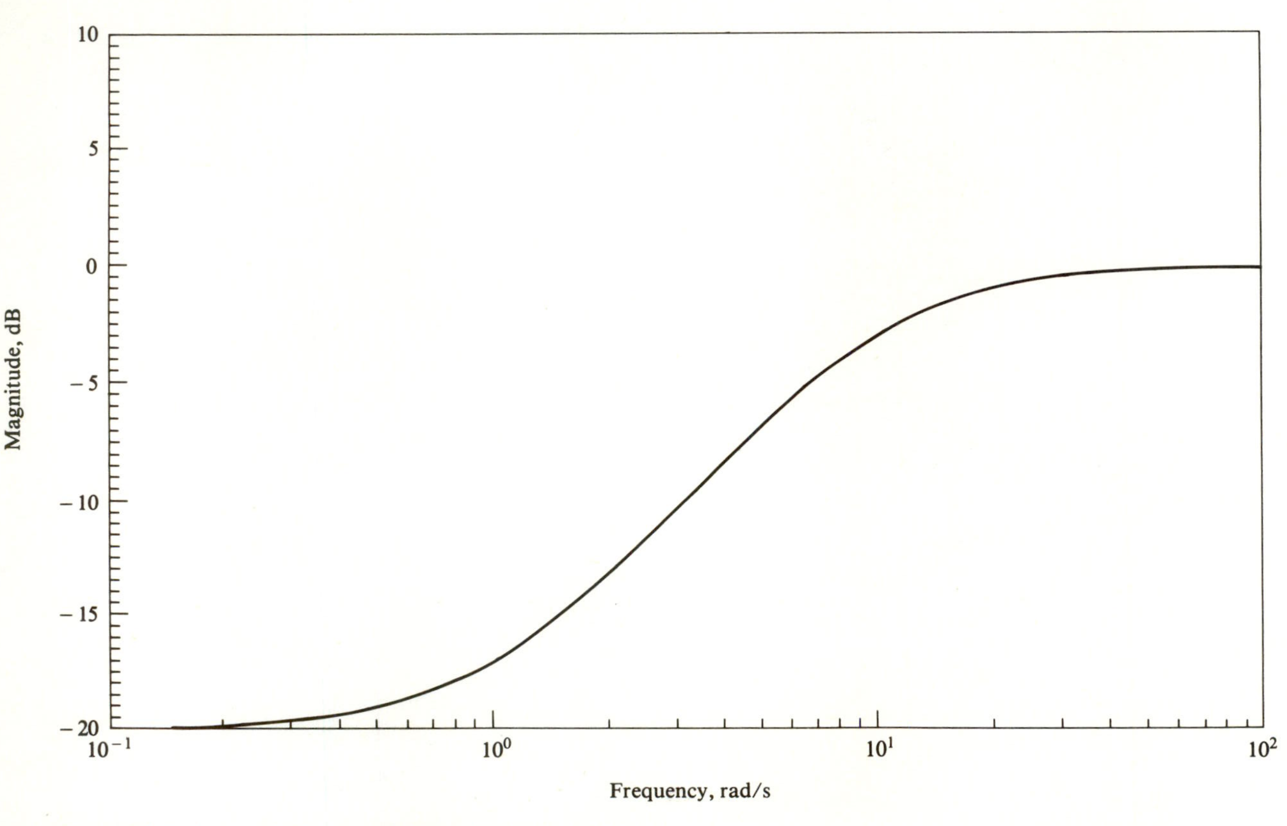

Figure 3-8 Bode magnitude plot for $H_1(s)$, $H_2(s)$, $H_3(s)$, $H_4(s)$.

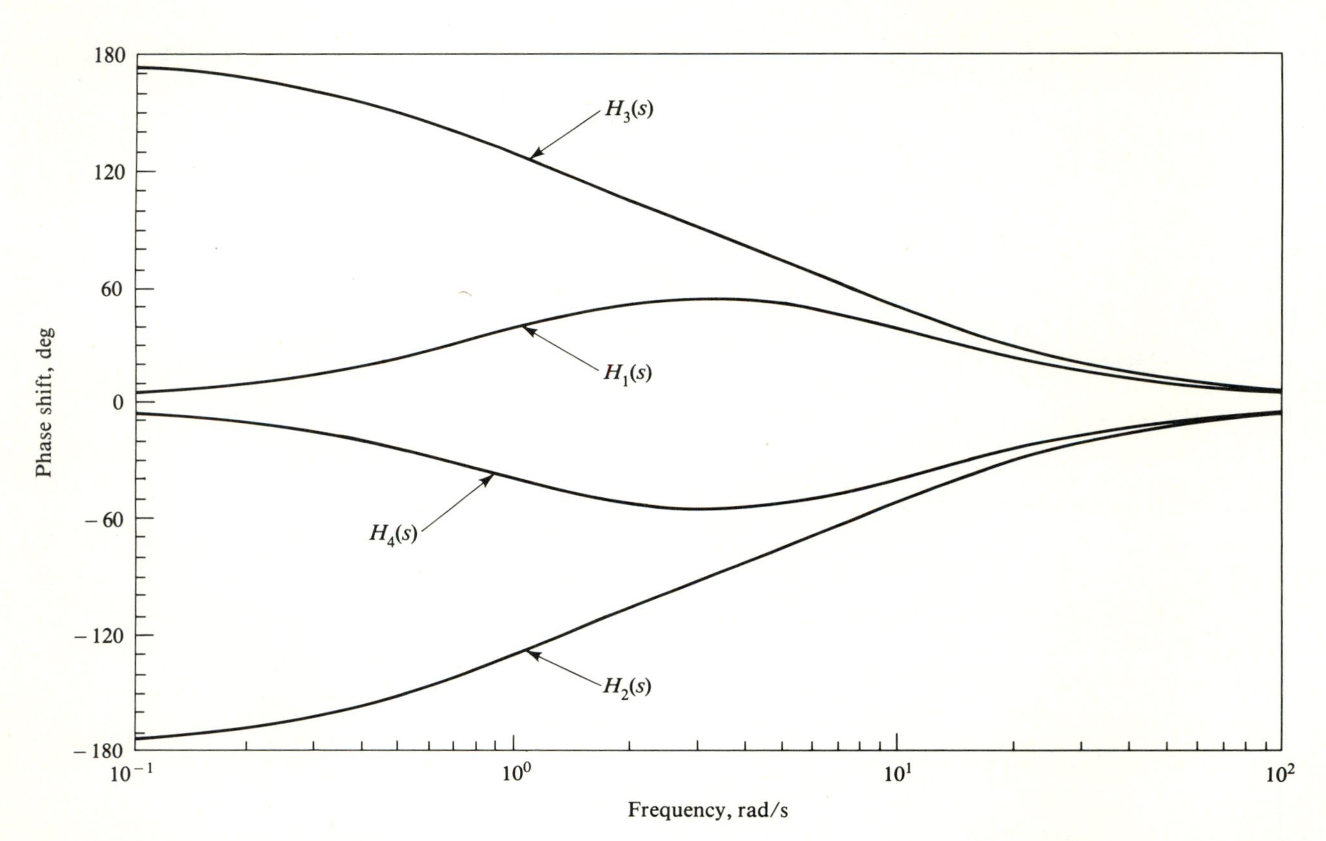

Figure 3-9 Bode phase-angle plots for $H_1(s)$, $H_2(s)$, $H_3(s)$, $H_4(s)$.

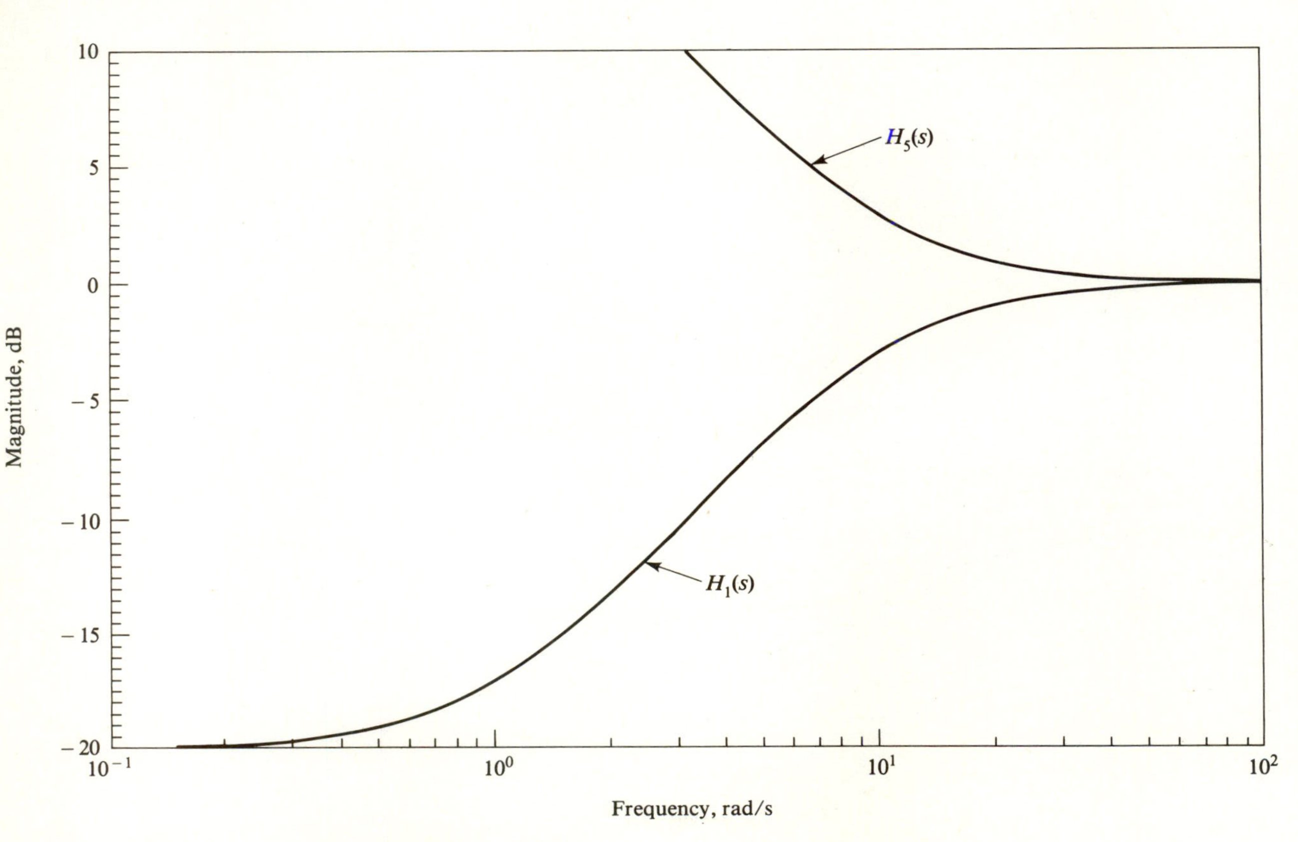

Figure 3-10 Bode magnitude plots for $H_1(s)$ and $H_5(s)$.

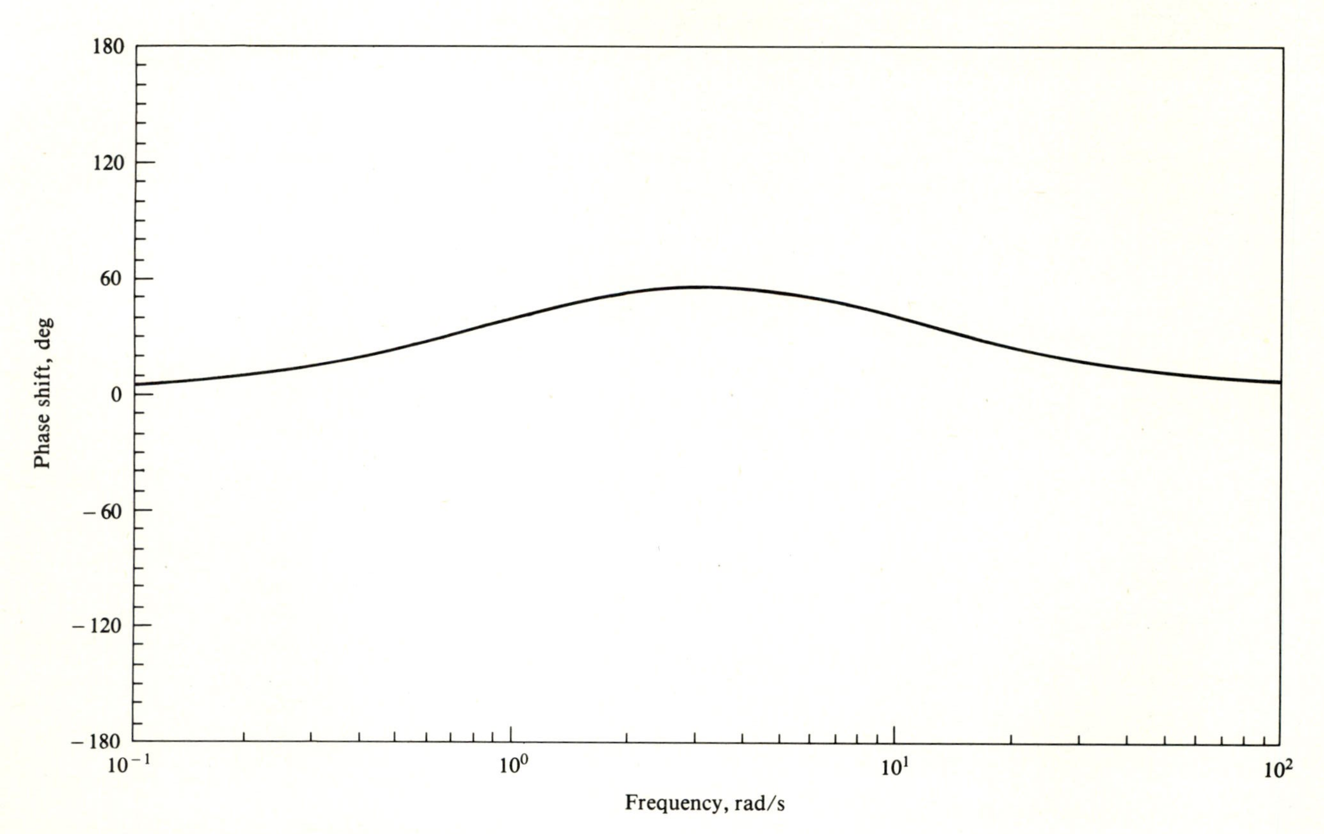

Figure 3-11 Bode phase-angle plot of $H_1(s)$ and $H_5(s)$.

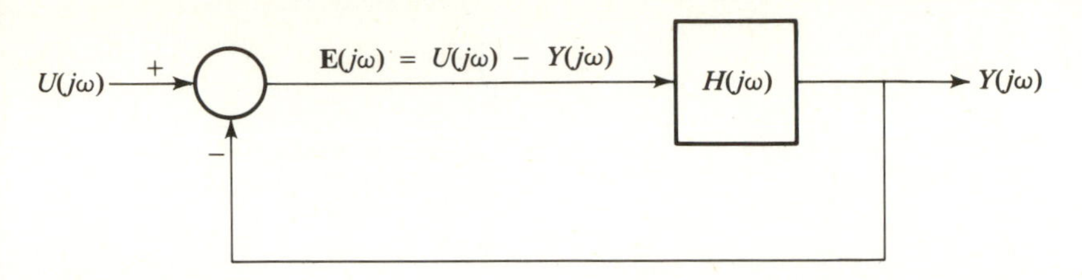

Figure 3-12 Feedback-control configuration.

exactly 180° out of phase with the input signal. If $|H(j\omega)| > 1$ at the frequency ω at which $\phi(\omega) = -180°$, then the system is unstable in a feedback configuration. This result is a simplified statement of the important Nyquist stability theorem [1, 2]. Similar results may be applied to a polar plot of the magnitude and phase angle plotted on the same plot. Such a polar plot is termed a *Nyquist plot*. See, e.g., [3] for greater discussion of the Nyquist stability theorem as applied to the topic of feedback control theory.

3-4 THE FOURIER TRANSFORM

Hypothetically assume that, instead of the pure sinusoid input, the system input to the stable linear time-invariant system is the complex exponential

$$u(t) = e^{j\omega t} = \cos(\omega t) + j\sin(\omega t) \tag{3-12}$$

By the methods of Chap. 2, it follows that the complex steady-state output would be

$$y_{ss}(t) = y_p(t) = H(j\omega)e^{j\omega t} \tag{3-13}$$

Alternatively, from the discussion of Chap. 1 it follows that the steady-state response is also given by the convolution integral

$$y_{ss}(t) = \int_{-\infty}^{t} h(t - \tau)e^{j\omega\tau}\, d\tau$$

The system is in steady state because it is assumed that the $e^{j\omega t}$ input persists all the way from $-\infty$. Since $h(t - \tau) \equiv 0$ for $\tau > t$, this convolution integral may also be rewritten as

$$y_{ss}(t) = \int_{-\infty}^{\infty} h(t - \tau)e^{j\omega\tau}\, d\tau$$

Making the change of variable $\xi = t - \tau$ then yields

$$y_{ss}(t) = \int_{-\infty}^{\infty} h(\xi)e^{j\omega(t-\xi)}\, d\xi$$

which may be rewritten as

$$y_{ss}(t) = \left(\int_{-\infty}^{\infty} h(\xi)e^{-j\omega\xi}\right)e^{j\omega t} \tag{3-14}$$

Comparing Eq. (3-13) to Eq. (3-14), both of which are true for all ω and all t, it is seen that the transfer function and the impulse response function are related by

$$H(j\omega) = \int_{-\infty}^{\infty} h(\xi)e^{-j\omega\xi}\, d\xi \tag{3-15}$$

This is a key integral relationship. It is known as the *Fourier transform;* that is, the transfer function, $H(j\omega)$, is the Fourier transform of the impulse response function, $h(t)$. This is written as

$$\boxed{H(j\omega) = \mathfrak{F}\{h(t)\} \equiv \int_{-\infty}^{\infty} h(\xi)e^{-j\omega\xi}\, d\xi} \tag{3-16}$$

The Fourier transform is a reversible transformation; that is, given $H(j\omega)$, the time-domain impulse response function, $h(t)$, may be found from the *inverse Fourier transform:**

$$\boxed{h(t) = \mathfrak{F}^{-1}\{H(j\omega)\} \equiv \frac{1}{2\pi}\int_{-\infty}^{\infty} H(j\omega)e^{j\omega t}\, d\omega} \tag{3-17}$$

The $1/2\pi$ is a normalizing factor.

Example 3-1 The impulse response function of the rocket-sled system was found in Chap. 2 to be

$$\begin{aligned} h(t) &= 0.90909e^{-0.2t} - 0.90909e^{-0.1t}\cos(1.044t) \\ &\quad + 0.087087e^{-0.1t}\sin(1.044t) \end{aligned} \tag{3-18}$$

The transfer function of the rocket-sled system is

$$\begin{aligned} H(s) &= \frac{1}{s^3 + 0.4s^2 + 1.14s + 0.22} \\ &= \frac{1}{(s + 0.2)(s^2 + 0.2s + 1.1)} \\ &= \frac{0.90909}{s + 0.2} + \frac{-0.90909(s + 0.1)}{(s + 0.1)^2 + 1.044^2} + \frac{(0.087087)(1.044)}{(s + 0.1)^2 + 1.044^2} \end{aligned} \tag{3-19}$$

where the third equality follows from the second by use of partial fraction

*Actually, there is only strict equality on the inverse Fourier transform if $h(t)$ is continuous. If $h(t)$ had a discontinuity at t_0 such that $h(t_0^-) = a$ and $h(t_0^+) = b$, then $h^*(t) \equiv \mathfrak{F}^{-1}\{H(j\omega)\}$ would be everywhere equal to $h(t)$ except at t_0, when $h^*(t_0) = (a + b)/2$. We do not discuss any further such discontinuities. See, e.g., Refs. 4 and 5 for reference.

expansions (see Appendix C). Thus

$$H(j\omega) = \frac{0.90909}{j\omega + 0.2} + \frac{-0.90909(j\omega + 0.1)}{(j\omega + 0.1)^2 + 1.044^2} + \frac{(0.087087)(1.044)}{(j\omega + 0.1)^2 + 1.044^2} \tag{3-20}$$

It therefore follows that

$$\begin{aligned} h(t) &= \mathfrak{F}^{-1}\{H(j\omega)\} \\ &= 0.90909e^{-0.2t} - 0.90909e^{-0.1t}\cos(1.044t) \\ &\quad + 0.087087e^{-0.1t}\sin(1.044t) \end{aligned} \tag{3-18}$$

This may be verified by checking the abbreviated table of Fourier transforms, Table 3-1. Note that one can go in either direction with this; one can take the Fourier transform of $h(t)$ to find $H(j\omega)$, or one can take the inverse Fourier transform of $H(j\omega)$ to find $h(t)$.

Table 3-1 Table of Fourier transforms

$f(t)$	$F(j\omega)$
1. $\delta(t)$	1
2. 1	$2\pi\,\delta(\omega)$
3. $u_{-1}(t)$	$\pi\,\delta(\omega) + 1/j\omega$
4. $\lvert t\rvert$	$-2/\omega^2$
5. $e^{-at}u_{-1}(t)$, $a > 0$	$1/(a + j\omega)$
6. $te^{-at}u_{-1}(t)$, $a > 0$	$1/(a + j\omega)^2$
7. $\cos\omega_0 t$	$\pi[\delta(\omega - \omega_0) + \delta(\omega + \omega_0)]$
8. $\sin\omega_0 t$	$j\pi[\delta(\omega - \omega_0) - \delta(\omega + \omega_0)]$
9. $\cos\omega_0 t u_{-1}(t)$	$\frac{\pi}{2}[\delta(\omega - \omega_0) + \delta(\omega - \omega_0)] + \frac{j\omega}{\omega_0^2 - \omega^2}$
10. $\sin\omega_0 t u_{-1}(t)$	$\frac{\pi}{2j}[\delta(\omega - \omega_0) - \delta(\omega + \omega_0)] + \frac{\omega_0}{\omega_0^2 - \omega^2}$
11. $g_\tau(t) \equiv \begin{cases} 1 & \lvert t\rvert < \tau/2 \\ 0 & \lvert t\rvert > \tau/2 \end{cases}$	$\tau\,\mathrm{sinc}\,(\omega\tau/2) \equiv \tau\frac{\sin(\omega\tau/2)}{(\omega\tau/2)}$
12. $\gamma_T(t) \equiv \sum_{k=-\infty}^{\infty} \delta(t - kT)$	$(2\pi/T)\sum_{n=-\infty}^{\infty} \delta[\omega - n(2\pi/T)] \equiv \omega_0\Gamma_{\omega_0}(\omega - n\omega_0)$
13. $e^{-at}\cos\omega_0 t u_{-1}(t)$	$\frac{j\omega + a}{(j\omega + a)^2 + \omega_0^2}$
14. $e^{-at}\sin\omega_0 t u_{-1}(t)$	$\frac{\omega_0}{(j\omega + a)^2 + \omega_0^2}$

Fourier Transform Generalization

Although the Fourier transformation is developed here in a natural way by equating the impulse-response/convolution-integral representation to the steady-state/particular solution for the hypothetical complex exponential input, $e^{j\omega t}$, the Fourier transformation is a completely general transformation; it finds many, many applications beyond its one use here to relate the time-domain impulse response function, $h(t)$, to the frequency-domain transfer function, $H(j\omega)$. Indeed, let $f(x)$ be any arbitrary function of the independent variable x, defined on the interval $-\infty < x < \infty$. Then the Fourier transform of $f(x)$ is defined by

$$\boxed{F(j\alpha) \equiv \int_{-\infty}^{\infty} f(\xi) e^{-j\alpha\xi} \, d\xi} \tag{3-21}$$

with inverse transform relation

$$\boxed{f(x) = \frac{1}{2\pi} \int_{-\infty}^{\infty} F(j\alpha) e^{j\alpha x} \, d\alpha} \tag{3-22}$$

The independent variable x does not have to be time, and the independent variable α does not have to be radians per time unit. For example, in certain applications the independent variable x might represent distance along an infinite rod. The $f(x)$ might represent the temperature at a particular location x. Then $F(j\alpha)$ represents the distribution of temperature variations in radians per unit of length for the rod. If the rod had a uniform temperature T it would have no variation. Then $F(j\alpha)$ would actually be a delta function, since

$$F(j\alpha) \equiv \int_{-\infty}^{\infty} f(\xi) e^{-j\alpha\xi} \, d\xi = \begin{cases} \displaystyle\int_{-\infty}^{\infty} T \, d\xi & \alpha = 0 \\ \displaystyle\int_{-\infty}^{\infty} T \, e^{-j\alpha\xi} \, d\xi & \alpha \neq 0 \end{cases}$$

or

$$F(j\alpha) = T\,\delta(\alpha) = \begin{cases} T \cdot \infty & \alpha = 0 \\ 0 & \alpha \neq 0 \end{cases} \tag{3-23}$$

This brings up an interesting point: not all functions have bounded Fourier transforms. More is discussed about this later. The following example gives a function which does happen to have a finite Fourier transform.

Example 3-2 Let

$$f(x) = \begin{cases} 100 & -1 < x < 1 \\ 0 & \text{otherwise} \end{cases}$$

where the distance x is measured in meters and $f(x)$ is temperature in degrees. The Fourier transform is defined by

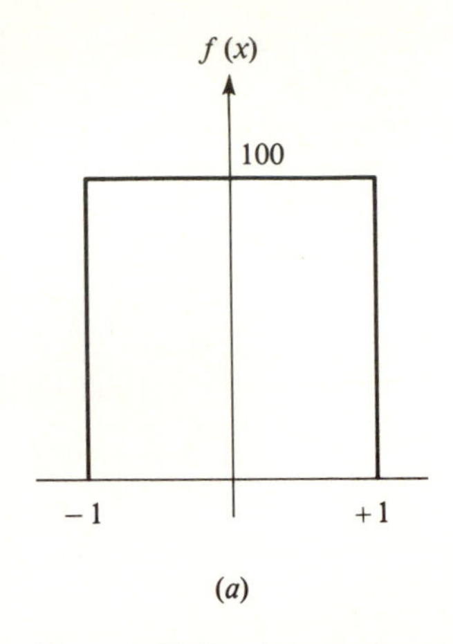

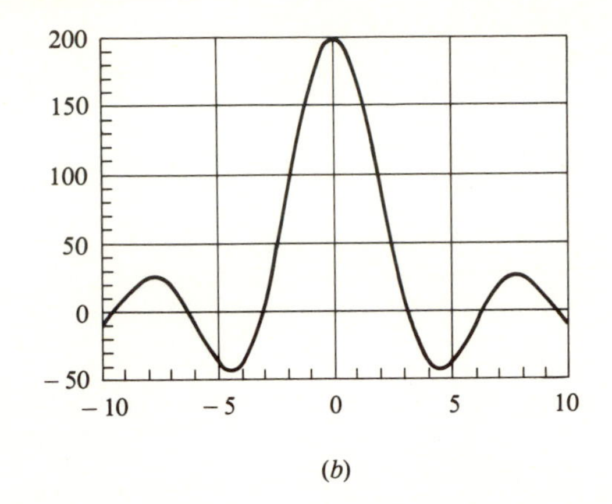

Figure 3-13 Fourier transform pair relation, Example 3-2.

(a) $f(x) = \begin{cases} 100 & \text{for } -1 < x < 1 \\ 0 & \text{for other} \end{cases}$

(b) $F(j\alpha) = \dfrac{200 \sin \alpha}{\alpha}$

$$F(j\alpha) = \int_{-\infty}^{\infty} f(\xi) e^{-j\alpha\xi} \, d\xi$$

$$= \int_{-1}^{1} 100 \, e^{-j\alpha\xi} \, d\xi$$

or

$$F(j\alpha) = \frac{100(e^{-j\alpha} - e^{j)})}{-j\alpha}$$

Using Euler's relation

$$\sin x = \frac{e^{jx} - e^{-jx}}{2j}$$

this becomes

$$F(j\alpha) = \frac{200 \sin \alpha}{\alpha}$$

The $f(x)$ and $F(j\alpha)$ are shown plotted in Fig. 3-13. The $f(x)$ is a special function called the *gate function*. The function $(\sin \alpha)/\alpha$ is also a special function called the *sinc function*. Both of these functions find particular significance in communications theory.

3-5 PROPERTIES OF THE FOURIER TRANSFORM

The Fourier transform has widespread application to linear systems theory in general. Many of these applications are derived because of its unique properties.

Linearity

The Fourier transformation is a linear transformation. Thus

$$\mathfrak{F}\{af_1(x) + bf_2(x)\} \equiv \int_{-\infty}^{\infty} [af_1(\xi) + bf_2(\xi)]e^{-j\alpha\xi}\, d\xi$$

$$= a\int_{-\infty}^{\infty} f_1(\xi)e^{-j\alpha\xi}\, d\xi + b\int_{-\infty}^{\infty} f_2(\xi)e^{-j\alpha\xi}\, d\xi$$

$$= a\mathfrak{F}\{f_1(x)\} + b\mathfrak{F}\{f_2(x)\} \tag{3-24}$$

and likewise for the inverse Fourier transform.

Time Shift

$$\mathfrak{F}\{f(x+\tau)\} \equiv \int_{-\infty}^{\infty} f(\xi+\tau)e^{-j\alpha\xi}\, d\xi$$

$$= \int_{-\infty}^{\infty} f(\zeta)e^{-j\alpha(\zeta-\tau)}\, d\zeta$$

or

$$\boxed{\mathfrak{F}\{f(x+\tau)\} = \mathfrak{F}\{f(x)\}e^{j\alpha\tau}} \tag{3-25}$$

Time Scaling

$$\mathfrak{F}\{f(ax)\} \equiv \int_{-\infty}^{\infty} f(a\xi)e^{-j\alpha\xi}\, d\xi$$

$$= \int_{-\infty}^{\infty} f(\zeta)e^{-(j\alpha/a)\zeta}\, d\zeta/|a|$$

or

$$\boxed{\mathfrak{F}\{f(ax)\} = F(j\alpha/a)/|a|} \tag{3-26}$$

where $F(j\alpha)$ is the Fourier transform of $f(x)$.

To better understand this scaling property, consider the *unit gate function*

$$g_1(x) \equiv \begin{cases} 0 & |x| > 1/2 \\ 1 & |x| < 1/2 \end{cases} \tag{3-27}$$

Letting $a = 1/3$ gives

$$g_3(x/3) = \begin{cases} 0 & |x| > 3/2 \\ 1 & |x| < 3/2 \end{cases}$$

The Fourier transform of the unit gate function gives the sinc function

$$\mathfrak{F}\{g_1(x)\} = \int_{-1/2}^{1/2} e^{-j\alpha\xi}\, d\xi = \frac{2\sin(\alpha/2)}{\alpha}$$

$$\mathfrak{F}\{g_3(x/3)\} = \int_{-3/2}^{3/2} e^{-j(\xi/3)\alpha}\, d\xi = \frac{6\sin(3\alpha/2)}{\alpha}$$

Figure 3-14 depicts this relation. Note that a time compression results in a frequency spreading in the transform domain and vice versa.

Differentiation

The function $f(x)$ is related to its Fourier transform $F(j\alpha) = \mathfrak{F}\{f(x)\}$ by

$$f(x) = \frac{1}{2\pi}\int_{-\infty}^{\infty} F(j\alpha)e^{j\alpha x}\, d\alpha \tag{3-22}$$

Differentiating $f(x)$ then yields

$$\frac{df(x)}{dx} = \frac{d}{dx}\left(\frac{1}{2\pi}\int_{-\infty}^{\infty} F(j\alpha)e^{j\alpha x}\, d\alpha\right)$$

$$= \frac{1}{2\pi}\int_{-\infty}^{\infty} (j\alpha)F(j\alpha)e^{j\alpha x}\, d\alpha$$

Thus it follows that

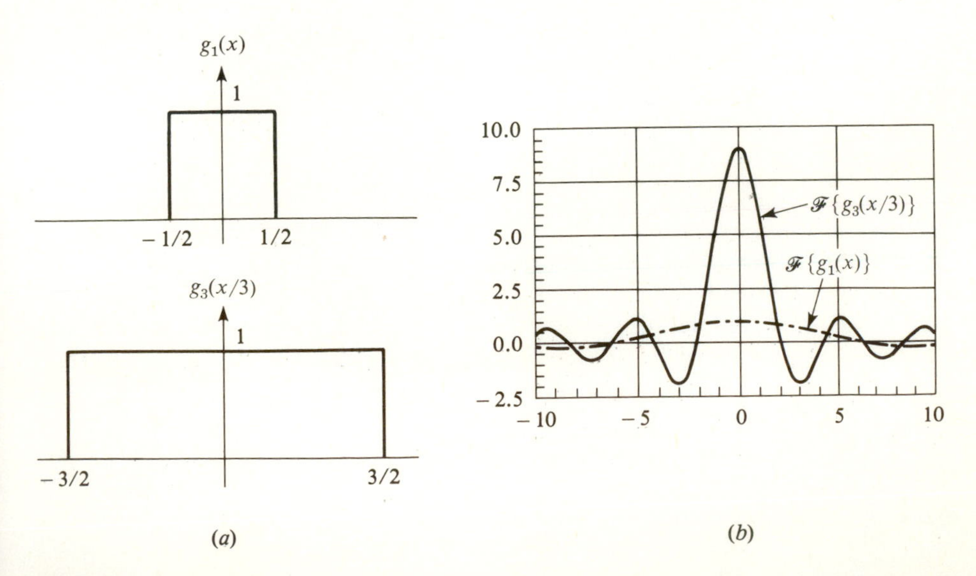

Figure 3-14 Illustration of time-scaling property of Fourier transform. (*a*) Functions $g(x)$ and $g(x/3)$. (*b*) Fourier transforms of $g(x)$ and $g(x/3)$.

$$\boxed{\mathfrak{F}\left\{\frac{df(x)}{dx}\right\} = (j\alpha)F(j\alpha)} \tag{3-28}$$

and, in general,

$$\boxed{\left\{\frac{d^n}{dx^n}f(x)\right\} = (j\alpha)^n F(j\alpha)} \tag{3-29}$$

Example 3-3 Using the differentiation property, the Fourier transform of the rocket-sled differential equation becomes

$$\mathfrak{F}\{\dddot{y}(t) + 0.4\ddot{y}(t) + 1.14\dot{y}(t) + 0.22y(t) = u(t)\}$$

or
$$[(j\omega)^3 + 0.4(j\omega)^2 + 1.14(j\omega) + 0.22]Y(j\omega) = U(j\omega)$$

Taking the ratio $Y(j\omega)/U(j\omega)$ then gives

$$\frac{Y(j\omega)}{U(j\omega)} = \frac{1}{(j\omega)^3 + 0.4(j\omega)^2 + 1.14(j\omega) + 0.22} = H(j\omega)$$

which is seen to be the system transfer function.

Determination of the Transfer Function

In Chap. 2 determination of the transfer function $H(s)$ is made on a fairly "mechanical" basis from the system differential equation by use of the operator variable s. By the time-differentiation property of the Fourier transform we see that we may legitimize this by taking the Fourier transform of both sides of the differential equation and solving for

$$\boxed{H(j\omega) = \frac{Y(j\omega)}{U(j\omega)}} \tag{3-30}$$

for the SISO system. Using the Fourier transform analysis, Eq. (3-30) is strictly true only for the steady state with a stable system (see Example 3-4). This procedure is used in Example 3-3 to find $H(j\omega)$ for the rocket-sled system. The same technique is generalized in the next chapter using the Laplace transformation instead of the Fourier transformation.

Integration

From the definition of the Fourier transform it is true that

$$f(x) = \frac{1}{2\pi}\int_{-\infty}^{\infty} F(j\alpha)e^{j\alpha x}\, d\alpha \tag{3-22}$$

Integrating both sides yields

$$\int f(x)\,dx = \frac{1}{2\pi}\int_{-\infty}^{\infty} F(j\alpha)\left(\int e^{j\alpha x}\,dx\right)d\alpha$$

where the order of integration is reversed on the right side of the equality. Carrying through the integration of $e^{j\alpha}$ and leaving the upper and lower limits of integration indefinite then yields

$$\int f(x)\,dx = \frac{1}{2\pi}\int_{-\infty}^{\infty} \frac{F(j\alpha)}{j\alpha} e^{j\alpha x}\,d\alpha$$

Thus it follows that

$$\boxed{\mathfrak{F}\left\{\int f(x)\,dx\right\} = \frac{F(j\alpha)}{j\alpha}} \tag{3-31}$$

Existence

As previously mentioned, the Fourier transform of many functions does not exist (remains finite). A function $f(x)$ defined on the interval $-\infty < x < \infty$ has a finite Fourier transform if it satisfies the condition

$$\int_{-\infty}^{\infty} \left|f(x)\right| dx < \infty \tag{3-32}$$

This is a sufficiency condition and not a necessary condition. It is interesting that this is the same condition as the stability condition for a linear time-invariant system:

$$\int_{0}^{\infty} \left|h(\tau)\right| d\tau = \int_{-\infty}^{\infty} \left|h(\tau)\right| d\tau < \infty \tag{3-33}$$

(See Sec. 1.5.) Thus only for stable linear time-invariant systems does the Fourier transform of the impulse response function necessarily exist. This problem is obviated by the Laplace transform as discussed in the next chapter.

Time Convolution

The convolution of two functions $f_1(x)$ and $f_2(x)$ is defined by

$$f_1(x) * f_2(x) \equiv \int_{-\infty}^{\infty} f_1(\tau) f_2(x-\tau)\,d\tau \tag{3-34a}$$

or equivalently

$$f_1(x) * f_2(x) \equiv \int_{-\infty}^{\infty} f_1(x-\tau) f_2(\tau)\,d\tau \tag{3-34b}$$

Using Eq. (3-34a) and taking the Fourier transform of both sides of the equation yields

$$\mathfrak{F}\{f_1(x) * f_2(x)\} \equiv \int_{-\infty}^{\infty} \left(\int_{-\infty}^{\infty} f_1(\tau) f_2(\xi - \tau)\, d\tau \right) e^{-j\alpha\xi}\, d\xi$$

$$= \int_{-\infty}^{\infty} f_1(\tau) \left(\int_{-\infty}^{\infty} f_2(\xi - \tau) e^{-j\alpha\xi}\, d\xi \right) d\tau$$

where the order of integration has been reversed.

Using the time-shift property, the term inside parentheses is seen to be $\mathfrak{F}\{f_2(x - \tau)\} = e^{-j\alpha\tau}\mathfrak{F}\{f_2(x)\}$. Thus

$$\mathfrak{F}\{f_1(x) * f_2(x)\} = \left(\int_{-\infty}^{\infty} f_1(\tau) e^{-j\alpha\tau}\, d\tau \right) \mathfrak{F}\{f_2(x)\}$$

$$= \mathfrak{F}\{f_1(x)\}\mathfrak{F}\{f_2(x)\}$$

or

$$\boxed{\mathfrak{F}\{f_1(x) * f_2(x)\} = F_1(j\alpha)F_2(j\alpha)} \tag{3-35}$$

Simply put, this says that the Fourier transform of convolution in the x (or time) domain yields multiplication in the transform (or frequency) domain. This property of the Fourier transform is extremely significant. It also holds for the Laplace transform and is discussed further in the next chapter.

Example 3-4 From Chap. 1 it is known the system steady-state response is given by

$$y_{ss}(t) = \int_{-\infty}^{t} h(t - \tau)u(\tau)\, d\tau$$

$$= \int_{-\infty}^{\infty} h(t - \tau)u(\tau)\, d\tau$$

$$= h(t) * u(t)$$

Then

$$\mathfrak{F}\{y_{ss}(t)\} = \mathfrak{F}\{h(t) * u(t)\}$$

$$= H(j\omega)U(j\omega)$$

where $H(j\omega)$ is the Fourier transform of $h(t)$ and $U(j\omega)$ is the Fourier transform of the input.

Now suppose that $u(t) = e^{j\omega_0 t}$. Then by Table 3-1

$$U(j\omega) = \mathfrak{F}\{e^{j\omega_0 t}\} = 2\pi\, \delta(\omega - \omega_0)$$

a delta function of strength 2π at ω_0. Taking the inverse Fourier transform of $\mathfrak{F}\{y_{ss}(t)\}$ yields $y_{ss}(t)$ again

$$y_{ss}(t) = \frac{1}{2\pi} \int_{-\infty}^{\infty} H(j\omega) 2\pi\, \delta(\omega - \omega_0) e^{j\omega t}\, d\omega$$

By the sampling property of the delta function this yields

$$y_{ss}(t) = H(j\omega_0)e^{j\omega_0 t}$$

This agrees with the Chap. 2 analysis of the steady-state response. Once again we have come full circle.

Frequency Convolution

By arguments directly analogous to deriving the time convolution property, it may be shown that

$$\boxed{\mathfrak{F}\{f_1(x)f_2(x)\} = F_1(j\alpha) * F_2(j\alpha)} \tag{3-36}$$

and

$$\boxed{\mathfrak{F}^{-1}\{F_1(j\alpha) * F_2(j\alpha)\} = f_1(x)f_2(x)} \tag{3-37}$$

In words, this says that multiplication in the time domain is convolution in the transform frequency domain. This is just the reverse of the time-convolution property.

Parseval's Theorem

Using the frequency-convolution property, an important result regarding signal energy may be proven. Consider an input signal, $u(t)$. Its signal energy is defined by

$$\mathcal{E} = \int_{-\infty}^{\infty} u(\tau)u(\tau)\,d\tau \tag{3-38}$$

Taking the Fourier transform of $u(t) \cdot u(t)$ and using the frequency-convolution property gives

$$\mathfrak{F}\{u(t)u(t)\} \equiv \int_{-\infty}^{\infty} u(\tau)u(\tau)e^{-j\omega\tau}\,d\tau = \int_{-\infty}^{\infty} U[j(\omega-\beta)]U(j\beta)\,d\beta$$

This Fourier transform relation is true for all $-\infty < \omega < \infty$, and so evaluating this expression at $\omega = 0$ gives

$$\int_{-\infty}^{\infty} u(\tau)u(\tau)\,d\tau = \int_{-\infty}^{\infty} U(-j\beta)U(j\beta)\,d\beta$$

The expression on the left is seen to be the signal energy. Since $U(-j\beta) = U^*(j\beta)$, the complex conjugate of $U(j\beta)$, the final expression of Parseval's theorem may be written as

$$\boxed{\mathcal{E} = \int_{-\infty}^{\infty} u(\tau)u(\tau)\,d\tau = \int_{-\infty}^{\infty} U^*(j\beta)U(j\beta)\,d\beta} \tag{3-39}$$

In words, this says that the signal energy as computed in the time domain is equal to the signal energy as computed in the frequency domain.

The expression

$$U^*(j\omega)U(j\omega) = U(j\omega)U^*(j\omega) = |U(j\omega)|^2 \tag{3-40}$$

also has a special name. It is called the energy spectrum of the signal, $u(t)$. The energy spectrum is real-valued, positive, and symmetric about $\omega = 0$. The energy spectrum of a signal may be used to ascertain useful characteristics about a signal in the same way that $|H(j\omega)|$ reveals useful characteristics about a system or a filter.

Example 3-5 The energy spectrum of the signal shown in Fig. 3-15*a* is shown in Fig. 3-15*b*. Although the signal appears very noisy, the power of the signal tends to drop off at high frequencies. The power at each frequency is measured by the energy spectrum of the signal; the example signal has nearly equal content of the low and middle frequencies, but it has low content of the high frequencies.

Time Correlation

The correlation between two functions $f_1(x)$ and $f_2(x)$ is defined by

$$f_1(x) \star f_2(x) \equiv \int_{-\infty}^{\infty} f_1(\tau) f_2(x+\tau)\, d\tau \tag{3-41a}$$

or equivalently by

$$f_1(x) \star f_2(x) \equiv \int_{-\infty}^{\infty} f_1(x+\tau) f_2(\tau)\, d\tau \tag{3-41b}$$

Using Eq. (3-41*a*) and taking the Fourier transform of both sides yields

$$\begin{aligned}\mathfrak{F}\{f_1(x) \star f_2(x)\} &= \int_{-\infty}^{\infty} \left(\int_{-\infty}^{\infty} f_1(\tau) f_2(\xi+\tau)\, d\tau \right) e^{-j\alpha\xi}\, d\xi \\ &= \int_{-\infty}^{\infty} f_1(\tau) \left(\int_{-\infty}^{\infty} f_2(\xi+\tau) e^{-j\alpha\xi}\, d\xi \right) d\tau\end{aligned}$$

Using the time-shift property of the Fourier transform, the term in parentheses becomes

$$e^{j\alpha\tau} F_2(j\alpha) = e^{-j\beta\tau} F_2^*(j\beta)$$

where we let $\alpha = -\beta$ and $F_2^*(j\beta)$ is the complex conjugate of $F_2(j\beta)$. Making this substitution results in

$$\boxed{\mathfrak{F}\{f_1(x) \star f_2(x)\} = F_1(j\beta) F_2^*(j\beta)} \tag{3-42}$$

This is similar to the convolution relationship and also proves to have important applications (see Sec. 3-7).

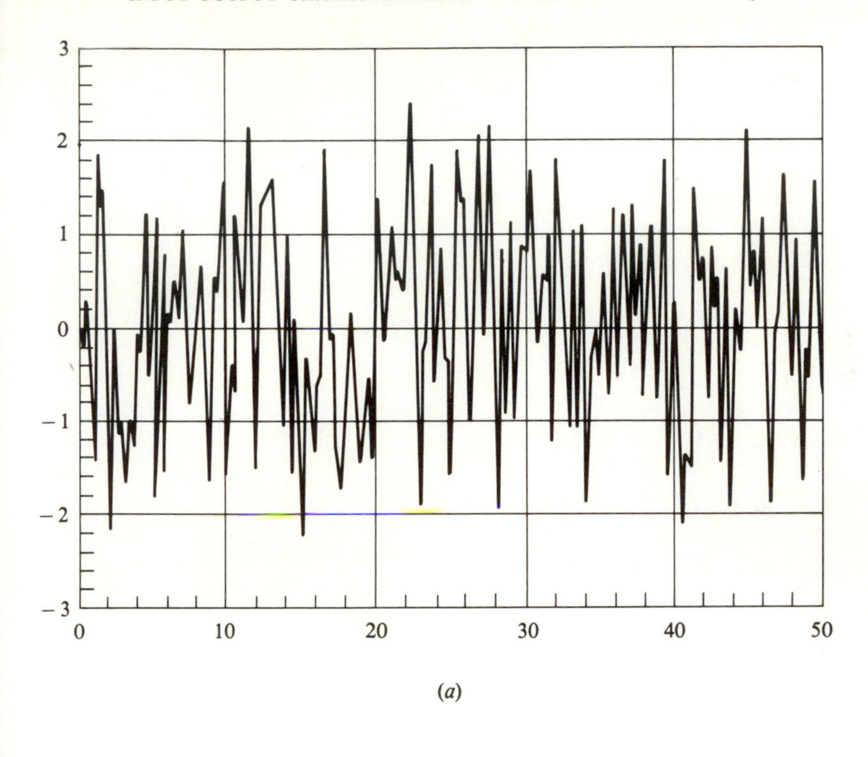

(a)

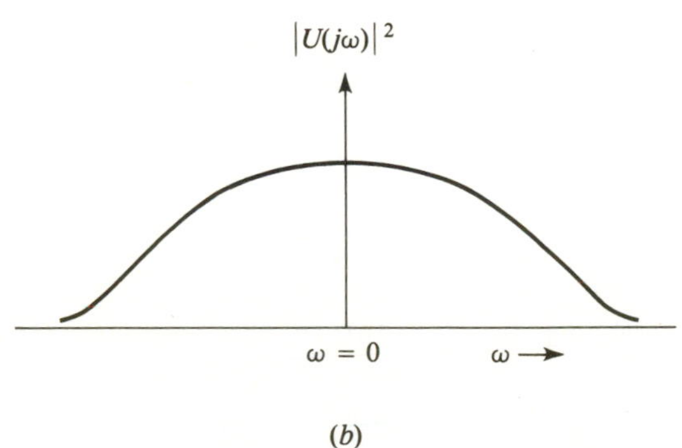

(b)

Figure 3-15 (a) Illustration of "noisy" signal. (b) Energy spectrum of noisy signal in a.

Alternative Forms of the Fourier Transform

The forms of the Fourier transform pair relation given by Eqs. (3-21) and (3-22) are convenient for our use here, but there are actually an infinite number of equivalent possible forms of the Fourier transform pair relations. In particular, we may write the general Fourier transform pair relation as

$$\mathfrak{F}\{f(x)\} = a \int_{-\infty}^{\infty} f(\zeta) e^{-j\alpha\zeta}\, d\zeta = F(j\alpha) \tag{3-43}$$

$$\mathfrak{F}^{-1}\{F(j\alpha)\} = b\int_{-\infty}^{\infty} F(j\alpha)e^{j\alpha x}\, d\alpha = f(x) \tag{3-44}$$

This is a valid transform pair relation provided $a \cdot b = 1/2\pi$. In Eqs. (3-21) and (3-22) we use $a = 1$, $b = 1/2\pi$.

Still another, often very useful, way to express the Fourier transform of a given $f(x)$ is via the Fourier transform expressed in cycles per time units (hertz) rather than radians per time unit:

$$\widetilde{\mathfrak{F}}\{f(x)\} \equiv \int_{-\infty}^{\infty} f(\zeta)e^{-2\pi\mathfrak{f}\zeta}\, d\zeta = \widetilde{F}(j\mathfrak{f}) \tag{3-45}$$

$$\widetilde{\mathfrak{F}}^{-1}\{\widetilde{F}(j\mathfrak{f})\} \equiv \int_{-\infty}^{\infty} \widetilde{F}(j\mathfrak{f})e^{j2\pi\mathfrak{f}t}\, d\mathfrak{f} = f(x) \tag{3-46}$$

The frequency $\mathfrak{f}$ in hertz (cycles per unit of x) is related to α in radians per unit of x via $\alpha = 2\pi\mathfrak{f}$. Thus $d\mathfrak{f} = d\alpha/2\pi$, and Eqs. (3-45) and (3-46) are seen to be exactly equivalent to Eqs. (3-21) and (3-22).

*3-6 TIME-DELAY SYSTEMS

In Sec. 1-8 the important class of systems—time-delay systems—was introduced. But even if it is linear and time-invariant, the time-delay system is exceedingly difficult to deal with in the time domain. The time-shift property of the Fourier transform, however, provides a fairly straightforward means to determine the transfer function of the linear time-invariant time-delay system. Consider, for example, the time-delay system described by the delay differential equation

$$\dot{y}(t) + 5y(t - 0.3) = u(t)$$

This is the same system for which the system impulse response was analyzed in Sec. 1-8. Taking the Fourier transform of this delay differential equation using zero initial conditions and using the time-shift property gives

$$(j\omega)Y(j\omega) + 5e^{-0.3j\omega}Y(j\omega) = U(j\omega)$$

Solving for $H(j\omega) = Y(j\omega)/U(j\omega)$ gives the transfer function of this time-delay system

$$H(j\omega) = \frac{1}{(j\omega) + 5e^{-0.3j\omega}}$$

This transfer function is in the form of an irrational polynomial because of the $e^{-0.3j\omega}$ factor. It is always the case that a time-delay system has an irrational transfer function. Indeed, it is true that a system has a rational polynomial transfer function if and only if it may be modeled by a linear time-invariant ordinary differential equation model.

Still, the time-delay transfer function has a magnitude, $|H(j\omega)|$, and phase angle, $\phi(\omega)$. A Bode plot could still be determined. The magnitude and phase angle still have the same significance regarding the steady-state system response to the sinusoid input, $u(t) = \sin \omega t$. Namely, it is still true that

$$y_{ss}(t) = |H(j\omega)| \sin [\omega t + \phi(\omega)]$$

for the persistent input, $u(t) = \sin \omega t$. Also true are the analyses of system feedback stability via the frequency-domain analysis.

It is also true that the impulse response function may be determined as the inverse Fourier transform of $H(j\omega)$:

$$h(t) = \mathfrak{F}^{-1}\{H(j\omega)\}$$

However, because of the irrational transfer function, this generally cannot be done in closed form. Instead, a rational polynomial approximation to $H(j\omega)$ can be found and an inverse transform of the rational approximation taken. A Padé approximation is often used for this purpose. See Ref. 6, for example.

*3-7 CORRELATION BETWEEN THE SYSTEM INPUT AND THE STEADY-STATE OUTPUT

The convolution and correlation properties of the Fourier transform may be used to derive an important experimental result regarding a linear time-invariant system. From Chap. 1 the steady-state output response for the input $u(t)$ is given by

$$y_{ss}(t) = \int_{-\infty}^{\infty} h(\xi)u(t - \xi)\, d\xi \tag{3-47}$$

Correlating the input with the steady-state output gives

$$\begin{aligned} y_{ss}(t) \star u(t) &\equiv \int_{-\infty}^{\infty} y_{ss}(t + \tau)u(\tau)\, d\tau \\ &= \int_{-\infty}^{\infty} \left[\int_{-\infty}^{\infty} h(\xi)u(t + \tau - \xi)\, d\xi \right] u(\tau)\, d\tau \end{aligned}$$

where Eq. (3-47) has been substituted for $y_{ss}(t + \tau)$. Reversing the order of integration gives

$$y_{ss}(t) \star u(t) = \int_{-\infty}^{\infty} h(\xi) \left[\int_{-\infty}^{\infty} u(t - \xi + \tau)u(\tau)\, d\tau \right] d\xi$$

But the term within brackets is $[u(t - \xi) \star u(t - \xi)]$, and so

$$y_{ss}(t) \star u(t) = \int_{-\infty}^{\infty} h(\xi)[u(t - \xi) \star u(t - \xi)]\, d\xi$$

The expression on the right is the convolution of $h(t)$ with $[u(t) \star u(t)]$, and so the final expression is

$$\boxed{y_{ss}(t) \star u(t) = h(t) * [u(t) \star u(t)]} \tag{3-48}$$

In words, Eq. (3-48) says that the correlation between the steady-state output response and the input producing that steady-state response is equal to the convolution of the impulse response with the *autocorrelation* function of the input, $[u(t) \star u(t)]$. Viewed another way, it says that the system forced by the input $[u(t) \star u(t)]$ has the steady-state output $[y_{ss}(t) \star u(t)]$. This is illustrated in Fig. 3-16. The term $[y_{ss}(t) \star u(t)]$ is called the *cross-correlation* of the input $u(t)$ with the steady-state output, $y_{ss}(t)$.

Equation (3-48) may be computed in the time domain, but a more significant result comes when the Fourier transform of both sides of the expression are taken to yield

$$Y_{ss}(j\omega)U^*(j\omega) = H(j\omega)[U(j\omega)U^*(j\omega)]$$

This expression may be solved for $H(j\omega)$ to give

$$\boxed{H(j\omega) = \frac{Y_{ss}(j\omega)U^*(j\omega)}{U(j\omega)U^*(j\omega)}} \tag{3-49}$$

This is a significant result. It says that one may determine the system transfer function, $H(j\omega)$, by finding the Fourier transform of the steady-state response, $Y_{ss}(j\omega)$, multiplying by the conjugate Fourier transform of the input, $U^*(j\omega)$, and then dividing by the power spectrum of the input, $U(j\omega)U^*(j\omega)$.

This is an experimental result. In the next section it is seen that all of the above operations may be done experimentally and analyzed on the computer by using the fast Fourier transform (FFT). Furthermore, if one chooses the input $u(t)$ to be pure white noise,† then the power spectrum of white noise is $U(j\omega)U^*(j\omega) = 1$. Then Eq. (3-49) becomes

$$\boxed{H(j\omega) = Y_{ss}(j\omega)U^*(j\omega) \qquad \text{for white noise } u(t)} \tag{3-50}$$

No physical process can ever have equal power, $U(j\omega)U^*(j\omega)$, at all frequencies. All physical processes tend to be band-limited; that is, their power spectrum will die off at frequencies beyond a certain bandwidth. Figure 3-17 shows the power spectrum of ideal white noise with unity power for all frequencies and a physical approximation to white noise with a wide but finite bandwidth. The bandwidth of a signal is defined as the frequency ω_c at which the power spectrum is diminished by 6 dB. This is the half-power point, since

$$20 \log_{10}\left(\frac{1}{2}\right) = -6.02 \text{ dB}$$

† We cannot go into a deep discussion of white noise because such a discussion involves an extensive probability theory background. Many physical processes closely approximate ideal white noise with a unity power spectrum. The noise you see on your TV screen when the picture goes out is a good example. Thermal noise in an electrical resistor is also very close to being pure white noise. White noise has equal power at all frequencies.

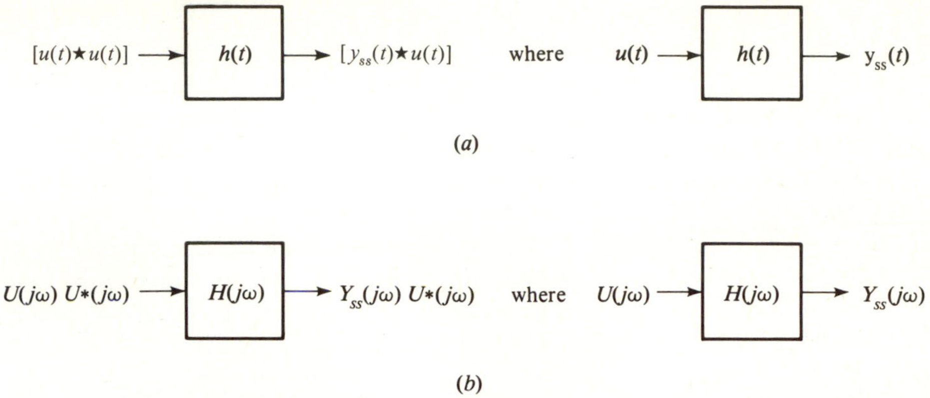

$[u(t)\star u(t)] \rightarrow \boxed{h(t)} \rightarrow [y_{ss}(t)\star u(t)]$ where $u(t) \rightarrow \boxed{h(t)} \rightarrow y_{ss}(t)$

(*a*)

$U(j\omega)\, U^*(j\omega) \rightarrow \boxed{H(j\omega)} \rightarrow Y_{ss}(j\omega)\, U^*(j\omega)$ where $U(j\omega) \rightarrow \boxed{H(j\omega)} \rightarrow Y_{ss}(j\omega)$

(*b*)

Figure 3-16 Time- and frequency-domain correlation relations of a linear time-invariant system. (*a*) Time-domain steady-state analysis. (*b*) Frequency-domain steady-state analysis.

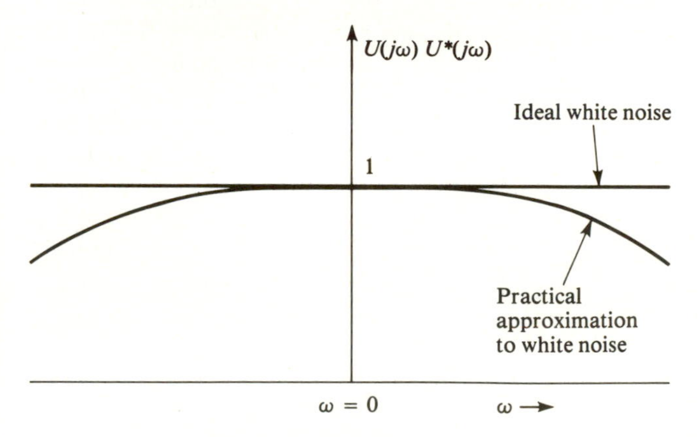

Figure 3-17 Illustration of power spectrum of ideal white noise.

Notice that the power spectrum $U^*(j\omega)U(j\omega)$ is symmetric about the origin; it is an even function of ω. Also notice that

$$|U(j\omega)|^2 = U(j\omega)U^*(j\omega)$$

Thus a 6-dB reduction in the power spectrum of the signal $u(t)$ corresponds to a 3-dB reduction in $|U(j\omega)|$.

*3-8 THE DISCRETE AND FAST FOURIER TRANSFORMS

One of the important reasons for the widespread use of the Fourier transform in signal analysis and design is because of the ease by which the Fourier transform of sampled data may be taken on the digital computer. The discrete time version of the Fourier transform is called the *discrete Fourier transform* (DFT). The straight DFT is fairly "expensive" computationally, requiring multiplications proportional in number to N^2, where N is the number of discrete samples of the given signal. However, Cooley and Tukey [7] devised an ingenious mechanization of the DFT that requires computations with a number of multiplications proportional to only $N \cdot \ln(N)$. Their algorithm is known as the fast Fourier transform (FFT). To illustrate the computational savings, for $N = 256$ the straight DFT would take some $N^2 = 65{,}536$ multiplications, while the FFT would only take some $N \cdot \ln(N) = 1420$ multiplications—about 2 percent of the DFT. For larger numbers of data points the savings become even more substantial. Furthermore, the FFT algorithm lends itself well to parallel computation. Special-purpose "signal processors" have thus been built which place the FFT at the heart of nearly all modern signal-processing applications.

In particular, the operations of signal convolution and signal correlation frequently arise in signal processing applications. For digitally sampled signals of length N data points each, the operations of discrete convolution and discrete correlation each take N^2 multiplications and additions. The convolution and correlation theorems still apply, though, to the discrete Fourier transform (DFT) and the FFT. Thus for large N it can be much more economical to take the FFT of both signals, multiply in the transform domain, and then take the inverse FFT. The FFT and inverse FFT only require $N \cdot \ln(N)$ operations each. For large N (say, $N = 256$ or $N = 512$) the computational savings by such an approach can be substantial.

The DFT and FFT are vast topic areas. It is beyond the scope of this text to go into the DFT and FFT. Fundamental to an understanding, though, of the DFT and FFT are the concepts of digitally sampled signals. The next section introduces one of the important results in the theory of digitally sampled signals—the Shannon sampling theorem.

*3-9 DIGITALLY SAMPLED SIGNALS, ALIASING, AND SHANNON'S SAMPLING THEOREM

A signal, $f(t)$, $-\infty < t < \infty$, sampled at an equal interval, T, may be represented by

$$f(kT) = \int_{-\infty}^{\infty} f(\tau)\,\delta(\tau - kT)\,d\tau \qquad |k| = 0, 1, 2, \ldots \tag{3-51}$$

Defining the *unit comb* function, $\gamma_T(t)$, of evenly spaced delta functions of interval T,

$$\gamma_T(t) \equiv \sum_{k=-\infty}^{\infty} \delta(t - kT) \tag{3-52}$$

the sequence of *all* $f(kT)$ samples may be expressed by

$$f(kT) = \int_{-\infty}^{\infty} f(\tau)\gamma_T(\tau)\,d\tau \qquad |k| = 0, 1, 2, \ldots \tag{3-53}$$

The Fourier transform* in hertz of the unit comb function is

$$\tilde{\mathfrak{F}}\{\gamma_T\} = \left(\frac{1}{T}\right)\sum_{n=-\infty}^{\infty}\delta(\mathfrak{f} - n\mathfrak{f}_0) \equiv \left(\frac{1}{T}\right)\tilde{\Gamma}_{\mathfrak{f}_0}(\mathfrak{f}) \tag{3-54}$$

where the base frequency $\mathfrak{f}_0$ is $\mathfrak{f}_0 = 1/T$ Hz. The summation on the right in Eq. (3-54) is a unit comb function in the frequency domain with frequency spacing $\mathfrak{f}_0 = 1/T$. Thus the shorter the sampling interval T in the time domain, the longer the frequency spacing $\mathfrak{f}_0 = 1/T$ in the transform domain. This is illustrated in Fig. 3-18.

By the convolution property of the Fourier transform (multiplication in the time domain is equivalent to convolution in the frequency domain) it is seen that

$$\begin{aligned}\tilde{\mathfrak{F}}\{f(t)\gamma_T(t)\} &= \tilde{\mathfrak{F}}\{f(t)\} * \tilde{\mathfrak{F}}\{\gamma_T(t)\}\\ &= \tilde{F}(j\mathfrak{f}) * \frac{1}{T}\tilde{\Gamma}_{\mathfrak{f}_0}(\mathfrak{f})\\ &\equiv \frac{1}{T}\int_{-\infty}^{\infty}\tilde{F}[j(\mathfrak{f}-\alpha)]\tilde{\Gamma}_{\mathfrak{f}_0}(\alpha)\,d\alpha\end{aligned} \tag{3-55}$$

Consider the convolution of any function, $g(t)$, with the unit comb function, $\gamma_T(t)$. First it is true that the convolution of $g(t)$ with $\delta(t)$ is

$$\int_{-\infty}^{\infty} g(t-\tau)\,\delta(\tau)\,d\tau = g(t) \qquad -\infty < t < \infty \tag{3-56}$$

This follows by the sampling property of the delta function. Next it is true that the convolution of $g(t)$ with $\delta(t - kT)$ is

$$\begin{aligned}\int_{-\infty}^{\infty} g(t-\tau)\,\delta(\tau - kT)\,d\tau &= g(t-\tau)\Big|_{\tau=kT}\\ &= g(t-kT) \qquad -\infty < t < \infty\end{aligned}$$

Therefore it follows that the convolution of $g(t)$ with the unit comb function, $\gamma_T(t)$, is given by

$$\begin{aligned}\int_{-\infty}^{\infty} g(t-\tau)\gamma_T(\tau)\,d\tau &= \int_{-\infty}^{\infty} g(t-\tau)\sum_{k=-\infty}^{\infty}\delta(\tau-kT)\,d\tau\\ &= \sum_{k=-\infty}^{\infty} g(t-kT) \qquad -\infty < t < \infty\end{aligned} \tag{3-57}$$

This is illustrated in Fig. 3-19. Notice that if $g(t) \equiv 0$ for $t > |T/2|$, the convolution of $g(t)$ with $\gamma_T(t)$ yields $g(t)$ perfectly repeated every T units of time. If

*The alternative form of the Fourier transform

$$\tilde{\mathfrak{F}}\{f(t)\} \equiv \int_{-\infty}^{\infty} f(\tau)e^{-j2\pi\mathfrak{f}\tau}\,d\tau$$

is being used here where the frequency $\mathfrak{f}$ is in hertz.

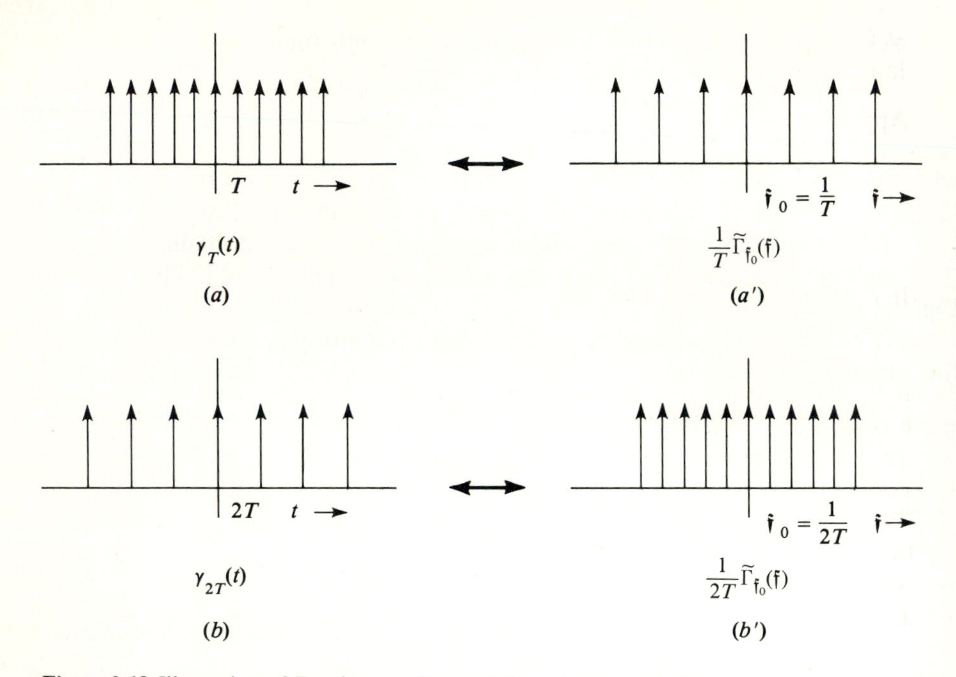

Figure 3-18 Illustration of Fourier-transform relation of unit comb function.

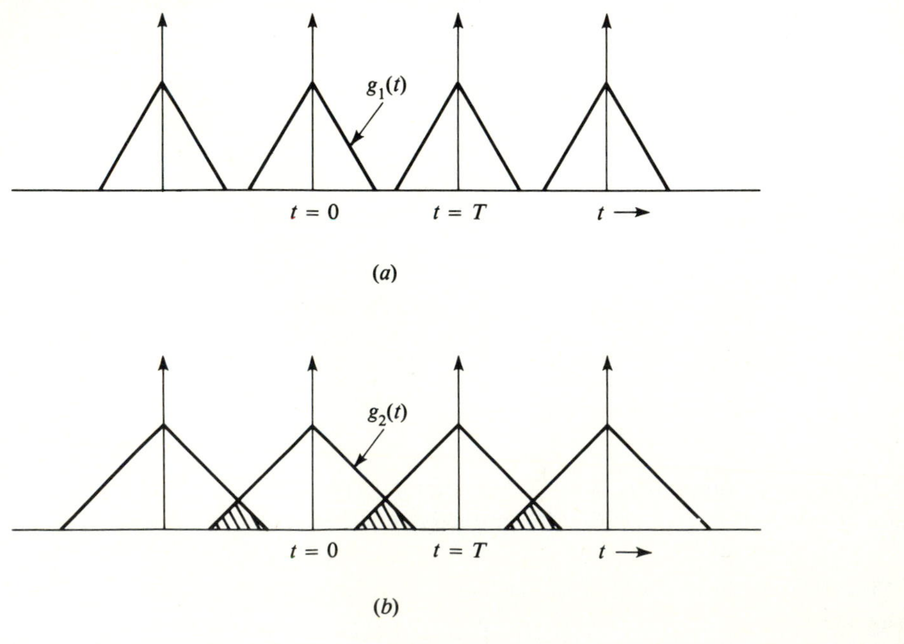

Figure 3-19 Illustration of time-domain aliasing. (*a*) No aliasing, $g_1(t) * \gamma_T(t)$. (*b*) Aliasing, $g_2(t) * \gamma_T(t)$.

$g(t) \neq 0$ for $T > |T/2|$, then the summation Eq. (3-57) distorts the signal at the overlap regions. This distortion of the overlapped signals is known as *aliasing*.

Applying the foregoing results in the frequency domain to Eq. (3-55) gives

$$\mathfrak{F}\{f(t)\gamma_T(t)\} = \widetilde{F}(j\mathfrak{f}) * \frac{1}{T}\widetilde{\Gamma}_{\mathfrak{f}_0}(\mathfrak{f})$$
$$= \frac{1}{T}\sum_{n=-\infty}^{\infty} \widetilde{F}[j(\mathfrak{f} - n/T)] \qquad -\infty < \mathfrak{f} < \infty \tag{3-58}$$

If $\widetilde{F}(j\mathfrak{f})$ satisfies

$$\widetilde{F}(j\mathfrak{f}) \equiv 0 \qquad |\mathfrak{f}| > \mathfrak{f}_0/2 = 1/2T\,\text{Hz} \tag{3-59}$$

i.e., if $f(t)$ is a band-limited signal with no frequency content above $1/2T$ Hz, then there is no frequency aliasing in the Fourier transform of $f(t)\gamma_T(t)$. If $f(t)$ does contain frequencies higher than $1/2T$ Hz, there will be frequency aliasing in the Fourier transform of $f(t)\gamma_T(t)$. This is illustrated in Fig. 3-20.

The foregoing discussion is the basis of an important result in the theory of digital sampling of signals. This result is *Shannon's sampling theorem*. This theorem states: Let a signal, $f(t)$, have highest frequency, $\mathfrak{f}_{\max}$ Hz, in its Fourier power spectrum; i.e.,

$$\widetilde{F}(j\mathfrak{f})\widetilde{F}^*(j\mathfrak{f}) \equiv 0 \qquad \text{for } |\mathfrak{f}| > \mathfrak{f}_{\max} \tag{3-60}$$

Then the sampled signal, $f(kT)$, $|k| = 0, 1, 2, \ldots$, contains all of the information equivalent to the original continuous signal, $f(t)$, provided that

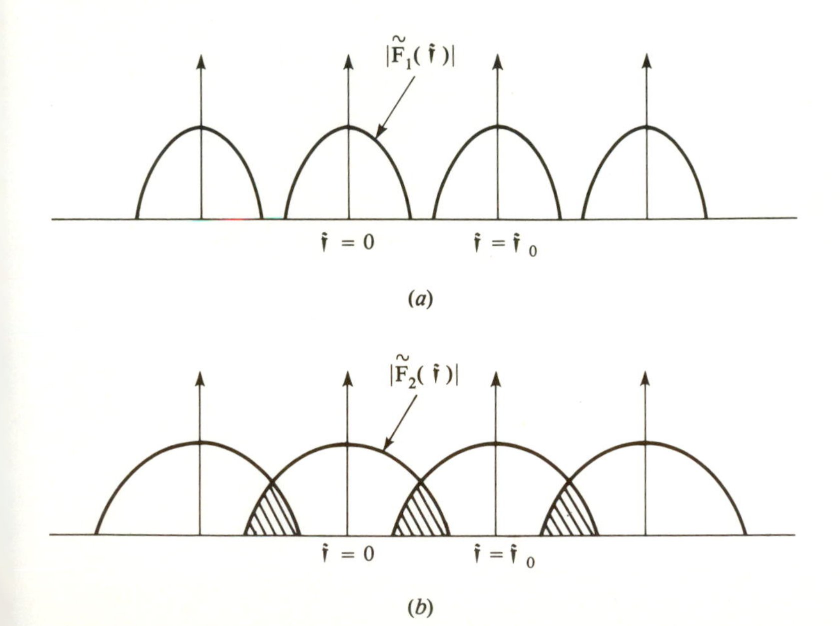

Figure 3-20 Illustration of frequency-domain aliasing. (*a*) No aliasing, $\widetilde{F}_1(\mathfrak{f}) * \widetilde{\Gamma}_{\mathfrak{f}_0}(\mathfrak{f})$; (*b*) Aliasing, $\widetilde{F}_2(\mathfrak{f}) * \widetilde{\Gamma}_{\mathfrak{f}_0}(\mathfrak{f})$.

$$T < 1/(2\mathfrak{f}_{max}) \tag{3-61}$$

The sampling frequency $2\mathfrak{f}_{max}$ is called the Nyquist or Shannon sampling frequency. As a result of this theorem one may exactly reconstruct the original $f(t)$ from the signal samples $f(kT)$.

The Shannon sampling theorem and the concepts of aliasing apply to many applications of digital signal analysis. They apply to discussion of the DFT/FFT, and they are encountered again in Chap. 5 in regard to discrete time analysis of the linear system and the z transform.

3-10 SUMMARY

Analysis of the linear system steady-state response provides a frequency-domain analysis of the linear time-invariant system. Equating the particular solution response to the convolution-integral response results in the fact that the impulse response function and the system transfer function are related by the Fourier transformation. The Fourier transformation has a great many applications in linear systems theory and signal analysis in general. The reason for this usefulness stems primarily from the special properties of the Fourier transformation. Most notable among these properties are the convolution and correlation properties. These convolution and correlation properties are also true of the discrete Fourier transform (DFT). A fast implementation of the DFT, the FFT, makes Fourier transform analysis a hallmark of modern digital signal processing. Finally, basic sampling concepts and the Shannon sampling theorem are discussed.

The next chapter examines a generalization of the Fourier transform—the Laplace transform. The Laplace transform possesses the same desirable properties as the Fourier transform. It is also quite valuable in the analysis of the linear time-invariant system.

PROBLEMS

3-1 Find $y_{ss}(t) = |H(j\omega)| \sin(\omega t + \phi)$ for the system $\ddot{y} + 0.4\dot{y} + 4y = \dot{u} + u$ with the inputs $u(t)$: (*a*) $\sin t$, (*b*) $\sin 2t$, (*c*) $\sin 8t$.

3-2 Find $y_{ss}(t) = |H(j\omega)| \sin(\omega t + \phi)$ for the system $\ddot{y} + 0.2\dot{y} + 4y = \dot{u} - u$ with inputs $u(t)$: (*a*) $\sin t$, (*b*) $\sin 2t$, (*c*) $\sin 8t$.

3-3 Find $|H(j\omega)| \sin(\omega t + \phi)$ for the following systems with the input $u(t) = \sin 2t$. What are the pole and zero locations for each of these systems? For which of these systems does this represent $y_{ss}(t)$?

(*a*) $\ddot{y} + 6\dot{y} + 5y = \dot{u} + 0.1u$
(*b*) $\ddot{y} + 6\dot{y} + 5y = \dot{u} - -0.1u$
(*c*) $\ddot{y} + 4\dot{y} - 5y = \dot{u} + 0.1u$
(*d*) $\ddot{y} + 4\dot{y} - 5y = \dot{u} - 0.1u$
(*e*) $\ddot{y} - 6\dot{y} + 5y = \dot{u} - 0.1u$

3-4 Repeat Prob. 3-3 for the following systems:

(*a*) $\ddot{y} + 6\dot{y} + 5y = \dot{u} + 10u$
(*b*) $\ddot{y} + 6\dot{y} + 5y = \dot{u} - 10u$

(*c*) $\ddot{y} - 4\dot{y} - 5y = \dot{u} + 10u$
(*d*) $\ddot{y} - 4\dot{y} - 5y = \dot{u} - 10u$
(*e*) $\ddot{y} - 6\dot{y} + 5y = \dot{u} - 10u$

3-5 Sketch the Bode magnitude plots for systems in Prob. 3-3. Which systems have the same magnitude plot?

3-6 Sketch the Bode magnitude plots for the systems in Prob. 3-4. Which systems have the same magnitude plot?

3-7 Sketch the Bode phase-angle plots for the systems in Prob. 3-3. Which systems have the same phase-angle plot?

3-8 Sketch the Bode phase-angle plots for the systems in Prob. 3-4. Which systems have the same phase-angle plot?

3-9 Are all five systems in Prob. 3-3 distinguishable by their Bode magnitude and phase-angle plots together? Explain.

3-10 Are all five systems in Prob. 3-4 distinguishable by their Bode magnitude and phase-angle plots together? Explain.

3-11 Using Table 3-1, find the Fourier transform of the following functions:

(*a*) $2e^{-3t}u_{-1}(t)$ (*b*) $4e^{-2t} \sin (8t)u_{-1}(t)$
(*c*) $3 \sin (4t) \quad -\infty < t < \infty$ (*d*) $2te^{-3t}u_{-1}(t)$

3-12 Using Table 3-1, find the Fourier transform of the following functions:

(*a*) $3 \cos (4t) \quad -\infty < t < \infty$ (*b*) $(5e^{-4t} - 3e^{-7t})u_{-1}(t)$
(*c*) $4e^{-2t} \cos (8t)u_{-1}(t)$ (*d*) $5\,\delta(t) + 4u_{-1}(t)$

3-13 Using Table 3-1, find $f(t)$ from the following $F(j\omega)$:

$$F(j\omega) = \frac{5j\omega + 5}{(j\omega)^2 + 2(j\omega) + 17}$$

3-14 Using Table 3-1,find $f(t)$ from the following $F(j\omega)$:

$$F(j\omega) = 40 \sin (2\omega)/2\omega + 5$$

3-15 Using a combination of the Fourier transform properties and Table 3-1, find the Fourier transform of the following expressions:

(*a*) $f(t) = \begin{cases} 8 & |t| < 5 \\ 0 & |t| > 5 \end{cases}$

(*b*) $f(t) = \begin{cases} 8 & |t| < 10 \\ 0 & |t| > 10 \end{cases}$

(*c*) $f(t) = e^{-2(t-t_0)}u_{-1}(t)$

(*d*) $f(t) = e^{-2t}u_{-1}(t - t_0)$

(*e*) $f(t) = e^{-2(t-t_0)}u_{-1}(t - t_0)$

(*f*) $f(t) = \int_{-\infty}^{\infty} [e^{-2(t-\tau)}u_{-1}(t - \tau)][e^{-5\tau}u_{-1}(\tau)]\, d\tau$
$= [e^{-2t}u_{-1}(t)] * [e^{-5t}u_{-1}(t)]$

3-16 Using a combination of the Fourier transform properties and Table 3-1, find the Fourier transform of the following expressions:

(*a*) $f(t) = \begin{cases} 10 & |t| < 1 \\ 0 & |t| > 1 \end{cases}$

(*b*) $f(t) = \begin{cases} 1 & |t| < 10 \\ 0 & |t| > 10 \end{cases}$

(*c*) $f(t) = \begin{cases} 10 & |t+2| < 1 \\ 0 & |t+2| > 1 \end{cases}$

(*d*) $f(t) = e^{-2(t-5)}u_{-1}(t-1)$

(*e*) $f(t) = \int_{-\infty}^{\infty} [e^{-3(t-\tau)}u_{-1}(t-\tau)][e^{-8\tau}u_{-1}(\tau)]\,d\tau$

$= [e^{-3t}u_{-1}(t)] * [e^{-8t}u_{-1}(t)]$

3-17 (*a*) Show that $f(t)$ in Prob. 3-15(f) may be written as

$$f(t) = e^{-2t}\int_0^t e^{-3\tau}\,d\tau$$

(*b*) Evaluate this integral and then take the Fourier transform.

(*c*) Should the results in (*b*) agree with the result from Prob. 3-15(f)?

3-18 (*a*) Show that $f(t)$ in Prob. 3-16(*e*) may be written as

$$f(t) = e^{-3t}\int_0^t e^{-5\tau}\,d\tau$$

(*b*) Evaluate this integral and then take the Fourier transform.

(*c*) Should the result in (*b*) agree with the result from Prob. 3-16(*e*)?

3-19 A signal has power spectrum given by

$$|U(j\omega)|^2 = \omega^2 e^{-\omega^2}$$

What is the signal energy, $\int_{-\infty}^{\infty} u^2(\tau)\,d\tau$?

3-20 A signal has power spectrum given by

$$\left|U(j\omega)\right|^2 = \frac{\sin^2(4\omega)}{\omega^2}$$

What is the signal energy, $\int_{-\infty}^{\infty} u^2(\tau)\,d\tau$?

3-21 What is $H(j\omega)$ for the system

$$\ddot{y}(t) + 2\dot{y}(t-1) + 3y(t) = \dot{u}(t) + 0.5u(t)$$

3-22 What is $H(j\omega)$ for the system

$$\ddot{y}(t) + 5\dot{y}(t) + 2y(t) - \dot{y}(t-0.2) = u(t-0.5)$$

3-23 According to Shannon's sampling theorem, what would be the highest frequency, in hertz, that could be present in $H(j\omega)$ to not have frequency aliasing in samples $h_T(k)$ taken with $T = 0.1$?

3-24 According to Shannon's sampling theorem, what would be the highest frequency, in hertz, that could be present in $H(j\omega)$ to not have frequency aliasing in samples $h_T(k)$ taken with $T = 0.2$?

3-25 What is the Nyquist sampling interval for a signal with frequency content up to 20 Hz?

3-26 What is the Nyquist sampling interval for a signal with frequency content up to 60 Hz?

NOTES AND REFERENCES

References 1 and 2 are classic references on the subject of frequency domain analysis of a linear dynamic system. See, e.g., Ref. 3 for a detailed treatment on the application of these results to feedback control theory.

References 4 and 5 provide general references on the Fourier transform and its applications. References 7 and 8 give further discussion of the topics of Sec. 3.7 on correlation methods. Finally, References 10 to 12 are general references on the theory and application of the fast Fourier transform (FFT).

1. Bode, H. W.: *Network Analysis and Feedback Amplifier Design,* Van Nostrand, Princeton, N.J., 1945, Chap. 8.
2. Nyquist, H.: "Regeneration Theory," *Bell Systems Technology J.,* vol. 11, 1932, pp. 126–147.
3. D'Azzo, J. J., and C. H. Houpis: *Linear Control System Analysis and Design, Conventional and Modern,* McGraw-Hill, New York, 2d ed., 1981, Chap. 8.
4. Bracewell, R. M.: *The Fourier Transform and Its Application,* McGraw-Hill, New York, 1965.
5. Papoulis, A.: *The Fourier Integral and Its Applications,* McGraw-Hill, New York, 1962.
6. Daniels, R. W.: *Approximation Methods for the Design of Passive, Active, and Digital Filters,* McGraw-Hill, New York, 1974.
7. Sage, A. P., and J. L. Melsa: *System Identification,* Academic, New York, 1971, pp. 13–18.
8. Graupe, D.: *Identification of Systems,* Van Nostrand, New York, 1972, pp. 75–91.
9. Cooley, J. W., and J. W. Tukey: "An Algorithm for Machine Calculation of Complex Fourier Series," *Math Computation,* vol. 19, April 1965, pp. 297–301.
10. Brigham, E. O.: *The Fast Fourier Transform,* Prentice-Hall, Englewood Cliffs, N.J., 1971.
11. Rabiner, L. R., and B. Gold: *Theory and Application of Digital Signal Processing,* Prentice-Hall, Englewood Cliffs, N.J., 1975.
12. Oppenheim, A. V., and R. W. Shafer: *Digital Signal Processing,* Prentice-Hall, Englewood Cliffs, N.J., 1975.

CHAPTER

FOUR

LAPLACE TRANSFORM ANALYSIS

The Laplace transform is a generalization of the Fourier transform. Like the Fourier transform it has proven to be an essential tool in the analysis of the linear time-invariant differential equation system. It possesses many of the same desirable properties as the Fourier transform and provides a frequency-domain analysis of a linear dynamic system.

Furthermore, the Laplace transform gives a convenient alternative to the Chap. 2 classic method for solving the differential equation. The Laplace transform solution method is seen to be midway between the classic solution method and the state-space solution method as developed in Chap. 7. The Laplace transform solution provides a decomposition of the total solution into the zero-input and zero-state portions as introduced in Chap. 1.

Finally, the convolution property of the Laplace transform (convolution in the time domain is equal to multiplication in the transform domain) underlies simple rules for the analysis of complex interconnections of linear subsystems. This is perhaps the foremost reason for the importance of the Laplace transform in the analysis of the linear time-invariant dynamic system.

4-1 DEFINITION AND RELATION TO FOURIER TRANSFORM

The Fourier transform pair of a given function, $f(t)$, is defined by

$$F(j\omega) \equiv \mathfrak{F}\{f(t)\} \equiv \int_{-\infty}^{\infty} f(\tau)e^{-j\omega\tau}\, d\tau \tag{4-1}$$

$$f(t) = \mathfrak{F}^{-1}\{F(j\omega)\} \equiv \frac{1}{2\pi} \int_{-\infty}^{\infty} F(j\omega) e^{j\omega\tau}\, d\omega \tag{4-2}$$

provided these integrals exist (remain finite). The *one-sided* Laplace transform* of $f(t)$ is obtained by replacing $j\omega$ with the complex variable $s = \sigma + j\omega$, constant $\sigma \geq 0$. Then $ds = j\,d\omega$ and the one-sided Laplace transform pair relations are

$$F(s) = \mathfrak{L}\{f(t)\} \equiv \int_{0^-}^{\infty} f(\tau) e^{-s\tau}\, d\tau \tag{4-3}$$

$$f(t) = \mathfrak{L}^{-1}\{F(s)\} \equiv \frac{1}{2\pi j} \int_{\sigma - j\infty}^{\sigma + j\infty} F(s) e^{st}\, ds \tag{4-4}$$

provided these integrals exist.

The lower limit on the Laplace transform of 0^- instead of 0 is so that any impulses in $f(t)$ at time zero are included within the limits of integration.

One advantage of the Laplace transform over the Fourier transform is that the Laplace variable $s = \sigma + j\omega$ gives

$$\int_{0^-}^{\infty} f(\tau) e^{-s\tau}\, d\tau = \int_{0^-}^{\infty} f(\tau) e^{-\sigma\tau} e^{-j\omega\tau}\, d\tau$$

Hence the condition for existence of the Laplace transform is now

$$\int_{0^-}^{\infty} \left| f(\tau) e^{-\sigma\tau} \right| d\tau < \infty \tag{4-5}$$

Even if $f(t) = e^{pt}, p > 0$, a positive exponential, the condition Eq. (4-5) is satisfied for all $\sigma > p$. Hence the Laplace transform of an unstable impulse response function will exist, whereas the Fourier transform will not. That we can find a $\sigma > p$ such that Eq. (4-5) is satisfied is called the condition of exponential dominance. All physical dynamic systems modeled by linear time-invariant ordinary differential equations have the property of exponential dominance; therefore, the impulse response function of such systems—whether stable or unstable—always has a Laplace transform. For our purposes we can thus ignore the issue of existence of the Laplace transform.

Table 4-1 gives the Laplace transform of several common functions. More extensive tables of Laplace transforms are available (e.g., Refs. 1 and 2), but this short table suffices for our purposes.

To see how some of the entries in the table come about, consider, for example, the first entry

$$\mathfrak{L}\{\delta(t)\} \equiv \int_{0^-}^{\infty} \delta(\tau) e^{-s\tau}\, d\tau$$

By the sampling property of the delta function this yields

*We present only the one-sided Laplace transform. In the one-sided Laplace transform it is assumed that $f(t) \equiv 0$ for $t < 0$. This is most convenient for differential equation solution, and for the impulse response, $h(t)$, because a causal system satisfies this condition.

$$\mathcal{L}\{\delta(t)\} = e^{-s\tau}|_{\tau=0} = 1$$

Next consider

$$\mathcal{L}\{u_{-1}(t)\} \equiv \int_{0^-}^{\infty} u_{-1}(\tau)e^{-s\tau}\,d\tau$$

or

$$\mathcal{L}\{u_{-1}(t)\} = \int_{0^-}^{\infty} e^{-s\tau}\,d\tau = \left.\frac{-e^{-s\tau}}{s}\right|_{\tau=0^-}^{\tau=\infty}$$

$$= \frac{1}{s}$$

Finally consider

$$\mathcal{L}\{e^{at}u_{-1}(t)\} \equiv \int_{0^-}^{\infty} e^{a\tau}u_{-1}(\tau)e^{s\tau}\,d\tau$$

$$= \int_{0^-}^{\infty} e^{(a-s)\tau}\,d\tau$$

$$= \left.\frac{e^{(a-s)\tau}}{a-s}\right|_{\tau=0^-}^{\tau=\infty}$$

$$= \frac{1}{s-a} \qquad \text{provided real } (s) > \text{real } a$$

Using the Laplace transform table avoids the evaluation of such integrals.

Example 4-1 Let

$$f(t) = [5te^{-2t} + 7e^{-3t}\cos(4t)]u_{-1}(t)$$

Multiplying by the unit step function, $u_{-1}(t)$, emphasizes that $f(t) \equiv 0$ for $t < 0$. By Table 4-1

$$F(s) = \frac{5}{(s+2)^2} + \frac{7(s+3)}{(s+3)^2 + 16}$$

Example 4-2 Let

$$F(s) = \frac{5(s-2)}{s^2 + 4s + 3}$$

Factoring the denominator polynomial and then expanding by partial fraction expansions (see Appendix C) gives

$$F(s) = \frac{5(s-2)}{(s+1)(s+3)}$$

$$= \frac{-15/2}{s+1} + \frac{25/2}{s+3}$$

Table 4-1 Table of Laplace transforms

$F(s)$	$f(t)$
1. 1	$\delta(t)$
2. $\frac{1}{s}$	$u_{-1}(t)$
3. $\frac{e^{-Ts}}{s}$	$u_{-1}(t - T)$
4. $\frac{1}{s^2}$	$tu_{-1}(t)$
5. $\frac{1}{s^n}$	$t^n u_{-1}(t)/(n - 1)!$, n positive integer
6. $\frac{1}{s + a}$	$e^{-at}u_{-1}(t)$
7. $\frac{1}{(s + a)^2}$	$te^{-at}u_{-1}(t)$
8. $\frac{1}{(s + a)^n}$	$t^{n-1}e^{-at}u_{-1}(t)/(n - 1)!$, n positive integer
9. $\frac{b}{s^2 + b^2}$	$\sin (bt)u_{-1}(t)$
10. $\frac{s}{s^2 + b^2}$	$\cos (bt)u_{-1}(t)$
11. $\frac{b}{(s + a)^2 + b^2}$	$e^{-at} \sin (bt)u_{-1}(t)$
12. $\frac{s + a}{(s + a)^2 + b^2}$	$e^{-at} \cos (bt)u_{-1}(t)$
13. $\frac{1}{s^2 + 2\zeta\omega_n s + \omega_n^2}$	$\frac{1}{\omega_n \sqrt{1 - \zeta^2}} e^{-\zeta\omega_n t} \sin (\omega_n \sqrt{1 - \zeta^2} t)$, for $0 < \zeta < 1$

Using Table 4-1 gives

$$f(t) = \mathcal{L}^{-1}\{F(s)\} = \left[-\frac{15}{2}e^{-t} + \frac{25}{2}e^{-3t} \right] u_{-1}(t)$$

4-2 PROPERTIES OF THE LAPLACE TRANSFORM

The properties of the Laplace transform are similar to the Fourier transform, and their proofs follow in an analogous fashion. Therefore, proofs are often omitted in the properties to follow. The notation $\leftrightarrow$ is used to indicate a Laplace transform pair relationship.

Linearity

The Laplace transform is a linear operator:

$$a[f_1(t) + f_2(t)] \longleftrightarrow a[F_1(s) + F_2(s)] \tag{4-6}$$

Scaling

For $a > 0$,

$$f(at) \longleftrightarrow \frac{1}{a}F\left(\frac{s}{a}\right) \tag{4-7}$$

Time Shifting

$$f(t - t_0)u_{-1}(t - t_0) \longleftrightarrow F(s)e^{-st_0} \tag{4-8}$$

The factor $u_{-1}(t - t_0)$ is used to indicate that the curve $f(t)$ is actually shifted by t_0 time units and does not initiate until time t_0. Note that $f(t - t_0)u_{-1}(t) \neq f(t - t_0)u_{-1}(t - t_0)$ and that they do not have the same Laplace transform. This is illustrated in Fig. 4-1.

Also note that the time shift property may be used to take the Laplace transform of some fairly complex time functions. For instance, consider the thrust input of the rocket-sled system as considered in Chap. 1. This is shown again in Fig. 4-2. The thrust input curve may be expressed as

$$\begin{aligned} u(t) = 2tu_{-1}(t) &- 2(t - 10)u_{-1}(t - 10) \\ &- 2(t - 30)u_{-1}(t - 30) + 2(t - 40)u_{-1}(t - 40) \end{aligned}$$

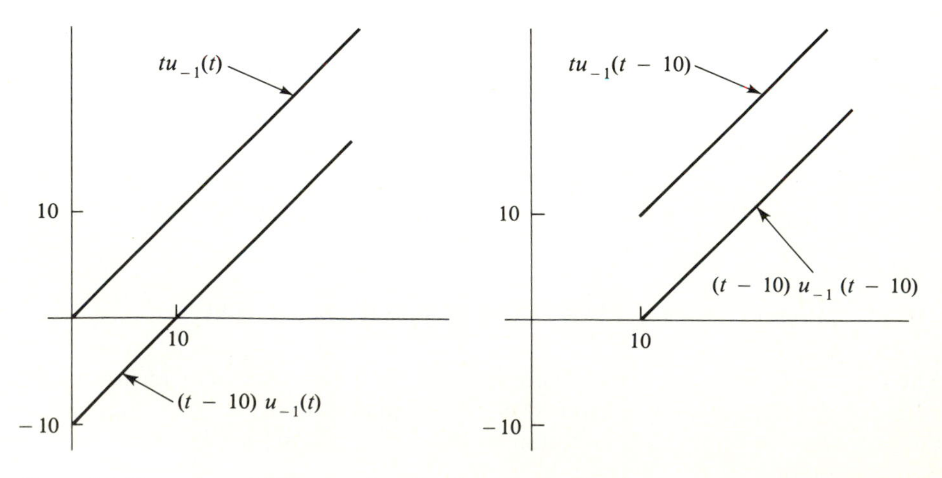

Figure 4-1 Illustration of differences in using the unit-step function.

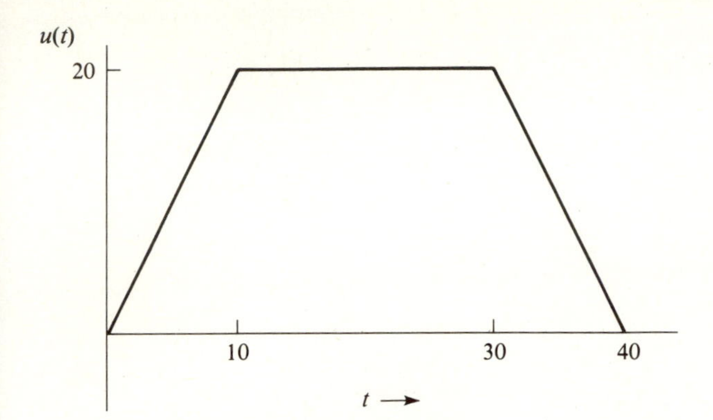

Figure 4-2 Thrust input curve of the rocket-sled system.

Using the time shift property, the Laplace transform of $u(t)$ is

$$U(s) = \frac{2}{s^2}(1 - e^{-10s} - e^{-30s} + e^{-40s})$$

Frequency Shifting

$$\boxed{f(t)e^{s_0 t} \longleftrightarrow F(s - s_0)} \tag{4-9}$$

Time Convolution

$$\boxed{f_1(t) * f_2(t) \longleftrightarrow F_1(s)F_2(s)} \tag{4-10}$$

This property that convolution in the time domain is equal to multiplication in the Laplace transform domain is of fundamental importance. To review the proof of this property consider

$$f_1(t) * f_2(t) \equiv \int_{-\infty}^{\infty} f_1(\tau) f_2(t - \tau)\, d\tau$$

$$= \int_{0^-}^{\infty} f_1(\tau) f_2(t - \tau)\, d\tau \tag{4-11}$$

where the lower limit of integration follows from the fact that both $f_1(\tau) = f_2(\tau) = 0$ for $\tau < 0$. Taking the Laplace transform of Eq. (4-11) yields

$$\mathcal{L}\{f_1(t) * f_2(t)\} \equiv \int_{0^-}^{\infty} \left(\int_{0^-}^{\infty} f_1(\tau) f_2(t - \tau)\, d\tau \right) e^{-st}\, dt$$

Changing the order of integration and using the time shift property gives

$$\mathcal{L}\{f_1(t) * f_2(t)\} = \int_{0^-}^{\infty} f_1(\tau)\left(\int_{0^-}^{\infty} f_2(t-\tau)e^{-st}\,dt\right)d\tau$$

$$= \int_{0^-}^{\infty} f_1(\tau)e^{-s\tau}F_2(s)\,d\tau$$

$$= F_1(s)F_2(s) \tag{4-12}$$

as claimed.

Frequency Convolution

$$\boxed{f_1(t)f_2(t) \longleftrightarrow \frac{1}{2\pi j}[F_1(s) * F_2(s)]} \tag{4-13}$$

Time Differentiation

$$\boxed{\frac{df(t)}{dt} \longleftrightarrow sF(s) - f(0^-)} \tag{4-14}$$

This may be derived as follows:

$$\mathcal{L}\left\{\frac{df(t)}{dt}\right\} \equiv \int_{0^-}^{\infty} \frac{df(\tau)}{d\tau}e^{-s\tau}\,d\tau$$

Integrating by parts yields

$$\mathcal{L}\left\{\frac{df(t)}{dt}\right\} = f(\tau)e^{-s\tau}\Big|_{0^-}^{\infty} + \int_{0^-}^{\infty} f(\tau)se^{-s\tau}\,d\tau$$

$$= -f(0^-) + sF(s)$$

which agrees with Eq. (4-14).

The same technique may be extended to higher-order derivatives. Thus

$$\boxed{\frac{d^nf(t)}{dt^n} \longleftrightarrow s^nF(s) - s^{n-1}f(0^-) - s^{n-2}f^{(1)}(0^-) - \cdots - f^{(n-1)}(0^-)} \tag{4-15}$$

This result is applied in the next section to provide a solution method to a given linear time-invariant differential equation.

Time Integration

$$\boxed{\int_{0^-}^{t} f(\tau)\,d\tau \longleftrightarrow \frac{F(s)}{s}} \tag{4-16}$$

To prove this result, consider

$$\mathcal{L}\left\{\int_{0^-}^{t} f(\tau)\,d\tau\right\} \equiv \int_{0^-}^{\infty}\left(\int_{0^-}^{t} f(\tau)\,d\tau\right)e^{-st}\,dt$$

Letting $u \equiv \int_{0^-}^{t} f(\tau)\,d\tau$, $du = f(t)\,dt$, $dv = e^{-st}\,dt$, and $v = -e^{-st}/s$, and integrating by parts, yields

$$\mathcal{L}\left\{\int_{0^-}^{t} f(\tau)\,d\tau\right\} = \left(\int_{0^-}^{t} f(\tau)\,d\tau\right)\frac{e^{-st}}{s}\Bigg|_{0^-}^{\infty} + \int_{0^-}^{\infty} f(t)\frac{e^{-st}}{s}\,dt$$

The first term is zero at both the upper limit $t = \infty$ and the lower limit $t = 0^-$. The second term is $F(s)/s$, and so Eq. (4-16) is proven.

Applying Eq. (4-16), it can also be shown that

$$\int_{-\infty}^{t} f(\tau)\,d\tau \longleftrightarrow \frac{F(s)}{s} + \frac{\int_{-\infty}^{0^-} f(\tau)\,d\tau}{s} \tag{4-17}$$

This follows from the fact that

$$\int_{-\infty}^{t} f(\tau)\,d\tau = \int_{-\infty}^{0^-} f(\tau)\,d\tau + \int_{0^-}^{t} f(\tau)\,d\tau \tag{4-18}$$

But the first term on the right in Eq. (4-18) is just a constant (assuming the integral exists). Taking the Laplace transform of both sides of Eq. (4-18) and using the result, Eq. (4-16) yields Eq. (4-17) as desired.

This result is reminiscent of the Chap. 1 discussion concerning the zero-input and zero-state response as given by

$$\begin{aligned} y(t) &= \int_{-\infty}^{t} h(t-\tau)u(\tau)\,d\tau \\ &= \int_{-\infty}^{0^-} h(t-\tau)u(\tau)\,d\tau + \int_{0^-}^{t} h(t-\tau)u(\tau)\,d\tau \\ &= y_{zi}(t) + y_{zs}(t) \end{aligned} \tag{1-24e}$$

The Laplace transform solution method leads naturally to this zero-input and zero-state decomposition.

Initial-Value Theorem

$$\boxed{f(0^+) = \lim_{s\to\infty} sF(s)} \tag{4-19}$$

provided that $f(0^+)$ and the limit both exist. To prove this, consider Eq. (4-14), the time differentiation property,

$$sF(s) - f(0^-) = \int_{0^-}^{\infty} \frac{df(\tau)}{d\tau} e^{-s\tau}\, d\tau$$

$$= \int_{0^-}^{0^+} \frac{df(\tau)}{d\tau} e^{-s\tau}\, d\tau + \int_{0^+}^{\infty} \frac{df(\tau)}{d\tau} e^{-s\tau}\, d\tau$$

$$= f(0^+) - f(0^-) + \int_{0^+}^{\infty} \frac{df(\tau)}{d\tau} e^{-s\tau}\, d\tau$$

since $e^{-s\tau}|_{\tau=0} = 1$. Thus

$$sF(s) = f(0^+) + \int_{0^+}^{\infty} \frac{df(\tau)}{d\tau} e^{-s\tau}\, d\tau \tag{4-20}$$

Then, since

$$\lim_{s\to\infty} e^{-st} = 0$$

the integral on the right in Eq. (4-20) is zero as the limit is taken. Eq. (4-19) thus results.

Final-Value Theorem

$$\boxed{f(\infty) = \lim_{s\to 0} sF(s)} \tag{4-21}$$

provided that $f(\infty)$ and the limit both exist.

To derive Eq. (4-21), take the limit of Eq. (4-20) as $s \to 0$. This gives

$$\lim_{s\to 0} sF(s) = f(0^+) + \int_{0^+}^{\infty} \frac{df(\tau)}{d\tau}\, d\tau$$

$$= f(0^+) + f(\infty) - f(0^+)$$

which is Eq. (4-21).

4-3 USING THE LAPLACE TRANSFORM TO SOLVE THE DIFFERENTIAL EQUATION

The differentiation property of the Laplace transform provides a convenient method to solve a given linear time-invariant differential equation. The Laplace transform of both sides of the differential equation are taken; algebraic manipulations are performed; and then the inverse Laplace transform yields the solution.

To develop this solution approach, we take a specific example system to illustrate the important points:

$$\ddot{y}(t) + 4\dot{y}(t) + 3y(t) = 5\dot{u}(t) + 10u(t) \tag{4-22}$$

with initial conditions $y(0^-) = 4$, $\dot{y}(0^-) = 0$, $u(0^-) = 1$, and $u(t) = 3t$, $t > 0$. Taking the Laplace transform of both sides of the differential equation yields

$$[s^2Y(s) - sy(0^-) - \dot{y}(0^-)] + 4[sY(s) - y(0^-)] + 3Y(s)$$
$$= 5[sU(s) - u(0^-)] + 10U(s)$$

Solving for $Y(s)$ yields

$$Y(s) = \frac{sy(0^-) + \dot{y}(0^-) + 4y(0^-) - 5u(0^-)}{s^2 + 4s + 3} + \frac{5s + 10}{s^2 + 4s + 3}U(s) \tag{4-23a}$$

Notice that the factor multiplying $U(s)$ in the second term is the transfer function of the system

$$H(s) = \frac{5s + 10}{s^2 + 4s + 3}$$

Also the denominator of the first term in Eq. (4-23*a*) is the characteristic polynomial

$$\Delta(s) = s^2 + 4s + 3$$

Thus Eq. (4-23*a*) may be written as

$$Y(s) = \frac{sy(0^-) + \dot{y}(0^-) + 4y(0^-) - 5u(0^-)}{\Delta(s)} + H(s)U(s) \tag{4-23b}$$

If the system had all zero initial conditions,* then Eq. (4-23*b*) would be

$$Y_{zs}(s) = H(s)U(s) \tag{4-24}$$

The notation $Y_{zs}(s)$ serves to emphasize that Eq. (4-24) is true when the system has zero initial conditions. Thus another interpretation of the transfer function, $H(s)$, for the SISO system is

$$H(s) = \frac{Y_{zs}(s)}{U(s)} \tag{4-25}$$

Equation (4-24) is true in the general multiple-input situation, but Eq. (4-25) only makes sense when $U(s) = \mathcal{L}\{u(t)\}$ is a scalar. It is not by accident that the notation s was chosen in Chap. 2 to develop the transfer-function concept.

The convolution property of the Laplace transform may also be used to observe another interesting fact about Eq. (4-24). Namely, from Chap. 1,

$$y_{zs}(t) = \int_0^t h(t - \tau)u(\tau)\, d\tau = h(t) * u(t) \tag{4-26}$$

*Note that the initial conditions include the initial conditions on the input. This set of initial conditions constitutes the initial state of the system. The state is always continuous for nonimpulsive input, and so it is immaterial whether these initial conditions are at 0^- or 0^+. Reasons for this are explained in Chap. 7.

Taking the Laplace transform of both sides of Eq. (4-26) gives Eq. (4-24), where

$$\boxed{H(s) = \mathcal{L}\{h(t)\}} \tag{4-27}$$

Thus it is also true that

$$\boxed{h(t) = \mathcal{L}^{-1}\{H(s)\}} \tag{4-28}$$

Equation (4-27) is equivalent to the relation $H(j\omega) = \mathfrak{F}\{h(t)\}$; but, using the Laplace transform, $H(s)$ always exists, regardless of whether or not the system is stable. We have come full circle another time. The basics never change; only the methods used to derive the basic results change.

Since the second term in Eq. (4-23*b*) is $Y_{zs}(s)$, the first term is $Y_{zi}(s)$:

$$Y_{zi}(s) = \frac{sy(0^-) + \dot{y}(0^-) + 4y(0^-) - 5u(0^-)}{\Delta(s)} \tag{4-29}$$

The terms of the numerator are thus equivalent to the initial state information on the system.

To solve the differential equation Eq. (4-22), substitution of the boundary conditions are made into Eq. (4-23*b*) along with the input $u(t)$:

$$U(s) = \mathcal{L}\{3t\} = \frac{3}{s^2}$$

This gives

$$Y(s) = \frac{4s + 11}{s^2 + 4s + 3} + \frac{3(5s + 10)}{s^2(s^2 + 4s + 3)} \tag{4-30}$$

The inverse transform of Eq. (4-30) must now be taken. In general this requires partial fraction expansions. The inherent linear equation problem encountered in the classic solution method for finding the arbitrary constants $c_1, c_2, \ldots, c_n$, has now been traded for a partial fraction expansion problem. Either method of solution yields the same total response, $y(t)$.

The partial fraction expansions of the two terms in Eq. (4-30) are found to be (see Appendix C)

$$Y_{zi}(s) = \frac{7/2}{s + 1} + \frac{1/2}{s + 3} \tag{4-31}$$

$$Y_{zs}(s) = \frac{-25/3}{s} + \frac{10}{s^2} + \frac{15/2}{s + 1} + \frac{15/18}{s + 3} \tag{4-32}$$

Taking the inverse Laplace transform of Eqs. (4-31) and (4-32) gives

$$y_{zi}(t) = 7/2\, e^{-t} + 1/2\, e^{-3t} \tag{4-33}$$

$$y_{zs}(t) = -25/3 + 10t + 15/2e^{-t} + 15/18e^{-3t} \tag{4-34}$$

Adding Eq. (4-33) and Eq. (4-34) gives the complete solution

$$y(t) = 11e^{-t} + 4/3\, e^{-3t} - 25/3 + 10t \tag{4-35}$$

If one used the classic method of solution to solve this same system, one would obtain

$$y_h(t) = 11\, e^{-t} + 4/3\, e^{-3t}$$

and
$$y_p(t) = -25/3 + 10t$$

The total solution is the same, but we see that $y_{\mathrm{zi}}(t) \neq y_h(t)$ and $y_{\mathrm{zs}}(t) \neq y_p(t)$. In general, the zero-state response includes natural transients that are caused by the application of the input to the system initially at the rest zero-state condition. For a persistent input, a stable system, and using the methods of Chap. 2

$$y_{\mathrm{ss}}(t) = y_p(t) = \lim_{t\to\infty} y_{\mathrm{zs}}(t)$$

The foregoing Laplace transform solution method may be applied to any system with forcing inputs that are Laplace-transformable. The steps taken are the same. One further point, though, is made before proceeding. This is how to handle boundary conditions at some $t_0^- \neq 0^-$. This is complicated if one attempts to apply the Laplace transform shift property. The easiest approach is to simply shift the time axis to $t' = t - t_0$. Then $t' = 0$ when $t = t_0$, and the solution can proceed as normal to find $y(t')$. The solution for $y(t)$ is then found via substitution of $t' = t - t_0$ into the solution for $y(t')$.

Example 4-3

$$\dot{y}(t) + 2y(t) = 3u(t)$$

with $y(t_0) = 4$ and

$$u(t) = [1 + 5e^{-(t-t_0)}]u_{-1}(t - t_0)$$

Let $t' = t - t_0$. Then the boundary conditions in terms of t' are $y(0) = 4$ and

$$u(t') = [1 + 5e^{-t'}]u_{-1}(t')$$

The differential equation in terms of t' is still

$$\dot{y}(t') + 2y(t') = 3u(t')$$

Taking the Laplace transform of both sides and solving for $Y(s)$ gives

$$Y(s) = \frac{y(0)}{s+2} + \frac{3}{s+2}U(s)$$

or
$$Y(s) = \frac{4}{s+2} + \frac{3}{s(s+2)} + \frac{(3)(5)}{(s+2)(s+1)}$$

Expanding via partial fraction expansions provides

$$Y(s) = \frac{4}{s+2} + \frac{3/2}{s} + \frac{-3/2}{s+2} + \frac{-15}{s+2} + \frac{15}{s+1}$$

Combining terms and taking the inverse Laplace transform gives

$$y(t') = [-25/2e^{-2t'} + 3/2 + 15e^{-t'}]u_{-1}(t')$$

Finally, making the substitution $t' = t - t_0$ on the right yields the final complete solution:

$$y(t) = [-25/2e^{-2(t-t_0)} + 3/2 + 15e^{-(t-t_0)}]u_{-1}(t - t_0)$$

The reader is invited to check that this is the solution by checking the boundary condition $y(t_0) = 4$ and by substitution of the solution $y(t)$ into the original differential equation using the input, $u(t)$.

4-4 UNIT STEP RESPONSE REVISITED

In Chaps. 1 and 2 the response to the unit step input played a major role in linear system analysis, and it was carefully considered. The Laplace transform solution method provides some additional insight into the nature of the important step response.

Consider a hypothetical system with transfer function

$$H(s) = \frac{5000(s^2 + 1.11s + 0.101)}{(s^4 + 15s^3 + 83s^2 + 359s + 290)}$$

$$= \frac{5000(s + 1.01)(s + 0.1)}{(s + 1)(s + 10)(s + 2 + 5j)(s + 2 - 5j)} \tag{4-36}$$

Since the Laplace transform of the unit step input is $1/s$, the unit step response is found by taking the inverse Laplace transform of

$$Y(s) = \frac{5000(s + 1.01)(s + 0.1)}{(s + 1)(s + 10)(s + 2 + 5j)(s + 2 - 5j)(s)}$$

$$= \frac{r_1}{s + 1} + \frac{r_2}{s + 10} + \frac{r_3}{s + 2 + 5j} + \frac{r_4}{s + 2 - 5j} + \frac{r_5}{s} \tag{4-37}$$

Taking the inverse Laplace transform of $Y(s)$ then gives the unit step response

$$y(t) = r_1 e^{-t} + r_2 e^{-10t} + r_3 e^{(-2-5j)t}$$

$$+ r_4 e^{(-2+5j)t} + r_5 \tag{4-38}$$

Letting $p_1 = -1$, $p_2 = -10$, $p_3 = (-2 - 5j)$, $p_4 = (-2 + 5j)$, $p_5 = 0$,* $z_1 = -1.01$, and $z_2 = -0.1$, the residues† are evaluated from

$$r_i = \left[5000 \prod_{k=1}^{2} (p_i - z_k)\right] \Big/ \left[\prod_{\substack{k=1 \\ k \neq i}}^{5} (p_i - p_k)\right] \qquad i = 1, 2, 3, 4, 5 \tag{4-39}$$

*Note that $p_5 = 0$ is an artificial pole corresponding to the unit step input.

†See Appendix C.

The solution for the residues is

$$r_1 = 0.19231 \qquad r_2 = 55.556$$

$$r_3 = -28.745 + 44.077j \qquad r_4 = -28.745 - 44.077j$$

$$r_5 = 1.7414 \tag{4-40}$$

The numerator factors in Eq. (4-39) are the directed distances between the pole p_i (including the artificial pole associated with the input—here $p_5 = 0$) and the system zeros. The denominator factors are the directed distances between the subject pole, p_i, and all of the other poles. These distances may be determined graphically as illustrated in Fig. 4-3. The angle of the residue is found by subtracting the denominator angles from the numerator angles. This procedure is one of simply treating the right-hand side of Eq. (4-39) in polar form.

Determination of the residues gives valuable insight into the nature of the time response for the subject input. For instance, the pole $p_2 = -10$ is fast compared to the other poles, and its influence decays quickly. However, its residue $r_2 = 55.556$, being large, shows that it will have a significant influence in the time response. This is shown by comparing the step time response to $H(s)$, Eq. (4-36), with the step response to

$$H_1(s) = \frac{500(s + 1.01)(s + 0.1)}{(s + 1)(s + 2 + 5j)(s + 2 - 5j)} \tag{4-41}$$

This comparison is shown in Fig. 4-4. The gain of $H_1(s)$ is reduced to 500 so that $H(s)$ and $H_1(s)$ have the same steady-state final value

$$y_{ss}(t) = H(0) = H_1(0) = r_5 = 1.7414$$

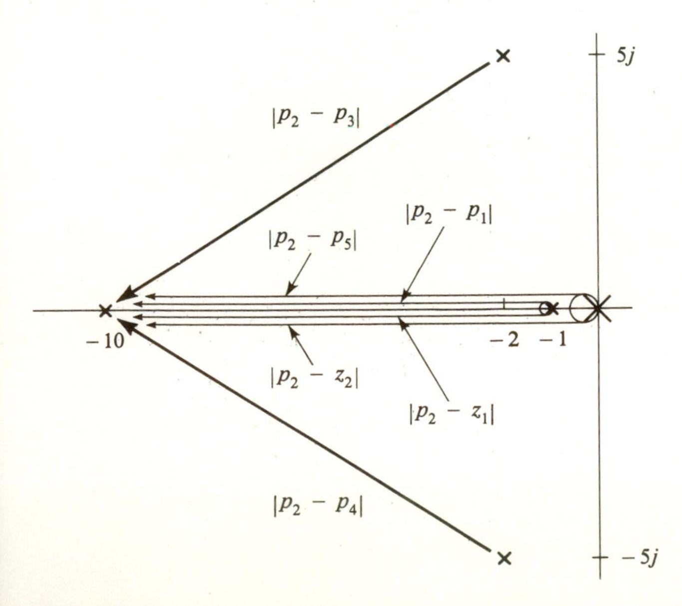

Figure 4-3 Illustration of graphical calculation of residues, Eq. (4-39).

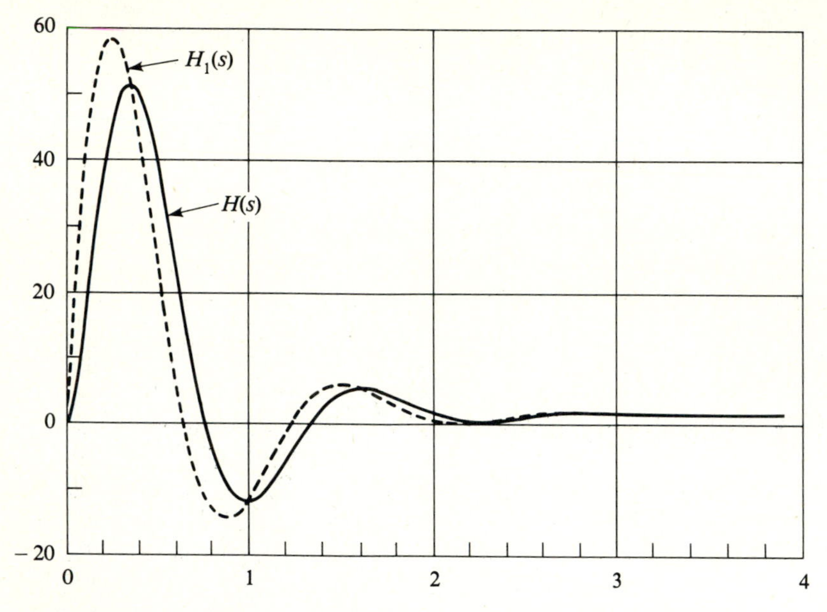

Figure 4-4 Step response $H(s)$, Eq. (4-36), versus $H_1(s)$, Eq. (4-41).

The pole $p_1 = -1$ is the slowest pole and would normally have a dominant influence on the time response. However, because of the closely coincident zero, $z_1 = -1.01$, the distance factor $(p_1 - z_1) = (0.01)$ in the residue equation for r_1 makes r_1 very small, $r_1 = 0.19231$. The zero effectively cancels the influence of the

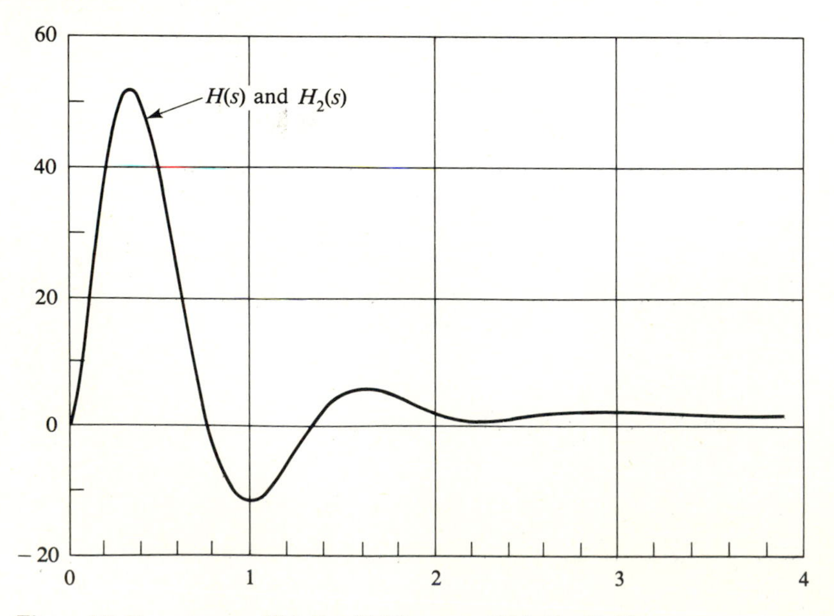

Figure 4-5 Step response $H(s)$, Eq. (4-36), versus $H_2(s)$, Eq. (4-42).

pole because they are so close together. The minor effect which this pole-zero combination has on the time response of the system is seen in Fig. 4-5, where the step response of

$$H_2(s) = \frac{5050(s + 0.1)}{(s + 10)(s + 2 + 5j)(s + 2 - 5j)} \tag{4-42}$$

is shown plotted versus the step response of the original $H(s)$, Eq. (4-36). Again the gain of $H_2(s)$ is adjusted so that $H_2(0) = H(0) = 1.7414$. A good third-order approximation to the fourth order $H(s)$ is thus $H_2(s)$. In fact,

$$H_3(s) = \frac{505(s + 0.1)}{(s + 2 + 5j)(s + 2 - 5j)} \tag{4-43}$$

makes a reasonably good second-order approximation to $H(s)$ where both the fast pole, $p_2 = -10$, and the pole-zero combination, $p_1 = -1$ and $z_1 = -1.01$, have been eliminated. A comparison plot is shown in Fig. 4-6.

Thus the lightly damped complex conjugate poles lying close to the imaginary axis, $p_{3,4} = -2 \pm 5j$, and the zero, $z_2 = -0.1$, which is also close to the imaginary axis, are the dominant factors in the time response. That they both dominate the response is seen in Fig. 4-7, in which the original $H(s)$ is shown plotted along with the step response of

$$H_4(s) = \frac{172.41(s + 1.01)(s + 0.1)}{(s + 1)(s + 10)} \tag{4-44}$$

and

$$H_5(s) = \frac{500(s + 1.01)}{(s + 1)(s + 10)(s + 2 + 5j)(s + 2 - 5j)} \tag{4-45}$$

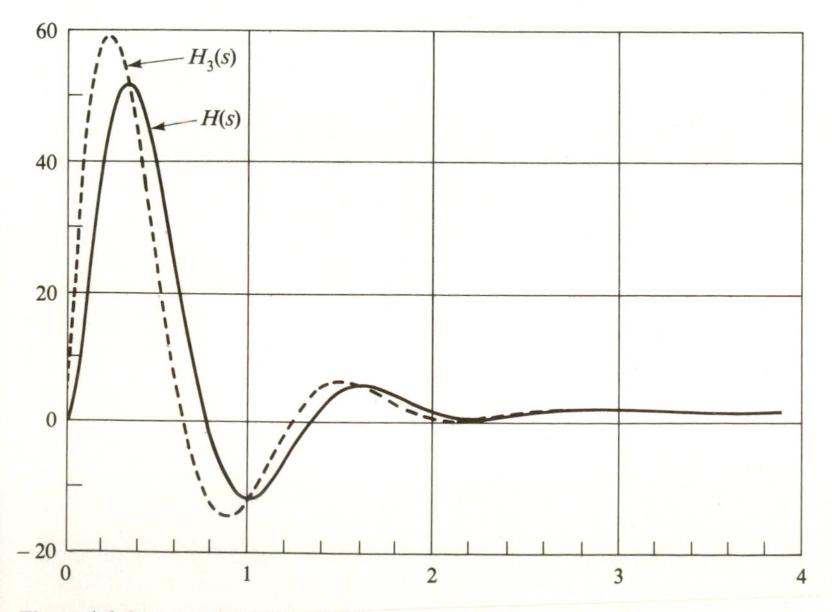

Figure 4-6 Step response $H(s)$, Eq. (4-36), versus $H_3(s)$, Eq. (4-43).

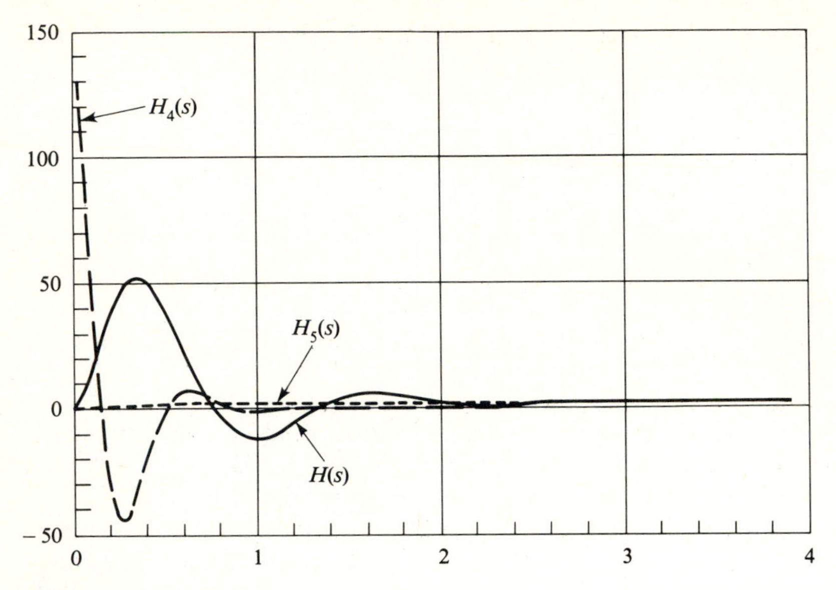

Figure 4-7 Step response $H(s)$, Eq. (4-36), versus $H_4(s)$, Eq. (4-44), versus $H_5(s)$, Eq. (4-45).

System $H_4(s)$, which eliminates the p_3 and p_4 poles, and system $H_5(s)$, which eliminates the zero z_2, bear little resemblance to the original $H(s)$.

While discussing the zero z_2, consider what happens as z_2 crosses the imaginary axis from -0.1 to $+0.1$. Figure 4-8 plots the step response for the original $H(s)$ and for

$$H_6(s) = \frac{5000(s + 1.01)(s - 0.1)}{(s + 1)(s + 10)(s + 2 + 5j)(s + 2 - 5j)} \tag{4-46}$$

The residues for $H_6(s)$ with the unit step $(1/s)$ input are

$$r_1 = 0.2350 \qquad r_2 = 56.679$$
$$r_3 = -27.586 + 45.668j \qquad r_4 = -27.586 - 45.668j$$
$$r_5 = -1.7414 \tag{4-47}$$

These residues are all very close to those for $H(s)$, Eq. (4-40), except that $r_5 = y_{ss}(t) = H_6(0) = -1.7414$ is now negative instead of positive. This is consistent with the analysis of the nonminimum-phase zero in Chap. 2, where it was shown that the initial slope of the step response is opposite in direction to the final steady-state value. Again, we see the same result, but derived from a different viewpoint.

4-5 INTERCONNECTION OF LINEAR SYSTEMS

One of the most useful properties of transform techniques is the facility they provide in the analysis of a complex intercoupling of linear time-invariant subsys-

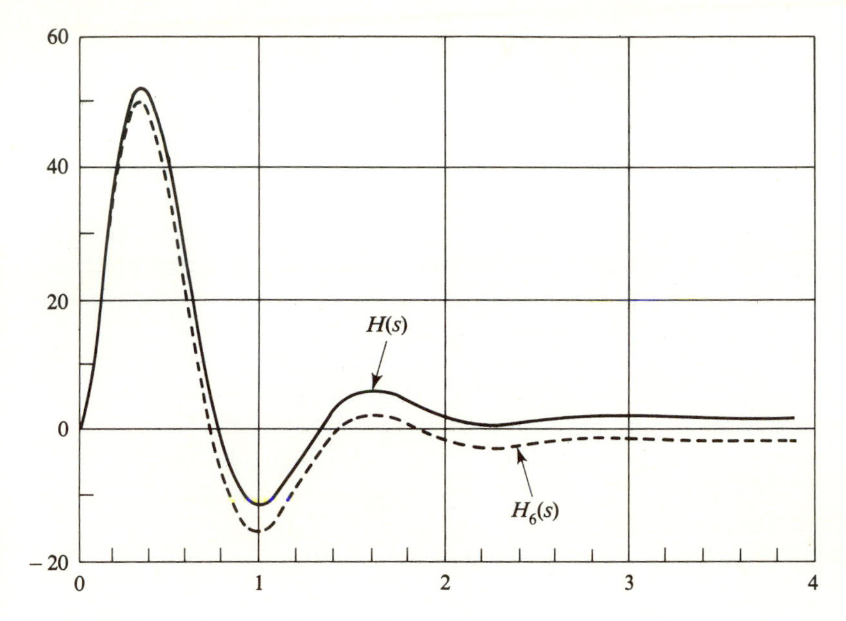

Figure 4-8 Step response $H(s)$, Eq. (4-36), versus nonminimum phase $H_6(s)$, Eq. (4-46).

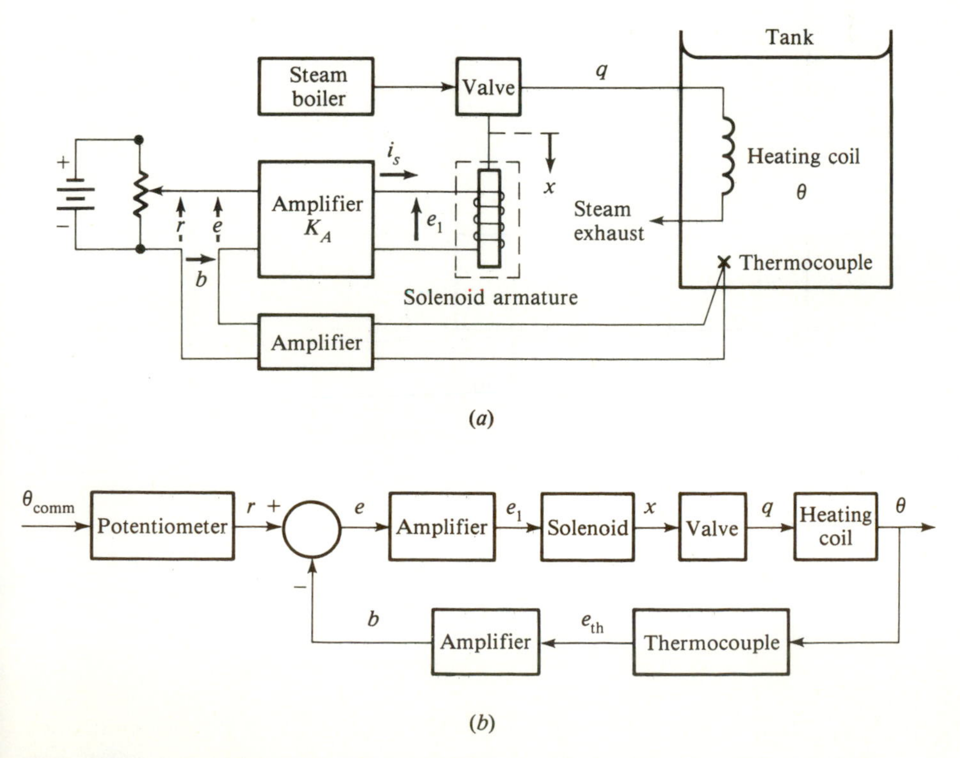

Figure 4-9 An industrial-process temperature-control system. (*From D'Azzo & Houpis,* 1981.)

tems. Figure 4-9 illustrates such an intercoupling for a hypothetical industrial heating system. Each element has an associated transfer function. This section reviews techniques for combining and simplifying a given transfer-function block diagram.

Consider first the two systems connected in series as shown in Fig. 4-10. The separate impulse responses are $h_1(t)$ and $h_2(t)$. The combined impulse response is the convolution, $h(t) = h_1(t) * h_2(t)$. However, in the transform domain the combined transfer function is simply the product of the two transfer functions, $H(s) = H_1(s)H_2(s)$. Then $h(t) = L^{-1}\{H(s)\}$. Transfer functions in series, such as in Fig. 4-10, may be multiplied and their order reversed as desired:

$$H(s) = H_1(s)H_2(s) = H_2(s)H_1(s) \tag{4-48}$$

This may be extended to any number of transfer functions connected in series.

For transfer functions connected in parallel, as in Fig. 4-11, the rule is

$$H(s) = H_1(s) + H_2(s) \tag{4-49}$$

For example, let $H_1(s) = 2/(s + 1)$ and $H_2(s) = -1/(s + 2)$. The parallel connection of $H_1(s)$ and $H_2(s)$ is

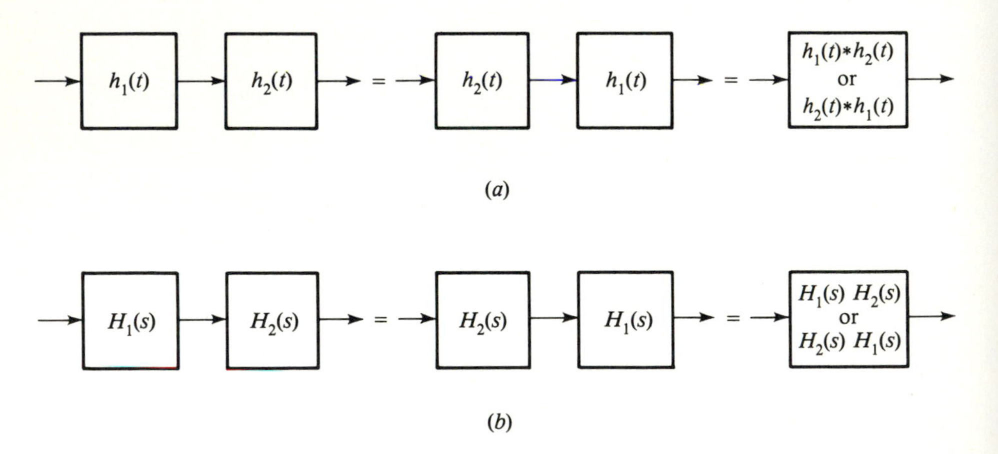

Figure 4-10 Illustration of linear time-invariant systems connected in series.

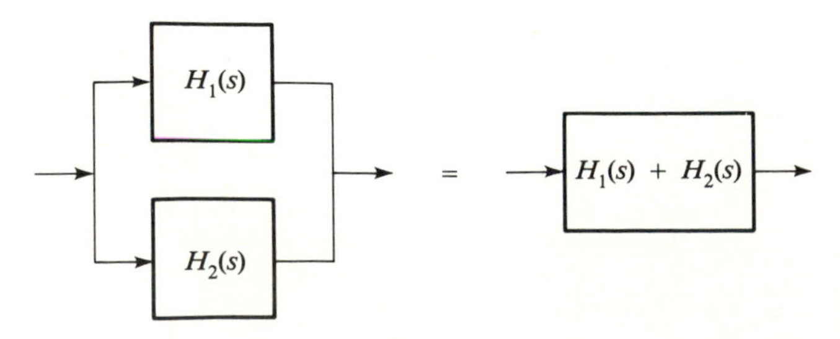

Figure 4-11 Illustration of linear time-invariant systems connected in parallel.

$$H(s) = \frac{2}{s+1} + \frac{-1}{s+2} = \frac{s+3}{(s+1)(s+2)}$$

This is partial fraction expansion in reverse.

For three transfer functions connected in the branch, as in Fig. 4-12*a*, the equivalent representation is shown in Fig. 4-12*b*.

The system of transfer functions in Fig. 4-13*a* is termed a *feedback* system; the output of the forward path, $Y(s)$, is fed back through a *compensator* with transfer function $H_2(s)$ and then subtracted from the input signal, $U(s)$, to form an *error signal*, $E(s)$. Such a feedback system can dramatically influence the overall *closed-loop transfer function* from input $U(s)$ to output $Y(s)$. How to design the proper feedback system to achieve a desired system response is the general topic of feedback control theory.

To gain some insight into how the closed-loop transfer function may be modified by a feedback system, consider the equivalent transfer function of Fig. 4-13*b*:

$$H(s) = \frac{H_1(s)}{1 + H_1(s)H_2(s)} \tag{4-50a}$$

To derive this, consider that $Y(s) = H_1(s)E(s)$. But $E(s) = U(s) - H_2(s)Y(s)$. Thus

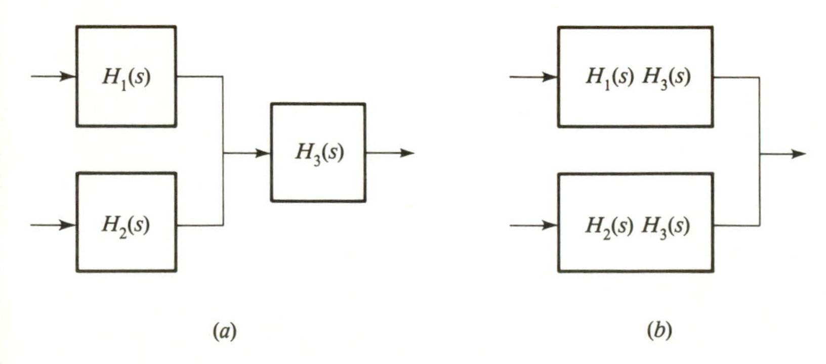

Figure 4-12 Illustration of branch network and its equivalent.

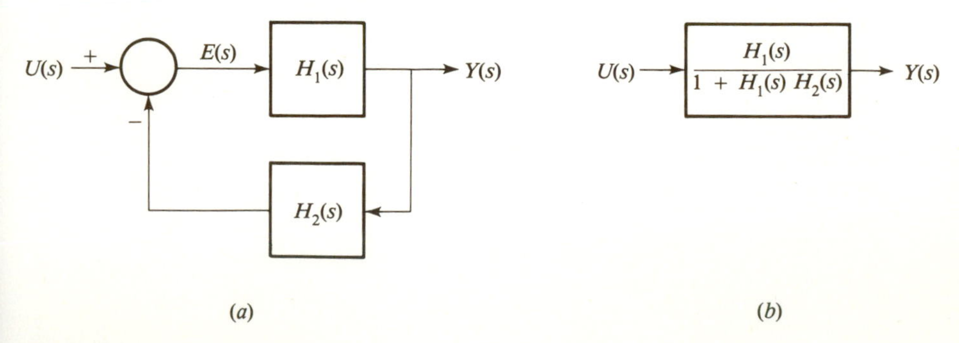

Figure 4-13 Illustration of feedback system, transform domain.

$$Y(s) = H_1(s)U(s) - H_1(s)H_2(s)Y(s)$$

Solving for $Y(s)$ gives

$$Y(s) = \left[\frac{H_1(s)}{1 + H_1(s)H_2(s)}\right]U(s)$$

Thus the equivalent closed-loop transfer function, $H(s) = Y(s)/U(s)$, is Eq. (4-50*a*) as claimed.

In a feedback situation, the return transfer function $H_2(s)$ is often a compensator for the forward-path *plant,* $H_1(s)$. To see more clearly how the poles and zeros of the compensator can alter the poles and zeros of the closed-loop plant, $H(s)$, let $H_1(s) = N_1(s)/\Delta_1(s)$ and $H_2(s) = N_2(s)/\Delta_2(s)$ be the respective zero-polynomial and characteristic-polynomial ratios. Making these substitutions into $H(s)$, Eq. (4-50*a*), gives

$$H(s) = \frac{N_1(s)\,\Delta_2(s)}{\Delta_2(s)\,\Delta_1(s) + N_2(s)N_1(s)} \tag{4-50b}$$

Hence the zeros of the forward-loop plant, along with the poles of $H_2(s)$, are the zeros of the closed-loop feedback plant. The poles of the feedback plant are the roots of

$$\Delta(s) = \Delta_2(s)\,\Delta_1(s) + N_2(s)N_1(s) \tag{4-51}$$

There is no simple relation of the closed-loop poles to the original system poles and/or zeros.

Example 4-4 Suppose an uncompensated plant has the transfer function

$$H_1(s) = \frac{s + 0.1}{s^2 + 2s + 29} = \frac{s + 0.1}{(s + 1 + 5.2915j)(s + 1 - 5.2915j)}$$

This system has an undesirable oscillatory response as shown in Fig. 4-14. The open-loop damping ratio is $\zeta = 0.1857$. The system also has a weak steady-state response to a step input of $y_{ss}(t) = H_1(0) = 0.1/29 = 0.00345$. A forward-loop compensator $H_2(s)$ along with unity feedback is added to the system. This is shown in Fig. 4-15. The compensator $H_2(s)$ is chosen as

$$H_2(s) = \frac{10(s + 5)}{s + 0.1}$$

This compensator is termed a *lag compensator* or a *lag filter*. It may be implemented by the passive electrical circuit as shown in Fig. 4-16. The transfer function of the circuit is

$$H_2(s) = \frac{V_{out}(s)}{V_{in}(s)}$$

$$= \frac{A(R_2Cs + 1)}{(R_1 + R_2)Cs + 1}$$

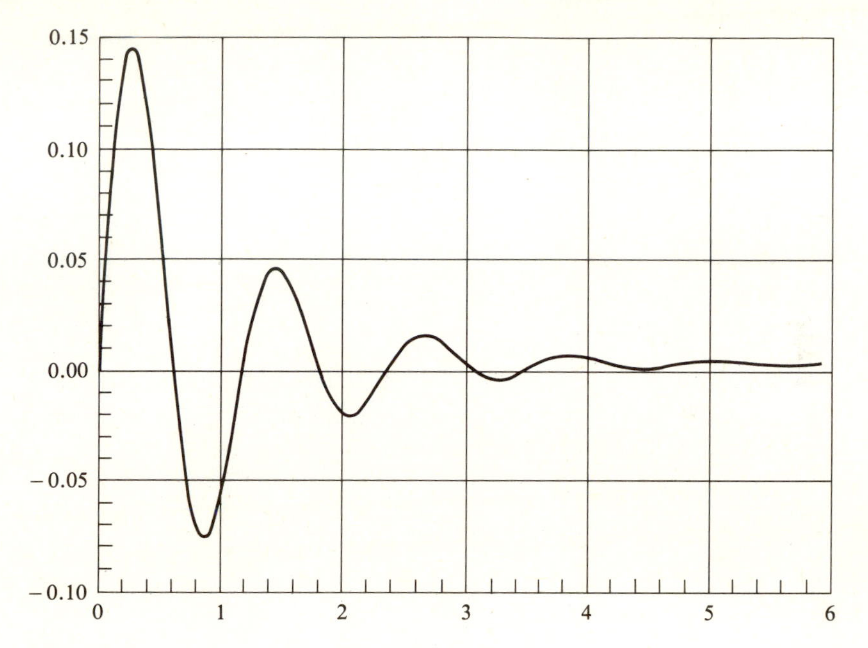

Figure 4-14 Step response, $H_1(s)$, Example 4-4.

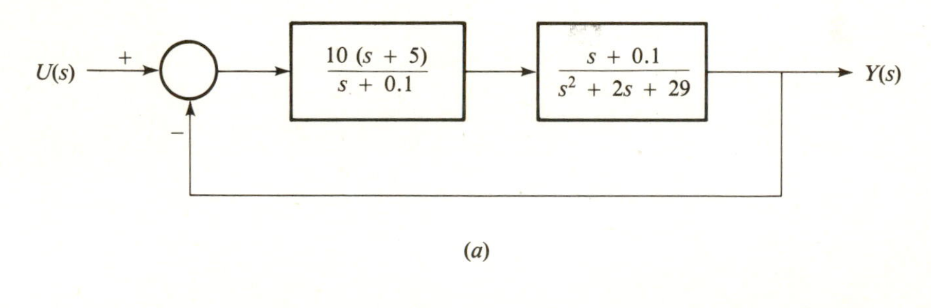

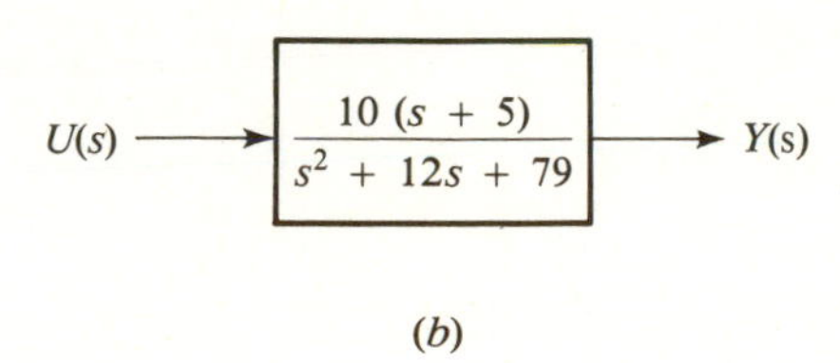

Figure 4-15 Compensator and feedback for Example 4-4. (*a*) Original system with forward loop "compensator" and negative feedback. (*b*) Equivalent "closed-loop" system.

where $A = 500$ is the amplifier gain and R_1, R_2, and C are chosen such that

$$R_2C = 1/5 \qquad \text{and} \qquad (R_1 + R_2)C = 10$$

Manipulating the variables then gives $H_2(s)$.

The pole of the forward-loop compensator is chosen to cancel the unde-

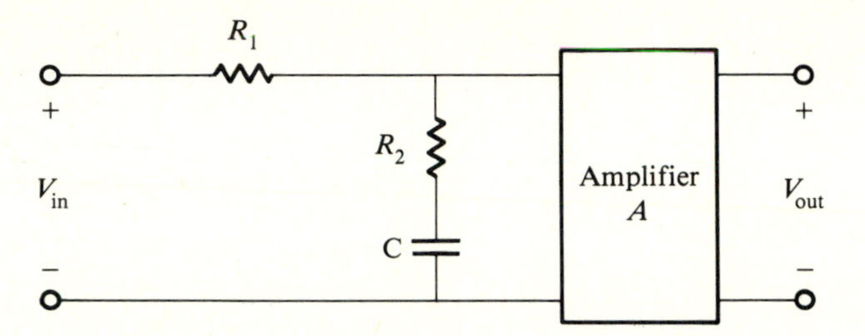

Figure 4-16 Passive electric network for lag compensator.

sirable zero at $z = -0.1$. The closed-loop transfer function is

$$H(s) = \frac{H_2(s)H_1(s)}{1 + H_2(s)H_1(s)}$$

$$= \frac{10(s+5)}{(s_2 + 2s + 29) + 10(s+5)}$$

$$= \frac{10(s+5)}{s^2 + 12s + 79}$$

The steady-state final value is $H(0) = 50/79 = 0.633$, and the closed-loop poles are $p_{1,2} = -6 \pm 6.557j$. The closed-loop step response is shown in Fig. 4-17. The closed-loop damping ratio is $\zeta = 0.675$, very satisfactory.

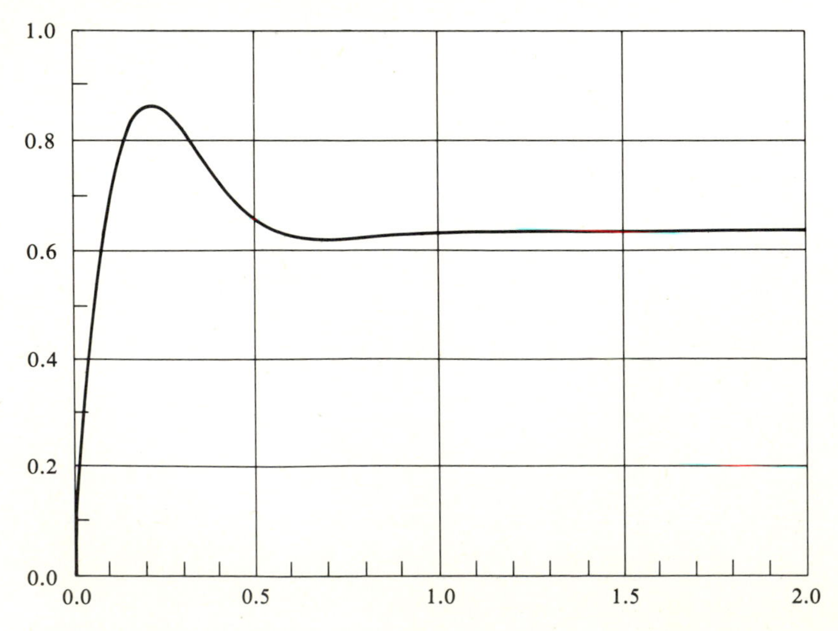

Figure 4-17 Step response $H(s)$ after feedback compensation, Example 4-4.

A well-designed compensator and feedback control system can dramatically improve an undesirable system response. The interested reader is referred to, for example, Refs. 3, 4, and 5. There is one further aspect, though, pertaining to system feedback which provides additional insight into the illusive system zero. This is explored briefly in the next section.

*4-6 UNITY FEEDBACK AND THE SYSTEM POLES AND ZEROS

This section briefly investigates some intrinsic properties of the system poles and zeros under feedback. This provides another interesting ramification of the system zero.

To be specific, consider the closed-loop plant as encountered in the previous example:

$$H(s) = \frac{KG(s)}{1 + KG(s)} = \frac{K(s+5)/(s^2+2s+29)}{1 + K(s+5)/(s^2+2s+29)} \tag{4-52a}$$

This is unity feedback around the plant

$$KG(s) = \frac{K(s+5)}{s^2+2s+29} = \frac{K(s+5)}{(s+1+10.583j)(s+1-10.583j)} \tag{4-53}$$

as shown in Fig. 4-18. The forward loop K is left as a variable so that the behavior of the closed-loop poles and zeros can be investigated as $K \to \infty$.

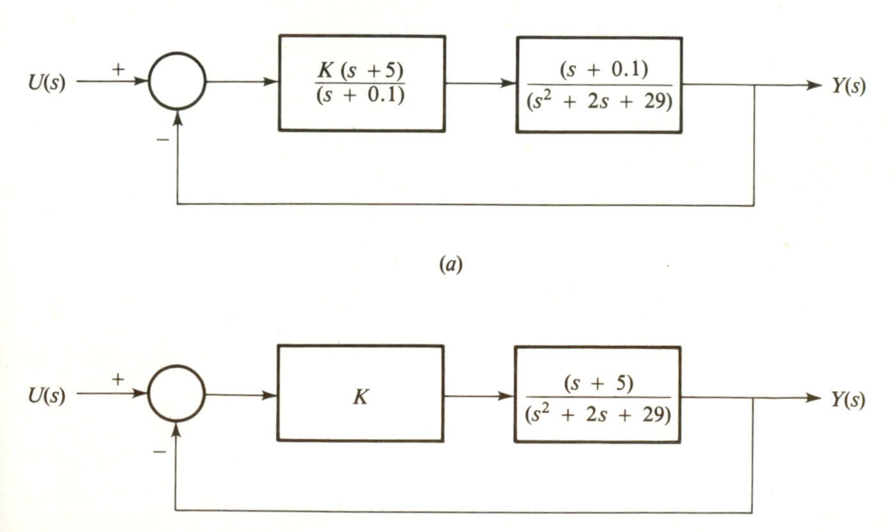

Figure 4-18 Explanation of $KG(s)$, Eq. (4-53). (*a*) Compensated system of Example 4-4. (*b*) Equivalent unity feedback system.

The zeros of $H(s)$ are the zeros of $G(s)$ regardless of the value of gain K. This is always true for the unity-feedback system.

To investigate what happens to the closed-loop poles, modify $H(s)$ by dividing numerator and denominator by K and multiply numerator and denominator by $(s^2 + 2s + 29)$.

$$H(s) = \frac{(s + 5)}{\frac{1}{K}(s^2 + 2s + 29) + (s + 5)} \tag{4-52b}$$

As K gets very large, one of the system poles approaches the zero at -5.

This is true in general: with unity feedback, m of the system poles will approach the m system zeros as the forward-loop gain, K, goes to infinity. This is another interesting property of the system zeros. It is also indicative of why the zero in the right half-plane causes so much difficulty in a feedback control system. Since one of the poles will migrate to the right half-plane zero as the gain is increased, the system will go unstable as soon as the pole crosses the imaginary axis.

In an nth-order system, m poles approach the m zeros as $K \to \infty$. Where do the $(n - m)$ remaining poles go? These remaining poles go to infinity along well-defined asymptotes. See, for example, Ref. 3 for a derivation of these facts along with a number of other rules associated with drawing the *root locus plot*—a plot of the n poles of the closed-loop transfer function

$$H(s) = \frac{KG(s)}{1 + KG(s)}$$

as K is varied from zero to $+\infty$.

Figure 4-19 depicts the root locus plot for the example system $G(s)$, Eq. (4-53). Notice that one pole approaches the zero at -5 and the other pole goes to $-\infty$ along the real axis.

4-7 SUMMARY

This chapter has introduced the one-sided Laplace transform as a convenient tool for the analysis of the linear time-invariant system. The Laplace transform is similar in many ways to the Fourier transform, and its properties are also quite similar. The differentiation property is used to provide an alternative to the classic method of system time solution as presented in Chap. 2. Further analysis on the system step response provides additional insights regarding the influence of the system poles and zeros.

The convolution property of the Laplace transform is again a key property of the transform domain analysis. Interconnections of linear subsystems are analyzed by algebraic manipulations of their transfer functions.

The methods presented in these first four chapters provide some powerful linear system analysis tools. But we still do not have an adequate method by

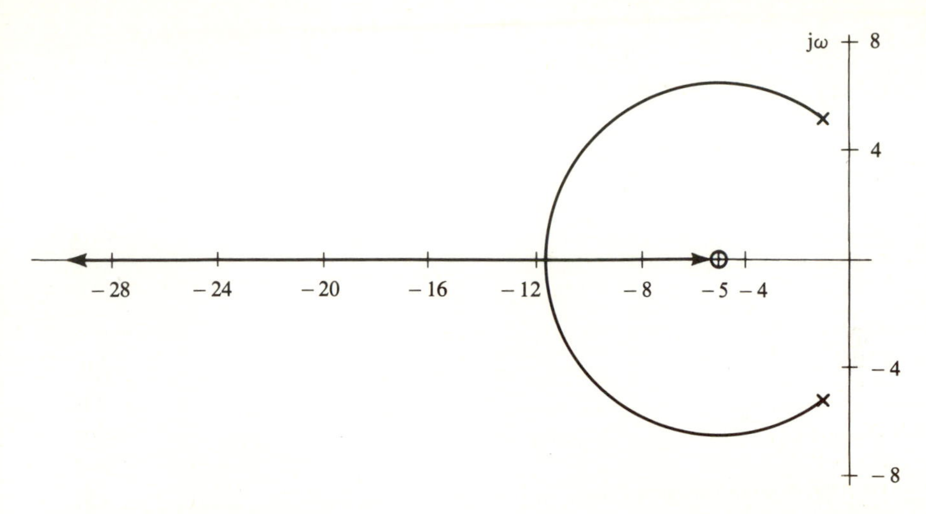

Figure 4-19 Example of root-locus plot.

which to analyze the system response to discontinuous types of inputs—particularly the digital pulse inputs that are typical with either digital simulation or on-line digital control systems. This need is partially answered by the material in the next chapter.

PROBLEMS

4-1 Find the Laplace transform of the following functions:

(*a*) $2e^{-3t}u_{-1}(t)$ (*b*) $4e^{-2t}\sin(8t)u_{-1}(t)$

(*c*) $3\sin(4t)u_{-1}(t)$ (*d*) $2te^{-3t}u_{-1}(t)$

4-2 Find the Laplace transform of the following functions:

(*a*) $3\cos(4t)u_{-1}(t)$ (*b*) $(5e^{-4t} - 3e^{-7t})u_{-1}(t)$

(*c*) $4e^{-2t}\cos(8t)u_{-1}(t)$ (*d*) $5\,\delta(t) + 4u_{-1}(t)$

4-3 Find the inverse Laplace transform of the following functions:

(*a*) $\dfrac{8s-1}{s^2+3s+2}$ (*b*) $\dfrac{4}{s^2+2s+17}$ (*c*) $\dfrac{s+1}{s^2+2s+5}$

4-4 Find the inverse Laplace transform of the following functions:

(*a*) $\dfrac{2}{s^2+7s+10}$ (*b*) $\dfrac{2}{s^2+2s+2}$ (*c*) $\dfrac{s+2}{s^2+2s+2}$

4-5 Use the properties of the Laplace transform to find the Laplace transform of the following functions. (*Hint:* draw a sketch of each of these.)

(*a*) $(t-5)u_{-1}(t)$ (*b*) $(t-5)u_{-1}(t-5)$

(*c*) $tu_{-1}(t-5)$ (*d*) $tu_{-1}(t)$

4-6 Repeat Prob. 4-5 for the functions:

(*a*) $5e^{(6-2t)}u_{-1}(t)$ (*b*) $5e^{(6-2t)}u_{-1}(t-3)$

(*c*) $5e^{-2t}u_{-1}(t-3)$ (*d*) $5e^{-2t}u_{-1}(t)$

4-7 A function $f(t)$ has Laplace transform $F(s) = 2/(s^3 + 7s^2 - 15s)$.

(*a*) What is $f(0^+)$?

(*b*) What is $f(\infty)$?

4-8 A function $f(t)$ has Laplace transform $F(s) = 7/(s^3 + 2s^2 + 9s + 7)$.

(*a*) What is $f(0^+)$?

(*b*) What is $f(\infty)$?

4-9 Solve the following differential equations using the method of Laplace transforms. Identify the zero-input and zero-state portions of the response.

(*a*) $\dot{y} + 7y = 2u$

$u(t) = u_{-1}(t)$

$y(0^-) = 0$

(*b*) $\dot{y} + 3y = \dot{u} + 2u$

$u(t) = 25\cos(4t)u_{-1}(t)$

$y(0^-) = -5 \qquad u(0^-) = 0$

(*c*) $\ddot{y} + 7\dot{y} + 12y = 2\dot{u} + 2u$

$u(t) = u_{-1}(t)$

$y(0^-) = 2 \qquad \dot{y}(0^-) = -3 \qquad u(0^-) = 0$

4-10 Solve the following differential equations using the method of Laplace transforms. Identify the zero-input and zero-state portions of the response.

(*a*) $\dot{y} + 5y = u$

$u(t) = (3 + t)u_{-1}(t)$

$y(0^-) = 1$

(*b*) $\dot{y} - y = 2\dot{u} + 4u$

$u(t) = u_{-1}(t)$

$y(0^-) = -1 \qquad u(0^-) = 0$

(*c*) $\ddot{y} + \dot{y} - 12y = \dot{u} + u$

$u(t) = (2e^{-t})u_{-1}(t)$

$y(0^-) = 0 \qquad \dot{y}(0^-) = -2 \qquad u(0^-) = 5$

4-11 Find $h(t)$ for the following systems by using $h(t) = \mathcal{L}^{-1}\{H(s)\}$:

(*a*) $H(s) = \dfrac{5}{s+2}$ (*b*) $H(s) = \dfrac{3s+1}{s^2+3s+2}$

(*c*) $H(s) = \dfrac{2}{s^2+0.4s+4.04}$ (*d*) $H(s) = \dfrac{s+2}{s+1}$

4-12 Find $h(t)$ for the following systems by using $h(t) = \mathcal{L}^{-1}\{H(s)\}$:

(*a*) $H(s) = \dfrac{2}{s+3}$ (*b*) $H(s) = \dfrac{2s-1}{s^2+5s+4}$

(*c*) $H(s) = \dfrac{1}{s^2-0.2s+4.01}$ (*d*) $H(s) = \dfrac{s+2}{s-1}$

4-13 Find the unit step response for the following systems which are initially at rest:

(*a*) $H(s) = 1/(s^2 + 4s + 4)$

(*b*) $H(s) = (s + 1)/(s^2 + 4s + 4)$

(*c*) $H(s) = (s - 1)/(s^2 + 4s + 4)$

4-14 Find the unit step response for the following systems which are initially at rest:

(*a*) $H(s) = 1/(s^2 + 3s + 2)$

(*b*) $H(s) = (10s + 1)/(s^2 + 3s + 2)$

(*c*) $H(s) = (10s - 1)/(s^2 + 3s + 2)$

4-15 Find the unit step response for the following systems which are initially at rest:

(*a*) $H(s) = \dfrac{1000(s/10 + 1)}{(s + 100)(s + 1)}$

(*b*) $H(s) = \dfrac{1000(s/100 + 1)}{(s^2 + 2s + 2)(s^2 + 2s + 101)}$

(*c*) $H(s) = \dfrac{10s + 110}{(s + 0.1)(s^2 + 2s + 5)}$

4-16 Find the unit step response for the following systems which are initially at rest:

(*a*) $H(s) = \dfrac{(s + 9.9)}{(s + 10)(s + 1)}$

(*b*) $H(s) = \dfrac{(s + 1.1)}{(s + 10)(s + 1)}$

(*c*) $H(s) = \dfrac{(s^2 + 2s + 100)}{(s^2 + 2s + 101)(s + 1)}$

4-17 Find the overall transfer function for the interconnected systems as shown in Fig. 4-20.

4-18 Find the overall transfer function for the interconnected systems as shown in Fig. 4-21.

4-19 Find the closed-loop poles and zeros of the feedback systems shown in Fig. 4-22.

4-20 Find the closed-loop poles and zeros of the feedback systems shown in Fig. 4-23.

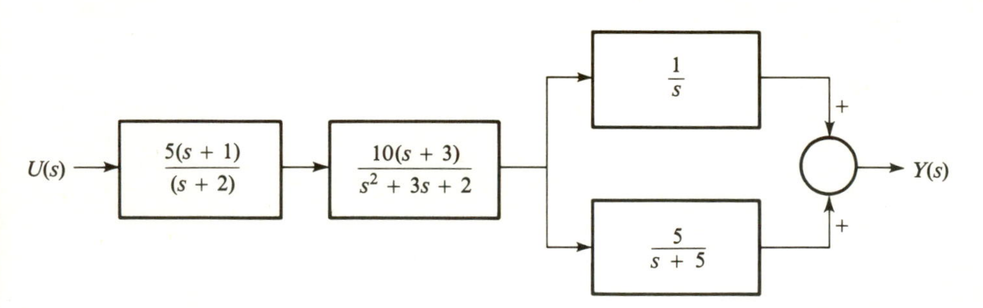

Figure 4-20 Interconnected system, Prob. 4-17.

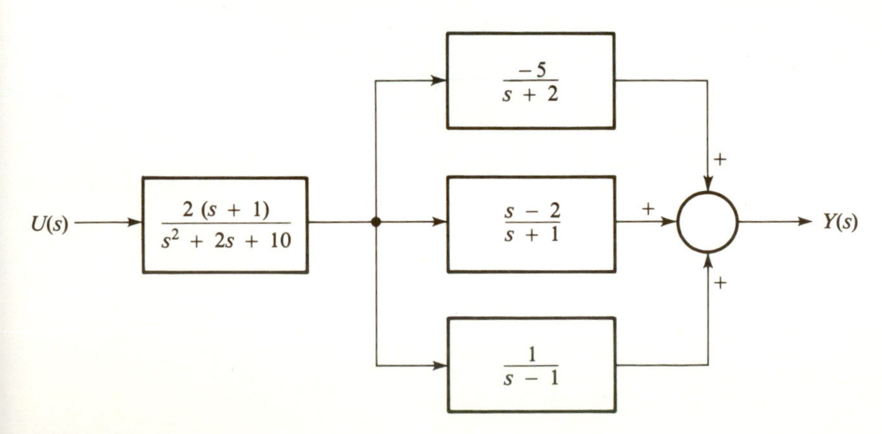

Figure 4-21 Interconnected system, Prob. 4-18.

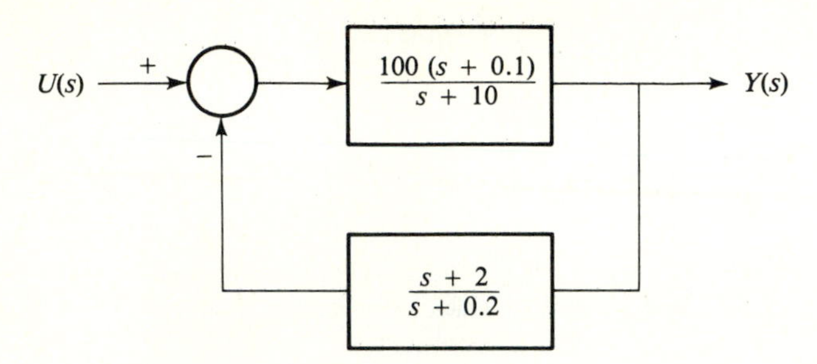

Figure 4-22 Feedback system, Prob. 4-19.

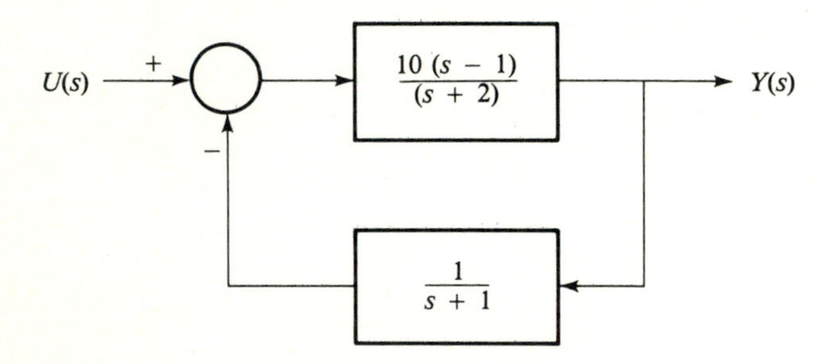

Figure 4-23 Feedback system, Prob. 4-20.

NOTES AND REFERENCES

References 1 and 2 are examples of sources for more extensive tables of Laplace transforms. References 3 through 5 are examples of sources to consult regarding feedback compensator design.

1. Selby, S., Ed.: *CRC Standard Mathematical Tables,* The Chemical Rubber Co., Cleveland, Ohio, 1980.
2. Abramowitz, M., and I. A. Segun, Ed.: *Handbook of Mathematical Functions,* Dover, New York, 1968.
3. D'Azzo, J. J., and C. H. Houpis: *Linear Control System Analysis and Design, Conventional and Modern,* McGraw-Hill, New York, 2d ed., 1981.
4. Truxal, J. G.: *Automatic Feedback Control System Synthesis,* McGraw-Hill, New York, 1955.
5. Fortmann, T. E., and K. L. Hitz: *An Introduction to Linear Control Systems,* Marcel Dekker, New York, 1977.

CHAPTER

FIVE

THE FINITE-DIFFERENCE EQUATION IN MODELING A CONTINUOUS SYSTEM

In Chap. 1 the discrete pulse-response model is presented as a method to handle the discrete sample-and-hold-type inputs. Such inputs might be associated with digital simulation of a continuous-input system or system analysis when the system actually has such inputs. The sample-and-hold-type inputs typically arise when an on-line digital computer is part of a system as a control element. However, computing the sampled output points via the discrete pulse-response model and discrete convolution is not particularly efficient because of the long "memory length" of past inputs—especially if the system is lightly damped.

The classic solution method and the Laplace transform method are not desirable for digital simulation because of the changes in boundary conditions that occur when the input makes discrete changes. The classic solution and Laplace transform methods are useful for analysis of continuous systems, but they are not particularly useful for discrete system analysis.

An efficient way to treat the discrete-time, linear, time-invariant, ordinary differential equation system is to use a finite-difference equation model of the system. This is still an input-output–based model, but it compresses the "infinite memory" of the pulse response model into a finite sum of n past input and n past output values. The form of the finite-difference equation bears striking resemblance to the nth order output differential equation model for the continuous time system. Indeed, many of the same analysis and solution techniques for the continuous-time differential equation are also applicable, in a modified form, to the finite-difference equation. For instance, the solution of the finite-difference equation may be separated into the sum of the homogeneous and particular solutions. The same basic concepts are used to find the homogeneous and particular solutions. Finally, transform techniques analogous to the Laplace transform may be used to solve and analyze the finite-difference equation. The appropriate transformation in this case is the z transform.

5-1 DERIVATION OF THE FINITE-DIFFERENCE EQUATION

The SISO system represented by the differential equation

$$y^{(n)}(t) + a_{n-1}y^{(n-1)}(t) + \cdots + a_0 y(t) = b_m u^{(m)}(t) + \cdots + b_0 u(t) \tag{5-1}$$

has transfer function

$$H(s) = \frac{b_m s^m + b_{m-1}s^{m-1} + \cdots + b_0}{s^n + a_{n-1}s^{n-1} + \cdots + a_0} \tag{5-2}$$

and impulse response function

$$h(t) = \mathcal{L}^{-1}\{H(s)\} \tag{5-3}$$

Initially it is assumed that the system Eq. (5-1) is proper so that $m < n$. Then $h(t)$ has no impulse at the origin and there is no direct feedforward of input to output.

If a sequence of discrete sample-and-hold-type inputs as depicted in Fig. 5-1 and as represented by

$$u_T(k) = u(t) \qquad t \in (kT - T, kT] \tag{5-4}$$

is input to the system Eq. (5-1), then the output at each sample time kT, $y(kT)$, is given by

$$\begin{aligned} y_T(k) &= \sum_{i=-\infty}^{k-1} h_T(k-i)u_T(i) \\ &= \sum_{i=-\infty}^{-1} h_T(k-i)u_T(i) + \sum_{i=0}^{k-1} h_T(k-i)u_T(i) \\ &= y_{\mathrm{zi}_T}(k) + y_{\mathrm{zs}_T}(k) \end{aligned} \tag{5-5}$$

Time $k = 0$ is arbitrarily selected as the initial time, so that the first summation from $-\infty$ to -1 represents the zero-input response while the second summation from 0 to $k - 1$ represents the zero-state response. The discrete pulse-response sequence $\{h_T(k), k = 1, 2, 3, \ldots\}$ is found via

$$h_T(k) \equiv \int_{(k-1)T}^{kT} h(\tau)\, d\tau \tag{5-6}$$

or via

$$h_T(k) = r(kT) - r(kT - T) \tag{5-7}$$

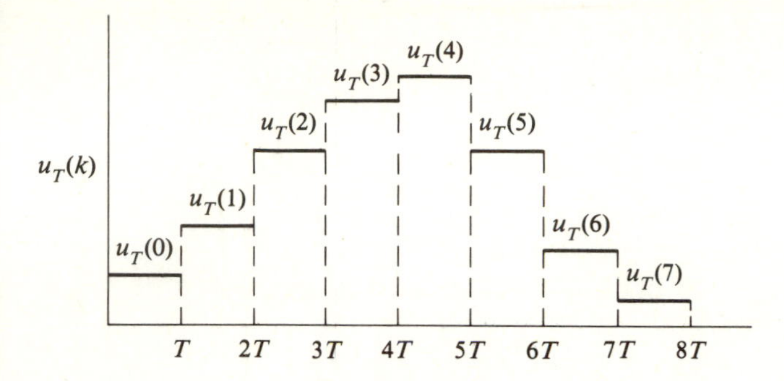

Figure 5-1 Piecewise constant sample-and-hold input.

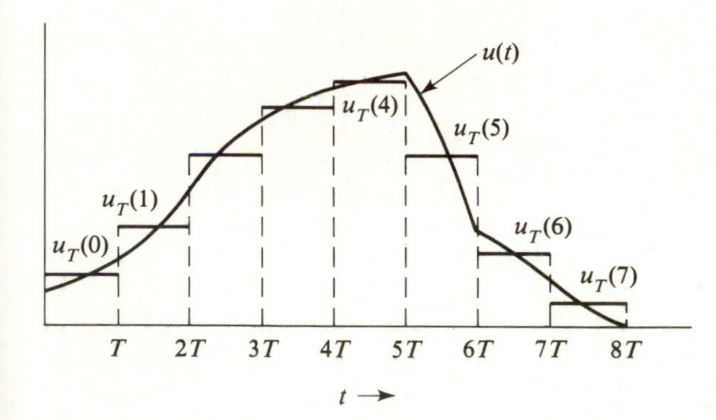

Figure 5-2 Piecewise constant approximation to continuous input, $u(t)$.

where $r(t)$ is the unit step response for the system. Using Eq. (5-7), the unit pulse response may be determined experimentally (see also Chap. 1).

The foregoing are a summary of some key results from Chaps. 1, 2, and 4. If the actual input is indeed the discrete sample-and-hold-type input, Eq. (5-4), then Eq. (5-5) is an exact representation of the output, $y_T(k) = y(kT)$, at the discrete sample points. There is no approximation, regardless of the sample duration T. If the input is actually a continuous-curve input as shown in Fig. 5-2, then Eq. (5-5) can still give a good approximation if T is short enough. The piecewise constant input, $u_T(k)$, then must be an approximation to the continuous-curve input. Possibilities are the sampled value $u_T(k) = u(kT)$, or, better, the average value

$$u_T(k) = \frac{u(kT + T) + u(kT)}{2} \tag{5-8}$$

The problem, of course, in the use of Eq. (5-5) for discrete time simulation is the long past memory of the inputs and the resulting computational burden of the discrete convolution summation. An alternative to the discrete pulse-response model is the discrete finite-difference equation*

*This is also termed an autoregressive moving average (ARMA) model in the statistical literature. An all-pole version is termed an autoregressive (AR) model. The discrete pulse response relation, Eq. (5-5), is also called a moving average (MA) model.

$$y_T(k+n) + a_{T_{n-1}} y_T(k+n-1) + a_{T_{n-2}} y_T(k+n-2) + \cdots + a_{T_0} y_T(k)$$
$$= b_{T_{n-1}} u_T(k+n-1) + b_{T_{n-2}} u_T(k+n-2) + \cdots + b_{T_0} u_T(k) \qquad \text{(5-9a)}$$

The $2n$ coefficients $\{a_{T_{n-1}}, \ldots, a_{T_0}\}$ and $\{b_{T_{n-1}}, \ldots, b_{T_0}\}$ are constant for the linear time-invariant system, but they are dependent on the sample duration T. They are not equal to the $(n + m + 1)$ coefficients $\{a_{n-1}, \ldots, a_0\}$ and $\{b_m, \ldots, b_0\}$ of the continuous-time model. The purpose of this first section is to see just how to derive the model Eq. (5-9a) from a given continuous-time model.

When properly computed, the representation Eq. (5-9a) is exact; it provides the exact sampled output $y(kT) = y_T(k)$ for the piecewise constant input $u_T(k)$. Thus Eq. (5-9a) is a desirable alternative to Eq. (5-5) for digital simulation of the continuous system Eq. (5-1). For digital simulation it is convenient to shift all the variables by n time units and solve for $y_T(k)$:

$$y_T(k) = -a_{T_{n-1}} y_T(k-1) - a_{T_{n-2}} y_T(k-2) - \cdots - a_{T_0} y_T(k-n)$$
$$+ b_{T_{n-1}} u_T(k-1) + b_{T_{n-2}} u_T(k-2) + \cdots + b_{T_0} u_T(k-n) \qquad \text{(5-9b)}$$

Simulation of Eq. (5-9b) on the digital computer is quite efficient. Computing $y_T(k) = y(kT)$ at each time k requires saving n past values of the output $\{y_T(k-1), y_T(k-2), \ldots, y_T(k-n)\}$ and n past values of the input $\{u_T(k-1), u_T(k-2), \ldots, u_T(k-n)\}$. This is equivalent to knowledge of the boundary conditions $\{y^{(n-1)}(t_0), y^{(n-2)}(t_0), \ldots, y(t_0)\}$ and $\{u^{(m-1)}(t_0), u^{(m-2)}(t_0), \ldots, u(t_0)\}$ in either the classic or Laplace transform continuous-time solution methods. It is also equivalent to knowledge of the system state, as will be seen in later chapters.

What is left unresolved is just how to compute the coefficients $\{a_{T_{n-1}}, a_{T_{n-2}}, \ldots, a_{T_0}\}$ and $\{b_{T_{n-1}}, b_{T_{n-2}}, \ldots, b_{T_0}\}$. There are any number of ways to derive an approximate-difference equation, Eq. (5-9), from the continuous-time model, Eq. (5-1), but one method that is especially illuminating is derived by considering the response to the unit pulse input

$$\delta_T(t) \equiv \begin{cases} 1 & 0 < t \leq T \\ 0 & \text{otherwise} \end{cases} \qquad \text{(5-10)}$$

Furthermore, this method is "exact" regardless* of the sample spacing, T.

If the digital pulse input, Eq. (5-10), is applied to the system, Eq. (5-9b), then $u_T(0) = 1$ and $u_T(k) = 0$, $k > 0$. Also the output to the unit pulse input is the discrete pulse-response sequence. Thus $y_T(k) = h_T(k)$, $k = 1, 2, 3, \ldots$. The discrete pulse-response sequence may be determined from Eq. (5-6) or by actually forcing the physical system with the digital pulse input.†

*There may be considerable computational sensitivity (small errors in the data may cause large errors in the computed model), but theoretically it is exact. More is discussed on this shortly.

†See Chap. 1 for further discussion. Also see Chap. 8 for state-variable methods to compute the discrete pulse-response sequence.

Stepping forward in time with Eq. (5-9b) and using the facts that $u_T(0) = 1$, $u_T(i) = 0$ for $i = 1, 2, \ldots$, and $h_T(k) = y_T(k)$ gives

$$h_T(1) = b_{T_{n-1}} \tag{5-11a}$$

$$h_T(2) = -a_{T_{n-1}} h_T(1) + b_{T_{n-2}} \tag{5-12a}$$

$$h_T(3) = -a_{T_{n-1}} h_T(2) - a_{T_{n-2}} h_T(3) + b_{T_{n-3}} \tag{5-13a}$$

....

$$h_T(n) = -a_{T_{n-1}} h_T(n-1) - \cdots - a_{T_1} h_T(1) + b_{T_0} \tag{5-14a}$$

$$h_T(n+1) = -a_{T_{n-1}} h_T(n) - \cdots - a_{T_1} h_T(2) - a_{T_0} h_T(1) \tag{5-15}$$

....

$$h_T(2n) = -a_{T_{n-1}} h_T(2n-1) - \cdots - a_{T_1} h_T(n+1) - a_{T_0} h_T(n) \tag{5-16}$$

There are $2n$ unknowns, $\{a_{T_0}, a_{T_1}, \ldots, a_{T_{n-1}}\}$ and $\{b_{T_0}, b_{T_1}, \ldots, b_{T_{n-1}}\}$, and $2n$ equations. These are represented by Eqs. (5-11a) to (5-16). The n equations represented by Eqs. (5-15) and (5-16) are conveniently set up as a linear equation problem:

$$\begin{bmatrix} h_T(1) & h_T(2) & \cdots & h_T(n) \\ h_T(2) & h_T(3) & \cdots & h_T(n+1) \\ \vdots & \vdots & & \vdots \\ h_T(n) & h_T(n+1) & \cdots & h_T(2n-1) \end{bmatrix} \begin{bmatrix} a_{T_0} \\ a_{T_1} \\ \vdots \\ a_{T_{n-1}} \end{bmatrix} = \begin{bmatrix} -h_T(n+1) \\ -h_T(n+2) \\ \vdots \\ -h_T(2n) \end{bmatrix} \tag{5-17}$$

This linear equation problem may be solved by use of readily available software algorithms (e.g., Refs. 1 and 2).

The $n \times n$ coefficient matrix in Eq. (5-17)

$$\mathcal{H}_T \equiv \begin{bmatrix} h_T(1) & h_T(2) & \cdots & h_T(n) \\ h_T(2) & h_T(3) & \cdots & h_T(n+1) \\ \cdots & \cdots & \cdots & \cdots \\ h_T(n) & h_T(n+1) & \cdots & h_T(2n-1) \end{bmatrix} \tag{5-18}$$

is actually a very special matrix with special system theoretic properties. It is called the system *Hankel matrix*. It is interesting that it arises in this application. More is discussed about the Hankel matrix in later chapters.*

*The Hankel matrix is encountered again in Chaps. 8 and 11. Also see Appendix E for discussion of numerical problems that might arise in solution of the linear equation problem, Eq. (5-17), because of "ill-conditioning" properties of the Hankel matrix.

The a_T coefficients may be solved from Eq. (5-17). Once these are found the b_T coefficients may be found from Eqs. (5-11a) through (5-14a). Namely,

$$b_{T_{n-1}} = h_T(1) \tag{5-11b}$$

$$b_{T_{n-2}} = h_T(2) + a_{T_{n-1}} h_T(1) \tag{5-12b}$$

$$b_{T_{n-3}} = h_T(3) + a_{T_{n-1}} h_T(2) + a_{T_{n-2}} h_T(1) \tag{5-13b}$$

$$\cdots\cdots\cdots\cdots\cdots\cdots\cdots\cdots\cdots\cdots$$

$$b_{T_0} = h_T(n) + a_{T_{n-1}} h_T(n-1) + \cdots + a_{T_1} h_T(1) \tag{5-14b}$$

In summary, all of the coefficients in the finite-difference equation, Eq. (5-9), may be determined directly from the discrete pulse-response sequence parameters, $h_T(k)$, $k = 1, 2, 3, \ldots$. It appears that we have come full circle yet another time. Furthermore, the coefficients make Eq. (5-9) an exact system representation for the digital pulse inputs. An optimum sample time, T, may also be determined to minimize errors in solving Eq. (5-17) for the a_T's. This is dependent on properties of the coefficient Hankel matrix, Eq. (5-18), as the sample time T is varied. These properties dependent on T are discussed in Chap. 11. Finally, it is important to remark that this method provides an experimental method to find the finite-difference equation; Chap. 1 discusses experimental methods to calculate the $h_T(k)$, and these are all that are required in Eq. (5-17) and in Eqs. (5-11b) through (5-14b) to find the finite-difference equation model.

Only one piece of information in Eq. (5-17) and Eqs. (5-11b) through (5-14b) is missing from a *pure* experimental determination of the finite-difference equation model from input-output measurements of an unknown linear dynamic system. This is what the system model order n should be. Interestingly, this key piece of information may be determined from the Hankel matrix, Eq. (5-18). The theory of this must await later chapters, as it is dependent on system properties known as controllability and observability. These properties are discussed in Chaps. 8 and 11 in connection with state-variable methods.

Example 5-1 The rocket-sled dynamic system introduced in Chap. 1 has the differential equation

$$\dddot{y}(t) + 0.4\ddot{y}(t) + 1.14\dot{y}(t) + 0.22y(t) = u(t)$$

and continuous impulse response function

$$h(t) = 0.90909e^{-0.2t} - 0.90909e^{-0.1t}\cos(1.044t) + 0.087078e^{-0.1t}\sin(1.044t)$$

Using a sample time $T = 1$, the discrete pulse-response sequence as determined in Chap. 1 has values

$$h_T(1) = 0.14287089 \qquad h_T(4) = 0.9600685$$
$$h_T(2) = 0.73603021 \qquad h_T(5) = 0.3299832$$
$$h_T(3) = 1.1658266 \qquad h_T(6) = -0.1493102$$

These values could be determined by integrating $h(t)$ and using Eq. (5-6), but usually it is more convenient to sample the unit step response, $r(t)$ and use the relation Eq. (5-7). More values are tabulated in Table 1-1, Chap. 1, but the first six values are all that are required to find the six unknown coefficients in the third-order discrete-time finite-difference equation

$$y_T(k) = -a_{T_2}y_T(k-1) - a_{T_1}y_T(k-2) - a_{T_0}y_T(k-3)$$
$$+ b_{T_2}u_T(k-1) + b_{T_1}u_T(k-2) + b_{T_0}u_T(k-3)$$

Using Eq. (5-17), the a_T's are found from

$$\begin{bmatrix} 0.14287089 & 0.73603021 & 1.1658266 \\ 0.73603021 & 1.1658266 & 0.9600685 \\ 1.1658266 & 0.9600685 & 0.3299832 \end{bmatrix} \begin{bmatrix} a_{T_0} \\ a_{T_1} \\ a_{T_2} \end{bmatrix} = \begin{bmatrix} -0.9600685 \\ -0.3299832 \\ 0.1493102 \end{bmatrix}$$

The coefficients are found to be

$$a_{T_0} = -0.67032471$$
$$a_{T_1} = 1.5636163$$
$$a_{T_2} = -1.7285310$$

The b_T's are then found from Eq. (5-11b) through (5-14b).

$$b_{T_2} = h_T(1) = 0.14287089$$
$$b_{T_1} = h_T(2) + a_{T_2}h_T(1) = 0.48907344$$
$$b_{T_0} = h_T(3) + a_{T_2}h_T(2) + a_{T_1}h_T(1) = 0.11697082$$

The third-order finite-difference model would give exactly the same discrete output points, $y_T(k) = y(kT)$, as the discrete convolution model

$$y_T(k) = \sum_{i=-\infty}^{k-1} h_T(k-i)u_T(i)$$

for a sequence of pulse inputs $u_T(i)$. Since the rocket-sled system is so lightly damped, it takes a long time before $|h_T(i)| \simeq 0$ for all $i > N$. Indeed, even for $T = 1$, $N = 64$, $h_T(64) = 0.0014$, which is still not insignificant. Thus the rocket sled has a long effective memory length, and the discrete convolution would not at all be efficient. The third-order finite-difference equation applicable to the third-order system requires only three past values of the input and three past values of the output to compute the next output. This is quite a compression of information from the discrete pulse-response model and is easily programmed on the digital computer.

To see that the finite-difference equation does indeed yield the sampled output at the discrete sample times kT

$$y_T(k) = y(kT)$$

for a piecewise constant sample-and-hold-type input $u_T(k)$, the rocket-sled finite-difference equation is used to find the system unit step response.

For the unit step response, the initial boundary conditions are all zero and $u_T(k) = 1$, $k = 0, 1, 2, \ldots$. Thus using the values obtained in Example 5-1, the finite-difference equation is solved forward in time

$$y_T(1) = b_{T_2} = 0.14287089$$

$$y_T(2) = -a_{T_2} y_T(1) + b_{T_2} + b_{T_1} = 0.87890110$$

$$\begin{aligned} y_T(3) &= -a_{T_2} y_T(2) - a_{T_1} y_T(1) + b_{T_2} + b_{T_1} + b_{T_0} \\ &= 2.0447277 \end{aligned}$$

$$\begin{aligned} y_T(4) &= -a_{T_2} y_T(3) - a_{T_1} y_T(2) - a_{T_0} y_T(1) + b_{T_2} + b_{T_1} + b_{T_0} \\ &= 3.0048238 \end{aligned}$$

and so on for all $k = 1, 2, 3, \ldots$. The result is shown plotted in Fig. 5-3 along with the continuous-time step response for the rocket-sled system. The results are seen to be "exact."

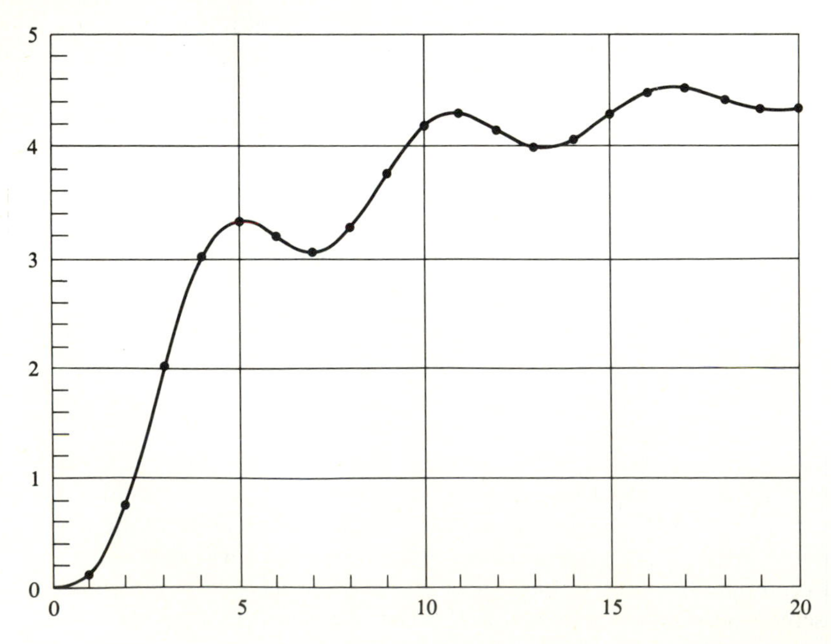

Figure 5-3 Rocket-sled step response and sampled values from finite-difference equation.

Improper, Feedforward Systems

To examine the improper feedforward system we consider a simple second-order system example

$$\ddot{y}(t) + 3\dot{y}(t) + 2y(t) = 10\ddot{u}(t) + 70\dot{u}(t) + 120u(t) \tag{5-19}$$

The transfer function is

$$H(s) = \frac{10s^2 + 70s + 120}{s^2 + 3s + 2}$$

By long division of polynomials, $H(s)$ is found to equal

$$H(s) = \frac{40s + 100}{s^2 + 3s + 2} + 10$$

With partial fraction expansions this becomes

$$H(s) = \frac{60}{s + 1} + \frac{-20}{s + 2} + 10$$

The impulse response function is

$$h(t) = 60e^{-t} - 20e^{-2t} + 10\delta(t)$$

with an impulse at time zero. In the output response the sampling property of the delta function gives

$$\begin{aligned} y(t) &= \int_{-\infty}^{t} h(t - \tau)u(\tau)\, d\tau \\ &= \int_{-\infty}^{t} [60e^{-(t-\tau)} - 20e^{-2(t-\tau)} + 10\delta(t - \tau)]u(\tau)\, d\tau \\ &= \int_{-\infty}^{t} [60e^{-(t-\tau)} - 20e^{-2(t-\tau)}]u(\tau)\, d\tau + 10u(t) \end{aligned}$$

The $10u(t)$ term is the direct feedforward of the input to the output.

This feedforward term also appears in the finite-difference equation. Thus for this second-order system the finite-difference equation has the form

$$\begin{aligned} y_T(k) = -a_{T_1}y_T(k-1) - a_{T_0}y_T(k-2) + b_{T_1}u_T(k-1) \\ + b_{T_0}u_T(k-2) + 10\begin{cases} u_T(k-1) \\ \text{or} \\ u_T(k) \end{cases} \end{aligned} \tag{5-20}$$

The last term is the feedforward term of the direct input to output. Really the feedforward term at instant $t = kT$ is $10u(kT)$. Thus it depends on whether

$$u(kT) = \begin{cases} u_T(k-1) \\ \text{or} \\ u_T(k) \end{cases}$$

as to which value should represent the feedforward term in calculating $y_T(k) = y(kT)$. This is usually a matter of convention, and which way it is implemented does not matter so long as there is consistency.

To calculate the a_T and b_T coefficients, the relations Eq. (5-17) and Eqs. (5-11*b*) through (5-14*b*) are used with the impulse response function *without* the impulse. Thus

$$h_T(k) = \int_{(k-1)T}^{kT} (60e^{-\tau} - 20e^{-2\tau})\, d\tau \qquad (5\text{-}21)$$

for $k = 1, 2, 3, 4$. Note that only $2n = 4$ values of the discrete pulse-response sequence are required to calculate the four unknown coefficients of the finite-difference equation.

Choosing $T = 0.2$ gives the discrete pulse-response sequence

$$h_T(1) = 7.579 \qquad h_T(3) = 5.809$$

$$h_T(2) = 6.695 \qquad h_T(4) = 4.976$$

The finite-difference equation corresponding to Eq. (5-21) is

$$y_T(k) = -a_{T_1} y_T(k-1) - a_{T_0} y_T(k-2) + b_{T_1} u_T(k-1) + b_{T_0} u_T(k-2)$$

The a_T coefficients are found from

$$\begin{bmatrix} 7.579 & 6.695 \\ 6.695 & 5.809 \end{bmatrix} \begin{bmatrix} a_{T_0} \\ a_{T_1} \end{bmatrix} = \begin{bmatrix} -5.809 \\ -4.976 \end{bmatrix}$$

Solving this linear equation problem yields

$$\begin{bmatrix} a_{T_0} \\ a_{T_1} \end{bmatrix} = \begin{bmatrix} 0.549 \\ -1.489 \end{bmatrix}$$

The b_T coefficients are then found from

$$b_{T_1} = h_T(1) = 7.579$$

$$b_{T_0} = h_T(2) + a_{T_1} h_T(1) = -4.591$$

Using these results, the finite-difference equation, Eq. (5-20), becomes

$$y_T(k) = 1.489 y_T(k-1) - 5.49 y_T(k-2) + 7.579 u_T(k-1)$$

$$- 4.591 u_T(k-2) + 10 \begin{cases} u_T(k-1) \\ \text{or} \\ u_T(k) \end{cases} \qquad (5\text{-}22)$$

The same procedure may be extended to any order improper system.

Zero-Input and Zero-State Response

The finite-difference equation, Eq. (5-9*b*), may be used to compute the discrete-time zero-input and zero-state responses, $y_{T_{zi}}(k)$ and $y_{T_{zs}}(k)$. Let $k = 0$ be the initial time and let $\{y_T(-1), y_T(-2), \ldots, y_T(-n)\}$ and $\{u_T(-1), u_T(-2), \ldots, u_T(-n)\}$ be prior output and input data. Then the zero-input response, $k = 1, 2, \ldots$, is computed from

$$y_{T_{zi}}(k) = -a_{T_{n-1}} y_{T_{zi}}(k-1) - a_{T_{n-2}} y_{T_{zi}}(k-2) - \cdots - a_{T_0} y_{T_{zi}}(k-n) + b_{T_{n-1}} u_T(k-1) + b_{T_{n-2}} u_T(k-2) + \cdots + b_{T_0} u_T(k-n) \quad (5\text{-}23)$$

where $u_T(i) \equiv 0$ for $i = 0, 1, 2, \ldots$. The zero-state response, $k = 1, 2, \ldots$, is computed from

$$y_{T_{zs}}(k) = -a_{T_{n-1}} y_{T_{zs}}(k-1) - a_{T_{n-2}} y_{T_{zs}}(k-2) - \cdots - a_{T_0} y_{T_{zs}}(k-n) + b_{T_{n-1}} u_T(k-1) + b_{T_{n-2}} u_T(k-2) + \cdots + b_{T_0} u_T(k-n) \quad (5\text{-}24)$$

where $y_{T_{zs}}(i) \equiv 0$ and $u_T(i) \equiv 0$ for $i = -1, -2, \ldots, -n$. Like the continuous-time system

$$y_T(k) = y_{T_{zi}}(k) + y_{T_{zs}}(k) \quad (5\text{-}25)$$

Again like the continuous-time system, the total solution may be decomposed into the homogeneous and particular solution responses

$$y_T(k) = y_{T_h}(k) + y_{T_p}(k) \quad (5\text{-}26)$$

This topic is discussed in the next section.

5-2 THE HOMOGENEOUS RESPONSE

The homogeneous finite-difference equation for the nth-order system corresponding to Eq. (5-9*a*) is

$$y_{T_h}(k+n) + a_{T_{n-1}} y_{T_h}(k+n-1) + \cdots + a_{T_0} y_{T_h}(k) = 0 \quad (5\text{-}27a)$$

Defining the operator variable z such that

$$z y_T(k) = y_T(k+1) \qquad z^2 y_T(k) = y_T(k+2) \qquad \text{etc.}$$

the homogeneous difference equation may be written as

$$(z^n + a_{T_{n-1}} z^{n-1} + \cdots + a_{T_0}) y_{T_h}(k) = 0 \quad (5\text{-}27b)$$

The polynomial

$$\Delta_T(z) = z^n + a_{T_{n-1}} z^{n-1} + \cdots + a_{T_0} \quad (5\text{-}28a)$$

is the characteristic polynomial of the finite-difference equation. Define p_{T_i} $i = 1, 2, \ldots, n$, as the n roots of the characteristic polynomial

$$\Delta_T(z) = \prod_{i=1}^{n} (z - p_{T_i})^n \quad (5\text{-}28b)$$

Then $\Delta_T(p_{T_i}) = 0$, and it is true that

$$y_{T_h}(k) = c_i p_{T_i}^k$$

is a solution of the homogeneous finite-difference equation, Eq. (5-27). Substitution of this claimed solution into Eq. (5-27) gives

$$c_i(p_{T_i}^{k+n} + a_{T_{n-1}} p_{T_i}^{k+n-1} + \cdots + a_{T_0} p_{T_i}^k)$$
$$= c_i p_{T_i}^k (p_{T_i}^n + a_{T_{n-1}} p_{T_i}^{n-1} + \cdots + a_{T_0} p_{T_i})$$

The term inside the parentheses is

$$\Delta_T(z)|_{z=p_{Ti}} = 0$$

so the claimed solution does indeed satisfy the homogeneous difference equation.

The most general form of the homogeneous solution for the case with distinct roots of $\Delta_T(z)$ is

$$y_{T_h}(k) = c_1 p_{T_1}^k + c_2 p_{T_2}^k + \cdots + c_n p_{T_n}^k \tag{5-29}$$

where $c_1, c_2, \ldots, c_n$ are arbitrary constants yet to be determined based on boundary conditions $y_T(0), y_T(-1), \ldots, y_T(-n+1), u_T(-1), \ldots, u_T(-n+1)$.

If root p_{T_1} is repeated, say, m times, then the general form of the homogeneous solution is

$$y_{T_h}(k) = c_1 p_{T_1}^k + c_2 k p_{T_1}^k + \cdots + c_m k^{m-1} p_{T_1}^k$$
$$+ c_{m+1} p_{T_{m+1}} + \cdots + c_n p_{T_n} \tag{5-30}$$

If p_{T_i} is complex, $p_{T_i} = \rho_i e^{\theta_i j}$, then $p_{T_{i+1}}$ is its complex conjugate, $p_{T_{i+1}} = \rho_i e^{-\theta_i j}$. Thus, as discussed in Chap. 2, there are several alternative forms of the solution

$$c_i p_{T_i}^k + c_{i+1} p_{T_{i+1}}^k = c_i' \rho_i^k \cos(\theta_i k) + c_{i+1}' \rho_i^k \sin(\theta_i k)$$
$$= c_i'' \rho_i^k \sin(\theta_i k + c_{i+1}'')$$

In the alternative forms c_i', c_{i+1}' and c_i'', c_{i+1}'' are both real-valued arbitrary constants, while c_i and c_{i+1} are complex conjugate pairs. These results are all exactly analogous to the results on the homogeneous solution from Chap. 2.

5-3 STABILITY

The total solution to the finite-difference equation Eq. (5-9) has the form

$$y_T(k) = y_{T_h}(k) + y_{T_p}(k) \tag{5-26}$$

where $y_{T_p}(k)$ is the particular solution forced by the input. Like the continuous-time case the homogeneous or "natural" solution determines whether or not the system is stable. From Eq. (5-29) or Eq. (5-30) it follows that

$$\lim_{k\to\infty} |y_{T_h}(k)| = 0$$

and the system is stable for all initial conditions if and only if $|p_{T_i}| < 1$ for all $i = 1, 2, \ldots, n$. If there is any one $|p_{T_i}| > 1$, then the limit of $p_{T_i}^k$ as $k \to \infty$ is infinite, and the system is unstable. If $|p_{T_i}| = 1$ and p_{T_i} is distinct, the limit of the corresponding $p_{T_i}^k$ term has a constant magnitude of unity. The system is said to be marginally stable. If $|p_{T_i}| = 1$ and the root is multiple, then the system is unstable.

In summary, the condition for stability of the finite-difference equation is that all roots of the characteristic polynomial lie inside the unit circle. This is equivalent to the stability condition that all poles be in the left half-plane for the continuous-time system.

5-4 RELATION BETWEEN THE CONTINUOUS POLES, p_i, AND THE DISCRETE POLES, p_{T_i}

The solution to the homogeneous finite-difference equation, Eq. (5-29), is equal to the sampled continuous time homogeneous solution

$$y_h^{(n)}(t) + a_{n-1}y_h^{(n-1)}(t) + \cdots + a_1 y_h^{(1)}(t) + a_0 y_h(t) = 0 \tag{5-31}$$

Suppose that the characteristic polynomial to Eq. (5-31) has n distinct poles $p_1, p_2, \ldots, p_n$; then it is true that

$$\begin{aligned} y_h(kT) &= c_1 e^{p_1 kT} + c_2 e^{p_2 kT} + \cdots + c_n e^{p_n kT} \\ &= y_{T_h}(k) \\ &= c_1 p_{T_1}^k + c_2 p_{T_2}^k + \cdots + c_n p_{T_n}^k \end{aligned} \tag{5-32}$$

This is true for all arbitrary constants $c_1, c_2, \ldots, c_n$. It thus follows that the roots of the finite-difference equation characteristic polynomial are related to the continuous system poles by

$$\boxed{p_{T_i} = e^{p_i T} \qquad i = 1, 2, \ldots, n} \tag{5-33}$$

This is also true in the repeated root case. Suppose that p_1 is repeated m times. The corresponding terms in the homogeneous solution are

$$y_h(t) = c_1 e^{p_1 t} + c_2 t e^{p_1 t} + \cdots + c_m t^{m-1} e^{p_1 t} \tag{5-34}$$

Sampling at $t = kT$ gives

$$y_h(kT) = c_1(e^{p_1 T})^k + (c_2 T)k(e^{p_1 T})^k + \cdots + (c_m T^{m-1})k^{m-1}(e^{p_1 T})^k$$

Defining the arbitrary constants $c_1, (c_2 T), \ldots, (c_m T^{m-1})$, this expression compares exactly equal to Eq. (5-30). Thus Eq. (5-33) is true regardless of whether or not the p_i are repeated.

Alternative Computation of the Finite-Difference Equation Coefficients

Eq. (5-33) provides an alternative way to find the a_T coefficients in the finite-difference equation. Suppose that the continuous system has the characteristic polynomial in factored form

$$\Delta(s) = \prod_{i=1}^{n} (s - p_i) \tag{5-35}$$

where all the p_i may not be distinct. Then

$$\boxed{\begin{aligned}\Delta_T(z) &= \prod_{i=1}^{n} [z - (e^{p_i T})] \\ &= z^n + a_{T_{n-1}} z^{n-1} + \cdots + a_{T_1} z + a_{T_0}\end{aligned}} \tag{5-36}$$

The b_T coefficients are still found from Eqs. (5-11*b*) through (5-14*b*).

Example 5-2 The rocket-sled system

$$\dddot{y}(t) + 0.4\ddot{y}(t) + 1.14\dot{y}(t) + 0.22y(t) = u(t)$$

has characteristic polynomial

$$\Delta(s) = (s + 0.2)(s + 0.1 + 1.044j)(s + 0.1 - 1.044j)$$

For $T = 1$ the discrete poles are

$$\begin{aligned}p_{T_1} &= e^{-0.2} = 0.81873075 \\ p_{T_2} &= e^{-0.1-1.044j} = e^{-0.1}[\cos(1.044) - \sin(1.044)j] \\ &= 0.45492203 - 0.78216155j \\ p_{T_3} &= e^{-0.1-1.044j} = e^{-0.1}[\cos(1.044) + \sin(1.044)j] \\ &= 0.45492203 + 0.78216155j\end{aligned}$$

Then

$$\begin{aligned}\Delta_T(z) &= (z - p_{T_1})(z - p_{T_2})(z - p_{T_3}) \\ &= z^3 - 1.7285748z^2 + 1.5636481z - 0.67032005\end{aligned}$$

Therefore, the a_T coefficients in the finite-difference equation

$$y_T(k+3) + a_{T_2} y_T(k+2) + a_{T_1} y_T(k+1) + a_{T_0} y_T(k) = b_{T_2} u_T(k+2) + b_{T_1} u_T(k+1) + b_{T_0} u_T(k)$$

are

$$a_{T_2} = -1.7285748 \qquad a_{T_1} = 1.5636481 \qquad a_{T_0} = -0.67032005$$

The homogeneous finite-difference equation is thus

$$y_{T_h}(k+3) - 1.7285748y_{T_h}(k+2) + 1.563481y_{T_h}(k+1) - 0.67032005y_{T_h}(k) = 0$$

with solution

$$y_{T_h}(k) = c_1(e^{-0.2})^k + c_2(e^{-0.1-1.044j})^k + c_3(e^{-0.1+1.044j})^k$$

Clearly $y_{T_h}(k)$ is equal to the sampled homogeneous response, $y_h(kT)$.

The b_T coefficients are still found from Eqs. (5-11*b*) through (5-15*b*) knowing $h_T(1)$, $h_T(2)$, $h_T(3)$. Using

$$h_T(1) = 0.14287089 \qquad h_T(2) = 0.73603021 \qquad h_T(3) = 1.1658266$$

the b_T's are found to be

$$b_{T_2} = h_T(1) = 0.14287089$$

$$b_{T_1} = h_T(2) + a_{T_2}h_T(1) = 0.48906719$$

$$b_{T_0} = h_T(3) + a_{T_2}h_T(2) + a_{T_1}h_T(1) = 0.11694311$$

Note that the values of the a_T's and the b_T's found in Example 5.2 are slightly different from those found in Example 5.1 for the same rocket-sled finite-difference equation. In Example 5.1 the a_T's are found via solution of the linear equation problem Eq. (5-17), while in Example 5.2 the a_T's are found from Eq. (5-36), the discrete characteristic polynomial, using the relation $p_{T_i} = e^{p_i T}$. The latter approach would generally turn out to be more accurate; the linear equation problem Eq. (5-17) frequently is very sensitive to small errors (as introduced here by rounding off). The reasons for this sensitivity are examined later,* but it illustrates an important point: when we are dealing with finite wordlength computations on a digital computer, some methods of computation are better than others. The main advantage of using Eq. (5-17) for finding the a_T's is that it can be based purely on an experimental determination of the $h_T(k)$. One does not need to know the continuous system poles, p_i, or a continuous-time math model. If one has a continuous math model and knows the system poles, then Eq. (5-37) would generally be a more accurate way to compute the a_T coefficients in the finite-difference equation. The b_T coefficients are found from Eqs. (5-11*b*) through (5-14*b*) regardless.

Experimental Determination of the Continuous System Poles

Equation (5-33) produces another interesting result. Equation (5-33) may be inverted to give†

*See Sec. 11-7 and Appendix E.

†See Probs. 5-15 and 5-16.

$$\boxed{p_i = \frac{\ln(p_{T_i})}{T}} \tag{5-37}$$

Since the p_{T_i} may be determined by an experimental determination of the discrete pulse-response model, the $h_T(i)$, $i = 1, 2, \ldots, n$, Eq. (5-37) provides a useful first guess of the continuous-time system poles. The terminology *first guess* is used because there could be considerable error introduced in the procedure.

Example 5-3 Suppose that experimental determination of the discrete pulse-response parameters of the rocket-sled system finds the $h_T(k)$ accurate to only three significant figures

$$h_T(1) = 0.143 \qquad h_T(4) = 0.960$$

$$h_T(2) = 0.736 \qquad h_T(5) = 0.330$$

$$h_T(3) = 1.17 \qquad h_T(6) = -0.149$$

instead of eight as used in Example 5.1. The three a_T coefficients are found from Eq. (5-17) as

$$\begin{bmatrix} 0.143 & 0.736 & 1.17 \\ 0.736 & 1.17 & 0.960 \\ 1.17 & 0.960 & 0.330 \end{bmatrix} \begin{bmatrix} a_{T_0} \\ a_{T_1} \\ a_{T_2} \end{bmatrix} = \begin{bmatrix} -0.960 \\ -0.330 \\ 0.149 \end{bmatrix}$$

Computed to four significant figures this gives

$$a_{T_0} = -0.6134 \qquad a_{T_1} = 1.479 \qquad a_{T_2} = -1.676$$

The characteristic polynomial of the finite-difference equation is

$$\Delta_T(z) = z^3 - 1.676z^2 + 1.479z - 0.6134$$

Factoring $\Delta_T(z)$ gives

$$\Delta_T(z) = (z - 0.7871)(z - 0.4445 + 0.7627j)(z - 0.4445 - 0.7627j)$$

The three roots are

$$p_{T_1} = 0.7871 \qquad p_{T_2} = 0.4445 - 0.7627j \qquad p_{T_3} = 0.4445 + 0.7627j$$

With $T = 1$ the estimated pole $\hat{p}_1$ is

$$\hat{p}_1 = \ln(0.7871) = -0.2394$$

The true pole p_1 is -0.2. To handle the complex conjugate poles p_{T_2}, p_{T_3}, they are put into polar form

$$p_{T_{2,3}} = 0.8828e^{\pm 1.043j}$$

Taking the logarithm gives

$$\hat{p}_{2,3} = \ln(0.8828) \pm \ln(e^{1.043j}) = -0.1247 \pm 1.043j$$

The true poles p_2 and p_3 are $-0.1 \pm 1.044j$.

The errors in the estimated poles are caused by the low accuracy in the "measured" discrete pulse response $h_T(k)$. However, they would still be a good first guess for more sophisticated estimation procedures.

Discrete System Stability Revisited

The result $p_{T_i} = e^{p_i T}$ provides additional insight into the stability condition on the finite-difference equation that $|p_{T_i}| < 1$ for all $i = 1, 2, \ldots, n$ roots of the characteristic polynomial, $\Delta_T(z)$. For a stable continuous system the poles p_i all lie in the left half-plane. But for $p_i = \sigma_i + j\omega_i$, the corresponding p_{T_i} root

$$p_{T_i} = e^{(\sigma_i T + j\omega_i T)}$$

has magnitude less than one if and only if $\sigma_i < 0$. Hence the finite-difference equation model is stable if and only if the continuous model is stable.

Analysis of the p_{T_i} Roots As T is Varied, $0 < T < \infty$

For real p_i poles, $p_{T_i} = e^{p_i T}$ is real and

$$\lim_{T\to 0} p_{T_i} = 1 \qquad \lim_{T\to\infty} p_{T_i} = 0$$

for a stable (negative) p_i.

For a complex pole, $p_i = \sigma_i + j\omega_i$, it is seen that

$$p_{T_i} = e^{\sigma_i T}[\cos(\omega_i T) + j \sin(\omega_i T)] \tag{5-38}$$

For $T = m\pi/2\omega_i$, $m = 1, 3, 5, \ldots$, the root p_{T_i} is purely imaginary and $|p_{T_i}| = e^{\sigma_i T} < 1$ if $\sigma_i < 0$.

For $T = m\pi/\omega_i$ the root $p_{T_i} = e^{\sigma_i T}(-1)^m$, $m = 1, 2, \ldots$, is real. In fact, the complex conjugate root

$$p_{T_{i+1}} = e^{\sigma_i T}[\cos(\omega_i T) - j \sin(\omega_i T)] \tag{5-39}$$

equals p_{T_i} at $T = m\pi/\omega_i$, and so the complex conjugate roots p_{T_i} and $p_{T_{i+1}}$ collapse to a single repeated root at $T = m\pi/\omega_i$. The natural oscillations of the homogeneous response for the continuous poles p_i and p_{i+1}

$$y_h(t) = c_i e^{\sigma_i t} e^{j\omega_i t} + c_{i+1} e^{\sigma_i t} e^{-j\omega_i t}$$
$$= c_i' e^{\sigma_i t} \cos(\omega_i t) + c_{i+1}' e^{\sigma_i t} \sin(\omega_i t)$$

are thus obscured at the sampling interval $T = m\pi/\omega_i$. This is illustrated in Fig. 5-4 for the rocket-sled impulse response with

$$T = \frac{\pi}{1.044} = 3.0091884$$

Figure 5-5 shows the movement of the rocket-sled discrete poles as T is varied from 0 to 5.

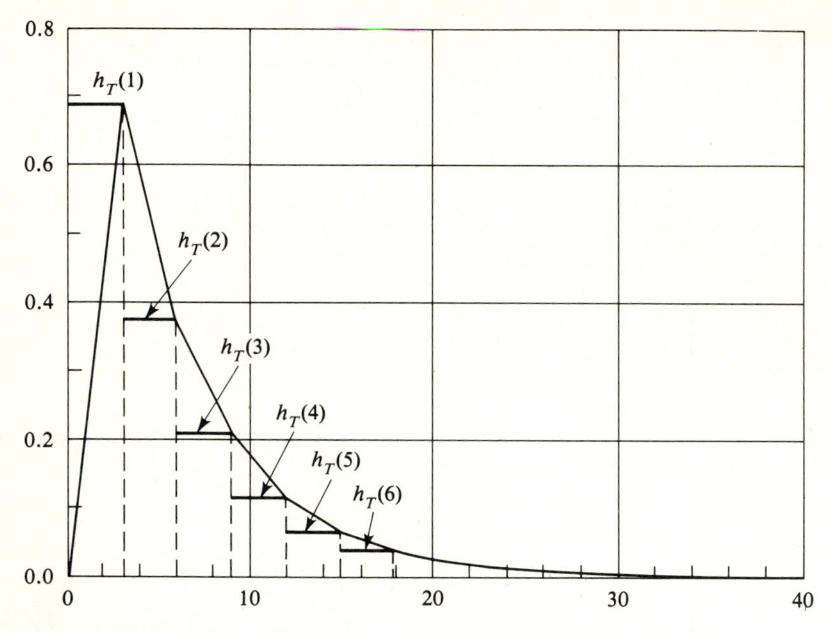

Figure 5-4 Discrete pulse-response sequence rocket sled system, $T = \pi/1.044$.

It is interesting to note that the sampling interval $T = \pi/2\omega_i$ is the Nyquist sampling interval for the frequency ω_i. Therefore, to accurately reproduce the homogeneous response, $y_h(t)$, from the sampled homogeneous response, $y_{T_h}(k) = y_h(kT)$, one should pick $T < \pi/2\omega_i$, for all poles $i = 1, 2, \ldots, n$. If this is satisfied, the roots of the finite-difference equation for a stable system satisfy

$$|p_{T_i} = e^{p_i T}| < 1 \qquad \text{and} \qquad \text{Re}\,\{p_{T_i} = e^{p_i T}\} > 0 \tag{5-40}$$

This is illustrated in Fig. 5-6.

5-5 THE PARTICULAR SOLUTION

Analogous to the continuous-time system, we find the particular solution to the finite-difference equation, Eq. (5-9), for only one piecewise constant input of the general form

$$u_T(k) = \alpha\, d^k \qquad k = 0, 1, 2, \ldots \tag{5-41}$$

where α is a real scalar multiplier and d is a given complex number. The particular solution must satisfy the finite-difference equation so that

$$y_{T_p}(k+n) + a_{T_{n-1}} y_{T_p}(k+n-1) + \cdots + a_{T_0} y_{T_p}(k)$$
$$= b_{T_{n-1}} \alpha\, d^{k+n} + b_{T_{n-2}} \alpha\, d^{k+n-1} + \cdots + b_{T_0} \alpha\, d^k \tag{5-42a}$$

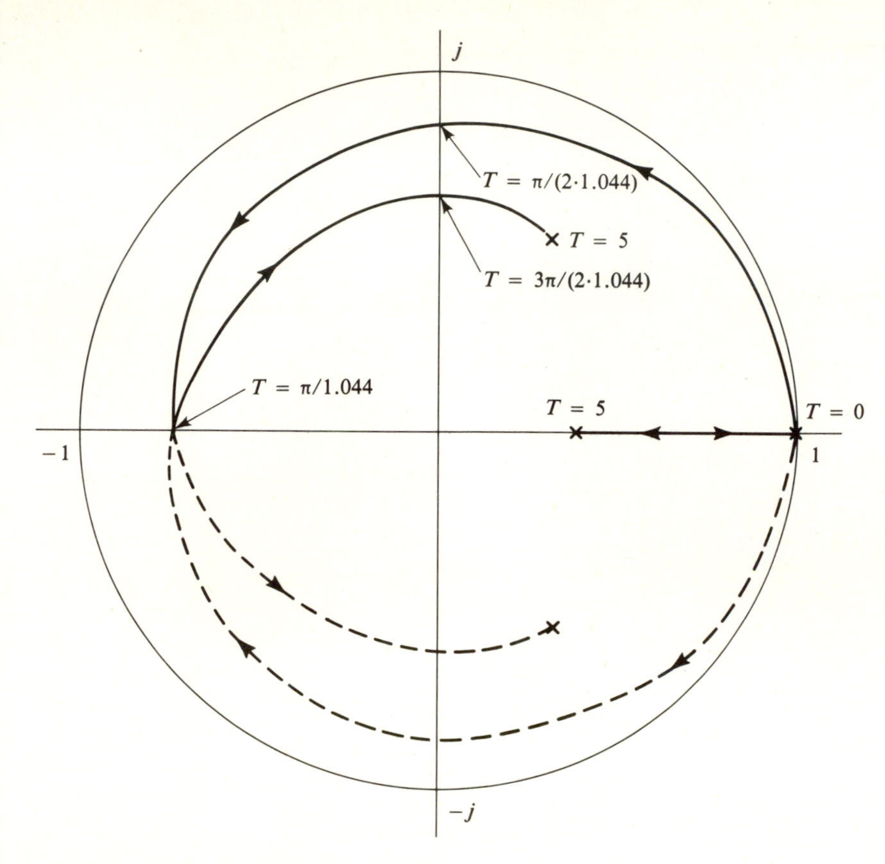

Figure 5-5 Locus of discrete poles, rocket-sled system, $0 \leq T \leq 5$.

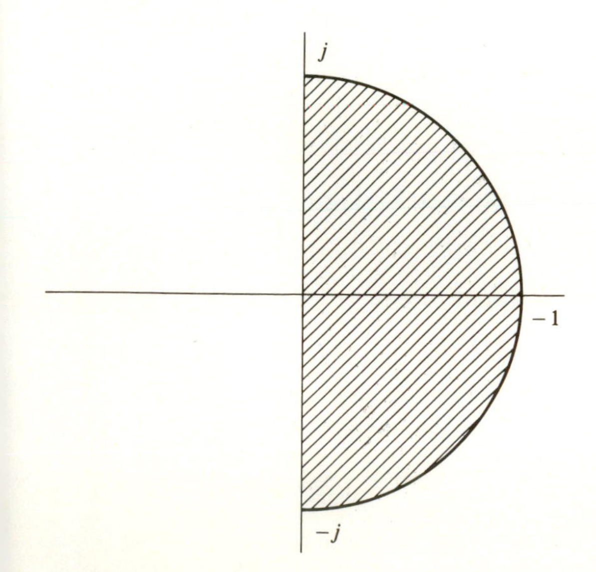

Figure 5-6 Discrete pole locations for stable system which satisfy Nyquist sampling criterion.

Using the z operator (see Sec. 5-2) this may be written as

$$(z^n + a_{T_{n-1}} z^{n-1} + \cdots + a_{T_0}) y_{T_p}(k) = (b_{T_{n-1}} z^{n-1} + b_{T_{n-2}} z^{n-2} + \cdots + b_{T_0}) \alpha\, d^k \tag{5-42b}$$

From Eq. (5-42b) it is apparent that

$$y_{T_p}(k) = \{H_T(z)\alpha z^k\}|_{z=d} \qquad k = 0, 1, 2, \ldots \quad d \neq p_{T_i} \quad i = 1, 2, \ldots, n \tag{5-43}$$

satisfies Eq. (5-42b) where

$$H_T(z) = \frac{b_{T_{n-1}} z^{n-1} + b_{T_{n-2}} z^{n-2} + \cdots + b_{T_0}}{z^n + a_{T_{n-1}} z^{n-1} + a_{T_{n-2}} z^{n-2} + \cdots + a_{T_0}} \tag{5-44}$$

is the finite-difference equation transfer function.

This is an exact analog of the results of Sec. 2-5 for the continuous input $u(t) = \alpha e^{qt}$ where $d = e^{qT}$. The major difference is that $u_T(k)$ represents a piecewise constant, sample-and-hold-type input. Thus

$$u_T(k) = u(kT) = \alpha e^{qkT}$$

at the sampling instants, but it is held constant between sampling instants.* Thus the particular solution $y_{T_p}(k)$ to the input $u_T(k) = \alpha e^{qkT}$ is not the sampled response $y_p(kT)$ to the input $u(t) = \alpha e^{qt}$. It would be the sampled particular solution response to piecewise constant input

$$u(t) \equiv \alpha e^{qkT} \qquad (k-1)T < t \leq kT \qquad k = 1, 2, \ldots$$

Linear Operations on the Input, $u_T(k) = \alpha\, d^k$

The particular solution to any input which may be written as

$$u_T(k) = L\{z^k\} \tag{5-45}$$

where $L\{\cdot\}$ is any linear operation is found to be

$$y_{T_p}(k) = L\{H_T(z) z^k\} \tag{5-46}$$

The derivation of this is directly analogous to that in Sec. 2-5 for the continuous-time particular solution. Furthermore, all of the linear operations considered in

*This is the way that we use the finite-difference equation here to represent the sampled response of an underlying continuous time system. There are situations where the finite-difference equation represents a pure discrete system in which there is no underlying continuous system. The results on the finite-difference equation still hold true but there is no applicable relation to an underlying continuous system.

Sec. 2-5 (e.g., differentiation, taking the real or imaginary parts, etc.) are directly applicable here.

Example 5-4 Consider the rocket-sled finite-difference equation for $T = 1$ found in Example 5-2:

$$y_T(k+3) - 1.7285748y_T(k+2) + 1.5636481y_T(k+1) - 0.67032005y_T(k)$$
$$= 0.14287089u_T(k+2) + 0.48906719u_T(k+1) + 0.11694311u_T(k)$$

and suppose that the input is the piecewise constant approximation to the input $u(t) = 3t$,

$$u_T(k) = \frac{3kT + 3(k+1)T}{2} \qquad k = 0, 1, 2, \ldots$$
$$= 3kT + \frac{T}{2}$$

This input may be written as

$$u_T(k) = 3T\frac{d}{dz}\{z^k\}\bigg|_{z=1} + \frac{T}{2}\{z^k\}\bigg|_{z=1}$$

Therefore, the particular solution is

$$y_{T_p}(k) = 3T\frac{d}{dz}\{H_T(z)z^k\}\bigg|_{z=1} + \frac{T}{2}\{H(z)z^k\}\bigg|_{z=1}$$
$$= 3TH_T(1)k + 3T\frac{d}{dz}\{H_T(z)\}\bigg|_{z=1} + \frac{T}{2}H_T(1)$$

where

$$H_T(z) = \frac{0.14287089z^2 + 0.73603021z + 1.1658266}{z^3 - 1.7285748z^2 + 1.5636481z - 0.67032005}$$

Performing the differentiation and evaluating at $z = 1$, $T = 1$ gives

$$y_{T_p}(k) = 13.636414k - 75.202185$$

To emphasize that $y_{T_p}(k)$ is not the sampled particular solution response to the continuous input $u(t) = 3t$, consider the continuous rocket-sled differential equation

$$\dddot{y}(t) + 0.4\ddot{y}(t) + 1.14\dot{y}(t) + 0.22y(t) = u(t)$$

with transfer function

$$H(s) = \frac{1}{s^3 + 0.4s^2 + 1.14s + 0.22}$$

By Table 2-2, the particular solution response to the input $u(t) = 3t$ is

$$y_p(t) = 3\frac{d}{ds}\{H(s)e^{st}\}\bigg|_{s=0}$$

$$= \frac{3}{0.22}t - \frac{(3)(1.14)}{0.22^2}$$

$$= 13.636364t - 70.661157$$

Sampling at kT, $T = 1$, gives

$$y_p(kT) = 13.636364k - 70.661157$$

The error between $y_{T_p}(k)$ and $y_p(kT)$ is that the former is found with the piecewise constant approximation to the $3t$ input, while the latter is the sampled particular solution to the actual continuous $3t$ input. If T is decreased and a new finite-difference equation is found for the new T, then the approximation between $y_{T_p}(k)$ and $y_p(kT)$ would get better and better.

Unit Step Input

One input for which the continuous input is exactly equal to its piecewise constant approximation is the unit step input

$$u_{-1}(kT) = u_T(k) = 1 = \{z^k\}|_{z=1} \qquad k = 0, 1, 2, \ldots \tag{5-47}$$

The particular solution response is

$$y_{T_p}(k) = \{H_T(z)z^k\}|_{z=1}$$

$$= H_T(1) \tag{5-48}$$

Like the continuous-time system, the particular solution is the steady-state response after the transients associated with the homogeneous response have decayed to zero. Therefore, for the unit step input it should be true that

$$y_p(t) = H(0) = \text{constant}$$

$$= y_{T_p}(k) = H_T(1) = \text{constant} \tag{5-49}$$

Unless there are errors in the calculations, this will be true.

Example 5-5 The rocket sled has transfer function

$$H(s) = \frac{1}{s^3 + 0.4s^2 + 1.14s + 0.22}$$

The particular solution to the unit step input is

$$y_p(t) = H(0) = \frac{1}{0.22} = 4.5454545$$

The finite-difference equation transfer function from Example 5-2 is

$$H_T(z) = \frac{0.14287089z^2 + 0.73603021z + 1.1658266}{z^3 - 1.7285748z^2 + 1.5636481z - 0.67032005}$$

The finite-difference equation particular solution to the unit step input is

$$y_{T_p}(k) = H_T(1) = 4.5454715$$

The two $y_p(t)$ and $y_{T_p}(k)$ are very nearly equal. This would indicate that the finite-difference equation model derived in Example 5-2 is very accurate. The particular solution value for the unit step input using the finite-difference equation derived in Example 5-1 is 4.5216692. This is not nearly so close to the actual value of 4.5454545.

*5-6 PARTICULAR SOLUTION TO DISCRETE SINUSOIDAL INPUT

The input

$$u_T(k) = \sin(\omega k T) = \operatorname{Im}\{(e^{j\omega T})^k\} \qquad k = 0, 1, 2, \ldots \tag{5-50}$$

is of a linear operation on d^k where $d = e^{j\omega T}$ and the linear operation is $L\{\cdot\} \equiv \operatorname{Im}\{\cdot\}$. This piecewise constant sample-and-hold input is illustrated in Fig. 5-7. Since $u_T(k)$ is of the form of a linear operation on d^k, the results of the previous section may be used to find the sampled particular solution (steady-state solution) response

$$y_p(kT) = y_{T_p}(k) = \operatorname{Im}\{H_T(z)z^k\}|_{z=e^{j\omega T}}$$

Since

$$\{H_T(z)\}|_{z=e^{j\omega T}} = H_T(e^{j\omega T})$$

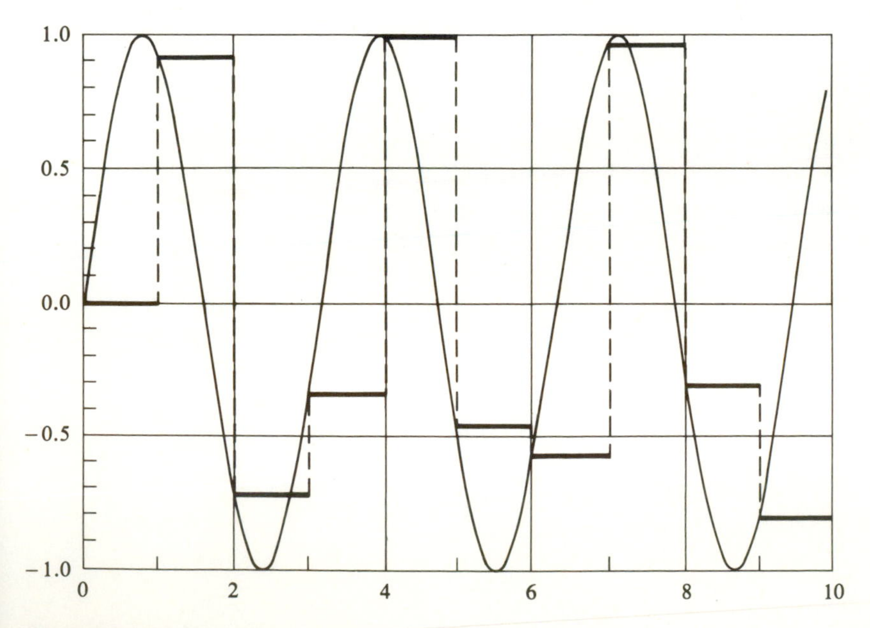

Figure 5-7 Piecewise constant sample-and-hold $u(t) = \sin(\pi k T)$, $T = 1$, input.

is simply a complex number, it may be written in terms of its magnitude and phase angle:

$$H_T(\beta) = |H_T(\beta)|e^{j\phi(\beta)}$$

where $\beta \equiv e^{j\omega T} = \cos(\omega T) + j\sin(\omega T)$

Then

$$y_p(kT) = y_{T_p}(k) = |H_T(\beta)| \sin[\omega kT + \phi(\beta)] \tag{5-51}$$

This form is exactly analogous to the form studied in Chap. 3 for the continuous sinusoid input. However, although $y_p(kT)$ is the sampled steady-state response to the piecewise constant sample-and-hold input as depicted in Fig. 5-7, the result Eq. (5-51) is not nearly so straightforward to analyze as the continuous-time case. Complete consideration of the frequency-response characteristics of the finite-difference equation is an involved topic area. The interested reader can find further discussion in Ref. 5.

The main complication arises in choice of the sample-and-hold time, T. Equation (5-51) is correct for any ω and any T, but the choice of the wrong T can obscure important phenomena.

5-7 THE COMPLETE SOLUTION

Again analogous to the classic solution method for the continuous-time system, the complete solution to the finite-difference equation Eq. (5-9) may be found as the sum of the homogeneous and particular solutions:

$$\begin{aligned} y_T(k) &= y_{T_h}(k) + y_{T_p}(k) \\ &= c_1 p_{T_1}^k + c_2 p_{T_2}^k + \cdots + c_n p_{T_n}^k + y_{T_p}(k) \end{aligned} \tag{5-52}$$

for the distinct root p_{T_i} case and similarly for the case with repeated roots. The particular solution corresponding to the forcing input $u_T(k)$, $k = 0, 1, 2, \ldots$, may be found as described in Sec. 5-5. The only remaining problem in finding the complete solution, Eq. (5-52), is to calculate the n arbitrary constants $c_1, c_2, \ldots, c_n$. These are found from boundary conditions on $\{y_T(0), y_T(-1), \ldots, y_T(-n+1)\}$ and $\{u_T(-1), u_T(-2), \ldots, u_T(-n+1)\}$. The procedure for doing this is illustrated in the following simple example. This procedure may be generalized to any problem for which one is able to find the particular solution.

Example 5-6 Consider the second-order finite-difference equation

$$y_T(k) = y_T(k-1) - 0.21y_T(k-2) + 3u_T(k-1) + 1.2u_T(k-2)$$

with discrete transfer function

$$H_T(z) = \frac{3z + 1.2}{z^2 - z + 0.21} = \frac{3z + 1.2}{(z - 0.3)(z - 0.7)}$$

Let $u_T(k) = (1)^k = 1$, $k = 0, 1, 2, \ldots$, the constant input. Also let boundary conditions on y_T and u_T be

$$y_T(0) = 3 \qquad y_T(-1) = -1 \qquad u_T(-1) = 4$$

Then the homogeneous solution is

$$y_{T_h}(k) = c_1(0.3)^k + c_2(0.7)^k$$

and the particular solution for $k = 1, 2, \ldots$, is

$$y_{T_p}(k) = H_T(1) = \frac{4.2}{0.21} = 20$$

Therefore,

$$y_T(k) = c_1(0.3)^k + c_2(0.7)^k + 20 \qquad k = 1, 2, \ldots$$

and the arbitrary constants c_1 and c_2 are yet to be determined.

The following procedure is used to find c_1 and c_2. Using the boundary conditions and the input $u_T(k)$, $k = 0, 1$, the finite-difference equation is stepped forward two steps to find $y_T(1)$ and $y_T(2)$. Only two steps are required, since there are only two unknown constants, c_1 and c_2. If the system were of a higher order n, then n steps forward would be computed. Using the known boundary conditions this gives

$$\begin{aligned} y_T(1) &= y_T(0) - 0.21y_T(-1) + 3u_T(0) + 1.2u_T(-1) \\ &= 11.01 \\ y_T(2) &= y_T(1) - 0.21y_T(0) + 3u_T(1) + 1.2u_T(0) \\ &= 14.58 \end{aligned}$$

These values for $y_T(1)$ and $y_T(2)$ are next equated to $y_T(1)$ and $y_T(2)$ as found by the sum of the homogeneous and particular solutions.

$$y_T(1) = 11.01 = c_1(0.3) + c_2(0.7) + 20$$

$$y_T(2) = 14.58 = c_1(0.3)^2 + c_2(0.7)^2 + 20$$

These may be formulated as a standard linear equation problem

$$\begin{bmatrix} 0.3 & 0.7 \\ 0.09 & 0.49 \end{bmatrix} \begin{bmatrix} c_1 \\ c_2 \end{bmatrix} = \begin{bmatrix} -8.99 \\ -5.42 \end{bmatrix}$$

Solving for c_1 and c_2 gives

$$c_1 = -7.275 \qquad c_2 = -9.725$$

The complete solution to the input $u_T(k) = 1$ for all $k = 1, 2, \ldots$ is thus

$$y_T(k) = -7.275(0.3)^k - 9.725(0.7)^k + 20$$

Using this solution, $y_T(k)$ may be solved for any value of k. However, for digital computations, the finite-difference equation itself is perhaps easier to solve on the computer than solving the sum of the homogeneous and particular solutions. The usefulness of the homogeneous and particular solutions lies in the analysis insights they provide and not particularly in digital simulation. The next section pursues another analysis technique applicable to the finite-difference equation—the z transform analysis.

5-8 THE z TRANSFORM

Analogous to the Laplace transform applied to the continuous linear time-invariant system, there is a transform theory applicable to the linear finite-difference equation. This is the z transform. The one-sided or unilateral z transform* of a discrete sequence, $f(k)$, $k = 0, 1, 2, \ldots$, is defined by

$$\mathcal{Z}\{f(k)\} \equiv F(z) = \sum_{k=0}^{\infty} f(k) z^{-k} \tag{5-53}$$

provided the summation converges. For the two-sided or bilateral z transform the summation is from $k = -\infty$ to $k = +\infty$. The region of convergence is the set of z in the complex plane in which the series Eq. (5-53) converges. This region is different for different $f(k)$.

Example 5-7 Let $f(k) = 1$, a constant for all $k > 0$. Then

$$\mathcal{Z}\{1\} \equiv \sum_{k=0}^{\infty} (1) z^{-k}$$

$$= 1 + \frac{1}{z} + \frac{1}{z^2} + \frac{1}{z^3} + \cdots$$

This is the geometric series, and it converges to

$$\mathcal{Z}\{1\} = \frac{1}{1 - (1/z)} = \frac{z}{z - 1} \qquad \text{for } |z| > 1$$

The region of convergence is for all complex z outside the unit circle.

Example 5-8 Let $u(k) = d^k$. This can represent any number of "exponential" sequences depending on the value of the constant d. If d is real and $d < 1$, then $d = e^{-qT}$ and $d^k = e^{-qkT}$ for $q > 0$; d^k represents a decaying exponential. If d is real and $d > 1$, then $d = e^{qT}$ and $d^k = e^{qkT}$ for $q > 0$; d^k represents a growing exponential. If d is complex, then d may be written in polar form as $d = \rho e^{j\phi}$ or in rectangular form as

$$d = \rho(\cos \phi + j \sin \phi)$$

Thus $$u(k) = d^k = \rho^k e^{jk\phi} = \rho^k(\cos k\phi + j \sin k\phi)$$

*Like the Fourier and Laplace transforms, the z transform is an invertible transform, but discussion of the inverse transform is quite complex. See Ref. 4, for example, for reference. For our purposes the simple table of transform pair relation, Table 5-1, suffices.

Whether the sequence is growing, decaying, or remaining constant depends on the magnitude ρ. This is the general form of the input as considered in the previous sections regarding the particular solution response.

Now consider the z transform of d^k:

$$\mathcal{Z}\{d^k\} \equiv \sum_{k=0}^{\infty} d^k z^{-k}$$

$$= 1 + \frac{d}{z} + \left(\frac{d}{z}\right)^k + \cdots$$

Again the series converges to

$$\mathcal{Z}\{d^k\} = \frac{1}{1-(d/z)} = \frac{z}{z-d}$$

if and only if $|d/z| < 1$. The region of convergence for z is all z outside the ball of radius $|d|$ in the complex plane.

Table 5-1 shows the z transform of some common sequences. This table may

Table 5-1 Table of z transforms

$f(k)$ with $f(k) = 0 \quad k < 0$	$F(z)$
1. $f(k) = \begin{cases} 1 & k = 0 \\ 0 & k > 0 \end{cases}$	1
2. $f(k) = 1 \quad k \geq 0$	$\dfrac{z}{z-1}$
3. $f(k) = k$	$\dfrac{z}{(z-1)^2}$
4. $f(k) = k^2$	$\dfrac{z(z+1)}{(z-1)^3}$
5. $f(k) = a^k$	$\dfrac{z}{z-a}$
6. $f(k) = 0 \quad k = 0;$ $f(k) = a^{k-1} \quad k > 0$	$\dfrac{1}{z-a}$
7. $f(k) = ka^{k-1}$	$\dfrac{z}{(z-a)^2}$
8. $\cos(kb)$	$\dfrac{z(z-\cos b)}{z^2 - 2z\cos b + 1}$
9. $\sin(kb)$	$\dfrac{z\sin b}{z^2 - 2z\cos b + 1}$
10. $a^k \cos(kb)$	$\dfrac{z(z - a\cos b)}{z^2 - 2az\cos b + e^{2b}}$
11. $a^k \sin(kb)$	$\dfrac{za\sin b}{z^2 - 2az\cos b + e^{2b}}$

be used to find the z transform of a given sequence, or conversely it may be used to find the sequence from a given z transformation.

5-9 PROPERTIES OF THE z TRANSFORM

Like the Laplace and Fourier transformations, the z transform has many desirable properties that make it useful in analysis of the discrete-time finite-difference equation. Reasons for using z transform analysis of the finite-difference equation are thus similar to reasons for using the Laplace transform to analyze continuous linear time-invariant systems. These properties are briefly reviewed as follows.

Linearity

Like the Fourier and Laplace transforms, the z transform is a linear operation. This is easily seen via

$$\begin{aligned} \mathcal{Z}\{c_1 f_1(k) + c_2 f_2(k)\} &\equiv \sum_{k=0}^{\infty} [c_1 f_1(k) + c_2 f_2(k)] z^{-k} \\ &= c_1 \sum_{k=0}^{\infty} f_1(k) z^{-k} + c_2 \sum_{k=0}^{\infty} f_2(k) z^{-k} \\ &= c_1 \mathcal{Z}\{f_1(k)\} + c_2 \mathcal{Z}\{f_2(k)\} \end{aligned} \tag{5-54}$$

The same applies to the inverse z transform.

Time Shift

If $F(z) = \mathcal{Z}\{f(k)\}$ then

$$\boxed{\mathcal{Z}\{f(k+1)\} = zF(z) - zf(0)} \tag{5-55}$$

This is similar to the differentiation property of the Laplace transform. To show this consider

$$\begin{aligned} \mathcal{Z}\{f(k+1)\} &\equiv \sum_{k=0}^{\infty} f(k+1) z^{-k} \\ &= z \sum_{k=0}^{\infty} f(k+1) z^{-(k+1)} \end{aligned}$$

Define a new dummy argument for summation $i = k + 1$. Then

$$\mathcal{Z}\{f(k+1)\} = z\sum_{i=1}^{\infty} f(i)z^{-i}$$

$$= z\left[\sum_{i=0}^{\infty} f(i)z^{-i} - f(0)\right]$$

Thus $\mathcal{Z}\{f(k+1)\} = zF(z) - zf(0)$ as claimed.

In a like fashion it may be shown that

$$\mathcal{Z}\{f(k+2)\} = z^2F(z) - z^2f(0) - zf(-1) \tag{5-56}$$

and, in general, for n shifts in the sequence

$$\boxed{\mathcal{Z}\{f(k+n)\} = z^nF(z) - z^nf(0) - z^{n-1}f(-1) - \cdots - zf(-n+1)} \tag{5-57}$$

This is almost a direct analogy to the Laplace transform result:

$$\mathcal{L}\{f^{(n)}(t)\} = s^nF(s) - s^{n-1}f(0^-) - s^{n-2}f^{(1)}(0^-) - \cdots - f^{(n-1)}(0^-) \tag{5-58}$$

Indeed, the form of this shift property makes the z transform directly applicable to solution of the finite-difference equation initial-value problem for the same reasons the Laplace transform could be used to solve the continuous differential equation. This technique using the z transform is examined in the next section.

Convolution

To be specific to our application, consider the convolution summation

$$y_{zs}(k) = \sum_{i=0}^{k-1} h_T(k-i)u_T(i)$$

$$= \sum_{i=0}^{\infty} h_T(k-i)u_T(i) \tag{5-59}$$

where h_T is the pulse response for a causal system so that $h_T(k) = 0$ for $k \leq 0$. The z transform of the convolution is

$$\mathcal{Z}\{y_{zs}(k)\} = \sum_{k=0}^{\infty} z^{-k}\left(\sum_{i=0}^{\infty} h_T(k-i)u_T(i)\right)$$

Reversing the order of summation gives

$$Y_{zs}(z) = \sum_{i=0}^{\infty}\left(\sum_{k=0}^{\infty} h_T(k-i)z^{-k}\right)u_T(i)$$

By the shift property of the z transform and the fact that $h_T(k) = 0$ for $k = 0, -1, -2, \ldots,$ the term within brackets is $z^{-i}H_T(z)$, where $H_T(z) \equiv \mathcal{Z}\{h_T(k)\}$. In the next section it is shown that this is the same $H_T(z)$ as found directly from the finite-difference equation in Sec. 5-5. Therefore, it follows that

$$Y_{zs}(z) = H_T(z)\left(\sum_{i=0}^{\infty} z^{-i}u_T(i)\right)$$

or
$$y_{zs}(z) = H_T(z)U_T(z) \tag{5-60}$$

This result applies in general for two causal sequences $f_1(k)$ and $f_2(k)$. The general discrete convolution property is

$$\boxed{\begin{aligned}\mathcal{Z}\{f_1(k) * f_2(k)\} &\equiv \mathcal{Z}\left\{\sum_{i=0}^{\infty} f_1(k-i)f_2(i)\right\} \\ &= F_1(z)F_2(z)\end{aligned}} \tag{5-61}$$

The z transform thus has the same desirable convolution property as did the Fourier and Laplace transforms.

The Relation $H_T(z) = z\{h_T(k)\}$

It is stated above and with considerable difficulty it may be shown that

$$\boxed{H_T(z) = \mathcal{Z}\{h_T(k)\}} \tag{5-62}$$

However, this is not a practical method for finding $H_T(z)$. It is quite difficult to demonstrate this for even a trivial example. There are other approximate ways for taking the z-transform of a continuous differential equation model. Since they are only approximations, we choose not to present them.

One further remark should be mentioned regarding the relation $H_T(z) = \mathcal{Z}\{h_T(k)\}$. Let $H(s) = H_1(s)H_2(s)$ be the transfer functions of two systems connected in series, and let $H_T(z)$, $H_{T_1}(z)$, and $H_{T_2}(z)$ be the finite-difference equation transfer functions for $H(s)$, $H_1(s)$, and $H_2(s)$, respectively. Then $H_T(z)$ is *not* equal to the product of $H_{T_1}(z)$ and $H_{T_2}(z)$. This may be somewhat surprising since the z transform has the desirable convolution property such that

$$\mathcal{Z}\{h_T(k) * u_T(k)\} = H_T(z)U_T(z)$$

Basically the reason that

$$H_T(z) \neq H_{T_1}(z)H_{T_2}(z)$$

is because

$$h_T(k) \neq h_{T_1}(k) * h_{T_2}(k)$$

even though

$$h(s) = h_1(s) * h_2(s)$$

The end result is that if one wishes to find the finite-difference equation representation of a complex interconnection of subsystems, one must first combine them in the continuous s domain. Then one may find the finite-difference equation representation of the combined result. Other discrete transforms such as the Tustin or bilateral transform partially alleviate this problem. See, for example, Ref. 6.

*5-10 THE z TRANSFORM USED TO SOLVE THE FINITE-DIFFERENCE EQUATION

By the shift property of the z transform, the z transform of the finite-difference equation may be taken, algebraic manipulations on both sides of the equation may be performed, and finally the inverse z transform may be used to find the solution $y_T(k)$ for any value k. This solution yields a natural separation between the zero-input and zero-state responses. The procedure is directly analogous to the Laplace transform solution method for continuous-time systems.

To describe the procedure for the general case is notationally cumbersome. We illustrate by considering only one example—the second-order finite-difference equation of Example 5-6.

Example 5-9 Consider the second-order finite-difference equation

$$y_T(k+2) - y_T(k+1) + 0.21y_T(k) = 3u_T(k+1) + 1.2u_T(k)$$

We find the system unit step response by using the z transform solution method. For the unit step input, $u_T(k) = (1)^k = 1, k = 0, 1, 2, \ldots$, and there are all zero initial conditions at time zero. Therefore, $y_T(k) = y_{T_{zs}}(k)$, since $y_{T_{zi}}(k) \equiv 0$.

Taking the z transform of the finite-difference equation with zero boundary conditions yields

$$(z^2 - z + 0.21)Y_T(z) = (3z + 1.2)U_T(z)$$

where
$$U_T(z) = \frac{z}{z-1}$$

Solving for $Y_T(z)$ gives

$$Y_T(z) = \frac{z(3z + 1.2)}{(z-1)(z^2 - z + 1.2)}$$

Expanding in partial fraction expansions gives

$$Y_T(z) = \frac{20z}{z-1} + \frac{7.5z}{z-0.3} - \frac{27.5z}{z-0.7}$$

Taking the inverse z transform yields the step response solution

$$y_T(k) = 20 + 7.5(0.3)^k - 27.5(0.7)^k$$

If the system had initial conditions, the time-shift property of the z transform can be used to also find the zero-input portion of the response. However, like the homogeneous and particular solution method, it would usually be much more convenient to simply propagate the finite-difference equation forward on the computer than to find a "closed-form" solution. For finding the time solution, neither the homogeneous and particular solution method nor the z transform solution method are especially useful. Rather, it is the analysis insights which they provide that are useful.

5-11 SUMMARY

This chapter has examined the discrete-time finite-difference equation in some depth. Such a model is convenient for digital simulation with piecewise constant sample-and-hold-type inputs. Useful relationships between the finite-difference-equation model and the discrete pulse-response model and the continuous-system poles are derived. Closed-form solution methods analogous to continuous-time solutions are next developed. The homogeneous and particular solutions are still applicable, and again they correspond to the transient and steady-state portions of the response. Analogous to the Laplace transform solution method is a z transform method. Finally, the discrete transfer function is the z transform of the discrete pulse-response sequence. This is directly analogous to the continuous transfer function being the Laplace transform of the continuous impulse-response function.

These first five chapters constitute what might be termed the *classic* methods of analysis of the linear time-invariant system. Analysis has been restricted to the single-input, single-output (SISO) system. The remainder of the text treats state-variable methods, or what might be termed the *modern* theory of linear systems. State-variable methods make it possible to conveniently examine the more complex multi-input, multi-output (MIMO) situation. Many of the same reoccurring themes encountered in these first five chapters reappear in the discussion of state-variable methods.

PROBLEMS

5-1 A second-order system has the sampled step response

$$r(T) = 2.0 \qquad r(2T) = 3.0 \qquad r(3T) = 3.4 \qquad r(4T) = 2.8$$

Find the finite-difference equation model [*Hint:* Use the sampled step response to find $h_T(1)$, $h_T(2)$, $h_T(3)$, $h_T(4)$].

5-2 Repeat Prob. 5-1 for the second-order system with sampled step response

$r(T) = 1.0 \qquad r(2T) = 1.8 \qquad r(3T) = 1.0$

$r(4T) = 0.6$

5-3 Using the finite-difference equation model, find $h_T(1)$, $h_T(2)$, $h_T(3)$, $h_T(4)$, $h_T(5)$, and $h_T(6)$ for the system in Prob. 5-1.

5-4 Repeat Prob. 5-3 for the system in Prob. 5-2.

5-5 Using the finite-difference equation model, find $r_T(1)$, $r_T(2)$, $r_T(3)$, $r_T(4)$, $r_T(5)$, and $r_T(6)$ for the system in Prob. 5-1.

5-6 Repeat Prob. 5-5 for the system in Prob. 5-2.

5-7 What is the form of the homogeneous solution for the system in Prob. 5-1?

5-8 What is the form of the homogeneous solution for the system in Prob. 5-2?

5-9 Is the system in Prob. 5-1 stable or unstable? Justify.

5-10 Is the system in Prob. 5-2 stable or unstable? Justify.

5-11 Determine whether the following systems are stable or unstable:

$$(a)\ H_T(z) = \frac{z + 0.5}{z^2 - 1.2z + 0.11}$$

$$(b)\ H_T(z) = \frac{2z - 1}{z^2 - 1.6z + 1.28}$$

5-12 Determine whether the following systems are stable or unstable:

$$(a)\ H_T(z) = \frac{z - 10}{z^2 - 1.7z + 0.72}$$

$$(b)\ H_T(z) = \frac{z + 2}{z^2 - \sqrt{3}z + 1}$$

5-13 Find the finite-difference equation model for the following systems using $T = 0.2$:

$$(a)\ H(s) = \frac{5}{s + 1} \qquad (b)\ H(s) = \frac{5}{s - 1}$$

$$(c)\ H(s) = \frac{s + 3}{x^2 + 3s + 2} \qquad (d)\ H(s) = \frac{s - 3}{s^2 + 3s + 2}$$

Hint: First find $r(t)$ and the sampled values $r(kT)$. Then find $h_T(1)$, $h_T(2)$. Then find the discrete characteristic polynomial via either Eq. (5-17) or Eq. (5-36). Then find the b_T coefficients via Eqs. (5-11*b*) through (5-14*b*).

5-14 Repeat Prob. 5-13 using $T = 0.1$ for the following systems:

$$(a)\ H(s) = \frac{5}{s + 2} \qquad (b)\ H(s) = \frac{5}{s - 2}$$

$$(c)\ H(s) = \frac{s + 3}{s^2 + 6s + 8} \qquad (d)\ H(s) = \frac{s - 3}{s^2 + 6s + 8}$$

5-15 For real discrete system pole p_{T_i} show that the corresponding continuous pole, p_i, is given by

$$p_i = \frac{\ln (p_{T_i})}{T}$$

5-16 For complex discrete system pole

$$p_{T_i} = \sigma_{T_i} + j\omega_{T_i} = \rho_{T_i} e^{j\phi_{T_i}}$$

where

$$\rho_{T_i} = (\sigma_{T_i}^2 + \omega_{T_i}^2)^{1/2} \qquad \text{and} \qquad \phi_{T_i} = \tan^{-1} \frac{\omega_{T_i}}{\sigma_{T_i}}$$

show that the corresponding continuous pole, p_i, is given by

$$p_i = \frac{\ln(\rho_{T_i})}{T} + \frac{\phi_{T_i}}{T} j$$

In other words, $p_i = \sigma_i + j\omega_i$ where

$$\sigma_i = \frac{\ln(\rho_{T_i})}{T} \quad \text{and} \quad \omega_i = \frac{\phi_{T_i}}{T}$$

5-17 Suppose $T = 0.2$. Estimate the continuous system poles for the system in Prob. 5-1 (*Hint:* Use the results of Probs. 5-15 and 5-16).

5-18 Suppose $T = 0.5$. Estimate the continuous system poles for the system in Prob. 5-2 (*Hint:* Use the results of Probs. 5-15 and 5-16).

5-19 (*a*) Find the particular solution for the unit step input with the systems in Prob. 5-11.
(*b*) Will your answers in (*a*) be the steady-state solution for the unit step input? Explain.

5-20 (*a*) Find the particular solution for the unit step input with the systems in Prob. 5-12.
(*b*) Will your answers in (*a*) be the steady-state solution for the unit step input? Explain.

5-21 Find the steady-state step response for the following systems:
(*a*) $y_T(k) - 0.5y_T(k-1) = 2u_T(k-1)$
(*b*) $y_T(k) - 0.5y_T(k-1) + 0.06y_T(k-2) = 2u_T(k-1) + u_T(k-2)$
(*c*) $y_T(k) - 0.4y_T(k-1) + 0.29y_T(k-2) = u_T(k-1) + 2u_T(k-2)$

5-22 Find the steady-state step response for the following systems:
(*a*) $y_T(k) - 0.1y_T(k-1) = 4u_T(k-1)$
(*b*) $y_T(k) - y_T(k-1) + 0.09y_T(k-2) = u_T(k-1) - u_T(k-2)$
(*c*) $y_T(k) - 0.2y_T(k-1) + 0.26y_T(k-2) = 2u_T(k-1) - u_T(k-2)$

5-23 For the systems in Prob. 5-21 find the homogeneous, particular, and complete solutions for the unit step input $u_T(k) = 1$, $k \geq 0$.

5-24 Repeat Prob. 5-23 for the systems in Prob. 5-22.

5-25 Compute the unit step response $r_T(k)$, $k = 1, 2, 3$, for the systems in Prob. 5-21 using the finite-difference equation itself. Compare these values to values you compute from the complete solution as found in Prob. 5-23.

5-26 Repeat Prob. 5-25 for the systems in Prob. 5-22.

5-27 For the system in Prob. 5-21*a* find the particular solution for the input $u_T(k) = k$, $k \geq 0$.

5-28 For the system in Prob. 5-22*a* find the particular solution for the input $u_T(k) = -k$, $k \geq 0$.

5-29 Find the particular solution for the system in Prob. 5-21*a* for the input $u_T(k) = (-1)^k$.

5-30 Find the particular solution for the system in Prob. 5-22*a* for the input $u_T(k) = (-1)^k$.

5-31 Repeat Prob. 5-29 for the input $u_T(k) = \text{Re}\{(1/\sqrt{2} + j/\sqrt{2})^k\}$.

5-32 Repeat Prob. 5-30 for the input $u_T(k) = \text{Im}\{(\sqrt{3}/2 + j/2)^k\}$.

5-33 Suppose $T = 0.2$. What are the continuous system poles for the systems in Prob. 5-21? Use the results of Probs. 5-15 and 5-16.

5-34 Suppose $T = 0.5$. What are the continuous system poles for the systems in Prob. 5-22? Use the results of Probs. 5-15 and 5-16.

5-35 Use the z transform solution method to find the unit step response for the system in Prob. 5.21*a*. Compare your solution to that obtained in Prob. 5-23.

5-36 Repeat Prob. 5-35 for the system in Prob. 5-22.

5-37 Show that Eq. (5-17) may be set up as an overdetermined linear equation problem of the form

$$\begin{bmatrix} h_T(1) & h_T(2) & \cdots & h_T(n) \\ h_T(2) & h_T(3) & \cdots & h_T(n+1) \\ \vdots & & & \vdots \\ h_T(N) & h_T(N+1) & \cdots & h_T(N+n-1) \end{bmatrix} \begin{bmatrix} a_{T_0} \\ a_{T_1} \\ \vdots \\ a_{T_{n-1}} \end{bmatrix} = \begin{bmatrix} -h_T(n+1) \\ -h_T(n+2) \\ \vdots \\ -h_T(N+n) \end{bmatrix}$$

for any $N \geq n$. This can be used to improve the accuracy of the computation. See Appendix E.

NOTE AND REFERENCES

Derivation of the linear equation problems Eq. (5-17) and Eqs. (5-11*b*) through (5-14*b*) is original to this author. More is discussed about the coefficient Hankel matrix, Eq. (5-18), and the conditioning of the inherent linear equation problem, Eq. (5-17), in Chaps. 8 and 11 and in Appendix E. It is noted that the accuracy involved in solving Eq. (5-17) may be improved by setting up an overdetermined linear equation problem (see Prob. 5-37) and solving the resulting linear least-squares problem (see Appendix E). Regardless, though, readers should avail themselves of expertly written software in solution of the inherent linear equation problem (e.g., Refs. 1 and 2).

References 3 through 6, for example, provide good general references on discrete-time system analysis and on the z transform analysis in particular.

1. Dongarra, J. J., et al.: *LINPACK User's Guide,* SIAM, Philadelphia, 1979.
2. *IMSL Reference Manual,* IMSL Inc., Houston, 1979.
3. Cadzow, J. A.: *Discrete Time Systems,* Prentice-Hall, New York, 1973.
4. Schwarz, R. J., and B. Friedland: *Linear Systems,* McGraw-Hill, New York, 1965.
5. Freeman, H.: *Discrete Time Systems,* Wiley, New York, 1966.
6. Houpis, C., and G. Lamont: *Digital Control Systems* (*Theory, Hardware, Software*), McGraw-Hill, New York, pending.

PART TWO

STATE-VARIABLE ANALYSIS

CHAPTER

SIX

DERIVING THE STATE-VARIABLE MODEL

This chapter develops two distinct methods for deriving a state-variable model of a given linear time-invariant system: first, deriving the state model from a given input-output transfer function; and second, deriving the state model directly from the original physical circuit relationships. The first method leads to several alternate canonical forms of the state-variable model. These different forms each have special applications and are fully equivalent; however, the individual states in these canonical forms have no special physical significance of and by themselves.

In the second, physical-variable state model, the states are selected to correspond directly to physical variables in the system. Theoretically, these physical variables could be measured even if they are not, explicitly, a measured output. Examples might be the velocity of a mass, the charge across a capacitor, the current through an inductor, etc. The state model so derived has no special standard form like the canonical forms do, but it has the advantage that the individual state variables have a particular physical significance.

One of the objectives for deriving several alternative, but equivalent, state-variable forms is to emphasize that the state-variable model is not unique. Indeed, there are actually an infinite number of state-variable model forms all describing the same system. The state-variable model is termed an *internal model* of the system, because the state-variable model can include dynamics that do not appear in the input-output, external transfer-function model of the system. This is demonstrated when considering the development of the physical variable-state model.

Lastly, this chapter introduces the multi-input, multi-output (MIMO) state model. One of the important advantages of state-variable methods over the classic methods of analysis is the facility by which the MIMO situation may be treated. Therefore, it is appropriate to consider how the MIMO state-space model may be developed.

6-1 PHASE-VARIABLE CANONICAL FORM

The next several sections develop general mathematical techniques for taking a given SISO nth-order transfer function

$$H(s) = \frac{K(b_m s^m + b_{m-1} s^{m-1} + \cdots + b_1 s + b_0)}{s^n + a_{n-1} s^{n-1} + a_{n-2} s^{n-2} + \cdots + a_1 s + a_0} \tag{6-1}$$

and deriving from this a set of simultaneous, linear, time-invariant, first-order, state-variable differential equations that have the general form

$$\dot{\mathbf{x}}(t) = A\mathbf{x}(t) + Bu(t) \tag{6-2}$$

$$y(t) = C\mathbf{x}(t) + Du(t) \tag{6-3}$$

where $\mathbf{x}(t)$ is the nth-order *state vector*

$$\mathbf{x}(t) = \begin{bmatrix} x_1(t) \\ x_2(t) \\ \vdots \\ x_n(t) \end{bmatrix} \tag{6-4}$$

and where $u(t)$ is the single input or forcing signal; $y(t)$ is the single output signal; A is termed the $n \times n$-dimensioned *plant* matrix; B is the $n \times 1$-dimensioned *input* matrix (really a column matrix for this single-input case); C is the $1 \times n$-dimensioned *output* matrix (really a row matrix for this single-output case); and D is the 1×1-dimensioned *feedforward* matrix (really a scalar for this general SISO system).

One of the most straightforward methods for finding a model of the form Eqs. (6-2) and (6-3) for a given SISO transfer function, Eq. (6-1), is by choice of *phase-variable states*.* The simplest case occurs when there are no system zeros.

No System Zeros

Consider the example transfer function

$$H(s) = \frac{K}{s^3 + 4s^2 + x - 6} \tag{6-7}$$

with no zeros. The corresponding output differential equation is

$$\dddot{y}(t) + 4\ddot{y}(t) + \dot{y}(t) - 6y(t) = Ku(t) \tag{6-8}$$

The phase-variable states are selected via defining the variables

$$x_1(t) \equiv y(t) \tag{6-9}$$

$$x_2(t) \equiv \dot{x}_1(t) = \dot{y}(t) \tag{6-10}$$

*The phase-variable canonical form is also referred to as the *companion* canonical form or the *controllability* canonical form.

$$x_3(t) \equiv \dot{x}_2(t) = \ddot{y}(t) \tag{6-11}$$

First-order differential equations for the variables $x_1(t)$, $x_2(t)$, and $x_3(t)$ are then given by

$$\dot{x}_1(t) = x_2(t) \tag{6-12}$$

$$\dot{x}_2(t) = x_3(t) \tag{6-13}$$

$$\dot{x}_3(t) = \dddot{y}(t) \tag{6-14}$$

Only the third differential equation presents a problem, but solving for $\dddot{y}(t)$ from Eq. (6-8) gives

$$\dot{x}_3(t) = -4\ddot{y}(t) - \dot{y}(t) + 6y(t) + Ku(t)$$

In terms of the state variables defined by Eqs. (6-9) through (6-11), this becomes

$$\dot{x}_3(t) = -4x_3(t) - x_2(t) + 6x_1(t) + Ku(t) \tag{6-15}$$

Combining Eqs. (6-9), (6-12), (6-13), and (6-15) into the matrix form of Eqs. (6-2) and (6-3) then yields

$$\begin{bmatrix} \dot{x}_1(t) \\ \dot{x}_2(t) \\ \dot{x}_3(t) \end{bmatrix} = \begin{bmatrix} 0 & 1 & 0 \\ 0 & 0 & 1 \\ 6 & -1 & -4 \end{bmatrix} \begin{bmatrix} x_1(t) \\ x_2(t) \\ x_3(t) \end{bmatrix} + \begin{bmatrix} 0 \\ 0 \\ K \end{bmatrix} u(t) \tag{6-16}$$

$$y(t) = [1 \quad 0 \quad 0] \begin{bmatrix} x_1(t) \\ x_2(t) \\ x_3(t) \end{bmatrix} \tag{6-17}$$

Notice that $D \equiv 0$; this will be true for any proper transfer function and represents the fact that there is no direct feedforward of the input to the output. Applying the same methods, the reader may easily verify that the phase-variable state equations for the no-zero, nth-order transfer function

$$H(s) = \frac{K}{s^n + a_{n-1}s^{n-1} + \cdots + a_1 s + a_0} \tag{6-20}$$

takes the form

$$\begin{bmatrix} \dot{x}_1(t) \\ \dot{x}_2(t) \\ \dot{x}_3(t) \\ \vdots \\ x_n(t) \end{bmatrix} = \begin{bmatrix} 0 & 1 & 0 & 0 & \cdots \\ 0 & 0 & 1 & 0 & \cdots \\ 0 & 0 & 0 & 1 & \cdots \\ \vdots & & & & \\ -a_0 & -a_1 & \cdots & \cdots & -a_{n-1} \end{bmatrix} \begin{bmatrix} x_1(t) \\ x_2(t) \\ x_3(t) \\ \vdots \\ x_n(t) \end{bmatrix} + \begin{bmatrix} 0 \\ 0 \\ 0 \\ \vdots \\ K \end{bmatrix} u(t) \tag{6-21}$$

$$y(t) = [1 \quad 0 \quad 0 \quad \cdots \quad 0] \begin{bmatrix} x_1(t) \\ x_2(t) \\ x_3(t) \\ \vdots \\ x_n(t) \end{bmatrix} \tag{6-22}$$

Notice that an nth-order transfer function produces n first-order, phase-variable state equations. Also notice the standard canonical form which each phase-variable matrix, also called a *companion matrix,* takes. These may, in fact, be written down by inspection from the no-zero transfer function.

Example 6-1 The transfer function of the rocket sled introduced in Chaps. 1 and 2 is

$$H(s) = \frac{1}{s^3 + 0.4s^2 + 1.14s + 0.22}$$

Therefore, by inspection, the phase-variable state model is

$$\begin{bmatrix} \dot{x}_1(t) \\ \dot{x}_2(t) \\ \dot{x}_3(t) \end{bmatrix} = \begin{bmatrix} 0 & 1 & 0 \\ 0 & 0 & 1 \\ -0.22 & -1.14 & -0.4 \end{bmatrix} \begin{bmatrix} x_1(t) \\ x_2(t) \\ x_3(t) \end{bmatrix} + \begin{bmatrix} 0 \\ 0 \\ 1 \end{bmatrix} u(t)$$

$$y(t) = [1 \quad 0 \quad 0] \begin{bmatrix} x_1(t) \\ x_2(t) \\ x_3(t) \end{bmatrix}$$

Note that the phase-variable state model is not complete unless the output relation is also included.

General Case, Proper Transfer Function

Consider now the third-order SISO transfer function

$$H(s) = \frac{K(s^2 + 2s - 15)}{s^3 + 4s^2 + s - 6} \tag{6-23}$$

The system is proper, since the zero polynomial is of lower order than the characteristic polynomial.

A direct application of the previous methods yields

$$\dot{x}'_1(t) = x'_2(t)$$

$$\dot{x}'_2(t) = x'_3(t)$$

$$\begin{aligned} \dot{x}'_3(t) = \dddot{y}(t) &= 6x'_1(t) - x'_2(t) - 4x'_3(t) \\ &\quad + K[\ddot{u}(t) + 2\dot{u}(t) - 15u(t)] \end{aligned}$$

which is a correct representation—only not very useful, because one would like to be rid of the $\ddot{u}(t)$ and $\dot{u}(t)$ terms in the model. These caused trouble in solving the output differential equation by classic methods, and they would cause difficulty here also if some method is not found to keep them from explicitly appearing in the state differential equation model.

Fortunately, this is easily accomplished. Consider the split of the complete transfer function into the factor of two parts:

$$H(s) = \frac{Y(s)}{U(s)} = \frac{Y(s)}{V(s)}\frac{V(s)}{U(s)} \tag{6-24}$$

where the following intermediate transfer functions are defined:

$$\frac{V(s)}{U(s)} \equiv \frac{K}{s^3 + 4s^2 + s - 6} \tag{6-25}$$

$$\frac{Y(s)}{V(s)} \equiv \frac{s^2 + 2s - 15}{1} \tag{6-26}$$

Then with respect to the dummy output $v(t)$, the $V(s)/U(s)$ transfer function, Eq. (6-25), may be used to give the phase-variable state equations

$$\begin{bmatrix} \dot{x}_1(t) \\ \dot{x}_2(t) \\ \dot{x}_3(t) \end{bmatrix} = \begin{bmatrix} 0 & 1 & 0 \\ 0 & 0 & 1 \\ 6 & -1 & -4 \end{bmatrix} \begin{bmatrix} x_1(t) \\ x_2(t) \\ x_3(t) \end{bmatrix} + \begin{bmatrix} 0 \\ 0 \\ K \end{bmatrix} u(t) \tag{6-27}$$

$$v(t) = [1 \quad 0 \quad 0] \begin{bmatrix} x_1(t) \\ x_2(t) \\ x_3(t) \end{bmatrix} \tag{6-28}$$

But $v(t)$ is not the real output of the system. However, the transfer function $Y(s)/V(s)$, Eq. (6-26), may be used to relate $y(t)$ to $v(t)$ via

$$Y(s) = (s^2 + 2s - 15)V(s) \tag{6-29}$$

or in the time domain

$$y(t) = \ddot{v}(t) + 2\dot{v}(t) - 15v(t) \tag{6-30}$$

But, by definition of the phase-variable states used in the formation of Eqs. (6-27) and (6-28), it is true that

$$v(t) \equiv x_1(t) \tag{6-31}$$

$$\dot{v}(t) \equiv x_2(t) \tag{6-32}$$

$$\ddot{v}(t) \equiv x_3(t) \tag{6-33}$$

Making these substitutions into Eq. (6-30) gives the output relation

$$y(t) = [-15 \quad 2 \quad 1] \begin{bmatrix} x_1(t) \\ x_2(t) \\ x_3(t) \end{bmatrix} \tag{6-34}$$

The complete phase-variable model for this proper transfer function is then Eqs. (6-27) and (6-34). Note that the intermediate variable, $v(t)$, Eq. (6-28), is not needed in the final state-space model, and $v(t)$ has no physical significance or importance.

The phase-variable state equations for the general, nth-order, proper transfer function

$$H(s) = \frac{K(b_{n-1}s^{n-1} + b_{n-2}s^{n-2} + \cdots + b_1 s + b_0)}{s^n + a_{n-1}s^{n-1} + a_{n-2}s^{n-2} + \cdots + a_1 s + a_0} \tag{6-35}$$

are given by

$$\begin{bmatrix} \dot{x}_1(t) \\ \dot{x}_2(t) \\ \vdots \\ \dot{x}_n(t) \end{bmatrix} = \begin{bmatrix} 0 & 1 & 0 & \cdots & \\ 0 & 0 & 1 & \cdots & \\ \vdots & & & & \\ -a_0(t) & -a_1(t) & -a_2(t) & \cdots & -a_{n-1}(t) \end{bmatrix} \begin{bmatrix} x_1(t) \\ x_2(t) \\ \vdots \\ x_n(t) \end{bmatrix} + \begin{bmatrix} 0 \\ 0 \\ \vdots \\ K \end{bmatrix} u(t) \tag{6-36}$$

$$y(t) = [b_0 \quad b_1 \quad \cdots \quad b_{n-1}] \begin{bmatrix} x_1(t) \\ x_2(t) \\ \vdots \\ x_n(t) \end{bmatrix} \tag{6-37}$$

where any of the a's or b's may be zero. Hence the phase-variable state-space model may be derived directly by inspection from the transfer function.

Notice that there is still no feedforward D matrix. This is because the system here is assumed to be proper.

General Case, Improper Transfer Function

Consider now the improper transfer function

$$H(s) = \frac{K(s^3 + 2s^2 - 15s)}{s^3 + 4s^2 + s - 6} \tag{6-38}$$

It is improper because the zero polynomial is of the same order as the characteristic polynomial (third-order, in this example).

Applying the previous approach gives the same state-variable differential equations Eq. (6-27), but the intermediate variable $v(t)$ has the relation to the output $y(t)$ given by

$$y(t) = \dddot{v}(t) + 2\ddot{v}(t) - 15\dot{v}(t) \tag{6-39}$$

Again Eqs. (6-31) through (6-33) apply, which takes care of $\ddot{v}(t)$ and $\dot{v}(t)$, but

$$\dddot{v}(t) = \dot{x}_3(t) = 6x_1(t) - x_2(t) - 4x_3(t) + Ku(t) \tag{6-40}$$

where this last relation is obtained from the state equation for $x_3(t)$. Making these substitutions into Eq. (6-39) yields

$$y(t) = [6x_1(t) - x_2(t) - 4x_3(t) + Ku(t)] + 2x_3(t) - 15x_2(t)$$

Collecting terms gives the final output expression

$$y(t) = [6 \quad -16 \quad -2]\begin{bmatrix} x_1(t) \\ x_2(t) \\ x_3(t) \end{bmatrix} + [K]u(t) \tag{6-41}$$

Note that the system now has a direct feedforward of the input to the output; the input, $u(t)$, is immediately felt in the output, $y(t)$, weighted by a factor $D = [K]$.

For the general case

$$H(s) = \frac{K(b_n s^n + \cdots + b_1 s + b_0)}{s^n + a_{n-1}s^{n-1} + \cdots + a_1 s + a_0} \tag{6-42}$$

the phase-variable state equations take exactly the same form as Eq. (6-36), but the output relation is now given by

$$y(t) = [(b_0 - b_n a_0) \quad (b_1 - b_n a_1) \quad \cdots \quad (b_{n-1} - b_n a_{n-1})]\begin{bmatrix} x_1(t) \\ x_2(t) \\ \vdots \\ x_n(t) \end{bmatrix} + [Kb_n]u(t) \tag{6-43}$$

This is the most general form for the SISO system transfer function, and it applies to all of the previous cases if one realizes that $b_n = 0$ for the proper systems.

Boundary Conditions on the Phase-Variable States

We now consider the problem of relating the state-variable boundary conditions

$$\mathbf{x}(t_0^-) = \begin{bmatrix} x_1(t_0^-) \\ x_2(t_0^-) \\ \vdots \\ x_n(t_0^-) \end{bmatrix} \tag{6-44}$$

to the initial conditions on y and its first $n - 1$ derivatives $[y(t_0^-), \ldots y^{(n-1)}(t_0^-)]$. The boundary conditions are taken at t_0^-, before the occurrence of any discontinuity changes in the input. As shall be seen, $\mathbf{x}(t_0^-) = \mathbf{x}(t_0^+)$ for finite inputs, and so this distinction really is not needed.

To begin the discussion, consider the no-zero example

$$H(s) = \frac{K}{s^3 + 4s^2 + s - 6} \tag{6-7}$$

whose phase-variable state equations are given by Eqs. (6-16) and (6-17). These equations are true for all t, and so for t_0^- in particular it follows that

$$y(t_0^-) = x_1(t_0^-) \tag{6-45}$$

Differentiating once yields

$$\dot{y}(t_0^-) = \dot{x}_1(t_0^-) = x_2(t_0^-) \tag{6-46}$$

and once again

$$\ddot{y}(t_0^-) = \dot{x}_2(t_0^-) = x_3(t_0^-) \tag{6-47}$$

Therefore, for this no-zero case the phase-variable states $\{x_1(t), x_2(t), x_3(t)\}$ are just $\{y(t), \dot{y}(t), \ddot{y}(t)\}$ for all t including t_0^-. Also, since there are no zeros, each of the variables is continuous for all finite $u(t)$ signals even if $u(t)$ is discontinuous itself.

Take now the second proper case, with two zeros,

$$H(s) = \frac{K(s^2 + 2s - 15)}{s^3 + 4s^2 + s - 6} \tag{6-23}$$

where the phase-variable state equations and output relation are given by Eqs. (6-27) and (6-34). From Eq. (6-34) it follows that

$$y(t_0^-) = -15x_1(t_0^-) + 2x_2(t_0^-) + x_3(t_0^-) \tag{6-48}$$

Differentiating once gives

$$\dot{y}(t_0^-) = -15\dot{x}_1(t_0^-) + 2\dot{x}_2(t_0^-) + \dot{x}_3(t_0^-)$$

Going back to the state differential equations Eq. (6-27) to evaluate these derivatives yields

$$\dot{y}(t_0^-) = -15x_2(t_0^-) + 2x_3(t_0^-) + [6x_1(t_0^-) - x_2(t_0^-) - 4x_3(t_0^-) + Ku(t_0^-)]$$

or

$$\dot{y}(t_0^-) = 6x_1(t_0^-) - 16x_2(t_0^-) - 2x_3(t_0^-) + Ku(t_0^-) \tag{6-49}$$

Finally, differentiating for a second time and repeating the same substitution from Eq. (6-27) yields

$$\ddot{y}(t_0^-) = -12x_1(t_0^-) + 8x_2(t_0^-) - 8x_3(t_0^-) - 2Ku(t_0^-) + K\dot{u}(t_0^-) \tag{6-50}$$

Equations (6-48) through (6-50) may be combined in the standard linear equation problem format

$$\begin{bmatrix} -15 & 2 & 1 \\ 6 & -16 & -2 \\ -12 & 8 & -8 \end{bmatrix} \begin{bmatrix} x_1(t_0^-) \\ x_2(t_0^-) \\ x_3(t_0^-) \end{bmatrix} = \begin{bmatrix} y(t_0^-) \\ \dot{y}(t_0^-) - Ku(t_0^-) \\ \ddot{y}(t_0^-) + 2Ku(t_0^-) - K\dot{u}(t_0^-) \end{bmatrix} \tag{6-51}$$

Specific values for $y(t_0^-)$, $\dot{y}(t_0^-)$, $\ddot{y}(t_0^-)$, $u(t_0^-)$, and $\dot{u}(t_0^-)$ may be used to solve for the initial-state vector. Notice that like the Laplace transform solution technique of Chap. 4, boundary conditions on both y and the input u are required in order to solve for the initial state.

General Procedure for Finding the Initial State

The procedure for setting up the boundary condition linear equation problem is quite systematic and is repeated in all of the different state-variable models to be looked at. The steps in this procedure are:

1. Write down the output relation to the states.
2. Differentiate once and make substitutions for the $\dot{x}$ terms to relate $\dot{y}$ to the x's and $u/\dot{u}$.
3. Continue differentiating up to $n - 1$ times, where at each step $y^{(i)}$ is related to the x's and u, $u^{(1)}, \ldots, u^{(i)}$.
4. Evaluate at t_0 using known boundary conditions on $\{y(t_0), \ldots, y^{(n-1)}(t_0)\}$ and $\{u(t_0), \ldots, u^{(n-1)}(t_0)\}$, as required. It is immaterial whether one uses t_0^- or t_0^+ as long as there is consistency. The state vector is always continuous, $\mathbf{x}(t_0^-) = \mathbf{x}(t_0^+)$, so long as the input is not an impulse.
5. Collect the relations in (4) into a linear equation problem for $\mathbf{x}(t_0)$ and solve.

This same procedure applies to the improper feedforward-system case.

Continuity of the State Variables

It is asserted that the state variables are always continuous even for discontinuous, but bounded, inputs. This may seem like a subtle point, but it is quite significant in regard to the simulation of the state-space model on the digital computer: the boundary conditions on the state variables are always continuous even for discontinuous, sample-and-hold-type inputs.

The reason for this fact is basically because the $\mathbf{x}(t)$ differential equation is driven only by $u(t)$ and *not* by any of its derivatives. This fact may also be verified by applying the analysis methods of Sec. 2-8.

Example 6-2

$$\ddot{y}(t) + 3\dot{y}(t) + 2y(t) = 4\dot{u}(t) + 5u(t)$$

The phase-variable state model is

$$\begin{bmatrix} \dot{x}_1(t) \\ \dot{x}_2(t) \end{bmatrix} = \begin{bmatrix} 0 & 1 \\ -2 & -3 \end{bmatrix} \begin{bmatrix} x_1(t) \\ x_2(t) \end{bmatrix} + \begin{bmatrix} 0 \\ 1 \end{bmatrix} u(t)$$

$$y(t) = [5 \quad 4] \begin{bmatrix} x_1(t) \\ x_2(t) \end{bmatrix}$$

The boundary-condition linear equation problem is

$$\begin{bmatrix} 5 & 4 \\ -8 & -7 \end{bmatrix} \begin{bmatrix} x_1(t_0) \\ x_2(t_0) \end{bmatrix} = \begin{bmatrix} y(t_0) \\ \dot{y}(t_0) - 4u(t_0) \end{bmatrix}$$

The claim is that $x_1(t_0^-) = x_1(t_0^+)$ and $x_2(t_0^-) = x_2(t_0^+)$. The only way that this can be true is for $y(t_0^-) = y(t_0^+)$ and $[\dot{y}(t_0^-) - 4u(t_0^-)] = [\dot{y}(t_0^+) - 4u(t_0^+)]$. Applying the same analysis as conducted in Sec. 2-8, the second-order differential equation for $\ddot{y}(t)$ may be integrated twice from t_0^- to t_0^+ to yield $y(t_0^-) = y(t_0^+)$. Integrating $\ddot{y}(t)$ once then gives

$$\dot{y}(t_0^+) - \dot{y}(t_0^-) + 3[\underbrace{y(t_0^+) - y(t_0^-)}_{\to 0}] = 4[u(t_0^+) - u(t_0^-)]$$

Rearranging gives

$$[\dot{y}(t_0^+) - 4u(t_0^+)] = [\dot{y}(t_0^-) - 4u(t_0^-)]$$

as claimed.

6-2 STATES DERIVED VIA THE SIMULATION DIAGRAM

An important objective of this chapter is to demonstrate that the state-variable model for a given transfer function is not unique—there are actually an infinite number of possible ways to describe the same transfer function by an internal state-variable model, and all would be correct. To this end a second, very systematic—and therefore often very useful—technique for finding the state-variable model is shown in this section.

Proper Transfer Function

Consider the third-order transfer function

$$H(s) = \frac{K(s^2 + 2s - 15)}{s^3 + 4s^2 + s - 6} \tag{6-23}$$

with corresponding output differential equation

$$\dddot{y} + 4\ddot{y} + \dot{y} - 6y = K\ddot{u} + 2K\dot{u} - 15Ku \tag{6-52}$$

where the t is deleted for the simplicity of notation. The procedure for deriving the first-order state-space differential equations is to integrate Eq. (6-52) a sufficient number of times to eliminate all derivatives, and then solve for y. This yields

$$y = -4\int y - \iint y + 6\iiint y + K\int u + 2K\iint u - 15K\iiint u \tag{6-53}$$

where three integrations are required to eliminate $\dddot{y}$.

The next step is to make a simulation diagram of the resulting expression for y. For Eq. (6-53) above, this is shown in Fig. 6-1.

The next step is to define a state variable as the immediate output of each of the three integrators. Then if x_i is the output of the ith integrator, the immediate input to the integrator must be $\dot{x}_i$. This is shown in Fig. 6-2.

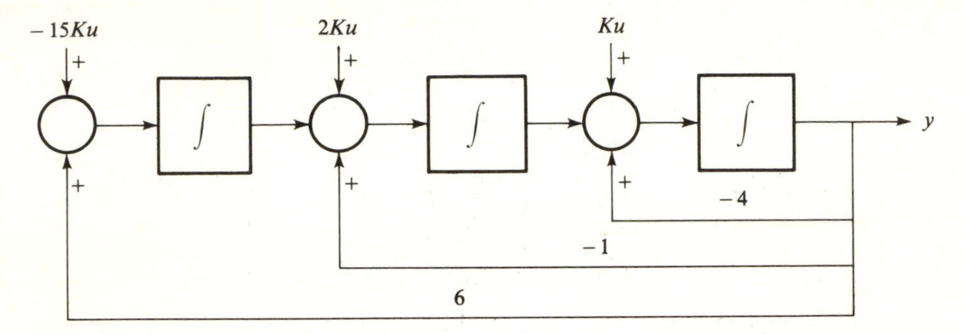

Figure 6-1 Simulation diagram of Eq. (6-53).

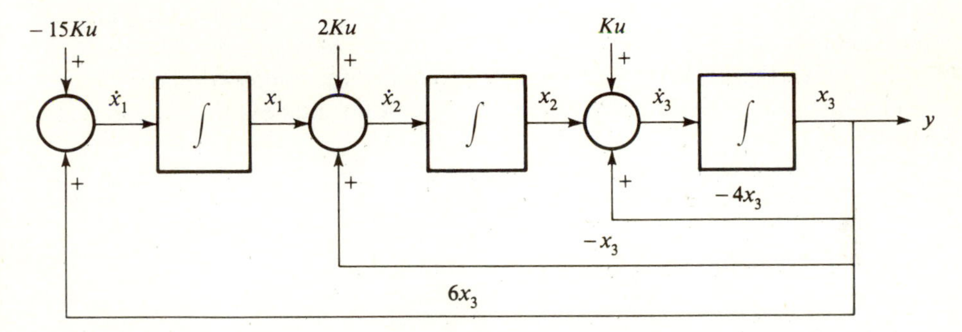

Figure 6-2 Simulation diagram of Eq. (6-53) with states labeled.

Here the states are numbered so that $x_3 = y$ (this is arbitrary). Hence the feedback path is really x_3. From this, the simulation diagram itself provides the needed expressions for $\dot{x}_1$, $\dot{x}_2$, and $\dot{x}_3$. Namely, by going to each summing junction,

$$\dot{x}_3 = -4x_3 + x_2 + Ku \tag{6-54}$$

$$\dot{x}_2 = -x_3 + x_1 + 2Ku \tag{6-55}$$

$$\dot{x}_1 = 6x_3 - 15Ku \tag{6-56}$$

or in matrix vector form

$$\begin{bmatrix} \dot{x}_1(t) \\ \dot{x}_2(t) \\ \dot{x}_3(t) \end{bmatrix} = \begin{bmatrix} 0 & 0 & 6 \\ 1 & 0 & -1 \\ 0 & 1 & -4 \end{bmatrix} \begin{bmatrix} x_1(t) \\ x_2(t) \\ x_3(t) \end{bmatrix} + \begin{bmatrix} -15K \\ 2K \\ K \end{bmatrix} u(t) \tag{6-57}$$

Finally, the output relation is simply

$$y(t) = [0 \quad 0 \quad 1] \begin{bmatrix} x_1(t) \\ x_2(t) \\ x_3(t) \end{bmatrix} \tag{6-58}$$

Feedforward Systems

Consider the system transfer function

$$H(s) = \frac{K(s^3 + 2s^2 - 15s)}{s^3 + 4s^2 + s - 6} \tag{6-38}$$

with corresponding output differential equation

$$\dddot{y} + 4\ddot{y} + \dot{y} - 6y = K\dddot{u} + 2K\ddot{u} - 15K\dot{u} \tag{6-59}$$

Integrating three times and solving for $y(t)$ then gives

$$y = -4\int y - \iint y + 6\iiint y + Ku + 2K\int u - 15K\iint u \tag{6-60}$$

The simulation diagram for Eq. (6-60) is shown in Fig. 6-3, in which the states are simultaneously labeled on the diagram. Notice that u is *fed forward* directly to the output—hence the name for the improper system of a *feedforward system*.

This fact can cause some confusion, though, in formulating the state differential equations. From the diagram it is seen that

$$\begin{aligned} \dot{x}_3 &= -4y + x_2 + 2Ku \\ &= -4(x_3 + Ku) + x_2 + 2Ku \\ &= -4x_3 + x_2 - 2Ku \end{aligned} \tag{6-61}$$

and likewise

$$\dot{x}_2 = -x_3 + x_1 - 16Ku \tag{6-62}$$

$$\dot{x}_1 = 6x_3 + 6Ku \tag{6-63}$$

Collecting Eqs. (6-61) through (6-63) into the standard state equation form yields

$$\begin{bmatrix} \dot{x}_1(t) \\ \dot{x}_2(t) \\ \dot{x}_3(t) \end{bmatrix} = \begin{bmatrix} 0 & 0 & 6 \\ 1 & 0 & -1 \\ 0 & 1 & -4 \end{bmatrix} \begin{bmatrix} x_1(t) \\ x_2(t) \\ x_3(t) \end{bmatrix} + \begin{bmatrix} 6K \\ -16K \\ -2K \end{bmatrix} u(t) \tag{6-64}$$

with output relation

$$y(t) = [0 \quad 0 \quad 1] \begin{bmatrix} x_1(t) \\ x_2(t) \\ x_3(t) \end{bmatrix} + [K]u(t) \tag{6-65}$$

Notice that the system A matrix is the same as the previous proper-system case; the A matrix is only dependent on the characteristic polynomial, which does not change between this case and the previous one. This property is the same for the phase-variable cases of the previous section. Also notice that the D matrix is the same in this example as it was for the previous phase-variable state model

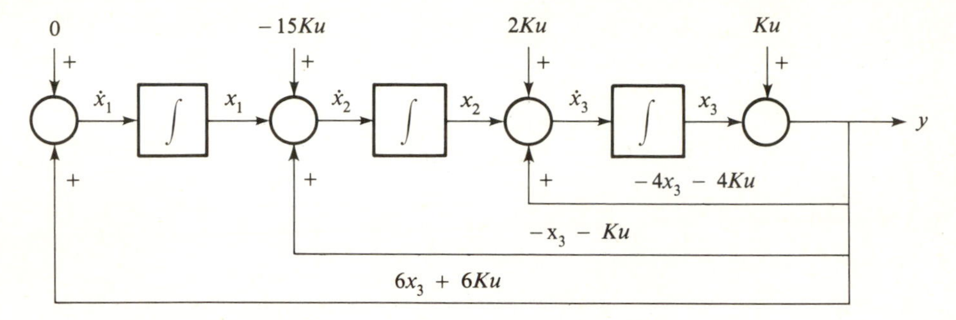

Figure 6-3 Simulation diagram of feed-forward system, Eq. (6-60).

with the same transfer function.* Since this part of the state-space model is a direct input-output property of the system, the D matrix for a given improper transfer function is invariant regardless of the state coordinates chosen.

Feedforward System Derived by Long Division

An alternative way to derive the state model for the improper transfer function is to apply long division of the characteristic polynomial into the zero polynomial. To illustrate, consider the previous example improper system

$$H(s) = \frac{K(s^3 + 2s^2 - 15s)}{s^3 + 4s^2 + s - 6} \tag{6-38}$$

By long division

$$\begin{array}{r|l} & \quad\quad\quad\quad\quad\quad\quad\quad\quad\quad K \\ \hline s^3 + 4s^2 + s - 6 & Ks^3 + 2Ks^2 - 15Ks + 0 \\ & \underline{Ks^3 + 4Ks^2 + \quad Ks - 6K} \\ & \quad\quad - 2Ks^2 - 16Ks + 6K \end{array}$$

Thus

$$H(s) = \frac{-2Ks^2 - 16Ks + 6K}{s^3 + 4s^2 + s - 6} + K \tag{6-38a}$$

The corresponding output equation is

$$y = -4\int y - \iint y + 6\iiint y - 2K\int u - 16K\iint u + 6K\iiint u + Ku \tag{6-66}$$

with corresponding simulation diagram shown in Fig. 6-4. Notice that this is the same simulation diagram as in Fig. 6-3 except that the feedforward $Ku(t)$ is added to $y(t)$ *after* the feedback of $x_3(t)$ rather than before the feedback of $y(t)$ as in Fig. 6-3. The state model Eqs. (6-64) and (6-65) may be derived now directly from Fig.

*Finally notice that comparing this model to the phase-variable model, $A = A_{PV}^T$, $B = C_{PV}^T$, $C = B_{PV}^T$. As will be discussed later, the two models are thus said to "dual".

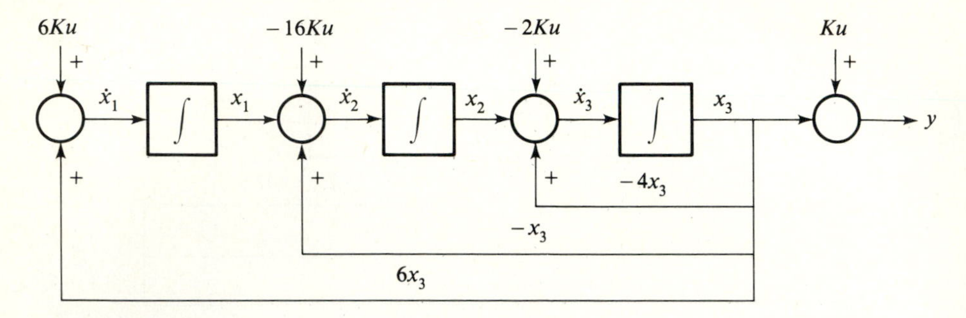

Figure 6-4 Simulation diagram of feed-forward system, Eq. (6-66).

6-4 without any complications. This method may also be applied to deriving the phase-variable state model or any other state model form for the improper feedforward system.

Boundary Conditions at t_0

The procedure for finding the boundary conditions on $\{x_1(t_0), x_2(t_0), x_3(t_0)\}$ related to $\{y(t_0), \dot{y}(t_0), \ddot{y}(t_0)\}$ is the same as for the phase-variable case: starting with the output relation on $y(t_0)$, differentiate a sufficient number of times to set up three equations in three unknowns [or n equations in n unknowns for the general nth-order $H(s)$]. This is illustrated for the first proper-system case, Eqs. (6-57) and (6-58).

From Eq. (6-58)

$$y(t_0) = x_3(t_0)$$

Differentiating and substitution from Eq. (6-57) gives

$$\dot{y}(t_0) = -4x_3(t_0) + x_2(t_0) + Ku(t_0)$$

Finally, a second differentiation gives

$$\ddot{y}(t_0) = -4[-4x_3(t_0) + x_2(t_0) + Ku(t_0)] + [-x_3(t_0) + x_1(t_0) + 2Ku(t_0)] + K\dot{u}(t_0)$$

Collecting terms then yields

$$\ddot{y}(t_0) = 15x_3(t_0) - 4x_2(t_0) + x_1(t_0) - 2Ku(t_0) + K\dot{u}(t_0)$$

Putting these into matrix vector form gives the linear equation problem

$$\begin{bmatrix} 0 & 0 & 1 \\ 0 & 1 & -4 \\ 1 & -4 & 15 \end{bmatrix} \begin{bmatrix} x_1(t_0) \\ x_2(t_0) \\ x_3(t_0) \end{bmatrix} = \begin{bmatrix} y(t_0) \\ \dot{y}(t_0) - Ku(t_0) \\ \ddot{y}(t_0) + 2Ku(t_0) - K\dot{u}(t_0) \end{bmatrix} \tag{6-67}$$

Equation (6-67) represents a special triangular form of the linear equation problem with ones on the cross diagonal. This form makes the linear equation problem Eq. (6-67) especially easy to solve by a method known as *back substitution*. Namely, working from the top down,

$$x_3(t_0) = y(t_0)$$
$$x_2(t_0) = \dot{y}(t_0) - Ku(t_0) + 4x_3(t_0)$$
$$x_1(t_0) = \ddot{y}(t_0) + 2Ku(t_0) - K\dot{u}(t_0) - 15x_3(t_0) + 4x_2(t_0)$$

At each step, all the required information is known to solve for the one remaining unknown.

It is also interesting that the right-hand side of Eq. (6-67) is the same as Eq. (6-51) for the phase-variable states. Like that case, the state $\mathbf{x}(t)$ is continuous for all t, even for discontinuous inputs, provided $u(t)$ is not an impulse. Again, this stems from the general form of the state equations as

$$\dot{\mathbf{x}}(t) = A\mathbf{x}(t) + Bu(t)$$

where there are no derivatives of $u(t)$ forcing $\dot{\mathbf{x}}(t)$.

6-3 DIAGONAL FORM, DISTINCT ROOTS

The third canonical form examined for a state-space model of the SISO transfer function $H(s)$ is the diagonal canonical form. This is one of the most important canonical forms for quite a number of applications.

Proper Transfer Function, Real Poles

Consider once more

$$H(s) = \frac{K(s^2 + 2s - 15)}{s^3 + 4s^2 + s - 6} = \frac{K(s^2 + 2s - 15)}{(s - 1)(s + 2)(s + 3)} \tag{6-23}$$

Expanding this transfer function in partial fraction form yields

$$H(s) = \frac{r_1}{(s - 1)} + \frac{r_2}{(s + 2)} + \frac{r_3}{(s + 3)}$$

where

$$r_1 = (s - 1)H(s)|_{s=1} = -K$$
$$r_2 = (s + 2)H(s)|_{s=-2} = 5K$$
$$r_3 = (s + 3)H(s)|_{s=-3} = -3K$$

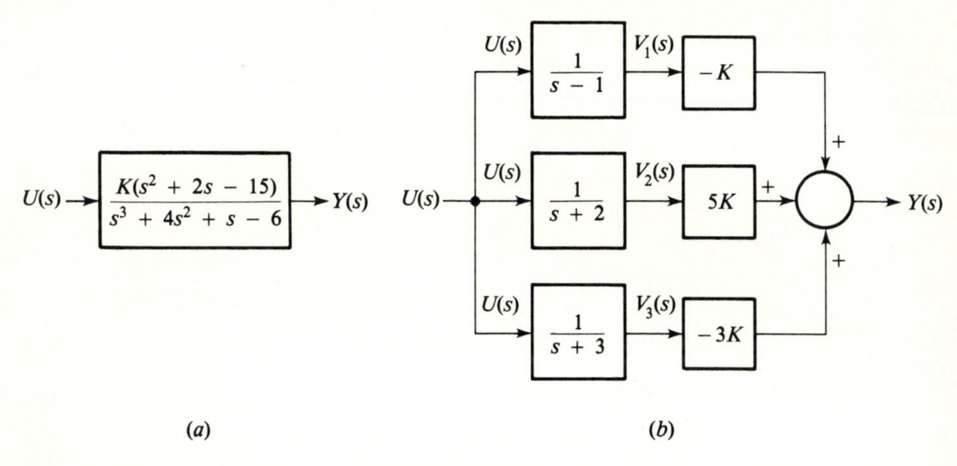

Figure 6-5 Transfer-function diagram of Eq. (6-23).

The transfer function diagram shown in Fig. 6-5*a* is thus equivalent to the parallel transfer functions in Fig. 6-5*b*. The intermediate outputs $V_i(s)$ are defined such that

$$y(t) = -Kv_1(t) + 5Kv_2(t) - 3Kv_3(t)$$

But each of the $V_i(s)/U(s)$ transfer blocks represents a first-order transfer function. Their state equation models become

$$\begin{aligned} \dot{x}_1(t) &= x_1(t) + u(t) \\ v_1(t) &= x_1(t) \\ \dot{x}_2(t) &= -2x_2(t) + u(t) \\ v_2(t) &= x_2(t) \\ \dot{x}_3(t) &= -3x_3(t) + u(t) \\ v_3(t) &= x_3(t) \end{aligned}$$

Hence the intermediate v's are really unnecessary, and so the final diagonal-form state-space model of $H(s)$ is

$$\begin{bmatrix} \dot{x}_1(t) \\ \dot{x}_2(t) \\ \dot{x}_3(t) \end{bmatrix} = \begin{bmatrix} 1 & 0 & 0 \\ 0 & -2 & 0 \\ 0 & 0 & -3 \end{bmatrix} \begin{bmatrix} x_1(t) \\ x_2(t) \\ x_3(t) \end{bmatrix} + \begin{bmatrix} 1 \\ 1 \\ 1 \end{bmatrix} u(t) \tag{6-68}$$

$$y(t) = [-K \quad 5K \quad -3K] \begin{bmatrix} x_1(t) \\ x_2(t) \\ x_3(t) \end{bmatrix} \tag{6-69}$$

Notice that the diagonal entries of the A matrix are the three distinct poles of the system transfer function. The diagonal elements of A are also termed the system *eigenvalues*.

Also note that the B matrix is chosen, for convenience, to be a column of 1s, and the C matrix has all the partial fraction *residues,* the r_i's. This, of course, could have been reversed—make the C matrix a row of 1s and put the residues in the B matrix. Actually there is an infinite number of possibilities here, as the only requirement is that the product of the B and C elements equal the residue values. Making the B matrix all 1s turns out to be desirable for later computational considerations.

The advantages of the diagonal-form states cannot all be understood just now, but one thing that is apparent is that each state, $x_i(t)$, is decoupled from the other states $x_j(t)$; that is, $\dot{x}_i(t)$ is only forced by the input and the state $x_i(t)$ itself. This is not the case in previous phase-variable and simulation diagram forms of the state equations, since the derivative relationships for the states are all intermeshed.

Proper Transfer Function, Complex Poles

Next consider

$$\begin{aligned} H(s) &= \frac{1}{s^3 + 3s^2 + 9s - 13} \\ &= \frac{1}{(s-1)(s^2+4s+13)} \\ &= \frac{1}{(s-1)(s+2-3j)(s+2+3j)} \end{aligned} \tag{6-70}$$

Partial fraction expansion gives

$$H(s) = \frac{r_1}{(s-1)} + \frac{r_2}{(s+2-3j)} + \frac{r_3}{(s+2+3j)}$$

where

$$r_1 = (s-1)H(s)\Big|_{s=1} = \frac{1}{18}$$

$$r_2 = (s+2-3j)H(s)\Big|_{s=-2+3j} = \frac{-1+j}{36}$$

$$r_3 = (s+2+3j)H(s)\Big|_{s=-2-3j} = \frac{-1-j}{36}$$

Therefore, by the previous approach it follows that the diagonal-form state equations are

$$\begin{bmatrix} \dot{x}_1(t) \\ \dot{x}_2(t) \\ \dot{x}_3(t) \end{bmatrix} = \begin{bmatrix} 1 & 0 & 0 \\ 0 & -2+3j & 0 \\ 0 & 0 & -2-3j \end{bmatrix} \begin{bmatrix} x_1(t) \\ x_2(t) \\ x_3(t) \end{bmatrix} + \begin{bmatrix} 1 \\ 1 \\ 1 \end{bmatrix} u(t) \tag{6-71}$$

$$y(t) = \begin{bmatrix} \frac{1}{18} & \frac{-1+j}{36} & \frac{-1-j}{36} \end{bmatrix} \begin{bmatrix} x_1(t) \\ x_2(t) \\ x_3(t) \end{bmatrix} \tag{6-72}$$

Procedures are examined in Chap. 10 to make the diagonal form in the complex case into the almost-diagonal *block diagonal* canonical form, thereby eliminating the complex variables in the A, B, and C matrices.

Boundary Conditions at t_0

The procedure for finding the boundary-condition relations on $\{x_1(t_0), x_2(t_0), \ldots, x_n(t_0)\}$ is again the same; starting with the output relation on $y(t_0)$, differentiate $n - 1$ times to establish n equations in n unknowns. Then form these simultaneous linear equations into a matrix-vector problem and solve.

To illustrate, consider the diagonal-form system in Eqs. (6-68) and (6-69). Equation (6-69) gives the first linear equation

$$y(t_0) = -Kx_1(t_0) + 5Kx_2(t_0) - 3Kx_3(t_0)$$

Differentiating once and substituting from Eq. (6-68) gives

$$\begin{aligned}\dot{y}(t_0) &= -K[x_1(t_0) + u(t_0)] + 5K[-2x_2(t_0) + u(t_0)] - 3K[-3x_3(t_0) + u(t_0)] \\ &= -Kx_1(t_0) - 10Kx_2(t_0) + 9Kx_3(t_0) + Ku(t_0)\end{aligned}$$

Repeating gives the third linear equation

$$\ddot{y}(t_0) = -Kx_1(t_0) + 20Kx_2(t_0) - 27Kx_3(t_0) - 2Ku(t_0) + K\dot{u}(t_0)$$

Combining these gives the linear equation problem in standard form:

$$\begin{bmatrix} -K & 5K & -3K \\ -K & -10K & 9K \\ -K & 20K & -27K \end{bmatrix} \begin{bmatrix} x_1(t_0) \\ x_2(t_0) \\ x_3(t) \end{bmatrix} = \begin{bmatrix} y(t_0) \\ \dot{y}(t_0) - Ku(t_0) \\ \ddot{y}(t_0) + 2Ku(t_0) - K\dot{u}(t_0) \end{bmatrix} \tag{6-73}$$

Notice the curious appearance of the same right-hand side of Eq. (6-73) as in Eqs. (6-51) and (6-67) of the previous phase-variable and simulation diagram state coordinate systems. Again the states are continuous for all t provided $u(t)$ is finite.

6-4 PHYSICAL-VARIABLE STATES

The previous three sections have demonstrated how three different canonical-form state-space models could be derived for a given nth-order SISO transfer function, $H(s)$: the phase-variable form, the form derived from the simulation diagram (this is sometimes referred to as the *observability canonical form*), and the diagonal or *decoupled* form. Yet in none of these canonical forms do the states necessarily correspond to physical variables of the system which might be directly measured.

This section examines a fourth alternative for deriving a state-variable model in which the states are specifically selected to correspond to physical variables of the system. In deriving such a physical-variable state-space model, several general principles are observed. First, a general linear network (be it electrical, mechanical, thermal, or whatever) will typically require one state variable for each independent energy-storage element. Energy-storage elements may be classified as either through-variable energy-storage elements or across-variable energy-storage elements (see Ref. 1 and Appendix A).

In passive electric systems, the through-variable energy-storage elements are inductors; they store energy proportional to the through variable, current. Across-variable energy-storage elements are capacitors; they store energy proportional to the across variable, voltage. Resistors are pure dissipators of energy, and they would have no state variable associated with them.

In simple mechanical translational systems, the through-variable energy-storage elements are springs; they store energy proportional to the through variable, force. The across-variable energy-storage elements are masses; they store energy (kinetic energy) proportional to the across variable, velocity.

The reader should consult Ref. 1 for a deeper explanation of these concepts and for a broader treatment of many other types of systems as well. In any event, the physical-variable states are generally selected as the value of the through variable (e.g., current or force) through independent through-variable energy-storage elements (e.g., inductors or springs) and as the value of the across variable (e.g., voltage or velocity) across the independent across-variable energy-storage elements (e.g., capacitors or masses).

These principles are illustrated in the examples to follow. But one further comment should be made regarding the concept of independent energy-storage elements before proceeding. Generally speaking, through-variable energy-storage elements connected in series are not independent. They may be combined into a single equivalent through-variable energy-storage element. This is illustrated in Fig. 6-6 for two inductors connected in series. Likewise, across-variable energy-storage elements connected in parallel are also not independent. They too may be combined into a single equivalent across-variable energy-storage element. This is illustrated in Fig. 6-7 for two capacitors connected in parallel.

What is often referred to, then, as a *minimal state model* is a state model in which only linearly independent energy-storage elements are represented by states. However, it is often difficult to determine linear independence of the energy-storage elements directly from the physical network. Furthermore, the state

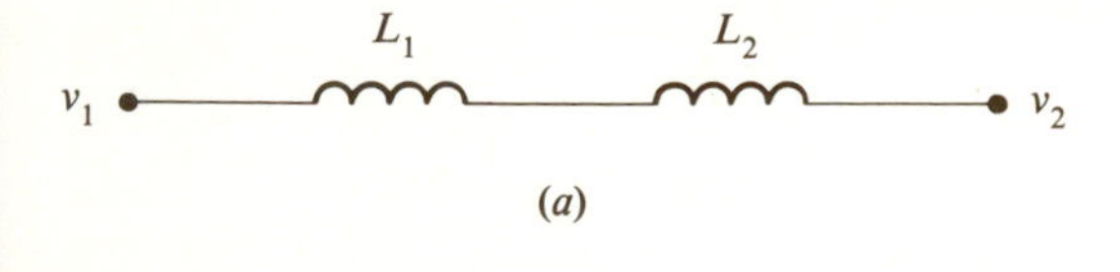

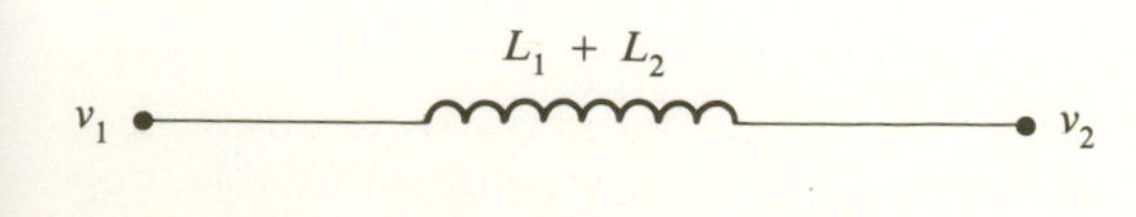

Figure 6-6 Illustration of equivalence of two inductors connected in series.

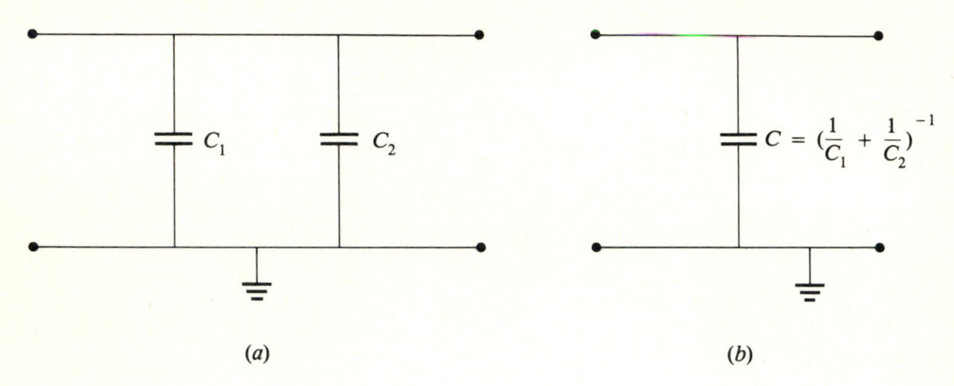

Figure 6-7 Illustration of equivalence of two capacitors connected in parallel.

model from an input-output viewpoint may be reducible regardless of whether or not the energy-storage elements are all independent. These situations are illustrated in Example 6-7. Fortunately, there are systematic state-variable analysis methods related to controllability and observability that will tell us when states are redundant and not necessary in a minimal representation. Additionally, derivation of the transfer function such as described in Appendix A can often be helpful in eliminating states unnecessary to the input-output properties of a given network.

Example 6-3 Consider the parallel *RLC* circuit of Fig. 6-8. A differential equation for the current through the inductor is given by a loop relation around the second loop:

$$L\frac{d}{dt}i_L(t) + \frac{v_C(t)}{R} = 0$$

where $v_C(t)/R$ is the current through the resistor branch.

A differential equation for the voltage drop across the capacitor C is given by a node relation:

$$i_{\text{in}}(t) = \frac{v_2(t)}{R} + i_L(t) + C\frac{d}{dt}v_C(t)$$

Defining $i_{\text{in}}(t) \equiv u(t)$, $x_1(t) \equiv i_L(t)$, and $x_2(t) \equiv v_C(t)$ gives the state-space differential equations

$$\begin{bmatrix} \dot{x}_1(t) \\ \dot{x}_2(t) \end{bmatrix} = \begin{bmatrix} 0 & -\dfrac{1}{LR} \\ -\dfrac{1}{C} & -\dfrac{1}{CR} \end{bmatrix} \begin{bmatrix} x_1(t) \\ x_2(t) \end{bmatrix} + \begin{bmatrix} 0 \\ \dfrac{1}{C} \end{bmatrix} u(t)$$

The output relation is then dependent on the desired output variable. For instance, if the current through the resistor is the desired output variable, then

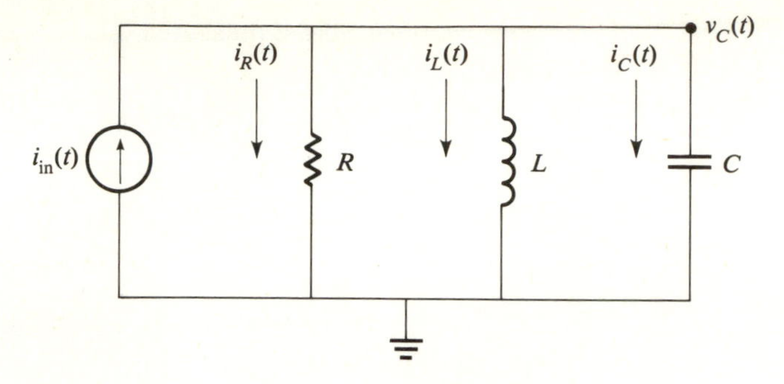

Figure 6-8 Parallel *RLC* network, Example 6-3.

$$y(t) \equiv i_R(t) = \frac{v_C(t)}{R}$$

$$= \begin{bmatrix} 0 & \frac{1}{R} \end{bmatrix} \begin{bmatrix} x_1(t) \\ x_2(t) \end{bmatrix}$$

Example 6-4 In a mechanical system, it is usually convenient to physically measure positions (integrated across-variables) and velocities (across-variables). Picking the position $x_1(t)$ as the first state and $x_2(t) = \dot{x}_1(t)$ as the second state (see Fig. 6-9) yields the force relation at the single node

$$m\dot{x}_2(t) + bx_2(t) + kx_1(t) = F(t)$$

This gives the two state model

$$\begin{bmatrix} \dot{x}_1(t) \\ \dot{x}_2(t) \end{bmatrix} = \begin{bmatrix} 0 & 1 \\ -\frac{k}{m} & -\frac{b}{m} \end{bmatrix} \begin{bmatrix} x_1(t) \\ x_2(t) \end{bmatrix} + \begin{bmatrix} 0 \\ 1 \end{bmatrix} u(t)$$

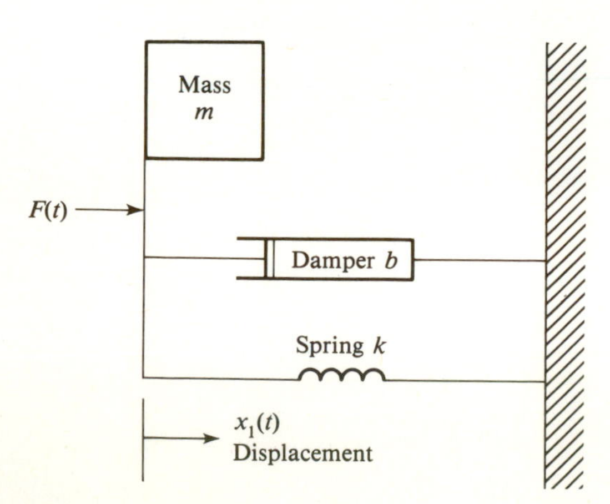

Figure 6-9 Parallel mass, damper, and spring system, Example 6-4.

where $u(t) = F(t)$ is the single force input. If the actual measured output were the force of the spring, then

$$y(t) = F_1(t) = [k \quad 0]\begin{bmatrix} x_1(t) \\ x_2(t) \end{bmatrix}$$

Notice that the reference point for measurement of the displacement $x_1(t)$ is chosen such that $F_1 = 0$ when $x_1 = 0$.

Example 6-5 Consider the rocket-sled system shown in Fig. 6-10. This system is introduced in Chap. 1 and is analyzed to find the output differential equation and transfer function in Chap. 2. Example 6-1 finds the state model in phase-variable form. Now a physical-variable model is derived.

Recalling Eqs. (2-2), (2-5), and (2-6) from Sec. 2.1,

$$m_2 \frac{d}{dt} v_2(t) + b_2 v_2(t) = F_k(t) \tag{2-2}$$

$$m_1 \frac{d}{dt} v_1(t) + b_1 v_1(t) + F_k(t) = u(t) \tag{2-5}$$

$$\frac{d}{dt} F_k(t) = k v_1(t) - k v_2(t) \tag{2-6}$$

Define the physical-variable states $x_1(t) \equiv v_2(t)$, $x_2(t) \equiv v_1(t)$, and $x_3(t) \equiv F_k(t)$. Then Eqs. (2-2), (2-5), and (2-6) may be formed into a standard state-variable model:

$$\begin{bmatrix} \dot{x}_1(t) \\ \dot{x}_2(t) \\ \dot{x}_3(t) \end{bmatrix} = \begin{bmatrix} -\dfrac{b_2}{m_2} & 0 & \dfrac{1}{m_2} \\ 0 & -\dfrac{b_1}{m_1} & -\dfrac{1}{m_1} \\ -k & k & 0 \end{bmatrix} \begin{bmatrix} x_1(t) \\ x_2(t) \\ x_3(t) \end{bmatrix} + \begin{bmatrix} 0 \\ \dfrac{1}{m_1} \\ 0 \end{bmatrix} u(t)$$

Since $y(t) \equiv v_2(t)$, the output relation is

$$y(t) = [1 \quad 0 \quad 0] \begin{bmatrix} x_1(t) \\ x_2(t) \\ x_3(t) \end{bmatrix}$$

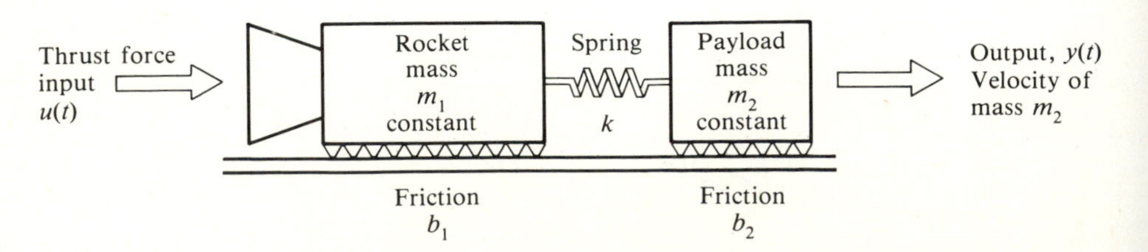

Figure 6-10 Rocket-sled system, Example 6-5.

Choosing the values $m_1 = 100$, $m_2 = 10$, $b_1 = 20$, $b_2 = 2$, and $k = 10$ then yields

$$\begin{bmatrix} \dot{x}_1(t) \\ \dot{x}_2(t) \\ \dot{x}_3(t) \end{bmatrix} = \begin{bmatrix} -0.2 & 0 & 0.1 \\ 0 & -0.2 & -0.01 \\ -10 & 10 & 0 \end{bmatrix} \begin{bmatrix} x_1(t) \\ x_2(t) \\ x_3(t) \end{bmatrix} + \begin{bmatrix} 0 \\ 1 \\ 0 \end{bmatrix} u(t)$$

where the input has been given an extra gain of 100. This physical-variable state model has the same input-output properties as the output differential equation, Eq. (2-12), derived in Sec. 2.1 or as the phase-variable state model derived in Sec. 6.1. From an input-output viewpoint they are all equivalent.

Example 6-6 Consider the series *RLC* circuit in Fig. 6-11. A loop relation yields

$$Ri_1(t) + L\frac{d}{dt}i_1(t) + \frac{1}{C}\int_0^t i_1(\tau)\,d\tau = v_{\text{in}}(t)$$

Then a node relation at $v_2(t)$ yields

$$i_1(t) = C\frac{d}{dt}v_2(t)$$

Integrating this it is seen that

$$v_2(t) = \frac{1}{C}\int_0^t i_1(\tau)\,d\tau$$

Then defining $u(t) \equiv v_{\text{in}}(t)$, $x_1(t) = i_1(t)$, and $x_2(t) = v_2(t)$, the equations above may be combined to give the state differential equations

$$\begin{bmatrix} \dot{x}_1(t) \\ \dot{x}_2(t) \end{bmatrix} = \begin{bmatrix} -R & -1 \\ \frac{1}{C} & 0 \end{bmatrix} \begin{bmatrix} x_1(t) \\ x_2(t) \end{bmatrix} + \begin{bmatrix} 1 \\ 0 \end{bmatrix} u(t)$$

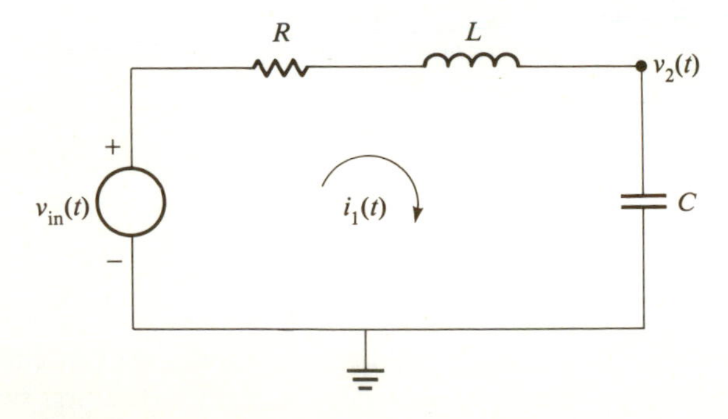

Figure 6-11 Series *RLC* circuit, Example 6-6.

The output relation depends on the particular output variable that is desired. If the voltage drop across the resistor is chosen, then

$$y(t) = [R \quad 0]\begin{bmatrix} x_1(t) \\ x_2(t) \end{bmatrix}$$

If the voltage drop across the inductor is chosen as the output, then

$$y(t) = [-R \quad -1]\begin{bmatrix} x_1(t) \\ x_2(t) \end{bmatrix} + v_{\text{in}}(t)$$

which is a feedforward system.

Example 6-7 There appear to be four energy-storage elements in the circuit of Fig. 6-12—three inductors L_1, L_2, and L_3, and capacitor C. Normally, four energy-storage elements would necessitate a four-state model. However, closer examination reveals that L_1 is really superfluous. The current at node v_1 is $i_{\text{in}}(t)$ regardless of the value of L_1 and R_1. Hence, a three-state model should be all that is required.

Normally one would like to choose as physical-variable states the current through the inductors and voltages across the capacitors. Differential equations can be found for $i_1(t)$, $i_2(t)$, and $v_3(t)$ as

$$\frac{di_1(t)}{dt} + 2i_1(t) = v_1(t)$$

$$\frac{di_2(t)}{dt} + 3i_2(t) = v_1(t) - v_3(t)$$

$$\frac{1}{2}\frac{dv_3(t)}{dt} = i_2(t)$$

These look straightforward except for one problem. We do not know $v_1(t)$. The voltage $v_1(t)$ is not the input.

There are systematic transfer-function-related techniques for solving this (see Appendix A), but one "brute-force" method is to use the facts that $i_1(t) = i_{\text{in}}(t) - i_2(t)$ and $i_2(t) = i_{\text{in}}(t) - i_1(t)$. Making these substitutions into the first two differential equations gives

$$\frac{d}{dt}(i_{\text{in}} - i_2) + 2(i_{\text{in}} - i_2) = v_1$$

$$\frac{d}{dt}(i_{\text{in}} - i_1) + 3(i_{\text{in}} - i_1) = v_1 - v_3$$

where the dependence upon t has been dropped for notational convenience. These give two separate relations for v_1, which may be used in the first two

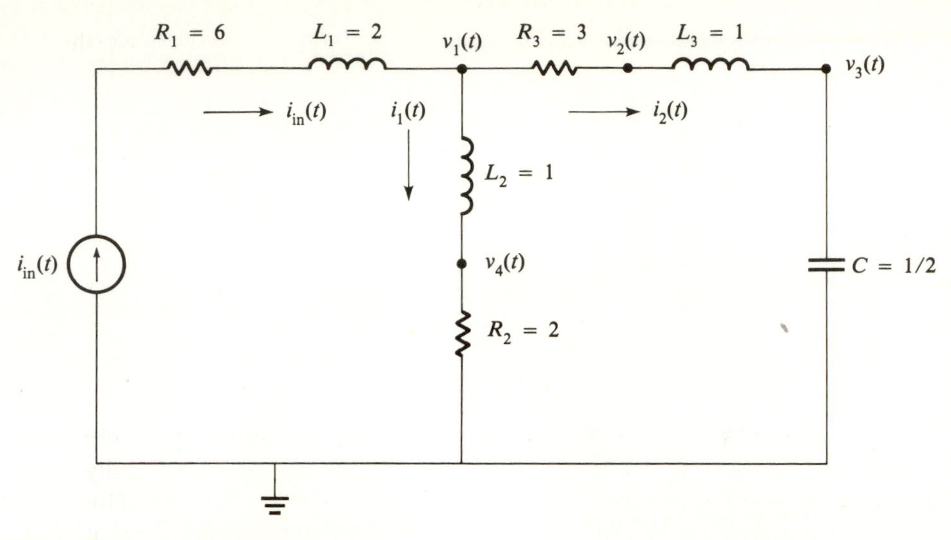

Figure 6-12 Circuit of Example 6-7.

differential equations for i_1 and i_2, respectively. Making these substitutions gives

$$\frac{d}{dt}i_1 + 2i_1 = \frac{d}{dt}(i_{in} - i_1) + 3(i_{in} - i_1) + v_3$$

$$\frac{d}{dt}i_2 + 3i_2 = \frac{d}{dt}(i_{in} - i_2) + 2(i_{in} - i_2) - v_3$$

These two differential equations are not suitable, though, as state equations because of the appearance of the derivatives of the input.

The derivative of the input may be eliminated by defining states $x_1 \equiv (i_1 - \frac{1}{2}i_{in})$ and $x_2 \equiv (i_2 - \frac{1}{2}i_{in})$. Namely, if one rearranges the differential equations above and performs some algebra, this yields

$$\frac{d}{dt}\left(i_1 - \frac{i_{in}}{2}\right) = -\frac{5}{2}\left(i_1 - \frac{i_{in}}{2}\right) + \frac{1}{2}v_3 + \frac{1}{4}i_{in}$$

$$\frac{d}{dt}\left(i_2 - \frac{i_{in}}{2}\right) = -\frac{5}{2}\left(i_2 - \frac{i_{in}}{2}\right) - \frac{1}{2}v_3 - \frac{1}{4}i_{in}$$

Finally, in terms of the variable $i_2 - i_{in}/2$, the v_3 differential equation is

$$\frac{d}{dt}v_3 = 2i_2 = 2\left(i_2 - \frac{i_{in}}{2}\right) + i_{in}$$

Choose as states $x_1 \equiv i_1 - i_{in}/2$, $x_2 \equiv i_2 - i_{in}/2$, and $x_3 \equiv v_3$; choose as output the voltage drop across the capacitor, $y \equiv v_3$; and choose as input

$u \equiv i_{\text{in}}$. These variables give the three-physical-variable state-space model

$$\begin{bmatrix} \dot{x}_1 \\ \dot{x}_2 \\ \dot{x}_3 \end{bmatrix} = \begin{bmatrix} -5/2 & 0 & 1/2 \\ 0 & -5/2 & -1/2 \\ 0 & 2 & 0 \end{bmatrix} \begin{bmatrix} x_1 \\ x_2 \\ x_3 \end{bmatrix} + \begin{bmatrix} 1/4 \\ -1/4 \\ 1 \end{bmatrix} u$$

$$y = [0 \quad 0 \quad 1] \begin{bmatrix} x_1 \\ x_2 \\ x_3 \end{bmatrix}$$

If the reader was observing carefully the derivation of this three-state model for this last example, some possible redundancy and simplification may have been noticed. Indeed, with little difficulty it may be seen that states x_1 and x_2 are redundant. In particular, the two inductors L_2 and L_3 are connected in series and thus may be combined into a single $L_{\text{eq}} = L_2 + L_3$. Hence they only provide a single state variable rather than two. Later analysis of the transfer function also shows that there is a pole-zero cancellation because of the particular values of L_2, R_2, L_3, R_3, and C. The three-state model derived here may thus actually be modeled by a single state model:

$$\dot{x}_1 = -\frac{1}{2}x_1 + u$$

$$y = x$$

Techniques are given in Chaps. 10 and 11 to discern the redundancy that exists in a given state model and eliminate it. See also the transfer-function methods of Appendix A. This raises an important point, though, about deriving a physical-variable state model rather than deriving a state model from a given transfer function. Namely, derivation of such a physical-variable state model can often lead to a state model that is of higher dimension than a minimum-dimension state model required to model the input-output transfer-function properties of the system. The higher-order state model is not incorrect, it simply has more states than are necessary for a minimal description of the system.

6-5 INTRODUCTION TO MIMO STATE EQUATIONS

The foregoing sections have illustrated the ease of deriving a state-space model to characterize the dynamic behavior of a given linear, time-invariant, SISO system. The MIMO case can add new complexities. The last example of the previous section demonstrates how a state-space model with physical-variable states could be generated having more than the minimal number of states necessary to model the SISO transfer function. Such redundant states may even be formed from the transfer function itself in the MIMO case.

Consider the two-input, two-output MIMO transfer function

$$H(s) = \begin{bmatrix} \dfrac{Y_1(s)}{U_1(s)} & \dfrac{Y_1(s)}{U_2(s)} \\ \dfrac{Y_2(s)}{U_1(s)} & \dfrac{Y_2(s)}{U_2(s)} \end{bmatrix}$$

$$= \begin{bmatrix} \dfrac{6s + 5}{s^2 + 3s + 2} & \dfrac{7}{s + 2} \\ \dfrac{5}{s + 2} & \dfrac{6s + 17}{s^2 + 5s + 6} \end{bmatrix} \tag{6-74}$$

as shown in Fig. 6-13. The ith row and jth column element of this transfer function represents the individual transfer function between the jth input alone and the ith output alone, $H_{ij}(s) = Y_i(s)/U_j(s)$.

If phase-variable states are chosen for each input-output pair, the following state-space model may be derived. For $H_{11}(s)$:

$$\begin{bmatrix} \dot{x}_1 \\ \dot{x}_2 \end{bmatrix} = \begin{bmatrix} 0 & 1 \\ -2 & -3 \end{bmatrix} \begin{bmatrix} x_1 \\ x_2 \end{bmatrix} + \begin{bmatrix} 0 \\ 1 \end{bmatrix} u_1$$

$$y_1 = [5 \quad 6] \begin{bmatrix} x_1 \\ x_2 \end{bmatrix}$$

For $H_{21}(s)$:

$$\dot{x}_3 = [-2]x_3 + u_1$$

$$y_2 = [5]x_3$$

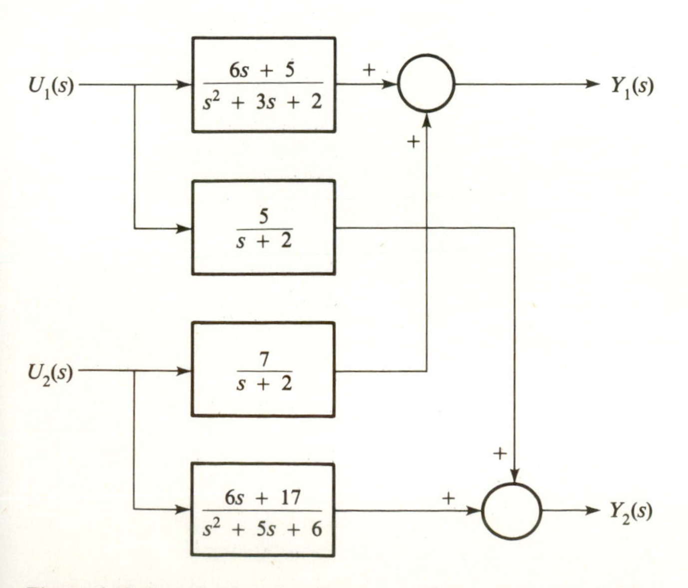

Figure 6-13 Transfer-function diagram, MIMO system, Eq. (6-74).

For $H_{12}(s)$:

$$\dot{x}_4 = [-2]x_4 + u_2$$

$$y_1 = [7]x_4$$

And finally for $H_{22}(s)$:

$$\begin{bmatrix} \dot{x}_5 \\ \dot{x}_6 \end{bmatrix} = \begin{bmatrix} 0 & 1 \\ -6 & -5 \end{bmatrix} \begin{bmatrix} x_5 \\ x_6 \end{bmatrix} + \begin{bmatrix} 0 \\ 1 \end{bmatrix} u_2$$

$$y_2 = [17 \quad 6] \begin{bmatrix} x_5 \\ x_6 \end{bmatrix}$$

Combined into a single matrix-vector state-space model this is

$$\dot{\mathbf{x}} = \begin{bmatrix} 0 & 1 & 0 & 0 & 0 & 0 \\ -2 & -3 & 0 & 0 & 0 & 0 \\ 0 & 0 & -2 & 0 & 0 & 0 \\ 0 & 0 & 0 & -2 & 0 & 0 \\ 0 & 0 & 0 & 0 & 0 & 1 \\ 0 & 0 & 0 & 0 & -6 & -5 \end{bmatrix} \mathbf{x} + \begin{bmatrix} 0 & 0 \\ 1 & 0 \\ 1 & 0 \\ 0 & 1 \\ 0 & 0 \\ 0 & 1 \end{bmatrix} \begin{bmatrix} u_1 \\ u_2 \end{bmatrix} \tag{6-75}$$

$$\begin{bmatrix} y_1 \\ y_2 \end{bmatrix} = \begin{bmatrix} 5 & 6 & 0 & 7 & 0 & 0 \\ 0 & 0 & 5 & 0 & 17 & 6 \end{bmatrix} \mathbf{x} \tag{6-76}$$

where $\mathbf{x}(t)$ is the six-dimensional state vector.

However, by predesign this same 2×2 transfer function may be modeled by

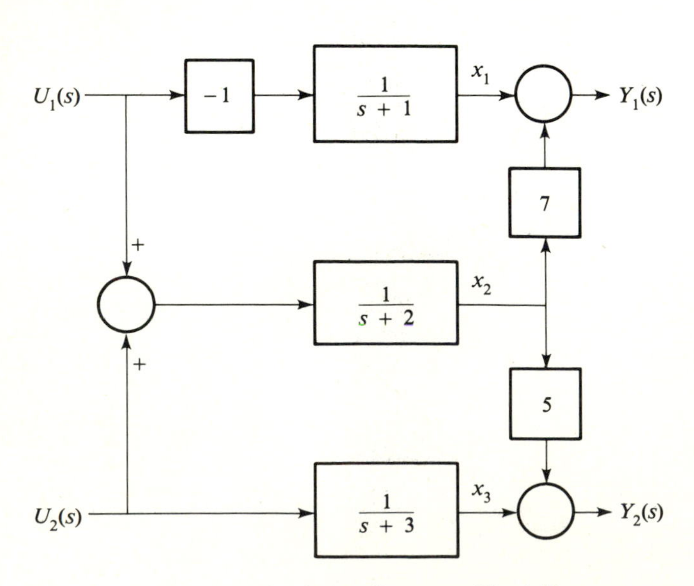

Figure 6-14 Equivalent simulation diagram to Fig. 6-13.

a three-state system model. A way to remove the redundancies can usually be achieved by using a combination of the diagonal form and the simulation diagram. Expanding each of the individual transfer-function blocks in partial fraction form gives

$$H_{11}(s) \equiv \left.\frac{Y_1(s)}{U_1(s)}\right|_{U_2 \equiv 0} = \frac{7}{s+2} + \frac{-1}{s+1} \tag{6-77}$$

$$H_{12}(s) \equiv \left.\frac{Y_1(s)}{U_2(s)}\right|_{U_1 \equiv 0} = \frac{7}{s+2} \tag{6-78}$$

$$H_{21}(s) \equiv \left.\frac{Y_2(s)}{U_1(s)}\right|_{U_2 \equiv 0} = \frac{5}{s+2} \tag{6-79}$$

$$H_{22}(s) \equiv \left.\frac{Y_2(s)}{U_2(s)}\right|_{U_1 \equiv 0} = \frac{5}{s+2} + \frac{1}{s+3} \tag{6-80}$$

The overall simulation block diagram may be constructed as shown in Fig. 6-14. The transfer functions

$$H_{11}(s) \equiv \left.\frac{Y_1(s)}{U_1(s)}\right|_{U_2 \equiv 0} \qquad H_{21}(s) \equiv \left.\frac{Y_2(s)}{U_1(s)}\right|_{U_2 \equiv 0}$$

and

$$H_{12}(s) \equiv \left.\frac{Y_1(s)}{U_2(s)}\right|_{U_1 \equiv 0} \qquad H_{22}(s) \equiv \left.\frac{Y_2(s)}{U_2(s)}\right|_{U_1 \equiv 0}$$

may be verified from Fig. 6-14.

The diagonal-form state model may then be derived from Fig. 6-14. Labeling the output of each integrator block as a state, x_1, x_2, x_3, gives three-state differential equations

$$\begin{bmatrix} \dot{x}_1 \\ \dot{x}_2 \\ \dot{x}_3 \end{bmatrix} = \begin{bmatrix} -1 & 0 & 0 \\ 0 & -2 & 0 \\ 0 & 0 & -3 \end{bmatrix} \begin{bmatrix} x_1 \\ x_2 \\ x_3 \end{bmatrix} + \begin{bmatrix} -1 & 0 \\ 1 & 1 \\ 0 & 1 \end{bmatrix} \begin{bmatrix} u_1 \\ u_2 \end{bmatrix} \tag{6-81}$$

$$\begin{bmatrix} y_1 \\ y_2 \end{bmatrix} = \begin{bmatrix} 1 & 7 & 0 \\ 0 & 5 & 1 \end{bmatrix} \begin{bmatrix} x_1 \\ x_2 \\ x_3 \end{bmatrix} \tag{6-82}$$

6-6 SUMMARY

This chapter has developed methods for deriving several different state-space models in canonical form to represent a given SISO input-output transfer function. The individual states do not necessarily have any physical significance in and of themselves. Then physical-variable state models are derived in which each state variable corresponds to a physically meaningful variable in the system. There is one state for each energy-storage element in the system. However, such a state model can be nonminimal—there are more states than the minimal number required to model the input-output transfer function. The nonminimal model could be reduced without sacrificing any important input-output characteristics of the system. Chapter 11 discusses this further.

The last section introduces the derivation of the MIMO state-space model. The technique of using the diagonal form in combination with the simulation diagram is seen to be a useful tool for avoiding model redundancy.

The next chapter considers solution of the state-space differential equations once they have been obtained. Many familiar topics from Part 1, Classic Analysis, reappear in the state-variable setting.

PROBLEMS

6-1 Find the phase-variable state-space model for the following systems:

(*a*) $H(s) = 3/(s + 2)$
(*b*) $H(s) = (3s + 8)/(s + 2)$
(*c*) $H(s) = 3/(s^2 + 3s + 2)$
(*d*) $H(s) = (3s + 9)/(s^2 + 3s + 2)$
(*e*) $H(s) = (3s^2 + 6s - 4)/(s^2 + 3s + 2)$

Assume $K = 1$ in each case.

6-2 Find the phase-variable state-space model for the following systems:

(*a*) $H(s) = 5/(s - 2)$
(*b*) $H(s) = (5s + 10)/(s - 2)$
(*c*) $H(s) = (5s + 15)/(s^2 + s)$
(*d*) $H(s) = (5s^2 + 4s - 10)/(s^2 + s)$

Assume $K = 1$ in each case.

6-3 Using the following boundary conditions, as applicable

$$y(0) = -1 \qquad \dot{y}(0) = 4 \qquad u(0) = -2 \qquad \dot{u}(0) = 1,$$

find the initial-state vector, $\mathbf{x}(0)$, for the state models in Prob. 6-1.

6-4 Repeat Prob. 6-3 using the same boundary conditions applied to Prob. 6-2.

6-5 Find the observability-form state model, or the states derived from the simulation diagram, for the systems in Prob. 6-1.

6-6 Repeat Prob. 6-5 for the systems of Prob. 6-2.

6-7 Repeat Prob. 6-3 for the state models of Prob. 6-5.

6-8 Repeat Prob. 6-4 for the state models of Prob. 6-6.

6-9 Find the diagonal-form state model for the systems in Prob. 6-1.

6-10 Repeat Prob. 6-9 for the systems in Prob. 6-2.

6-11 Repeat Prob. 6-3 for the state models of Prob. 6-9.

6-12 Repeat Prob. 6-4 for the state models of Prob. 6-10.

6-13 Find physical-variable state models of the electrical circuit shown in Fig. 6-15.

6-14 Find physical-variable state models of the electrical circuit shown in Fig. 6-16.

6-15 Find physical-variable state models of the mechanical network shown in Fig. 6-17.

6-16 Find physical-variable state models of the mechanical network shown in Fig. 6-18.

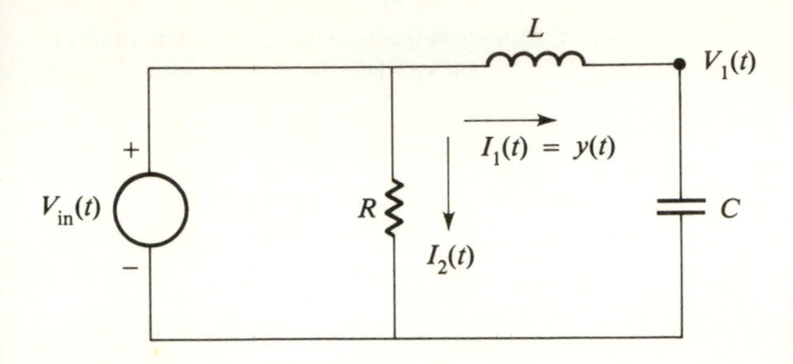

Figure 6-15 Electric circuit for Prob. 6-13.

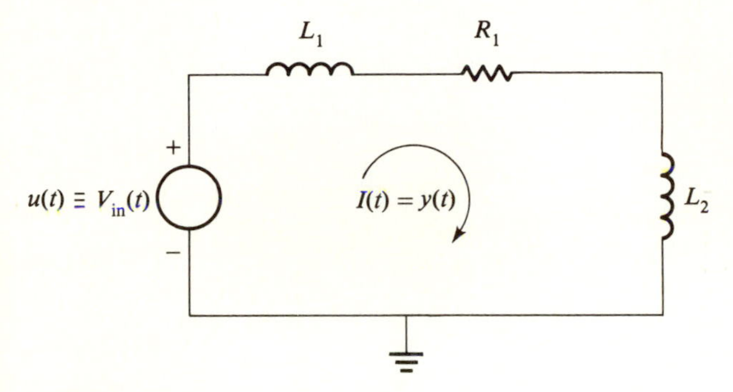

Figure 6-16 Electric circuit for Prob. 6-14.

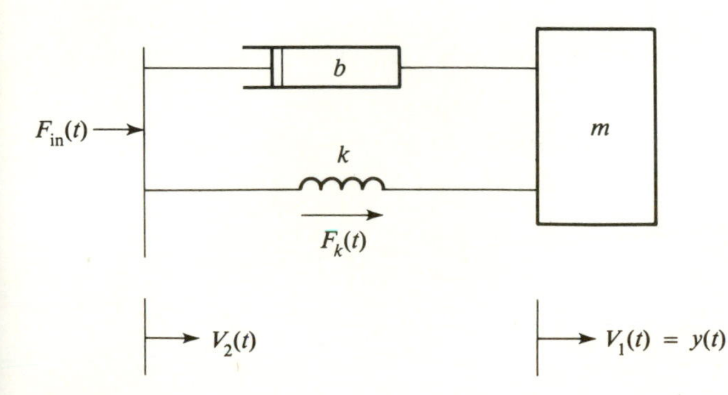

Figure 6-17 Mechanical network for Prob. 6-15.

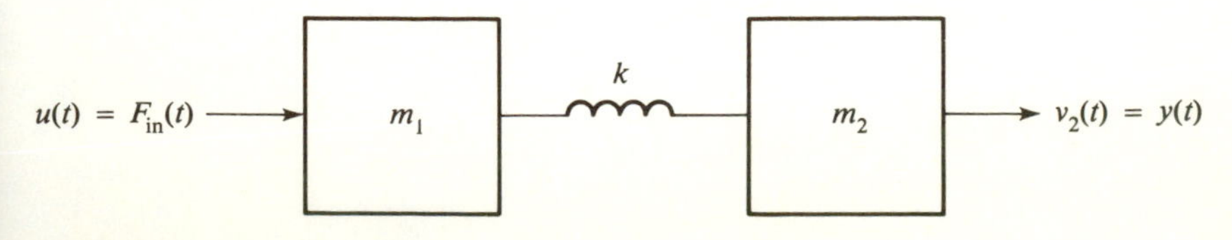

Figure 6-18 Mechanical network for Prob. 6-16.

6-17 By defining phase-variable states for each subtransfer function (each input-output pair combination), derive a state-variable model for each of the following MIMO transfer functions:

$$(a)\ H(s) = \left[\frac{s+2}{s^2+6s+5}, \frac{3}{s+5}\right]$$

$$(b)\ H(s) = \begin{bmatrix} \dfrac{s+2}{s^2+6s+5} \\ \dfrac{3}{s+5} \end{bmatrix}$$

$$(c)\ H(s) = \begin{bmatrix} \dfrac{s+2}{s^2+6s+5} & \dfrac{2}{s^2+6s+5} \\ \dfrac{5}{s^2+6s+5} & \dfrac{3}{s+5} \end{bmatrix}$$

6-18 By defining phase-variable states for each subtransfer function (each input-output pair combination), derive a state-variable model for each of the following MIMO transfer functions:

$$(a)\ H(s) = \begin{bmatrix} \dfrac{5}{s+2} \\ \dfrac{s+1}{s^2+s-2} \end{bmatrix}$$

$$(b)\ H(s) = \left[\frac{5}{s+2} \quad \frac{s+1}{s^2+s-2}\right]$$

$$(c)\ H(s) = \begin{bmatrix} \dfrac{s+1}{s^2+s-2} & \dfrac{7}{s+2} \\ \dfrac{4}{s^2+3s+2} & \dfrac{s}{s^2-1} \end{bmatrix}$$

6-19 For each of the MIMO transfer functions in Prob. 6.17 accomplish the following:

1. Expand each subtransfer function in partial fraction expansion form.
2. Draw a simulation diagram of the MIMO system using the results of *a* to try to eliminate redundant poles.
3. Use the results of *b* to derive a "minimal" state-variable model in diagonal form. (Note: The form you end up with is not unique, so it may disagree in the B and C matrices with the answers in Appendix F.)

6-20 Repeat Prob. 6.19 for the MIMO systems of Prob. 6.18.

NOTES AND REFERENCES

Reference 1 provides a deeper discussion of energy-storage elements for a wide class of dynamic systems. See also Ref. 2.

1. Shearer, J. L., A. T. Murphy, and H. H. Richardson: *Introduction to System Dynamics,* Addison-Wesley, Reading, Mass., 1971.
2. Ogata, K.: *Modern Control Engineering,* Prentice-Hall, Englewood Cliffs, N.J., 1979.

CHAPTER

SEVEN

STATE-SPACE ANALYSIS: CONTINUOUS TIME

The previous chapter developed several alternatives for deriving a general state-space model of a given linear time-invariant system. This chapter derives the time solution to this general state-space model. The state-space solution is thus seen to be an alternative to the classic solution approach of Chap. 2 or the Laplace transform solution approach of Chap. 4. New analysis insights are gained, and, in particular, the MIMO situation is handled with facility. Yet in spite of the fundamental differences between the state-space method and the classic methods, some familiar concepts reappear—only dressed differently now in new state-space notation and terminology.

7-1 SIMPLEST CASE: FIRST-ORDER STATE MODEL

This section considers solution and analysis of the first-order scalar state-space model

$$\dot{x}(t) = ax(t) + bu(t) \tag{7-1}$$

$$y(t) = cx(t) + du(t) \tag{7-2}$$

where $x(t)$, $u(t)$ and $y(t)$ are scalar variables and a, b, c, and d are each scalars. This is the simplest case, but all of the methods applied to this scalar system easily generalize to the nth-order, multi-input, multi-output situation.

Let $x(0)$ be the initial state at the assumed initial time $t_0 = 0$. Taking the Laplace transform of Eq. (7-1) and solving for $X(s)$ gives

$$X(s) = (s - a)^{-1}x(0) + (s - a)^{-1}bU(s) \tag{7-3}$$

The first term on the right in Eq. (7-3) is the zero-input portion of the state:

$$X_{zi}(s) = (s - a)^{-1}x(0) \tag{7-4}$$

Taking the inverse Laplace transform gives

$$x_{zi}(t) = e^{at}x(0) \tag{7-5}$$

The second term in Eq. (7-3) is the zero-state portion of the state

$$X_{zs}(s) = (s - a)^{-1}bU(s) \tag{7-6}$$

Using the convolution property of the Laplace transform (multiplication in the Laplace domain equals convolution in the transform domain—see Sec. 4-2) the inverse Laplace transform of Eq. (7-6) yields

$$x_{zs}(t) = \int_0^t e^{a(t-\tau)}bu(\tau)\,d\tau \tag{7-7a}$$

$$= \int_0^t e^{a\xi}bu(t - \xi)\,d\xi \tag{7-7b}$$

The dummy variables of integration τ and ξ are related by $\tau = t - \xi$, and either form of the convolution integral, Eq. (7-7*a*) or Eq. (7-7*b*), are equivalent.

The additive decomposition of the zero-input and zero-state response is always applicable to this linear system model, and so the complete state solution is

$$x(t) = x_{zi}(t) + x_{zs}(t)$$

$$= e^{at}x(0) + \int_0^t e^{a(t-\tau)}bu(\tau)\,d\tau \tag{7-8}$$

The first form of the convolution integral Eq. (7-7*a*) is chosen because the undesirable convolution integral may be transformed to an ordinary integral by using the property that

$$e^{a(t-\tau)} = e^{at}e^{-a\tau} \tag{7-9}$$

Thus the state solution Eq. (7-8) may be written in the more desirable form

$$x(t) = e^{at}x(0) + e^{at}\int_0^t e^{a\tau}bu(\tau)\,d\tau \tag{7-10}$$

This is the more desirable form here for expressing the continuous-time solution, but in the next chapter it is seen that the equivalent form Eq. (7-7*b*) is more convenient for expressing the discrete state model solution.

Using Eq. (7-10), the state $x(t)$ may be determined for any $t \geq 0$ if the specific $x(0)$, initial state, and future input $u(\tau)$, $0 < \tau < t$, are known. Notice that there is no concern about knowing $x(0)$ at 0^- or 0^+, as the state is always continuous for nonimpulsive inputs; $x(0^-) = x(0^+)$ always, provided $u(t) \neq \delta(t)$.

Verification of the State Solution

It is instructive to verify that Eq. (7-10) is indeed the state solution for $t \geq 0$. To do this, two things must be true. First, the boundary condition, $x(0)$, must be satisfied. Second, the solution must satisfy the state differential equation, Eq. (7-1).

For the boundary condition, evaluation of Eq. (7-10) at $t = 0$ gives

$$x(0) = e^{a \cdot 0}x(0) + e^{a \cdot 0} \int_{0^-}^{0^+} e^{-a\tau}bu(\tau)\, d\tau$$

If $u(t) \neq \delta(t)$ is not an impulse, the integral from 0^- to 0^+ is zero. Hence the boundary condition checks.

To verify the differential equation, the solution Eq. (7-10) is put into Eq. (7-1) to give

$$\frac{d}{dt}\left[e^{at}x(0) + e^{at} \int_0^t e^{-a\tau}bu(\tau)\, d\tau\right] \stackrel{?}{=} a\left[e^{at}x(0) + e^{at} \int_0^t e^{-a\tau}bu(\tau)\, d\tau\right] + bu(t)$$

Differentiation of e^{at} is no trouble:

$$\frac{d}{dt}e^{at} = ae^{at}$$

The Leibnitz rule may be used to differentiate the integral

$$\frac{d}{dt}\left[\int_0^t e^{-a\tau}bu(\tau)\, d\tau\right] = e^{-at}bu(t)$$

These results then provide

$$\frac{d}{dt}\left[e^{at}x(0) + e^{at} \int_0^t e^{-a\tau}bu(\tau)\, d\tau\right]$$

$$= ae^{at}x(0) + ae^{at} \int_0^t e^{-a\tau}bu(\tau)\, d\tau + e^{at}e^{-at}bu(t)$$

Collecting terms then gives

$$\frac{d}{dt}\left[\quad\right] = a\left[\quad\right] + bu(t)$$

where the term in brackets is the claimed solution, $x(t)$, Eq. (7-10). This verifies that Eq. (7-10) satisfies both the boundary condition $x(0)$ and the state differential equation, Eq. (7-1).

Generalization to $t_0 \neq 0$

By an approach similar to that applied to the Laplace transform solution when $t_0 \neq 0$, it may be shown that the solution to

$$\dot{x}(t) = ax(t) + bu(t) \tag{7-1}$$

subject to the boundary condition $x(t_0)$ is

$$x(t) = e^{a(t-t_0)}x(t_0) + e^{at}\int_{t_0}^{t} e^{-a\tau}bu(\tau)\,d\tau \tag{7-11}$$

Verification that Eq. (7-11) is the solution to Eq. (7-1) proceeds in the same way as verification of Eq. (7-10). Evaluation of Eq. (7-11) at $t = t_0$ gives $x(t_0) = e^{a\cdot 0}x(t_0) + 0 = x(t_0)$. Substitution of Eq. (7-11) into the differential equation, Eq. (7-1), again shows equality (see Prob. 7-3). Thus Eq. (7-11) is indeed the general state model solution for this scalar state case. In the next section it is seen that this same exact form is repeated for the general case. The scalars a, b, c, and d then simply become matrices A, B, C, and D, and the scalar variables $x(t)$, $y(t)$, and $u(t)$ simply become vector-state, output, and input variables.

Determination of the Transfer Function

Recall from Chap. 4 that $H(s)$ is given by

$$H(s) = \frac{Y_{zs}(s)}{U(s)} \tag{7-12}$$

But from Eq. (7-2) it is true that

$$Y_{zs}(s) = cX_{zs}(s) + dU(s) \tag{7-13}$$

Substitution of Eq. (7-6) for $X_{zs}(s)$ then gives

$$\begin{aligned} Y_{zs}(s) &= c(s-a)^{-1}bU(s) + dU(s) \\ &= [c(s-a)^{-1}b + d]U(s) \end{aligned} \tag{7-14}$$

From Eq. (7-12) it is seen that the term inside brackets is $H(s)$:

$$\begin{aligned} H(s) &= c(s-a)^{-1}b + d \\ &= \frac{cb}{s-a} + d \\ &= \frac{ds + (cb - da)}{s-a} \end{aligned} \tag{7-15}$$

This transfer function corresponds to the output differential equation model

$$\dot{y}(t) - ay(t) = d\dot{u}(t) + (cb - da)u(t) \tag{7-16}$$

This is the output differential equivalent to the state-space model Eqs. (7-1) and (7-2).

The transfer function Eq. (7-15) and output differential-equation model Eq. (7-16) for a given state model, Eqs. (7-1) and (7-2), are always unique; however, going the other way (from a transfer function to a state model) is not unique. Indeed, there are an infinite number of state models all corresponding to the same transfer function. For example, an alternative state model corresponding to the transfer function, Eq. (7-15), is

$$\dot{x}'(t) = ax'(t) + u(t) \tag{7-17}$$

$$y(t) = (cb)x'(t) + du(t) \tag{7-18}$$

That this is so may be checked by taking the Laplace transform of Eqs. (7-17) and (7-18) and solving for $H(s) = Y_{zs}(s)/U(s)$:

$$X'_{zs}(s) = (s - a)^{-1}U(s)$$

$$Y_{zs}(s) = (cb)X_{zs}(s) + dU(s)$$

$$= (s - a)^{-1}(cb)X_{zs}(s) + d \cdot U(s)$$

or

$$H(s) = \frac{cb}{s - a} + d$$

as before. Notice that in the $x'(t)$ state model, Eqs. (7-17) and (7-18), the variables $u(t)$ and $y(t)$ do not change. These are invariant in any state coordinate transformations because they are input-output, external variables, and not internal variables.

Example 7-1

$$\dot{x}(t) = -2x(t) + 5u(t)$$

$$y(t) = 7x(t)$$

with initial state $x(3) = 8$. The state solution for $t \geq 3$ is

$$x(t) = e^{-2(t-3)}\,8 + 5e^{-2t}\int_3^t e^{2\tau}u(\tau)\,d\tau$$

with output

$$y(t) = 7x(t)$$

For any given input $u(t)$, $t \geq 3$, the integral may be solved to give the complete state solution. For example, suppose $u(t) = \sin(4t)$. Then, by a table of integrals,

$$\int_3^t e^{2\tau}\sin(4\tau)\,d\tau = \frac{e^{2\tau}[2\sin(4\tau) - 4\cos(4\tau)]}{20}\bigg|_3^t$$

$$= \frac{e^{2t}[2\sin(4t) - 4\cos(4t)]}{20} - \frac{e^{6}[2\sin(12) - 4\cos(12)]}{20}$$

Substitution of this into the expression for $x(t)$ give $x(t)$ for the input $u(t) = \sin(4t)$. The output $y(t)$ may then be found for $y(t) = 7x(t)$.

Even for this simple example it is seen that the state-space solution by hand is not one of the high points of the state-space methods. The real power of state-space methods comes with digital analysis and digital simulation as studied in the next chapter. Also, the form of the state-space solution is quite valuable for analysis and design, particularly control-system design.

7-2 GENERAL STATE-SPACE MODEL

Consider the general nth-order, ℓ-input, m-output state-space model

$$\dot{\mathbf{x}}(t) = A\mathbf{x}(t) + B\mathbf{u}(t) \tag{7-19}$$

$$\mathbf{y}(t) = C\mathbf{x}(t) + D\mathbf{u}(t) \tag{7-20}$$

with initial state $\mathbf{x}(0)$ at $t_0 = 0$. The matrices $\{A, B, C, D\}$ are $n \times n$, $n \times \ell$, $m \times n$, and $m \times \ell$, respectively. Analysis and solution of this general state model proceeds in direct analogy to the scalar state model of the previous section.

Taking the Laplace transform of Eq. (7-19) gives

$$sX(s) - \mathbf{x}(0) = AX(s) + BU(s)$$

Solving for $X(s)$ then gives

$$X(s) = [sI - A]^{-1}\mathbf{x}(0) + [sI - A]^{-1}BU(s) \tag{7-21}$$

Again this may be separated into the zero-input and zero-state portions:

$$X_{zi}(s) = [sI - A]^{-1}\mathbf{x}(0) \tag{7-22}$$

$$X_{zs}(s) = [SI - A]^{-1}BU(s) \tag{7-23}$$

Equations (7-21) through (7-23) may be used in the Laplace transform of Eq. (7-20) to find corresponding expressions for the output:

$$Y(s) = C[sI - A]^{-1}\mathbf{x}(0) + \{C[sI - A]^{-1}B + D\}U(s) \tag{7-24}$$

$$Y_{zi}(s) = C[sI - A]^{-1}\mathbf{x}(0) \tag{7-25}$$

$$Y_{zs}(s) = \{C[sI - A]^{-1}B + D\}U(s) \tag{7-26}$$

Inverse Laplace transform yields the time solution, as it did in the scalar case. Additionally, Eq. (7-26) may be used to identify $H(s)$ from the fact that

$$Y_{zs}(s) = H(s)U(s) \tag{7-27}$$

Therefore, it follows that the $m \times \ell$ transfer-function matrix $H(s)$ is related to the state-space model $\{A, B, C, D\}$ via

$$\boxed{H(s) = C[sI - A]^{-1}B + D} \tag{7-28}$$

Again this corresponds to the scalar state case; however, the only factor in all of these expressions that would appear to give much difficulty is the $n \times n$ *resolvent* matrix, $[sI - A]^{-1}$. The other factors are all constant matrices B, C, D, and so they are treated as constant multipliers similar to the scalar multipliers. The only difficulty in working with matrices is that the basic rules of matrix multiplication must be observed.

7-3 THE RESOLVENT MATRIX AND e^{At}

In analysis of the state-space model, Eqs. (7-19) and (7-20), by Laplace transform techniques the $n \times n$ resolvent matrix, $[sI - A]^{-1}$, plays a fundamental role. Then to find the time solution from Eqs. (7-21) through (7-23) for the state or Eqs. (7-24) through (7-26) for the output, one must take the inverse Laplace transform of $[sI - A]^{-1}$. The result is the $n \times n$ matrix exponential function, e^{At}. This section examines these issues so that the results may be applied to the equations of Sec. 7-2.

First of all, $[sI - A]$ is an $n \times n$ matrix. For example, consider the 2×2

$$A = \begin{bmatrix} 0 & 1 \\ -2 & -3 \end{bmatrix}$$

Then

$$[sI - A] = s\begin{bmatrix} 1 & 0 \\ 0 & 1 \end{bmatrix} - \begin{bmatrix} 0 & 1 \\ -2 & -3 \end{bmatrix}$$

$$= \begin{bmatrix} s & -1 \\ 2 & s + 3 \end{bmatrix}$$

To invert $[sI - A]$, Cramer's rule may be used (see Appendix B):

$$[sI - A]^{-1} = \frac{\text{adj}\,[sI - A]}{\det\,[sI - A]} \tag{7-29}$$

For the example $[sI - A]$,

$$\det\,[sI - A] = s^2 + 3s + 2$$

and

$$\text{adj}\,[sI - A] = \begin{bmatrix} s + 3 & 1 \\ -2 & s \end{bmatrix}$$

Notice that the example A matrix is the phase-variable A matrix for a system with characteristic polynomial

$$\Delta(s) = s^2 + 3s + 2$$

This is equal to $\det\,[sI - A]$. Indeed, it is always true that

$$\det\,[sI - A] = \Delta(s) = s^n + a_{n-1}s^{n-1} + \cdots + a_0 \tag{7-30}$$

regardless of the state coordinate system. For example, the poles of $\Delta(s)$ are found from

$$\Delta(s) = s^2 + 3s + 2 = (s + 2)(s + 1)$$

Hence the diagonal-form state matrix equivalent to A would be

$$A' = \begin{bmatrix} -1 & 0 \\ 0 & -2 \end{bmatrix}$$

and $$\det[sI - A'] = \det\begin{bmatrix} s+1 & 0 \\ 0 & s+2 \end{bmatrix} = (s+1)(s+2) = \Delta(s)$$

which is invariant.

Using the results for the adjoint matrix and determinant of the example $[sI - A]$ in Cramer's rule, Eq. (7-29), the resolvent matrix is found to be

$$[sI - A]^{-1} = \frac{\begin{bmatrix} s+3 & 1 \\ -2 & s \end{bmatrix}}{(s^2 + 3s + 2)}$$

$$= \begin{bmatrix} \dfrac{s+3}{s^2+3s+2} & \dfrac{1}{s^2+3s+2} \\ \dfrac{-2}{s^2+3s+2} & \dfrac{s}{s^2+3s+2} \end{bmatrix}$$

The inverse Laplace transform of $[sI - A]^{-1}$ is then found by taking the inverse Laplace transform of each element of the $n \times n$ $[sI - A]^{-1}$ matrix. Thus for the example

$$\mathcal{L}^{-1}\{[sI - A]^{-1}\} = \begin{bmatrix} \mathcal{L}^{-1}\left\{\dfrac{s+3}{s^2+3s+2}\right\} & \mathcal{L}^{-1}\left\{\dfrac{1}{s^2+3s+2}\right\} \\ \mathcal{L}^{-1}\left\{\dfrac{-2}{s^2+3s+2}\right\} & \mathcal{L}^{-1}\left\{\dfrac{s}{s^2+3s+2}\right\} \end{bmatrix}$$

$$= \begin{bmatrix} 2e^{-t} - e^{-2t} & e^{-t} - e^{-2t} \\ -2e^{-t} + 2e^{-2t} & -e^{-t} + 2e^{-2t} \end{bmatrix}$$

Notice that each element of the $n \times n$ matrix $\mathcal{L}^{-1}\{[sI - A]^{-1}\}$ is made up of linear combinations of $e^{p_i t}$ where p_i, $i = 1, 2, \ldots, n$, are the roots of the characteristic polynomial, or system poles (assuming here n distinct poles). For the example A the poles are $p_1 = -1$ and $p_2 = -2$, and so the elements of $\mathcal{L}^{-1}\{[sI - A]^{-1}\}$ are made up of linear combinations of e^{-t} and e^{-2t}.

This fact is always true regardless of the state coordinate system for the A matrix. For the example diagonal form A', it is true that

$$\mathcal{L}^{-1}\{[sI - A']\} = \mathcal{L}^{-1}\left\{\begin{bmatrix} \dfrac{1}{s+1} & 0 \\ 0 & \dfrac{1}{s+2} \end{bmatrix}\right\}$$

$$= \begin{bmatrix} e^{-t} & 0 \\ 0 & e^{-2t} \end{bmatrix}$$

Again, the elements are linear combinations of e^{-t} and e^{-2t}—very simple linear combinations. This is another desirable feature of the diagonal state coordinate system—it makes finding $\mathcal{L}^{-1}\{[sI - A]^{-1}\}$ very, very easy. In fact, it may be done by inspection.

Example 7-2 Let

$$A = \begin{bmatrix} -1 & 0 & 0 \\ 0 & -2+3j & 0 \\ 0 & 0 & -2-3j \end{bmatrix}$$

Then, by inspection,

$$\mathcal{L}^{-1}\{[sI - A]^{-1}\} = \begin{bmatrix} e^{-t} & 0 & 0 \\ 0 & e^{(-2+3j)t} & 0 \\ 0 & 0 & e^{(-2-3j)t} \end{bmatrix}$$

Notice that it makes no difference that the elements on the diagonal of A are complex; the form is the same, regardless. The inspection method does not work for a general-form A matrix, as $\mathcal{L}^{-1}\{[sI - A]^{-1}\}$ can be made up of complicated linear combinations of the $e^{p_i t}$ terms.

Example 7-3 Let

$$A = \begin{bmatrix} 0 & 1 & 0 \\ 0 & 0 & 1 \\ -13 & -17 & -5 \end{bmatrix}$$

This is the phase-variable representation of the A matrix in the previous example, since

$$\begin{aligned} \Delta(s) &= (s+1)(s+2+3j)(s+2-3j) \\ &= s^3 + 5s^2 + 17s + 13 \end{aligned}$$

Then $\quad \det[sI - A] = s^3 + 5s^2 + 17s + 13$

and

$$\operatorname{adj}[sI - A] = \operatorname{adj}\begin{bmatrix} s & -1 & 0 \\ 0 & s & -1 \\ 13 & 17 & s+5 \end{bmatrix}$$

$$= \begin{bmatrix} s^2+5s+17 & s+5 & 1 \\ -13 & s^2+5s & s \\ -13s & -(17s+13) & s^2 \end{bmatrix}$$

Thus

$$[sI - A]^{-1} = \begin{bmatrix} s^2+5s+17 & s+5 & 1 \\ -13 & s^2+5s & s \\ -13s & -(17s+13) & s^2 \end{bmatrix} \Big/ (s^3 + 5s^2 + 17s + 13)$$

After considerable labor in partial fraction expansions, it is found that

$$\mathcal{L}^{-1}\{[sI - A]^{-1}\} = \begin{bmatrix} 13/10 & 4/10 & 1/10 \\ -13/10 & -4/10 & -1/10 \\ 13/10 & 17/10 & 5/10 \end{bmatrix} e^{-t}$$

$$+ \begin{bmatrix} -13/10 & -4/10 & -1/10 \\ 13/10 & 4/10 & 1/10 \\ -13/10 & -17/10 & -5/10 \end{bmatrix} e^{-2t} \cos(3t)$$

$$+ \begin{bmatrix} 7/30 & 6/30 & -1/30 \\ 13/30 & 21/30 & 11/30 \\ -143/30 & -187/30 & -55/30 \end{bmatrix} e^{-2t} \sin(3t)$$

By this example the clear advantage of the diagonal-form A matrix is seen for this particular application. One would hope for a better method for finding $\mathcal{L}^{-1}\{[sI - A]^{-1}\}$ in the general case. Appendix D presents a recursive method known as Fadeev's method. The algorithm of Appendix D would be suitable for digital computation. Other methods are also considered in Chap. 10.

The Matrix Exponential Function

By appealing to the analogy in the scalar case

$$\mathcal{L}^{-1}\{(s - a)^{-1}\} = e^{at} \tag{7-30}$$

the inverse Laplace transform of the $n \times n$ resolvent matrix is defined as the $n \times n$ matrix exponential function, e^{At}:

$$\boxed{\mathcal{L}^{-1}\{[sI - A]^{-1}\} \equiv e^{At}} \tag{7-31}$$

The e^{At} is an $n \times n$ matrix made up of linear combinations of the $e^{p_i t}$ terms (it is, simply, the $n \times n$ $\mathcal{L}^{-1}\{[sI - A]^{-1}\}$ matrix).

For the example A matrix

$$e^{At} = \exp\left(\begin{bmatrix} 0 & 1 \\ -2 & -3 \end{bmatrix} t\right) = \begin{bmatrix} 2e^{-t} - e^{-2t} & e^{-t} - e^{-2t} \\ -2e^{-t} + 2e^{-2t} & -e^{-t} + 2e^{-2t} \end{bmatrix}$$

or for the A' diagonal matrix

$$e^{A't} = \exp\left(\begin{bmatrix} -1 & 0 \\ 0 & -2 \end{bmatrix} t\right) = \begin{bmatrix} e^{-t} & 0 \\ 0 & e^{-2t} \end{bmatrix}$$

7-4 PROPERTIES OF THE MATRIX EXPONENTIAL FUNCTION

In Chap. 10 sophisticated matrix-theory techniques are used to analyze and calculate e^{At}, but for now one may take Eq. (7-31) as both a definition and a method for calculation of e^{At}. However, it is quite informative to consider some further analogies between the scalar e^{at} and the matrix e^{At}. Nearly all properties which are true of the scalar e^{at} are also true of the matrix e^{At} when properly interpreted in matrix expressions. These properties of e^{At} prove to be invaluable in the analysis of the state-space model.

One of the most important analogies is the power-series expression for e^{At}. For the scalar a it is true that

$$e^{at} = 1 + (at) + \frac{(at)^2}{2!} + \frac{(at)^3}{3!} + \cdots \tag{7-32}$$

This infinite series converges for any (at), but it converges most quickly if $|(at)| < 1$. Equation (7-32) may actually be considered as the definition of e^{at}; most all of the properties of e^{at} may be derived from Eq. (7-32).

In a like fashion, the matrix exponential function satisfies the power-series relation

$$\boxed{e^{At} = I + At + \frac{A^2t^2}{2!} + \frac{A^3t^3}{3!} + \cdots} \tag{7-33}$$

Again this converges for any At, but it converges quickly only if the norm of At is less than 1. More is discussed about the issue of matrix norms and the rapid convergence of Eq. (7-33) in Chap. 8. Indeed, the relation Eq. (7-33) is the foundation of computational methods for discrete state simulation studied in Chap. 8. Again Eq. (7-33) may be taken as a definition of e^{At}, but it is not particularly useful for finding a closed-form expression for e^{At} with variable t. In contrast, the alternative definition Eq. (7-31) more easily finds e^{At} with variable t.

Equation (7-33) is most useful here to derive other important properties of e^{At}. For instance, to find the derivative of e^{At}, differentiation of Eq. (7-33) gives

$$\begin{aligned}
\frac{d}{dt}e^{At} &= \frac{d}{dt}\left(I + At + \frac{A^2t^2}{2!} + \frac{A^3t^3}{3!} + \cdots\right) \\
&= A + A^2t + \frac{A^3t^2}{2!} + \frac{A^4t^3}{3!} + \cdots \\
&= A\left(I + At + \frac{A^2t^2}{2!} + \frac{A^3t^3}{3!} + \cdots\right) \\
&= \left(I + At + \frac{A^2t^2}{2!} + \frac{A^3t^3}{3!} + \cdots\right)A
\end{aligned}$$

The term inside parentheses is e^{At}, and so it has been proven that

$$\boxed{\frac{d}{dt}e^{At} = Ae^{At} = e^{At}A} \tag{7-34}$$

This is a matrix differential equation for the $n \times n$ matrix e^{At}. This differential equation may be used to calculate e^{At}. The boundary condition on the differential equation at $t = 0$ is

$$e^{A \cdot 0} = I + A \cdot 0 + \frac{A \cdot 0^2}{2!} + \cdots$$

or

$$\boxed{e^{A \cdot 0} = I} \tag{7-35}$$

Example 7-4 Let

$$A = \begin{bmatrix} 0 & 1 \\ -2 & -3 \end{bmatrix}$$

Then denote the 2×2 e^{At} matrix by

$$e^{At} = \Phi(t) = \begin{bmatrix} \Phi_{11}(t) & \Phi_{12}(t) \\ \Phi_{21}(t) & \Phi_{22}(t) \end{bmatrix}$$

The differential equation for e^{At} gives

$$\frac{d}{dt}\begin{bmatrix} \Phi_{11}(t) & \Phi_{12}(t) \\ \Phi_{21}(t) & \Phi_{22}(t) \end{bmatrix} = \begin{bmatrix} 0 & 1 \\ -2 & -3 \end{bmatrix}\begin{bmatrix} \Phi_{11}(t) & \Phi_{12}(t) \\ \Phi_{21}(t) & \Phi_{22}(t) \end{bmatrix}$$

with boundary condition at $t = 0$

$$\Phi(0) = \begin{bmatrix} 1 & 0 \\ 0 & 1 \end{bmatrix}$$

The differential equation for $\Phi(t)$ may be written as four coupled linear differential equations for each $\Phi_{ij}(t)$ element

$$\begin{aligned}
\dot{\Phi}_{11}(t) &= \Phi_{21}(t) & \Phi_{11}(0) &= 1 \\
\dot{\Phi}_{21}(t) &= -2\Phi_{11}(t) - 3\Phi_{21}(t) & \Phi_{21}(0) &= 0 \\
\dot{\Phi}_{12}(t) &= \Phi_{22}(t) & \Phi_{12}(0) &= 0 \\
\dot{\Phi}_{22}(t) &= -2\Phi_{12}(t) - 3\Phi_{22}(t) & \Phi_{22}(0) &= 1
\end{aligned}$$

Because these differential equations are all intercoupled (that is, $\dot{\Phi}_{11}$ is dependent upon Φ_{21}, and so forth) their solution by hand is quite difficult. However, by numerical integration on the digital computer the integration of a set of coupled first-order differential equations presents no difficulty. This provides an alternative numerical procedure for finding $e^{At} = \Phi(t)$. Indeed, this method for finding the $\Phi(t)$ matrix may be generalized to the time-varying $A(t)$ matrix. Then $\Phi(t)$ is denoted by $\Phi(t, t_0)$, where t_0 is an initial time. $\Phi(t, t_0)$ is termed the *system state transition matrix* corresponding to the $A(t)$ plant matrix. See Chap. 12. The state transition matrix is fundamental to the internal state-space analysis in the same way that the impulse response function, $h(t)$, is fundamental to the external input-output analysis. The state transition matrix for the constant A matrix is $\Phi(t) = e^{At}$.

Continuing with the properties of e^{At}, we may find an expression for the integral of e^{At} by integrating Eq. (7-33):

$$\int_0^t e^{A\tau}\,d\tau = \int_0^t \left(I + A\tau + \frac{A^2\tau^2}{2!} + \cdots\right) d\tau$$

Integrating and evaluating from 0 to t gives

$$\boxed{\int_0^t e^{A\tau}\,d\tau = It + \frac{At^2}{2!} + \frac{A^2t^3}{3!} + \cdots} \tag{7-36}$$

Premultiplying (or postmultiplying) both sides of this expression by A and adding I then gives

$$\begin{aligned} A\int_0^t e^{A\tau}\,d\tau + I &= \int_0^t e^{A\tau}\,d\tau A + I \\ &= I + At + \frac{A^2t^2}{2!} + \frac{A^3t^3}{3!} + \cdots \\ &= e^{At} \end{aligned}$$

Solving for $\int_0^t e^{A\tau}\,d\tau$ (assuming A^{-1} exists—that is, A is nonsingular) then gives

$$\boxed{\begin{aligned} \int_0^t e^{A\tau}\,d\tau &= A^{-1}(e^{At} - I) \\ &= (e^{At} - I)A^{-1} \end{aligned}} \tag{7-37}$$

Example 7-5 Let

$$A = \begin{bmatrix} 0 & 1 \\ -2 & -3 \end{bmatrix}$$

Then

$$A^{-1} = \begin{bmatrix} -3/2 & -1/2 \\ 1 & 0 \end{bmatrix}$$

and from before

$$e^{At} = \begin{bmatrix} 2e^{-t} - e^{-2t} & e^{-t} - e^{-2t} \\ -2e^{-t} + 2e^{-2t} & -e^{-t} + 2e^{-2t} \end{bmatrix}$$

Then

$$\int_0^t e^{A\tau}\,d\tau = A^{-1}(e^{At} - I)$$

or

$$\int_0^t e^{A\tau}\,d\tau = \begin{bmatrix} -3/2 & -1/2 \\ 1 & 0 \end{bmatrix} \begin{bmatrix} 2e^{-t} - e^{-2t} - 1 & e^{-t} - e^{-2t} \\ -2e^{-t} + 2e^{-2t} & -e^{-t} + 2e^{-2t} - 1 \end{bmatrix}$$

The integral relation of e^{At} proves quite valuable in later analysis and, in particular, for consideration of the discrete state simulation as presented in the next chapter. For this application the power-series Eq. (7-36) is actually more useful than Eq. (7-37).

Like the scalar relation $e^{a(t_1+t_2)}$, it is true that

$$\boxed{e^{A(t_1+t_2)} = e^{At_1}e^{At_2}} \tag{7-38}$$

This is called the *semigroup* property and may again be derived from the power-series relation Eq. (7-33). (See Prob. 7-19.) The analogy to the scalar e^{at} breaks down slightly, though, because

$$\boxed{\begin{aligned} e^{(A_1+A_2)t} &= e^{A_1t}e^{A_2t} \\ \text{if and only if} \quad A_1A_2 &= A_2A_1 \end{aligned}} \tag{7-39}$$

(See Prob. 7-20.)

The properties and relations involving e^{At} are summarized in Table 7-1. The last entry, number 7, is quite important in regard to the relation between e^{At} and system stability. Recall that a linear time-invariant system is stable if and only if all poles lie in the left half-plane (have negative real parts). But the $n \times n$ ele-

Table 7-1 Analogy of e^{at} and e^{At}

1. $e^{at} \equiv 1 + (at) + \dfrac{(at)^2}{2!} + \cdots$	1. $e^{At} \equiv I + (At) + \dfrac{(A^2t^2)}{2!} + \cdots$
2. $e^{a(t_1+t_2)} = e^{at_1}e^{at_2}$	2. $e^{A(t_1+t_2)} = e^{At_1}e^{At_2}$
3. $e^{(a_1+a_2)t} = e^{a_1t}e^{a_2t}$	3. $e^{(A_1+A_2)t} = e^{A_1t}e^{A_2t}$ *if and only if* $A_1A_2 = A_2A_1$
4. $\dfrac{d}{dt}e^{at} = ae^{at} = e^{at}a$ with boundary condition $e^{a0} = 1$	4. $\dfrac{d}{dt}e^{At} = Ae^{At} = e^{At}A$ with boundary condition $e^{A\cdot 0} = I$
5. $\displaystyle\int_0^t e^{a\tau}\,d\tau = t + \frac{at^2}{2!} + \frac{a^2t^3}{3!} + \cdots = \frac{e^{at}-1}{a}$ if $a \neq 0$	5. $\displaystyle\int_0^t e^{A\tau}\,d\tau = It + \frac{At^2}{2!} + \frac{A^2t^3}{3!} + \cdots = A^{-1}[e^{At} - I] = [e^{At} - I]A^{-1}$ if A^{-1} exists
6. $\mathcal{L}\{e^{at}\} = \dfrac{1}{s-a}$	6. $\mathcal{L}\{e^{At}\} = [sI - A]^{-1}$
7. $\lim\limits_{t\to\infty} e^{at} = 0$ when $\text{Re}\,\{a\} < 0$	7. $\lim\limits_{t\to\infty} e^{At} = 0$ when $\text{Re}\,\{p_i\} < 0$ for all eigenvalues, or poles of A

ments of e^{At} are made up of linear combinations of the $e^{p_i t}$ where the p_i are the n roots of the characteristic polynomial, $\det[sI - A] = 0$. These decay to zero as $t \to \infty$ for "stable" poles.

In Chap. 10 it is seen that the roots of the characteristic polynomial, $\det[sI - A] = 0$, are quite basic to a given $n \times n$ matrix, A. They may be considered without having some underlying state-space model. The roots of the characteristic polynomial for a given A matrix are then called the *eigenvalues* of A. It is customary to denote these eigenvalues as λ_i, $i = 1, 2, \ldots, n$, and to write the characteristic polynomial as $\det[\lambda I - A] = 0$.

7-5 THE GENERAL STATE SOLUTION

Knowing that

$$\mathcal{L}^{-1}\{[sI - A]^{-1}\} = e^{At} \tag{7-31}$$

and how to calculate this, the inverse Laplace transform of Eqs. (7-21) through (7-26) may be made to give

$$\mathbf{x}_{\text{zi}}(t) = e^{At}\mathbf{x}(0) \tag{7-40}$$

$$\mathbf{x}_{\text{zs}}(t) = e^{At}\int_0^t e^{-A\tau}B\mathbf{u}(\tau)\,d\tau \tag{7-41}$$

$$\mathbf{x}(t) = \mathbf{x}_{\text{zi}}(t) + \mathbf{x}_{\text{zs}}(t)$$

$$= e^{At}\mathbf{x}(0) + e^{At}\int_0^t e^{-A\tau}B\mathbf{u}(\tau)\,d\tau \tag{7-42}$$

$$\mathbf{y}_{\text{zi}}(t) = Ce^{At}\mathbf{x}(0) \tag{7-43}$$

$$\mathbf{y}_{\text{zs}}(t) = Ce^{At}\int_0^t e^{-A\tau}B\mathbf{u}(\tau)\,d\tau \tag{7-44}$$

$$\mathbf{y}(t) = \mathbf{y}_{\text{zi}}(t) + \mathbf{y}_{\text{zs}}(t) \tag{7-45}$$

In deriving the zero-state results, use is made of the fact that multiplication in the Laplace domain equals convolution in the time domain and also the semigroup property

$$e^{A(t-\tau)} = e^{At}e^{-A\tau} \tag{7-46}$$

Equations (7-40) through (7-45) are the general state-space solution analogs to Eqs. (7-5), (7-7), and (7-8). Like the scalar case, it may be verified that Eq. (7-43) is the solution to the state differential equation

$$\dot{\mathbf{x}}(t) = A\mathbf{x}(t) + B\mathbf{u}(t) \tag{7-19}$$

with boundary condition $\mathbf{x}(0)$. Use is made of the facts that

$$e^{A\cdot 0} = I \qquad \text{and} \qquad \frac{d}{dt}e^{At} = Ae^{At} \tag{7-47}$$

Table 7-2 Summary of state-space solution

System
$\dot{\mathbf{x}}(t) = A\mathbf{x}(t) + B\mathbf{u}(t)$ with $\mathbf{x}(t_0)$ $\mathbf{y}(t) = C\mathbf{x}(t) + D\mathbf{u}(t)$
Zero-input response
$\mathbf{x}_{zi}(t) = e^{A(t-t_0)}\mathbf{x}(t_0)$ $\mathbf{y}_{zi}(t) = Ce^{A(t-t_0)}\mathbf{x}(t_0)$
Zero-state response
$\mathbf{x}_{zs}(t) = e^{At}\int_{t_0}^{t} e^{-A\tau}B\mathbf{u}(\tau)\,d\tau = \int_0^{t-t_0} e^{A\xi}B\mathbf{u}(t-\xi)\,d\xi$ $\mathbf{y}_{zs}(t) = Ce^{At}\int_{t_0}^{t} e^{-A\tau}B\mathbf{u}(\tau)\,d\tau + D\mathbf{u}(t) = C\int_0^{t-t_0} e^{A\xi}B\mathbf{u}(t-\xi)\,d\xi + D\mathbf{u}(t)$
Total solution
$\mathbf{x}(t) = \mathbf{x}_{zi}(t) + \mathbf{x}_{zs}(t)$ $\mathbf{y}(t) = \mathbf{y}_{zi}(t) + \mathbf{y}_{zs}(t)$

Using these facts the proof follows exactly the same steps as made for the scalar case in Sec. 7-1. (See Prob. 7-24.)

Again, like the scalar case, the general state-space solution may be generalized to the case when the state boundary condition is $\mathbf{x}(t_0)$, $t_0 \neq 0$. The results are summarized in Table 7-2. Applying these results by hand is extremely tedious in all but the most trivial example. The power of the state-space methods is clearly not in the hand-solution capability they provide for finding the state or output as a function of time; their real power comes with the digital computer and the analysis insights that result for such design problems as feedback control.

7-6 FURTHER CONSIDERATION OF THE FUNDAMENTAL CONCEPT OF STATE

In Sec. 1-2 it is remarked that the state of a dynamic system at any time t is a summary of all past information that would allow one to predict all future output response knowing only the future inputs and the system model. This is true of deterministic systems; for random, probabilistic, or stochastic systems this concept of state must be generalized further, but the basic concept still applies.

In Chap. 1 it is seen that *all* past input information provides equivalent knowledge to system state, but this is clearly not a minimal or even efficient representation of state. Now with the general state-space solution as summarized in Table 7-2 and expressed by

$$\mathbf{y}(t) = Ce^{A(t-t_0)}\mathbf{x}(t_0) + \int_{t_0}^{t} Ce^{A(t-\tau)}B\mathbf{u}(\tau)\,d\tau + D\mathbf{u}(t)$$

we see very explicitly that $\mathbf{x}(t_0)$ is a summary of past information that allows us to predict $\mathbf{y}(t)$ for $t \geq t_0$ knowing $\mathbf{u}(\tau)$, $t_0 \leq \tau \leq t$ and the state model $\{A, B, C, D\}$. Since t_0 is arbitrary, this is true of any $\mathbf{x}(t)$.

Although it is not usually possible to write a closed-form solution for $\mathbf{y}(t)$ for the general nonlinear state-space model

$$\dot{\mathbf{x}}(t) = \mathbf{f}(\mathbf{x}, \mathbf{u}, t)$$

$$\mathbf{y}(t) = \mathbf{h}(\mathbf{x}, \mathbf{u}, t)$$

where $\mathbf{f}$ and $\mathbf{h}$ are n-dimensional and m-dimensional nonlinear vector functions of the n-dimensional state $\mathbf{x}(t)$, the ℓ-dimensional input $\mathbf{u}(t)$, and perhaps explicitly the independent variable, t, the state $\mathbf{x}(t_0)$ is generally all the information that one needs (assuming existence and uniqueness of the given nonlinear differential-equation model) to find the future output $\mathbf{y}(t)$. Hence the nomenclature of a *state model* refers to a model with a set of n first-order differential equations.

State Dimension

If the first-order differential equations for $\mathbf{x}(t)$ are ordinary differential equations, the state dimension is n. However, if the state differential equations are differential delay equations or partial differential equations, the state dimension is infinite-dimensional. These notions were introduced in Sec. 1-8.

Here they provide a good illustration of the concept of state. The simple delay state equation model

$$\dot{x}(t) = -5x(t - 0.3) + u(t)$$

$$y(t) = x(t)$$

has an infinite-dimensional state; to find the solution to this delay differential for $t \geq t_0$, one must know the whole line segment of $x(t)$, $t_0 - 0.3 < t \leq t_0$. Namely, to calculate the derivative $\dot{x}(t)$ one must utilize $x(t - 0.3)$. This shifts as t shifts, and so the state of this delay differential-equation system at any time, t, is the line segment $x(\tau)$, $t - 0.3 < \tau \leq t$. This line segment has an infinite number of points, and so the delay-system state model is referred to as being infinite-dimensional.

Minimal State Model

In the previous chapter there is some discussion of deriving a minimal-dimension state model. States are associated with the independent energy-storage elements of a system. This is an important point of view where it is applicable. However, there are many situations where one would like to derive and utilize a state model where the concept of an energy-storage element is difficult to deal with. Economic systems, for example, may not have a good counterpart to through- and across-variable energy-storage elements. Furthermore, in dealing with a system on an experimental input-output basis, one may have no idea regarding the internal energy-storage elements, or the independence of such energy-storage elements. The concepts of a minimal state model and state-model order reduction are complex topics. Chapter 11 addresses these topics further.

7-7 STATE-SPACE SOLUTION TO UNIT SINGULARITY INPUTS

In Part I, Classic Analysis, the unit impulse response and unit step response played fundamental roles. With the general state-space solution as presented in Sec. 7.5, we may express the impulse and step responses in terms of the state-space model.

We consider the general nth-order MIMO state-space system

$$\dot{\mathbf{x}}(t) = A\mathbf{x}(t) + B\mathbf{u}(t) \tag{7-19}$$

$$\mathbf{y}(t) = C\mathbf{x}(t) + D\mathbf{u}(t) \tag{7-20}$$

with zero initial state, $\mathbf{x}(0) = 0$. Since $\mathbf{u}(t)$ is, in general, an ℓ-input vector function

$$\mathbf{u}(t) = \begin{bmatrix} u_1(t) \\ u_2(t) \\ \vdots \\ u_\ell(t) \end{bmatrix}_{\ell \times 1} \tag{7-48}$$

we consider only one input at a time. This is denoted by $u_j(t)$, $j = 1, 2, \ldots, \ell$. Then, for example, we let $u_j(t) = \delta(t)$ and hold all other input channels identically zero. The output response of the m output channels will be the jth-column vector of the impulse response function matrix:

$$[h(t)]_{m \times \ell} = [\mathbf{h}_1(t), \mathbf{h}_2(t), \ldots, \mathbf{h}_\ell(t)] \tag{7-49}$$

The input vector for producing the jth column of $[h(t)]$ is

$$\mathbf{u}_j(t) = \begin{bmatrix} 0 \\ \vdots \\ \delta(t) \\ \vdots \\ 0 \end{bmatrix}_{\ell \times 1} \tag{7-50}$$

where $\delta(t)$ is at the jth-row entry. Denoting the ℓ columns of the input B matrix by $\mathbf{b}_1, \mathbf{b}_2, \ldots, \mathbf{b}_\ell$ such that

$$B = [\mathbf{b}_1, \mathbf{b}_2, \ldots, \mathbf{b}_\ell]_{n\times\ell} \tag{7-51}$$

the term $B\mathbf{u}_j(t)$ with $\mathbf{u}_j(t)$ given by Eq. (7-50) becomes

$$B\mathbf{u}_j(t) = \mathbf{b}_j\, \delta(t) \qquad j = 1, 2, \ldots, \ell \tag{7-52}$$

This same discussion applies to each of the inputs. This notation is used with respect to each of the inputs to follow. The same is done regarding the $m \times \ell$-dimension feedforward D matrix; thus the notation $\mathbf{d}_j$ is used to denote the jth column of D.

The Unit Impulse Response

Let the jth input channel to the MIMO system Eqs. (7-19) and (7-20) be the unit impulse function input, $\delta(t)$, and all other inputs be identically zero. Let the system start from rest, $\mathbf{x}(0) = 0$. Then, applying Eq. (7-44), the unit impulse response vector corresponding to that jth input channel is

$$\mathbf{h}_j(t) = Ce^{At} \int_0^t e^{-A\tau}\mathbf{b}_j\, \delta(\tau)\, d\tau + \mathbf{d}_j\, \delta(t)$$

Using the sampling property of the delta function and the fact that $e^{-A\cdot 0} = I$ then gives

$$\boxed{\mathbf{h}_j(t) = Ce^{At}\mathbf{b}_j + \mathbf{d}_j\, \delta(t)} \tag{7-53}$$

Note that $\mathbf{h}_j(t)$ denotes an m-channel output response for a jth input channel

$$\mathbf{h}_j(t) = \begin{bmatrix} h_{1j}(t) \\ h_{2j}(t) \\ \vdots \\ h_{mj}(t) \end{bmatrix}_{m\times 1} \qquad j = 1, 2, \ldots, \ell \tag{7-54}$$

Forming the matrix of such vector impulse response functions gives the $m \times \ell$-dimension impulse response function matrix

$$\boxed{\begin{aligned} [h(t)]_{m\times\ell} &= [\mathbf{h}_1(t), \mathbf{h}_2(t), \ldots, \mathbf{h}_\ell(t)] \\ &= Ce^{At}B + D\,\delta(t) \end{aligned}} \tag{7-55}$$

It is interesting to take the Laplace transform of the $m \times \ell[h(t)]$. Using the facts that

$$\mathcal{L}\{e^{At}\} = [sI - A]^{-1} \qquad \mathcal{L}\{\delta(t)\} = 1$$

gives

$$\boxed{\mathcal{L}\{[h(t)]\} = C[sI - A]^{-1}B + D} \tag{7-56}$$

This agrees with the previous result regarding $H(s)$, Eq. (7-28), thus showing that even for the MIMO situation it is true that

$$H(s) = \mathcal{L}\{[h(t)]\} \tag{7-57}$$

The Unit Step Response

Let $\mathbf{u}_j(t)$ denote the unit step input $u_{-1}(t)$ coming in the jth input channel with all other inputs identically zero. Then the corresponding output step response is

$$\mathbf{r}_j(t) = Ce^{At} \int_0^t e^{-A\tau}\mathbf{b}_j u_{-1}(\tau)\, d\tau + \mathbf{d}_j$$

Using the equivalent form of the convolution integral this may be written as

$$\mathbf{r}_j(t) = C \int_0^t e^{A\xi}\mathbf{b}_j u_{-1}(t - \xi)\, d\xi + \mathbf{d}_j$$

Then since $u_{-1}(t - \xi) = 1$ for all $0 < \xi < t$ and $\mathbf{b}_j$ is constant, this becomes

$$\mathbf{r}_j(t) = C \int_0^t e^{A\xi}\, d\xi \mathbf{b}_j + \mathbf{d}_j$$

Finally, using the integration property of e^{At} [see Eq. (7-37)], this becomes

$$\boxed{\begin{aligned} \mathbf{r}_j(t) &= CA^{-1}[e^{At} - I]\mathbf{b}_j + \mathbf{d}_j \\ &= C[e^{At} - I]A^{-1}\mathbf{b}_j + \mathbf{d}_j \end{aligned}} \tag{7-58}$$

provided A^{-1} exists and is nonsingular.

Forming the matrix of step responses gives

$$\boxed{\begin{aligned} [r(t)] &\equiv [\mathbf{r}_1(t), \mathbf{r}_2(t), \ldots, \mathbf{r}_\ell(t)]_{m\times\ell} \\ &= CA^{-1}[e^{At} - I]B + D \\ &= C[e^{At} - I]A^{-1}B + D \end{aligned}} \tag{7-59}$$

Since the unit step input is the integral of the impulse input

$$u_{-1}(t) = \int_0^t \delta(\tau)\, d\tau \tag{7-60}$$

it follows that

$$[r(t)] = \int_0^t [h(\tau)]\, d\tau \tag{7-61}$$

This is indeed true, which may be checked by integrating Eq. (7-55).

The Steady-State Step Response

Taking the limit as $t \to \infty$ of the unit step response, $[r(t)]$, gives the steady-state response to the step input:

$$[r_{ss}] \equiv \lim_{t\to\infty} [r(t)]$$

$$= \lim_{t\to\infty} \{CA^{-1}[e^{At} - I]B + D\} \tag{7-62}$$

Assuming a stable system

$$\lim_{t\to\infty} e^{At} = 0$$

Then

$$\boxed{[r_{ss}] = -CA^{-1}B + D} \tag{7-63}$$

It is interesting that this agrees with the Chap. 2 result that $r_{ss} = y_p(t) = H(0)$ for the unit step input. Since

$$H(s) = C[sI - A]^{-1}B + D \tag{7-28}$$

it is seen that

$$H(s)|_{s=0} = -CA^{-1}B + D \tag{7-64}$$

An Alternative Point of View on the Steady-State Step Response

By another point of view, the equilibrium or steady-state condition on the state differential equation for a constant input $\mathbf{u}(t) = \mathbf{u}$ is

$$0 = A\mathbf{x}_{ss} + B\mathbf{u} \tag{7-65}$$

This condition says that the state derivative relation at state $\mathbf{x}_{ss}$ is zero for the constant input $\mathbf{u}$; if the state derivative is zero, the state does not change in time and hence it is in a steady-state or equilibrium condition. Solving for $\mathbf{x}_{ss}$ gives

$$\mathbf{x}_{ss} = -A^{-1}B\mathbf{u} \tag{7-66}$$

assuming A is nonsingular (has no poles at the origin). Then

$$\mathbf{y}_{ss} = -CA^{-1}B\mathbf{u} + \mathbf{Du} \tag{7-67}$$

Applying this to the jth unit step input function gives

$$\mathbf{r}_{j_{ss}} = -CA^{-1}\mathbf{b}_j + \mathbf{d}_j \qquad j = 1, 2. \ldots, \ell \tag{7-68}$$

Forming a matrix of these ℓ-column vectors then gives Eq. (7-63) as before.

7-8 FURTHER CONSIDERATION OF H(s)

In Secs. 6-4 and 6-5 the problem is encountered that a state-space model might be derived in which there are more states than a minimal state model of the transfer function would yield. The statement was made that the higher-order-than-necessary state model is not incorrect, because it still correctly models the input-output transfer function. With the relation

$$H(s) = C[sI - A]^{-1}B + D \tag{7-28}$$

this fact may now be verified.

To demonstrate, consider a two-input, single-output system with transfer function

$$H(s) = \begin{bmatrix} \dfrac{Y(s)}{U_1(s)} & \dfrac{Y(s)}{U_2(s)} \end{bmatrix}$$
$$= \begin{bmatrix} \dfrac{3}{s+1} & \dfrac{4s-1}{s^2+3s+2} \end{bmatrix}$$

The transfer-function diagram for this system is shown in Fig. 7-1*a*.

Expanding $Y(s)/U_2(s)$ in partial fraction expansion

$$\frac{Y(s)}{U_2(s)} = \frac{4s-1}{s^2+3s+2} = \frac{-5}{s+1} + \frac{9}{s+2}$$

and choosing the diagonal-form state-space model, it is seen that the two-state model is

$$\begin{bmatrix} \dot{x}_1 \\ \dot{x}_2 \end{bmatrix} = \begin{bmatrix} -1 & 0 \\ 0 & -2 \end{bmatrix} \begin{bmatrix} x_1 \\ x_2 \end{bmatrix} + \begin{bmatrix} 3 & -5 \\ 0 & 9 \end{bmatrix} \begin{bmatrix} u_1 \\ u_2 \end{bmatrix}$$

$$y = \begin{bmatrix} 1 & 1 \end{bmatrix} \begin{bmatrix} x_1 \\ x_2 \end{bmatrix}$$

This state model is derived from the equivalent simulation diagram shown in Fig. 7-1*b*.

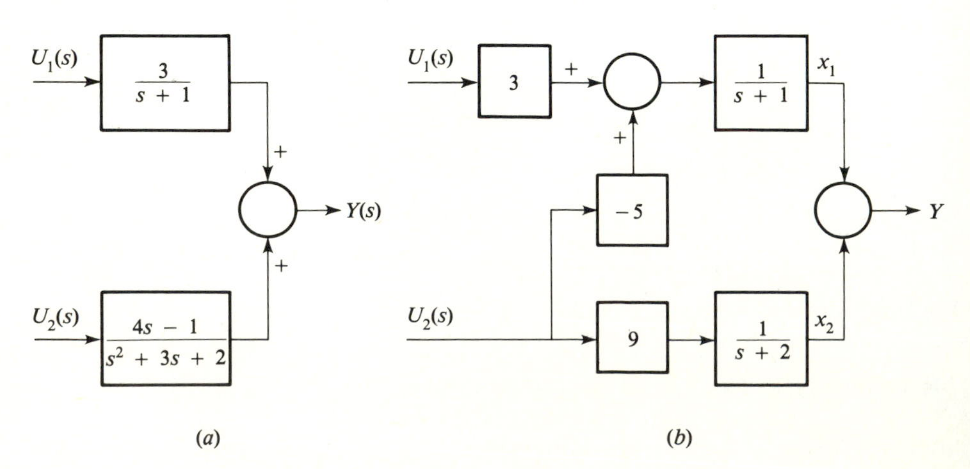

Figure 7-1 Transfer-function diagram and its equivalent.

To verify $H(s)$ we calculate

$$H(s) = [1 \quad 1]\begin{bmatrix} s+1 & 0 \\ 0 & s+2 \end{bmatrix}^{-1}\begin{bmatrix} 3 & -5 \\ 0 & 9 \end{bmatrix}$$

$$= [1 \quad 1]\begin{bmatrix} \dfrac{1}{s+1} & 0 \\ 0 & \dfrac{1}{s+2} \end{bmatrix}\begin{bmatrix} 3 & -5 \\ 0 & 9 \end{bmatrix}$$

or

$$H(s) = \left[\frac{3}{s+1} \quad \frac{-5}{s+1} + \frac{9}{s+2}\right]$$

This is the same as the original $H(s)$, only in partial fraction expansion form.

A three-state model can also be formed by using a phase-variable state model for each subtransfer function:

$\dfrac{3}{s+1}$:

$$\dot{x}_1 = -x_1 + 3u$$

$$v_1 = x_1$$

$\dfrac{4s-1}{s^2+3s+2}$:

$$\begin{bmatrix} \dot{x}_2 \\ \dot{x}_3 \end{bmatrix} = \begin{bmatrix} 0 & 1 \\ -2 & -3 \end{bmatrix}\begin{bmatrix} x_2 \\ x_3 \end{bmatrix} + \begin{bmatrix} 0 \\ 1 \end{bmatrix}u$$

$$v_2 = [-1 \quad 4]\begin{bmatrix} x_2 \\ x_3 \end{bmatrix}$$

Then

$$y = v_1 + v_2$$

The overall three-state model thus derived is

$$\begin{bmatrix} \dot{x}_1 \\ \dot{x}_2 \\ \dot{x}_3 \end{bmatrix} = \begin{bmatrix} -1 & 0 & 0 \\ 0 & 0 & 1 \\ 0 & -2 & -3 \end{bmatrix}\begin{bmatrix} x_1 \\ x_2 \\ x_3 \end{bmatrix} + \begin{bmatrix} 3 & 0 \\ 0 & 0 \\ 0 & 1 \end{bmatrix}\begin{bmatrix} u_1 \\ u_2 \end{bmatrix}$$

$$y = [1 \quad -1 \quad 4]\begin{bmatrix} x_1 \\ x_2 \\ x_3 \end{bmatrix}$$

This model is not incorrect. It still correctly models the input-output properties specified. It simply has one more state than is needed in a minimal representation.

To verify this, form

$$H(s) = [1 \quad -1 \quad 4]\begin{bmatrix} s+1 & 0 & 0 \\ 0 & s & -1 \\ 0 & 2 & s+3 \end{bmatrix}^{-1}\begin{bmatrix} 3 & 0 \\ 0 & 0 \\ 0 & 1 \end{bmatrix}$$

After carrying out all of the algebra it is seen that the $H(s)$ so calculated indeed agrees with the original $H(s)$ input-output transfer function.

7-9 SUMMARY

This chapter has derived the state-space solution for the general nth-order, multi-input, multi-output system. Fundamental to this state-space solution is the state transition matrix or matrix exponential function, $\Phi(t) = e^{At}$. Laplace transform analysis provides a connection between a given state-space model $\{A, B, C, D\}$ and the MIMO transfer function $H(s)$. Finally, the state-space solution to the unit impulse response and the unit step response inputs are derived. Relations derived in the classic analysis of Part I are seen to hold in the MIMO analysis using state-variable methods.

What is less than pleasing is the amount of labor involved in solving for the state continuous-time solution. The next chapter shows that one real power of the state-space methods comes with the discrete time representation and discrete simulation on the digital computer.

This power with regard to digital simulation is based on two fundamental facts as brought out in this chapter:

1. The state is everywhere continuous, even for discontinuous (but nonimpulsive) sample-and-hold-type inputs.
2. The state, $\mathbf{x}(t)$, at any time, t, summarizes all information regarding the past behavior of the system that is necessary to predict the future, knowing also the future input.

These facts are basic to the state-space approach and play a key role in the methods of the next chapter.

PROBLEMS

7-1 Write down the form of the complete solution for $x(t)$ and $y(t)$ for the following system with $x(1) = 7$:

$$\dot{x}(t) = -2x(t) + 3u(t)$$

$$y(t) = 5x(t) + 4u(t)$$

7-2 Repeat Prob. 7-1 for the following system with $x(-1) = 7$:

$$\dot{x}(t) = -x(t) + 2u(t)$$

$$y(t) = -3x(t) + 4u(t)$$

7-3 Verify that the solution to Prob. 7-1 is the correct solution by showing that $x(t)$ satisfies both the boundary condition and the state differential equation.

7-4 Verify that the solution to Prob. 7-2 is the correct solution by showing that $x(t)$ satisfies both the boundary condition and the state differential equation.

7-5 For the input $u(t) = u_{-1}(t)$ and $x(1) = 7$, find the solution to the system in Prob. 7-1 for $t \geq 1$.

7-6 For the input $u(t) = u_{-1}(t)$ and $x(2) = 5$, find the solution to the system in Prob. 7-2 for $t \geq 2$.

7-7 Resolve Prob. 7-5 for the input $u(t) = u_{-1}(t - 1)$, the unit step input with step at $t = 1$. Is there any difference in the solution? Explain.

7-8 Resolve Prob. 7-6 for the input $u(t) = u_{-1}(t - 2)$, the unit step input with step at $t = 2$. Is there any difference in the solution? Explain.

7-9 Find the transfer functions, $H(s)$, for the system in Prob. 7-1.

7-10 Find the transfer functions, $H(s)$, for the system in Prob. 7-2.

7-11 For the following matrices find both $[sI - A]^{-1}$ and e^{At}:

$$(a)\ A = \begin{bmatrix} -1+2j & 0 \\ 0 & -1-2j \end{bmatrix} \qquad (b)\ A = \begin{bmatrix} -1 & 2 \\ -2 & -1 \end{bmatrix}$$

$$(c)\ A = \begin{bmatrix} 0 & 1 \\ -5 & -2 \end{bmatrix} \qquad (d)\ A = \begin{bmatrix} 0 & -5 \\ 1 & -2 \end{bmatrix}$$

7-12 For the following matrices find both $[sI - A]^{-1}$ and e^{At}:

$$(a)\ A = \begin{bmatrix} -2+j & 0 \\ 0 & -2-j \end{bmatrix} \qquad (b)\ A = \begin{bmatrix} -2 & 1 \\ -1 & -2 \end{bmatrix}$$

$$(c)\ A = \begin{bmatrix} 0 & 1 \\ -5 & -4 \end{bmatrix} \qquad (d)\ A = \begin{bmatrix} 0 & -5 \\ 1 & -4 \end{bmatrix}$$

7-13 Find $H(s)$ for each of the following systems. Note that the A matrices are the same ones as in Prob. 7-11.

$$(a)\ \dot{\mathbf{x}} = \begin{bmatrix} -1+2j & 0 \\ 0 & -1-2j \end{bmatrix}\mathbf{x} + \begin{bmatrix} 1 \\ 1 \end{bmatrix}u$$

$$y = [2-3j,\ \ 2+3j]\mathbf{x}$$

$$(b)\ \dot{\mathbf{x}} = \begin{bmatrix} -1 & 2 \\ -2 & -1 \end{bmatrix}\mathbf{x} + \begin{bmatrix} 1 \\ 1 \end{bmatrix}u$$

$$y = [5 \quad -1]\mathbf{x}$$

$$(c)\ \dot{\mathbf{x}} = \begin{bmatrix} 0 & 1 \\ -5 & -2 \end{bmatrix}\mathbf{x} + \begin{bmatrix} 0 \\ 1 \end{bmatrix}u$$

$$y = [16 \quad 4]\mathbf{x}$$

$$(d)\ \dot{\mathbf{x}} = \begin{bmatrix} 0 & -5 \\ 1 & -2 \end{bmatrix}\mathbf{x} + \begin{bmatrix} 16 \\ 4 \end{bmatrix}u$$

$$y = [0 \quad 1]\mathbf{x}$$

(*e*) Are these four systems related to each other? How?

7-14 Find $H(s)$ for each of the following systems. Note that the A matrices are the same as in Prob. 7-12.

$$(a)\ \dot{\mathbf{x}} = \begin{bmatrix} -2+j & 0 \\ 0 & -2-j \end{bmatrix}\mathbf{x} + \begin{bmatrix} 5-3j \\ 5+3j \end{bmatrix}u$$

$$y = [1 \quad 1]\mathbf{x}$$

$$(b)\ \dot{\mathbf{x}} = \begin{bmatrix} -2 & 1 \\ -1 & -2 \end{bmatrix}\mathbf{x} + \begin{bmatrix} 2 \\ 8 \end{bmatrix}u$$

$$y = [1 \quad 1]\mathbf{x}$$

$$(c)\ \dot{\mathbf{x}} = \begin{bmatrix} 0 & 1 \\ -5 & -4 \end{bmatrix}\mathbf{x} + \begin{bmatrix} 0 \\ 2 \end{bmatrix}u$$

$$y = [13 \quad 5]\mathbf{x}$$

$$(d)\ \dot{\mathbf{x}} = \begin{bmatrix} 0 & -5 \\ 1 & -4 \end{bmatrix}\mathbf{x} + \begin{bmatrix} 13 \\ 5 \end{bmatrix}u$$

$$y = [0 \quad 2]\mathbf{x}$$

(*e*) Are these four systems related to each other? How?

7-15 Find $\int_0^t e^{A\tau}\, d\tau$ for the A matrix of Prob. 7-11*a*.

7-16 Repeat Prob. 7-15 for the A matrix of Prob. 7-12*b*.

7-17 Find four scalar differential equations and boundary conditions for the four elements of $\Phi(t)$ using the A matrix of Prob. 7-11*a*. Don't attempt solving; just write down what the differential equations and boundary conditions would be.

7-18 Repeat Prob. 7-17 for the A matrix of Prob. 7-12*b*.

7-19 Use the power-series definition of e^{At}, Eq. (7-33), to demonstrate the validity of Eq. (7-38).

7-20 Use the power-series definition of e^{At}, Eq. (7-33), to demonstrate the validity of Eq. (7-39).

7-21 Suppose the system in Prob. 7-13*b* is at the initial state

$$x(-3) = \begin{bmatrix} 20 \\ -10 \end{bmatrix}$$

and the input is $u(t) = 4 \sin (7t) u_{-1}(t + 3)$. Write down expressions for the complete state and output solutions for $t \geq -3$. Don't try to solve the integral expressions; just write them down. Use the results from Prob. 7-11 regarding e^{At} in doing this.

7-22 Suppose the system in Prob. 7-14*c* is at the initial state

$$x(13) = \begin{bmatrix} 0 \\ -10 \end{bmatrix}$$

and the input $u(t) = 5tu_{-1}(t - 13)$ is applied. Write down expressions for the complete state and output solutions for $t \geq 13$. Don't try to solve the integral expressions; just write them down. Use the results from Prob. 7-12 regarding e^{At} in doing this.

7-23 Verify the general state solution from Table 7-2 by showing that the solution satisfies both the boundary condition, $\mathbf{x}(0)$, and the state differential equation, Eq. (7-19).

7-24 Verify the general state solution as presented in Table 7-2 by showing that the solution satisfies both the boundary condition, $\mathbf{x}(t_0)$, and the state differential equation, Eq. (7-19).

7-25 (*a*) Use Eq. (7-55) to find the impulse response functions for the systems in Prob. 7-13. Use the results of Prob. 7-11 for the expressions for e^{At}.

(*b*) Take the Laplace transform of $h(t)$ and verify the results against the results of Prob. 7-13.

7-26 (*a*) Use Eq. (7-55) to find the impulse response functions for the systems in Prob. 7-14. Use the results of Prob. 7-12 for the expressions for e^{At}.

(*b*) Take the Laplace transform of $h(t)$ and verify the results against the results of Prob. 7-14.

7-27 (*a*) Use Eq. (7-63) to determine the steady-state step response for the systems in Prob. 7-13.

(*b*) Compare the result of (*b*) to $r_{ss} = H(0)$ where $H(s)$ is found from Prob. 7-13.

7-28 Repeat Prob. 7-27 for the systems in Prob. 7-14.

7-29 Consider the single-input state model

$$\dot{\mathbf{x}} = A\mathbf{x} + \mathbf{b}u$$

$$\mathbf{y} = C\mathbf{x}$$

Suppose $\mathbf{x}(0^-) = 0$ and the unit impulse input, $U(t) = \delta(t)$, is applied. Show that $\mathbf{x}(0^+) = \mathbf{b}$.

7-30 Consider the single-input, two-output system with transfer function

$$H(s) = \begin{bmatrix} \dfrac{s + 2}{s^2 + 4s + 3} \\ \dfrac{5}{s + 1} \end{bmatrix}$$

(*a*) Expand the elements of $H(s)$ in partial fraction expansion; draw a simulation diagram; and thereby verify that

$$\begin{bmatrix} \dot{x}_1 \\ \dot{x}_2 \end{bmatrix} = \begin{bmatrix} -3 & 0 \\ 0 & -1 \end{bmatrix} \begin{bmatrix} x_1 \\ x_2 \end{bmatrix} + \begin{bmatrix} 1 \\ 1 \end{bmatrix} u$$

$$\begin{bmatrix} y_1 \\ y_2 \end{bmatrix} = \begin{bmatrix} \frac{1}{2} & \frac{1}{2} \\ 0 & 5 \end{bmatrix} \begin{bmatrix} x_1 \\ x_2 \end{bmatrix}$$

is a valid two-state model of the system.

(*b*) Find $H(s) = C[sI - A]^{-1}B$ for the model of (*a*), and also verify that the two-state model is a valid input-output representation of $H(s)$.

(*c*) Find $H(s) = C[sI - A]^{-1}B$ for the three-state model

$$\dot{\mathbf{x}} = \begin{bmatrix} 0 & 1 & 0 \\ -3 & -4 & 0 \\ 0 & 0 & -1 \end{bmatrix} \mathbf{x} + \begin{bmatrix} 0 \\ 1 \\ 1 \end{bmatrix} u$$

$$\begin{bmatrix} y_1 \\ y_2 \end{bmatrix} = \begin{bmatrix} 2 & 1 & 0 \\ 0 & 0 & 5 \end{bmatrix} \begin{bmatrix} x_1 \\ x_2 \\ x_3 \end{bmatrix}$$

and thereby validate that this three-state model in phase-variable form correctly represents the input-output properties of the system.

NOTES AND REFERENCES

This chapter presents the basics regarding the continuous-time state solution and analysis. Chapter 10 considers the important matrix exponential function, e^{At}, in greater mathematical detail. Alternative references for most of the material are Refs. 1 and 2. Reference 3 also provides good additional examples and problems.

The Leibnitz rule for differentiation of an integral is a result in advanced calculus, and Ref. 4 provides one possible reference.

1. Zadeh, L. A., and C. A. DeSesoer: *Linear Systems Theory,* McGraw-Hill, New York, 1963.
2. DeRusso, P. M., R. J. Roy, and C. M. Close: *State Variables for Engineers,* Wiley, New York, 1965.
3. Wiberg, D. M.: *State Space and Linear Systems Theory,* Schaum's Outline Series, McGraw-Hill, New York, 1971.
4. Apostel, T. M.: *Mathematical Analysis,* Addison-Wesley, Reading, Mass., 1957, p. 99.

CHAPTER

EIGHT

STATE SPACE ANALYSIS: DISCRETE TIME

One of the important reasons for going to a state-space description of a linear dynamic system is to be able to readily apply the tool of the digital computer. There is very little that can be done in hand analysis with state-space methods, but with the digital computer the state-space methods are ideally suited.

This chapter develops methods applicable to the discrete-time, digital-computer, state-space representation of the linear dynamic system. It is interesting, though, that many of the same techniques and concepts introduced in the classic analysis reappear here and are quite helpful. This reemphasizes the point that they systems analyst of today must be conversant in both the classic techniques and the modern, or state-variable, techniques.

8-1 DISCRETE TIME SIMULATION: FIRST-ORDER SYSTEM

One important application of the state-space solution method is to discrete time simulation on the digital computer. This section introduces the fundamentals of this application for the very simplest case: the linear, time-invariant, first-order, SISO system. However, the basic concepts of this case extend naturally to the most general nth-order state-space model.

Consider the scalar state-space model

$$\dot{x}(t) = ax(t) + bu(t) \tag{8-1}$$

$$y(t) = cx(t) + du(t) \tag{8-2}$$

where a, b, c, and d are all scalars. Let $x(0)$ be the initial state at time $t_0 = 0$ (note that, since the system model is time-invariant, there is no loss in generality for the assumption that $t_0 = 0$). Next let the input signal $u(t)$, $t > 0$, be the sequence of digital pulses of equal duration T as shown in Fig. 8-1.

By the notation shown in Fig. 8-1,

$$u(t) = u_T(k) \qquad t \,\varepsilon\, (kT, (k + 1)T] \tag{8-3}$$

Such a constant-pulse-type input might be typical if a digital computer were actually in the loop and supplying control input signals to the system. Or it might just be the average, or *mean value,* of a continuous input signal which is being simulated on the digital computer. This is illustrated in Fig. 8-2 for the hypothetical continuous $u(t)$. The approximation to the continuous $u(t)$ might be made as the sampled value, $u(kT)$, or as the *average sampled value*

$$u_T(k) = u(kT) + \frac{u(kT + T)}{2} \tag{8-4}$$

where $u(kT)$ is the value of $u(t)$ sampled at the instant $t = kT$. Such a discrete approximation can then be made as accurate as desired by picking shorter and shorter T intervals.

In any event, the piecewise constant input signal $u(t)$ given by Eq. (8-3) and

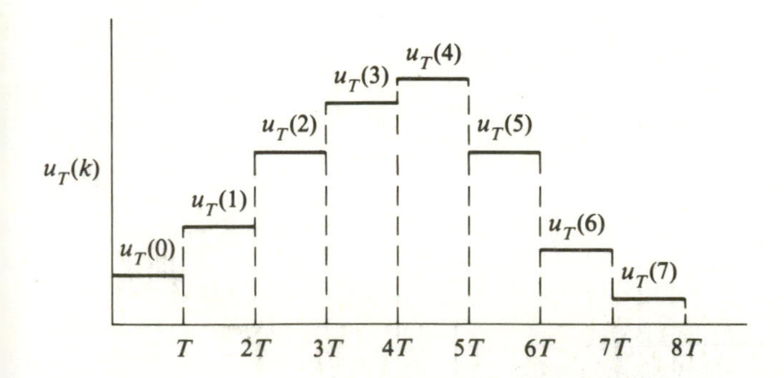

Figure 8-1 Piecewise-constant sample-and-hold input.

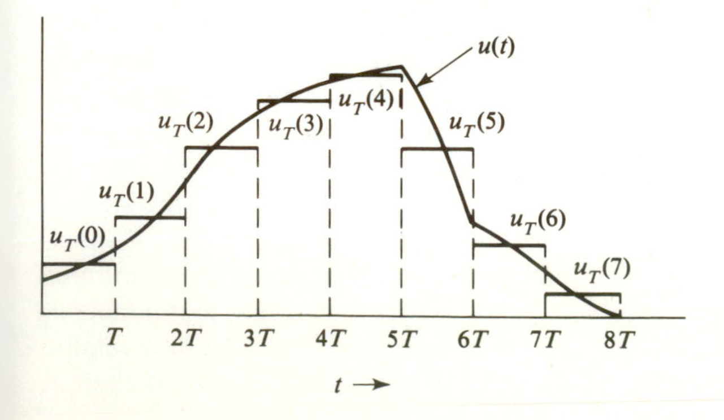

Figure 8-2 Piecewise-constant approximation to continuous input, $u(t)$.

shown in Fig. 8-1 is assumed to be the input to the continuous system Eqs. (8-1) and (8-2). From the state-space solution Eq. (7-10) it follows that

$$x(T) = e^{aT}x(0) + \int_0^T e^{a\xi}bu(T - \xi)\,d\xi \tag{8-5}$$

Notice that the second form of the convolution expression Eq. (7-7*b*) is used here. But the convolution is still avoided since $u(T - \xi) = u_T(0)$ for all $0 < \xi \leq T$ (this is the piecewise constant-input assumption). Then defining

$$x_T(k) \equiv x(kT) \qquad a_T \equiv e^{aT} \qquad b_T \equiv \left(\int_0^T e^{a\xi}\,d\xi\right)b \tag{8-6}$$

Eq. (8-5) is

$$x_T(1) = a_T x_T(0) + b_T u_T(0) \tag{8-7}$$

This is similar to the discrete finite-difference equation studied in Chap. 5. The relation between the discrete state model and the discrete finite-difference equation is examined in Sec. 8.4.

Propagating forward to find $x_T(2) = x(2T)$ by using $x_T(1)$ as the boundary condition and $u_T(1) = u(2T - \xi)$, $T < \xi \leq 2T$, as the input gives

$$x_T(2) = e^{a(2T-T)}x_T(1) + \left(\int_0^T e^{a\xi}\,d\xi\right)bu_T(1)$$

Then $e^{a(2T-T)} = e^{aT} = a_T$, and the integral is derived by letting $\xi = 2T - \tau$ in

$$\int_T^{2T} e^{a(2T-\tau)}bu(\tau)\,d\tau = \int_0^T e^{a\xi}bu(2T - \xi)\,d\xi = \int_0^T e^{a\xi}\,d\xi bu_T(1)$$

Using these results, the expression for $x_T(2)$ is

$$x_T(2) = a_T x_T(1) + b_T u_T(1) \tag{8-8}$$

This process may be continued by using $x_T(2)$ and $u_T(2)$ to find $x_T(3)$ and so forth. The general update expression always has the same recursive form:

$$\boxed{x_T(k) = a_T x_T(k - 1) + b_T u_T(k - 1)} \tag{8-9}$$

Equation (8-9) is the discrete time-state representation of the continuous-state differential Eq. (8-1). It is convenient for digital computations because of its recursive form: to propagate the state forward in time as far as one wants, one only has to propagate the discrete state propagation equation Eq. (8-9). Since the state at any time $x_T(k)$ summarizes all the information required about the past to calculate the future, all of the past states may be discarded. There is only the minimal computer storage of the current state $x_T(k)$ and the discrete model parameters a_T and b_T. Notice that a_T and b_T have to be computed just one time, and they remain invariant for as long as T does not change and the model itself does not change.

If at any time $t = kT$ it is desired to know the output, then this may be calculated from Eq. (8-2) as

$$\boxed{y(kT^-) = y_T(k^-) = c_T x_T(k) + d_T u_T(k-1)} \tag{8-10}$$

where $c_T = c$ and $d_T = d$. Notice that this is described at $t = kT^-$ because the piecewise constant input switches from $u_T(k-1)$ to $u_T(k)$ at $t^+ = kT^+$. This would not be a problem if there were no feedforward term in the output relation.

Also note that if the input $u(t)$ is indeed the piecewise constant input Eq. (8-3), and not an approximation as in Eq. (8-4), the discrete system model Eq. (8-9) is *exact*. Stated another way, the only approximation made in the discrete time representation Eq. (8-9) to the system Eq. (8-1) is the approximation (if any) which is made in representing the input signal by the piecewise constant input $u_T(k)$. This approximation may be made as close as desired by finer and finer sample spacings on T. Alternately, higher-order polynomial expansions of $u(t)$ are possible rather than the rectangular stairstep approximation with a corresponding change required in the discrete state update model (see Prob. 8-13). For now it is assumed that $u_T(k)$ is an adequate discrete time representation of the input signal.

Example 8-1 Find $x(0.1)$, $x(0.2)$, $x(0.3)$, $x(0.4)$ for the system

$$\dot{x}(t) = -5x(t) + 3u(t)$$
$$y(t) = x(t)$$

with piecewise constant input sequence $\{u_T(0), u_T(1), u_T(2), u_T(3)\} = \{1, -2, 2, -1\}$, where $T = 0.1$ and $x(0) = 10$.

From Eq. (8-6) it is seen that

$$a_T = e^{-5(0.1)} = e^{-0.5} = 0.6065$$

$$b_T = \left(\int_0^{-0.1} e^{-5\xi}\, d\xi\right)3 = 0.2361$$

Then

$$\begin{aligned}
x_T(1) &= a_T x_T(0) + b_T u_T(0) \\
&= a_T 10 + b_T 1 = 6.3014 \\
x_T(2) &= a_T x_T(1) + b_T u_T(1) = 3.3498 \\
x_T(3) &= a_T x_T(2) + b_T u_T(2) = 2.5039 \\
x_T(4) &= a_T x_T(3) + b_T u_T(3) = 1.2826
\end{aligned}$$

The drawback of this discrete system approach is naturally that one only obtains the state solution (and thus the system output) at the discrete instants of time, kT. But one can never do better than this with the inherently discrete digital computer. If one wants a tighter spacing, one simply makes T shorter.

Zero-Input and Zero-State Response

The discrete state solution separates quite naturally into the zero-input and zero-state portions of the response. The zero-input solution is seen to be

$$x_{T_{zi}}(1) = a_T x_T(0)$$
$$x_{T_{zi}}(2) = a_T x_{T_{zi}}(1) = a_T^2 x_T(0)$$
$$x_{T_{zi}}(3) = a_T x_{T_{zi}}(2) = a_T^3 x_T(0)$$

and in general

$$x_{T_{zi}}(k) = a_T^k x_T(0) \tag{8-11}$$
$$y_{T_{zi}}(k) = c_T a_T^k x_T(0) \tag{8-12}$$

For the discrete zero-state response,

$$x_{T_{zs}}(1) = b_T u_T(0)$$
$$x_{T_{zs}}(2) = a_T x_{T_{zs}}(1) + b_T u_T(1)$$
$$= a_T b_T u_T(0) + b_T u_T(1)$$
$$x_{T_{zs}}(3) = a_T x_{T_{zs}}(2) + b_T u_T(2)$$
$$= a_T^2 b_T u_T(0) + a_T b_T u_T(1) + b_T u_T(2)$$

and in general

$$x_{T_{zs}}(k) = \sum_{i=0}^{k-1} a_T^{(k-1-i)} b_T u_T(i) \tag{8-13}$$

$$y_{T_{zs}}(k) = \sum_{i=0}^{k-1} c_T a_T^{(k-1-i)} b_T u_T(i) \tag{8-14}$$

Then
$$x_T(k) = x_{T_{zi}}(k) + x_{T_{zs}}(k) \tag{8-15}$$
$$y_T(k) = y_{T_{zi}}(k) + y_{T_{zs}}(k) \tag{8-16}$$

the same as the continuous-state system.

Expression (8-14) looks very reminiscent of the discrete convolution relation

$$y_{T_{zs}}(k) = \sum_{i=0}^{k-1} h_T(k-i) u_T(i) \tag{1-20a}$$

developed in Chap. 1. Indeed, it is the same. Section 8-6 examines this issue further.

Example 8-1 (cont'd) For the discrete state model of Example 8-1, it is true that

$$y_{T_{zi}}(k) = x_{T_{zi}}(k) = (0.6065)^k \cdot 10$$

and

$$x_{T_{zs}}(1) = (0.2361)(1) = 0.2361$$

$$x_{T_{zs}}(2) = (0.6065)(0.2361) + (0.2361)(-2) = -0.3290$$

$$x_{T_{zs}}(3) = (0.6065)(-0.3290) + (0.2361)(2) = 0.2727$$

$$x_{T_{zs}}(4) = (0.6065)(0.2727) + (0.2361)(-1) = -0.0707$$

The total solution $x_T(k)$ is then found by adding $x_{T_{zi}}(k)$ and $x_{T_{zs}}(k)$. Note how much easier it is to perform the recursive update of the state than it is to compute the growing-memory-length discrete convolution, Eq. (8-13).

8-2 DISCRETE TIME SIMULATION: GENERAL MIMO SYSTEM

The discrete time results of Sec. 8-1 carry over to the general MIMO state-space model in a natural way. Consider the nth-order MIMO state-space model

$$\dot{\mathbf{x}}(t) = A\mathbf{x}(t) + B\mathbf{u}(t) \tag{8-17}$$

$$\mathbf{y}(t) = C\mathbf{x}(t) + D\mathbf{u}(t) \tag{8-18}$$

where $\mathbf{x}(t)$ is n-dimensional, $\mathbf{u}(t)$ is ℓ-dimensional, $\mathbf{y}(t)$ is m-dimensional, and $\{A, B, C, D\}$ are dimensioned compatibly. By following the development of Sec. 8-1 (see Prob. 8-5), the discrete time state-space model corresponding to Eqs. (8-17) and (8-18) is

$$\boxed{\begin{aligned} \mathbf{x}_T(k) &= A_T\mathbf{x}_T(k-1) + B_T\mathbf{u}_T(k-1) \quad (8\text{-}19)\\ \mathbf{y}_T(k^-) &= C_T\mathbf{x}_T(k) + D_T\mathbf{u}_T(k-1) \quad (8\text{-}20)\end{aligned}}$$

where

$$\mathbf{u}_T(k) = \mathbf{u}(t) \qquad t \in (kT, kT+T] \tag{8-21}$$

$$A_T \equiv e^{AT} \qquad B_T \equiv \left(\int_0^T e^{A\xi}\, d\xi\right)B \qquad C_T = C \qquad D_T = D \tag{8-22}$$

If the piecewise constant input, $\mathbf{u}_T(k)$, is exact then

$$\mathbf{x}_T(k) = \mathbf{x}(kT) \qquad \mathbf{y}_T(k^-) = \mathbf{y}(kT^-) \tag{8-23}$$

Example 8-2 Consider the MIMO system in diagonal form

$$\begin{bmatrix} \dot{x}_1(t) \\ \dot{x}_2(t) \\ \dot{x}_3(t) \end{bmatrix} = \begin{bmatrix} -1 & 0 & 0 \\ 0 & -2 & 0 \\ 0 & 0 & -3 \end{bmatrix} \begin{bmatrix} x_1(t) \\ x_2(t) \\ x_3(t) \end{bmatrix} + \begin{bmatrix} 1 & 0 \\ 1 & 0 \\ 0 & 1 \end{bmatrix} \begin{bmatrix} u_1(t) \\ u_2(t) \end{bmatrix}$$

$$\begin{bmatrix} y_1(t) \\ y_2(t) \end{bmatrix} = \begin{bmatrix} 2 & -1 & 0 \\ 0 & 0 & 1 \end{bmatrix} \begin{bmatrix} x_1(t) \\ x_2(t) \\ x_3(t) \end{bmatrix} + \begin{bmatrix} 2 & 0 \\ 0 & 0 \end{bmatrix} \begin{bmatrix} u_1(t) \\ u_2(t) \end{bmatrix}$$

Picking $T = 0.1$, the discrete state model has the form of Eqs. (8-19) and (8-20) where

$$A_T = \begin{bmatrix} e^{-0.1} & 0 & 0 \\ 0 & e^{-0.2} & 0 \\ 0 & 0 & e^{-0.3} \end{bmatrix} = \begin{bmatrix} 0.9048 & 0 & 0 \\ 0 & 0.8187 & 0 \\ 0 & 0 & 0.7408 \end{bmatrix}$$

$$B_T = \begin{bmatrix} \int_0^{0.1} e^{-\tau}\, d\tau & 0 \\ \int_0^{0.1} e^{-2\tau}\, d\tau & 0 \\ 0 & \int_0^{0.1} e^{-3\tau}\, d\tau \end{bmatrix} = \begin{bmatrix} 0.9516 & 0 \\ 0.9063 & 0 \\ 0 & 0.8639 \end{bmatrix}$$

The matrices C and D are used in Eq. (8-20) without change. Equation (8-19) can be used to propagate the initial state $\mathbf{x}_T(0)$ forward in time to $\mathbf{x}_T(k)$, for any k, given the discrete pulse input sequence $\{\mathbf{u}_T(0), \mathbf{u}_T(1), \ldots, \mathbf{u}_T(k-1)\}$. Equation (8-20) then provides the output at any desired discrete time $t = kT$. Notice that Eq. (8-20) is not used in the state propagation relation—only the current input and the current state are used to propagate to the next state. All of the necessary memory about the past is contained in the present state, and so all past state values and past input values can be discarded. This is fundamental to the state-space approach to digital simulation of the linear time-invariant system.

This example uses the very simplest state model form—the diagonal state-space model. But, as seen in Chap. 6, there are many possible state-space models of which the diagonal form is only one special case. The problem, then, is how to find A_T and B_T for general A matrices. This topic is treated in the next section.

8-3 CALCULATION OF A_T AND B_T

By Table 7-1 the A_T and B_T matrices may be computed by the power-series relations

$$A_T = I + AT + \frac{A^2T^2}{2!} + \frac{A^3T^3}{3!} + \cdots \tag{8-24}$$

$$B_T = \left\{ IT + \frac{AT^2}{2!} + \frac{A^2T^3}{3!} + \cdots \right\} B \tag{8-25}$$

Indeed, both A_T and B_T may be computed from a single power series

$$E \equiv IT + \frac{AT^2}{2!} + \frac{A^2T^3}{3!} + \cdots \tag{8-26}$$

Then

$$\begin{aligned} A_T &= I + AE \\ B_T &= EB \end{aligned} \tag{8-27}$$

Equations (8-26) and (8-27) may be used for *any* $n \times n$ A matrix, *any* $n \times \ell$ B matrix, and *any* T pulse sample time. However, convergence is slow and accuracy poor in the power series Eq. (8-26) unless T is "small enough."

To examine this issue, consider the power series expression for the scalar e^{aT}:

$$e^{aT} = 1 + (aT) + \frac{(aT)^2}{2!} + \frac{(aT)^3}{3!} + \cdots \tag{8-28}$$

If $|aT| < 1$, then

$$\sum_{k=0}^{\infty} (aT)^k$$

is a convergent series itself, and so Eq. (8-28) converges very quickly; round-off and truncation errors are minimal. If $|aT| > 1$, then $(aT)^k$ is divergent. The series Eq. (8-28) eventually converges, regardless, but round-off and truncation errors can cause unacceptable errors for digital computation.

A convenient technique for reducing these errors is by a procedure known as *squaring*. One finds an N such that $|aT/2^N| < 1$, computes

$$e^{(aT/2^N)} = 1 + \left(\frac{aT}{2^N}\right) + \frac{\left(\frac{aT}{2^N}\right)^2}{2!} + \cdots \tag{8-29}$$

by the rapidly converging power series, and then calculates

$$e^{(aT/2^{N-1})} = (e^{(aT/2^N)})^2 \tag{8-30}$$

$$e^{(aT/2^{N-2})} = (e^{(aT/2^{N-1})})^2 \tag{8-31}$$

. . . .

$$e^{(aT)} = (e^{(aT/2)})^2 \tag{8-32}$$

Example 8-3 Calculate e^{-100T}, $T = 0.1$. The power series

$$e^{-10} = 1 - 10 + 100 - \frac{1000}{2!} + \frac{10{,}000}{3!} - \cdots$$

eventually converges, but the round-off errors would be enormous.

By the squaring approach, choose $N = 4$ so that

$$|-10/2^4| = |-10/16| = 0.625 < 1$$

Then compute by power series

$$e^{-0.625} = 1 - 0.625 + \frac{(0.625)^2}{2!} - \frac{(0.625)^3}{3!} + \cdots$$

After 12 terms in the power series, this is computed on the calculator as

$$e^{-0.625} \simeq 0.5352614285$$

Then computing

$$(\{[(e^{-0.625})^2]^2\}^2)^2 = e^{-10}$$

gives

$$e^{-10} = 0.0000453999$$

The same squaring technique applies to computing the matrix power series for $e^{A \cdot T}$, Eq. (8-24). Only the scalar bound $|(aT/2^N)| < 1$ must be put into an equivalent matrix bound. For the diagonal $A = \Lambda$ matrix

$$A = \Lambda \equiv \text{diagonal}\,[\lambda_1, \lambda_2, \ldots, \lambda_n] \tag{8-33}$$

it is true that

$$A^2 = \Lambda^2 = \text{diagonal}\,[\lambda_1^2, \lambda_2^2, \ldots, \lambda_n^2]$$

$$A^3 = \Lambda^3 = \text{diagonal}\,[\lambda_1^3, \lambda_2^3, \ldots, \lambda_n^3]$$

and so forth. Thus the appropriate bound for this diagonal matrix case is

$$\left|\left(\frac{\lambda_{\max} T}{2^N}\right)\right| < 1 \tag{8-34}$$

where $\lambda_{\max}$ is the eigenvalue λ_i, having the largest absolute value.

Use of the bound Eq. (8-34) is applicable for the general A matrix, but it is inconvenient to have to calculate the system eigenvalues or poles for the general A matrix. A more convenient measure of the "size," or "magnitude," of a given A matrix is the *Froebinius norm* of the matrix:

$$\|A\|_F \equiv \left(\sum_{i=1}^{n} \sum_{j=1}^{n} a_{ij}^2\right)^{1/2} \tag{8-35}$$

where a_{ij} is the ith-row and jth-column element of A. This is similar to the Euclidean norm of a vector as studied in Chap. 9. It is convenient to use the Froebinius norm in this squaring application because it is easy to compute and it is always true that

$$\|A\|_F \geq |\lambda_{\max}| > 0 \tag{8-36}$$

Then the N is found such that

$$\frac{\|A\|_F T}{2^N} < 1 \tag{8-37}$$

This N is used in the squaring application to calculate $e^{(AT/2^N)}$ by power series. Then e^{AT} is computed by squaring the result N times.

A squaring technique may also be devised for calculation of B_T. In fact, both the squaring technique for A_T and the one for B_T may be combined into a single recursive algorithm, as shown in Algorithm 8-1. This algorithm requires only a single matrix power series calculation. The derivation of this recursive algorithm follows from the relations of Table 7-1 and the previous results discussed here. The reader is taken through the steps to derive this algorithm in Problem 8-12.

Algorithm 8.1: Squaring Algorithm for A_T and B_T

1. Given $A_{n\times n}$, $B_{n\times l}$, and T, find N such that
$$\frac{\|A\|_F T}{2^N} < 1$$
2. Calculate E_0 from the power series
$$E_0 \equiv I\left(\frac{T}{2^N}\right) + \frac{A\left(\frac{T}{2^N}\right)^2}{2!} + \frac{A^2\left(\frac{T}{2^N}\right)^3}{3!} + \cdots$$
(usually 10 to 12 terms in the power series is sufficient).
3. Form
$$F_0 \equiv I + AE_0 = e^{(AT/2^N)}$$
4. Calculate the recursion $M = 1, 2, 3, \ldots, N$
$$E_M = E_{M-1} + F_{M-1}E_{M-1}$$
$$F_M = F_{M-1}F_{M-1}$$
5. Then
$$A_T \equiv e^{AT} = F_N$$
$$B_T \equiv \left(\int_0^T e^{A\xi}\, d\xi\right)B = E_N B$$

Example 8-4 The continuous state-space model of the rocket-sled system in phase-variable form is (see Example 6-1)

$$\begin{bmatrix} \dot{x}_1(t) \\ \dot{x}_2(t) \\ \dot{x}_3(t) \end{bmatrix} = \begin{bmatrix} 0 & 1 & 0 \\ 0 & 0 & 1 \\ -0.22 & -1.14 & -0.22 \end{bmatrix} \begin{bmatrix} x_1(t) \\ x_2(t) \\ x_3(t) \end{bmatrix} + \begin{bmatrix} 0 \\ 0 \\ 1 \end{bmatrix} u(t)$$

$$y(t) = [1 \quad 0 \quad 0] \begin{bmatrix} x_1(t) \\ x_2(t) \\ x_3(t) \end{bmatrix}$$

Then

$$\|A\|_F = [1 + 1 + (0.22)^2 + (1.14)^2 + (0.22)^2]^{1/2}$$
$$= 1.843$$

Using $T = 1$, it is seen that

$$\frac{\|A\|_F T}{2^1} = 0.9215 < 1$$

Therefore, the previous squaring algorithm would use $N = 1$. For even more rapid convergence one could use $N = 2$. The bound $\|A\|_F$ is conservative, as the largest-magnitude pole of the rocket-sled system is

$$|0.1 + 1.044j| = 1.049 < 1.843 = \|A\|_F$$

In general, though, one would not know the poles (eigenvalues) of a given A, and so Eq. (8-37) is a convenient bound to use.

The actual calculation of A_T and B_T on the computer for this rocket-sled system with $T = 1$ is found to be

$$A_T = \begin{bmatrix} 0.96857 & 0.82896 & 0.39887 \\ -0.08775 & 0.51386 & 0.66942 \\ -0.14727 & -0.85088 & 0.24610 \end{bmatrix}$$

$$B_T = \begin{bmatrix} 0.14287 \\ 0.39887 \\ 0.66942 \end{bmatrix}$$

Notice that while the original A and B are in the special phase-variable canonical form, the corresponding A_T and B_T are in no special form at all and are actually full of all elements. Only in the diagonal Λ matrix case, Eq. (8-33), is the A_T matrix also diagonal:

$$A_{T_\Lambda} = \begin{bmatrix} e^{\lambda_1 T} & 0 & \cdots & 0 \\ 0 & e^{\lambda_2 T} & & \\ \cdots & \cdots & \cdots & \cdots \\ 0 & & & e^{\lambda_n T} \end{bmatrix} \tag{8-38}$$

Sections 8-5 and 8-6 consider how to take a given $\{A_T, B_T, C_T, D_T\}$ model and transform it to special canonical forms by an intermediate calculation of the discrete finite-difference equation. Transforming to a canonical form for A_T and B_T

can cut down on the number of multiplications and additions required in the discrete state update equation:

$$\mathbf{x}_T(k) = A_T\mathbf{x}_T(k-1) + B_T\mathbf{u}_T(k-1) \tag{8-19}$$

In terms of each element, $x_{T_i}(k)$, $i = 1, 2, \ldots, n$, the state update equation is

$$x_{T_i}(k) = \sum_{j=1}^{n} A_{T_{ij}} x_{T_j}(k-1) + \sum_{j=1}^{\ell} B_{T_{ij}} u_{T_j}(k-1) \tag{8-39}$$

Having A_T and B_T in special canonical forms, in which many elements, $A_{T_{ij}}$ and $B_{T_{ij}}$, are zero can thus substantially reduce the computations required for each state update.

8-4 STABILITY AND THE POLES OF A_T

For the diagonal

$$A = \Lambda = \text{diagonal}\,[\lambda_1, \lambda_2, \ldots, \lambda_n] \tag{8-33}$$

where the λ_i are the n eigenvalues or poles of A (assumed here to be distinct), it is seen that

$$A_{T_\Lambda} = e^{\Lambda T} = \text{diagonal}\,[e^{\lambda_1 T}, e^{\lambda_2 T}, \ldots, e^{\lambda_n T}] \tag{8-40}$$

Hence the n eigenvalues or poles of the discrete plant matrix A_{T_Λ} are $(e^{\lambda_i T})$, $i = 1, 2, \ldots, N$.

This is true, in general, no matter what the form of A or A_T (again assuming distinct λ_i—the case of repeated λ_i is more complicated and is treated in Chap. 10). This result is exactly the same result as found for the discrete time finite-difference equation studied in Chap. 5. Note that the λ_i and p_i are one and the same and that

$$e^{\lambda_i T} = e^{p_i T} \equiv p_{T_i} \tag{8-41}$$

Stability results of the discrete state model Eq. (8-19) are thus compatible with the stability results of the discrete time finite-difference equation. The n eigenvalues of A_T, the $(e^{\lambda_i T})$, $i = 1, 2, \ldots, n$, must all have absolute value less than one:

$$|e^{\lambda_i T}| < 1 \tag{8-42}$$

for the discrete state model to be stable. This is true for $T > 0$ if and only if all of the system eigenvalues (poles) are in the left half-plane; real $(\lambda_i) < 0$ for all $i = 1, 2, \ldots, n$.

The discrete state model, Eqs. (8-19) and (8-20), shares many other similarities with the finite-difference equation for the special case of the SISO system. Indeed, the finite-difference equation may be derived from the discrete state model $\{A_T, B_T, C_T, D_T\}$, and vice versa. These relations are examined in the next section.

8-5 RELATION BETWEEN THE DISCRETE STATE MODEL AND FINITE-DIFFERENCE EQUATION FOR THE SISO SYSTEM

Consider the SISO system output differential equation

$$y^{(n)}(t) + a_{n-1}y^{(n-1)}(t) + \cdots + a_0 y(t) = b_m u^{(m)}(t) + \cdots + b_0 u(t) \quad (8\text{-}43)$$

In Chap. 5 a finite-difference equation model

$$\begin{aligned} y_T(k) = &-a_{T_{n-1}} y_T(k-1) - \cdots - a_{T_0} y_T(k-n) \\ &+ b_{T_{n-1}} u_T(k-1) + \cdots + u_T(k-n) \end{aligned} \quad (8\text{-}44)$$

is found which models the sampled time response, $y(kT) = y_T(k)$, for the system Eq. (8-43) driven by the piecewise constant, sample-and-hold inputs, $u_T(k)$.

A relation may be found going from Eq. (8-44) to a discrete state model, Eqs. (8-19) and (8-20), in the same way that a continuous state model could be derived from the continuous-output differential equation. This, in fact, yields canonical forms for the state model Eqs. (8-19) and (8-20). We do this only for the proper SISO system (no feedforward and $D \equiv 0$).

Phase-Variable Canonical Form

Given the finite-difference equation model Eq. (8-44), the discrete state model in phase-variable form is

$$\begin{aligned} \mathbf{x}_T(k) = &\begin{bmatrix} 0 & 1 & 0 & \cdots & 0 \\ 0 & 0 & 1 & & \\ \vdots & & & & 1 \\ -a_{T_0} & -a_{T_1} & -a_{T_2} & \cdots & -a_{T_{n-1}} \end{bmatrix} \mathbf{x}_T(k-1) \\ &+ \begin{bmatrix} 0 \\ 0 \\ \vdots \\ 1 \end{bmatrix} u_T(k-1) \end{aligned} \quad (8\text{-}45)$$

$$y_T(k) = [b_{T_0} \quad b_{T_1} \quad \cdots \quad b_{T_{n-1}}]\mathbf{x}_T(k) \quad (8\text{-}46)$$

This may be derived in exactly the same way as done in Sec. 6-1.

The phase-variable canonical form is sometimes called the *controllability* canonical form.

Example 8-5 For the rocket-sled system and $T = 1$, the finite-difference equation is found in Example 5-2 to be

$$\begin{aligned} y_T(k) = &-a_{T_2} y_T(k-1) - a_{T_1} y_T(k-2) - a_{T_0} y_T(k-3) \\ &+ b_{T_2} u_T(k-1) + b_{T_1} u_T(k-1) + b_{T_0} u_T(k-2) \end{aligned}$$

where

$$a_{T_2} = -1.7285748 \qquad a_{T_1} = 1.5636481 \qquad a_{T_0} = -0.6703200$$

$$b_{T_2} = 0.14287089 \qquad b_{T_1} = 0.48906719 \qquad b_{T_0} = 0.11694311$$

The phase-variable or controllability canonical-form discrete state model is, therefore,

$$\mathbf{x}_T(k) = \begin{bmatrix} 0 & 1 & 0 \\ 0 & 0 & 1 \\ -a_{T_0} & -a_{T_1} & -a_{T_2} \end{bmatrix} \mathbf{x}_T(k-1) + \begin{bmatrix} 0 \\ 0 \\ 1 \end{bmatrix} u_T(k-1)$$

$$y_T(k) = [b_{T_0} \quad b_{T_1} \quad b_{T_2}]\mathbf{x}_T(k)$$

where $\mathbf{x}_T(k)$ is a three-dimensional state vector.

States Derived from Simulation Diagram

The simulation diagram of Eq. (8-44) using the unit delay operation rather than the integral operation is shown in Fig. 8-3. Defining a state as the output of each delay operator as shown in Fig. 8-3 then gives the state-space model

$$\mathbf{x}_T(k) = \begin{bmatrix} 0 & 0 & \cdots & \cdots & -a_{T_0} \\ 1 & 0 & \cdots & \cdots & -a_{T_1} \\ 0 & 1 & \cdots & \cdots & -a_{T_2} \\ \cdots & \cdots & \cdots & \cdots & \cdots \\ 0 & 0 & & 1 & -a_{T_{n-1}} \end{bmatrix} \mathbf{x}_T(k-1) + \begin{bmatrix} b_{T_0} \\ b_{T_1} \\ b_{T_2} \\ \vdots \\ b_{T_{n-1}} \end{bmatrix} u_T(k-1) \tag{8-47}$$

$$y_T(k) = [0 \quad 0 \quad \cdots \quad 1]\mathbf{x}_T(k) \tag{8-48}$$

This state model is also called the *observability* canonical form.

Example 8-6 For the rocket-sled system, the states derived from the simulation diagram or the observability canonical form is

$$\tilde{\mathbf{x}}_T(k) = \begin{bmatrix} 0 & 0 & -a_{T_0} \\ 1 & 0 & -a_{T_1} \\ 0 & 1 & -a_{T_2} \end{bmatrix} \tilde{\mathbf{x}}_T(k-1) + \begin{bmatrix} b_{T_0} \\ b_{T_1} \\ b_{T_2} \end{bmatrix} u_T(k-1)$$

$$y_T(k) = [0 \quad 0 \quad 1]\tilde{\mathbf{x}}_T(k)$$

The notation $\tilde{\mathbf{x}}_T(k)$ is used to differentiate this observability canonical form from the phase-variable, or controllability, canonical form.

Diagonal-Form State-Space Model

Expand the transfer function of the finite-difference equation in partial fraction expansion form

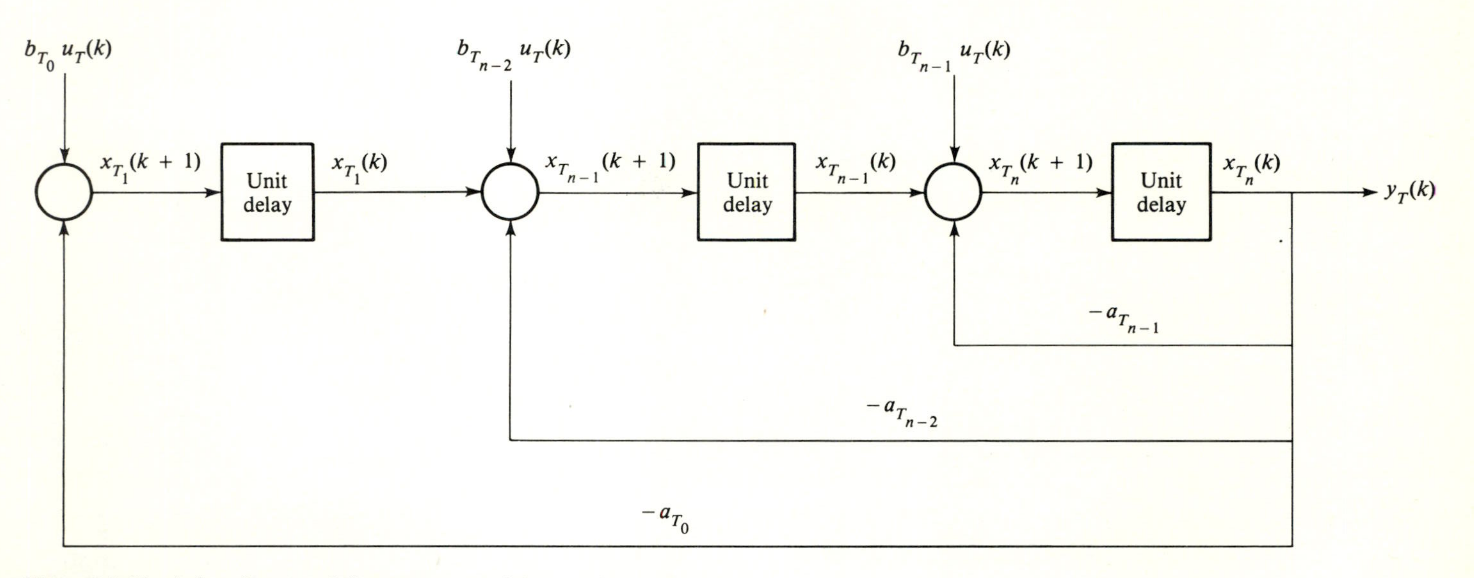

Figure 8-3 Simulation diagram of discrete-state model, Eqs. (8-47) and (8-48).

$$H_T(z) = \frac{b_{T_{n-1}}z^{n-1} + \cdots + b_{T_0}}{z^n + a_{T_{n-1}}z^{n-1} + \cdots + a_{T_0}}$$

$$= \frac{b_{T_{n-1}}z^{n-1} + \cdots + b_{T_0}}{(z - p_{T_1})(z - p_{T_2}) \cdots (z - p_{T_n})}$$

$$= \sum_{i=1}^{n} \frac{r_{T_i}}{z - p_{T_i}} \tag{8-49}$$

This assumes the discrete poles, p_{T_i}, are all distinct. If they are not, the appropriate partial fraction expansion may be performed (see Appendix C).

Analogous to the continuous-time case, the discrete state model in diagonal form is then

$$\begin{bmatrix} x_{T_1}(k) \\ x_{T_2}(k) \\ \vdots \\ x_{T_n}(k) \end{bmatrix} = \begin{bmatrix} p_{T_1} & 0 & \cdots & 0 \\ 0 & p_{T_2} & \cdots & \\ \cdots & \cdots & \cdots & \cdots \\ 0 & \cdots & \cdots & p_{T_n} \end{bmatrix} \begin{bmatrix} x_{T_1}(k-1) \\ x_{T_2}(k-1) \\ \vdots \\ x_{T_n}(k-1) \end{bmatrix} + \begin{bmatrix} 1 \\ 1 \\ \vdots \\ 1 \end{bmatrix} u_T(k-1) \tag{8-50}$$

$$y_T(k) = [r_{T_1} \quad r_{T_2} \quad \cdots \quad r_{T_n}] \begin{bmatrix} x_{T_1}(k) \\ x_{T_2}(k) \\ \vdots \\ x_{T_n}(k) \end{bmatrix} \tag{8-51}$$

Again the p_{T_i} poles are the same as the eigenvalues of the diagonal Λ_T matrix,

$$p_{T_i} = e^{\lambda_i T} \qquad i = 1, 2, \ldots, n$$

Example 8-7 For the rocket-sled system with $T = 1$, the discrete poles are found in Example 5-2 to be

$$p_{T_1} = e^{-0.2} = 0.81873075$$

$$p_{T_2} = e^{-0.1}e^{-1.044j} = 0.45492203 - 0.78216155j$$

$$p_{T_3} = e^{-0.1}e^{1.044j} = 0.45492203 + 0.78216155j$$

Partial fraction expansion of $H_T(z)$ gives

$$H_T(z) = \frac{b_{T_2}z^2 + b_{T_1}z + b_{T_0}}{(z - p_{T_1})(z - p_{T_2})(z - P_{T_3})}$$

$$= \frac{r_{T_1}}{z - p_{T_1}} + \frac{r_{T_2}}{z - p_{T_2}} + \frac{r_{T_3}}{z - p_{T_3}}$$

where $\qquad r_{T_1} = 0.824 \qquad$ and $\qquad r_{T_{2,3}} = 0.088 \pm 0.578.$

The diagonal-form discrete state model for the rocket-sled system with $T = 1$ is thus

$$\hat{\mathbf{x}}_T(k) = \begin{bmatrix} p_{T_1} & 0 & 0 \\ 0 & p_{T_2} & 0 \\ 0 & 0 & p_{T_3} \end{bmatrix} \hat{\mathbf{x}}_T(k-1) + \begin{bmatrix} 1 \\ 1 \\ 1 \end{bmatrix} u_T(k-1)$$

$$y_T(k) = [r_{T_1} \quad r_{T_2} \quad r_{T_3}]\hat{\mathbf{x}}_T(k)$$

The notation $\hat{\mathbf{x}}_T$ is used to differentiate this diagonal canonical form from the previous model forms. They all represent the same input-output system.

Calculation of $H_T(z)$ from $\{A_T, B_T, C_T, D_T\}$

Analogous to the continuous-time result

$$H(s) = C[sI - A]^{-1}B + D$$

it is true that

$$H_T(z) = C_T[zI - A_T]^{-1}B_T + D_T \tag{8-52}$$

This may be derived by taking the z transform of the discrete state model Eqs. (8-19) and (8-20) and solving for

$$Y_T(z) = H_T(z)U_T(z) \tag{8-53}$$

Like the continuous-time case, Eq. (8-52) is also true for the MIMO situation. Then $H_T(z)$ is an $m \times \ell$ discrete transfer-function matrix.

Eq. (8-52) provides a method to find the finite-difference equation model from a given state-space model $\{A_T, B_T, C_T, D_T\}$. Computational methods in finding the resolvent matrix $[zI - A_T]^{-1}$ are just like computational methods in computing $[sI - A]^{-1}$ (see Appendix D). Also computed with facility is the discrete pulse-response sequence $h_T(k)$, $k = 1, 2, 3, \ldots$ from the discrete state model $\{A_T, B_T, C_T, D_T\}$.

8-6 CALCULATION OF THE DISCRETE PULSE RESPONSE FROM THE DISCRETE STATE MODEL

Fundamental to the classic discrete input-output analysis of a linear time-invariant system is the discrete pulse response model $h_T(k)$, $k = 1, 2, 3, \ldots$. The discrete pulse response model for the SISO system is defined as the zero-state system response to the discrete pulse input

$$\delta_T(k) \equiv \begin{cases} 1 & k = 0 \\ 0 & k = 1, 2, \ldots \end{cases} \tag{8-54}$$

The discrete pulse response model is used to find the discrete output via the convolution summation

$$y_{zs}(kT) = \sum_{i=0}^{k-1} h_T(k - i)u_T(i) \tag{1-20a}$$

Also, $h_T(k)$ is directly related to the finite-difference equation as studied in Chap. 5 and may actually be used to find the finite-difference equation. (See Secs. 5-1 and 5-4.) The finite-difference equation can, in turn, be used to find useful canonical forms for the discrete state-space model. Methods for doing this have been presented in the previous section.

Between these different methods there is a complete circuit from an initial continuous-time transfer function to a discrete state model in a desired special canonical form. These interconnections are illustrated in Fig. 8-4. It is an interesting and useful part of linear systems theory that all of these concepts are so closely intermeshed.

Now to find $h_T(k)$ from the model $\{A_T, B_T, C_T, D_T\}$, consider forcing the system, Eqs. (8-19) and (8-20), with the discrete pulse input, $\delta_T(k)$, Eq. (8-54), starting from the zero initial state, $x_T(0) = 0$. Since $\delta_T(1) = 1$, Eqs. (8-19) and (8-20) yield

$$x_T(1) = B_T$$
$$y_T(1^-) = C_T x(1) + D_T = C_T B_T + D_T$$

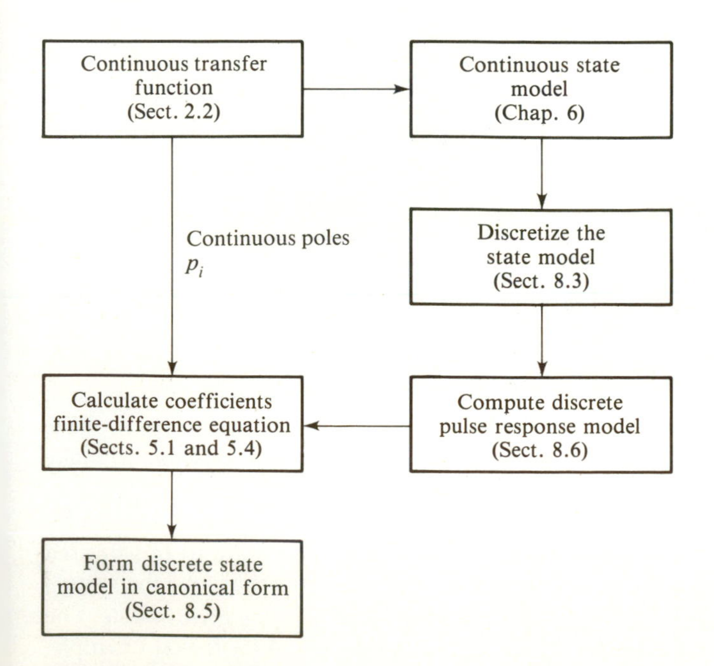

Figure 8-4 Interconnections of the classic and state-space approaches.

Then $\delta_T(k) = 0$, $k = 1, 2, \ldots$, and so the state may be propagated forward to $\mathbf{x}_T(k)$ using $\mathbf{x}_T(k-1)$ as the boundary condition:

$$\begin{aligned}
\mathbf{x}_T(2) &= A_T\mathbf{x}_T(1) = A_TB_T \\
y_T(2) &= C_TA_TB_T \\
\mathbf{x}_T(3) &= A_T\mathbf{x}_T(2) = A_T^2B_T \\
y_T(3) &= C_TA_T^2B_T \\
\mathbf{x}_T(4) &= A_T\mathbf{x}_T(3) = A_T^3B_T \\
y_T(4) &= C_TA_T^3B_T
\end{aligned}$$

Since $h_T(k) = y_T(k)$, the general form is thus seen to be

$$\boxed{\begin{aligned}
h_T(1^-) &= C_TB_T + D_T \\
h_T(k) &= C_TA_T^{k-1}B_T \qquad k = 2, 3, \ldots
\end{aligned}} \tag{8-55}$$

Eq. (8-55) is a calculation that obviously needs to be done on the computer; but so too is the process of finding the discrete $\{A_T, B_T, C_T, D_T\}$ model from a continuous $\{A, B, C, D\}$ state model. The calculations are quite direct, though, and so Eq. (8-55) is a useful relation. This relation is also true for calculating the MIMO pulse-response model, $[h_T(k)]$, $k = 1, 2, \ldots$, where the pulse-response model has dimension $m \times \ell$ (ℓ inputs, m outputs).

8-7 INTRODUCTION TO DISCRETE CONTROLLABILITY

One of the important applications of state-variable methods, particularly in the discrete time, is to the design of feedback control laws for use with on-line digital control systems. This topic area is quite broad and generally beyond our scope. However, fundamental to concepts of state-variable control laws is the system theory concept of controllability. If a system has states that are uncontrollable, then it is impossible for the control input to do anything that will influence these states.

To introduce the concept of discrete system controllability, consider the following hypothetical problem: We would like to transfer the system from the zero initial state, $\mathbf{x}_T(0) = 0$, to some desired state $\mathbf{x}_T(N) = \mathbf{x}_d$ by using the appropriate sequence of control inputs $\{\mathbf{u}_T(0), \mathbf{u}_T(1), \ldots, \mathbf{u}_T(N-1)\}$. If it is possible to make this transfer within some finite number of steps, N, then the state $\mathbf{x}_d \in R^n$ (i.e., $\mathbf{x}_d$ is an n-dimensional state vector) is said to be *controllable;* we *can* find a sequence of unconstrained controls that take the system from the rest state to this desired $\mathbf{x}_d$. If it is impossible to find such a sequence of control inputs that take the system to $\mathbf{x}_T(N) = \mathbf{x}_d$, no matter how large N, the state $\mathbf{x}_d$ is said to be *uncontrollable.* The state $\mathbf{x}_d$ simply cannot be reached with the control inputs. As a trivial example, consider the hypothetical discrete state model

$$\begin{bmatrix} x_{T_1}(k) \\ x_{T_2}(k) \end{bmatrix} = \begin{bmatrix} 0.8 & 0 \\ 0 & 0.5 \end{bmatrix} \begin{bmatrix} x_{T_1}(k-1) \\ x_{T_2}(k-1) \end{bmatrix} + \begin{bmatrix} 1 \\ 0 \end{bmatrix} u_T(k-1) \tag{8-56}$$

The state $\mathbf{x}_d = [0 \quad 1]^T$ simply cannot be reached from the zero initial state because the input never influences the x_{T_2} element of the state vector.

If *any* initial state in n-dimensional state space, R^n, can be reached for some given $N < \infty$, then the state-space model is said to be *completely controllable.* For example, the state model

$$\begin{bmatrix} x_{T_1}(k) \\ x_{T_2}(k) \end{bmatrix} = \begin{bmatrix} 0.8 & 0 \\ 0 & 0.5 \end{bmatrix} \begin{bmatrix} x_{T_1}(k-1) \\ x_{T_2}(k-1) \end{bmatrix} + \begin{bmatrix} 1 \\ 1 \end{bmatrix} u_T(k-1) \tag{8-57}$$

is completely controllable because any desired state $\mathbf{x}_d \ \varepsilon \ R^2$ can be reached in at least two steps ($N = 2$).

To demonstrate, suppose that we would like to reach $\mathbf{x}_d = [-3, 4]^T$ in two steps. A linear equation problem may be formed as follows: Since $\mathbf{x}_T(0) = 0$, $\mathbf{x}_T(1)$ is computed as

$$\mathbf{x}_T(1) = \begin{bmatrix} 1 \\ 1 \end{bmatrix} u_T(0) \tag{8-58}$$

Then $\mathbf{x}_T(2)$ is propagated forward from $\mathbf{x}_T(1)$ via

$$\mathbf{x}_T(2) = \begin{bmatrix} 0.8 & 0 \\ 0 & 0.5 \end{bmatrix} \mathbf{x}_T(1) + \begin{bmatrix} 1 \\ 1 \end{bmatrix} u_T(1) \tag{8-59}$$

Combining the expressions for $\mathbf{x}_T(1)$ and $\mathbf{x}_T(2)$, and using the fact that we want $\mathbf{x}_T(2) = \mathbf{x}_d$, then gives

$$\begin{bmatrix} -3 \\ 4 \end{bmatrix} = \begin{bmatrix} 0.8 \\ 0.5 \end{bmatrix} u_T(0) + \begin{bmatrix} 1 \\ 1 \end{bmatrix} u_T(1) \tag{8-60}$$

This may be written as a linear equation problem for the two unknowns, $u_T(0)$ and $u_T(1)$:

$$\begin{bmatrix} 1 & 0.8 \\ 1 & 0.5 \end{bmatrix} \begin{bmatrix} u_T(1) \\ u_T(0) \end{bmatrix} = \begin{bmatrix} -3 \\ 4 \end{bmatrix} \tag{8-61}$$

The solution for the two inputs $u_T(0)$ and $u_T(1)$ to transfer the system from the zero initial state to $\mathbf{x}_T(2) = \mathbf{x}_d$ is $u_T(0) = -23.333$ and $u_T(1) = 15.667$. The reader is invited to verify this solution by propagating the system forward from the zero state to $\mathbf{x}_T(2)$ using the two inputs, $u_T(0) = -23.333$ and $u_T(1) = 15.667$.

The results of this example may be generalized. In general, the discrete time controllability problem may be formulated as a linear equation problem involving the zero-state solution

$$\mathbf{x}_{T_{zs}}(N) = \sum_{i=0}^{N-1} A_T^{N-1-i} B_T \mathbf{u}_T(i) \tag{8-62}$$

We wish to find the sequence of inputs $\{\mathbf{u}_T(0), \mathbf{u}_T(1), \ldots, \mathbf{u}_T(N-1)\}$ such that

$\mathbf{x}_{T_{zs}}(N) = \mathbf{x}_d$. Equation (8-62) may be written as a matrix times the vector of inputs:

$$\mathbf{x}_{T_{zs}}(N) = [B_T, A_T B_T, \ldots, A_T^{N-1} B_T] \begin{bmatrix} \mathbf{u}_T(N-1) \\ \mathbf{u}_T(N-2) \\ \vdots \\ \mathbf{u}_T(0) \end{bmatrix}_{N\ell \times 1} \tag{8-63}$$

The coefficient matrix is dimension $n \times N\ell$ and is termed the *discrete controllability matrix:*

$$M_{T_c} \equiv [B_T, A_T B_T, \ldots, A_T^{N-1} B_T] \tag{8-64}$$

If the sequence of inputs $\{\mathbf{u}_T(0), \mathbf{u}_T(1), \ldots, \mathbf{u}_T(N-1)\}$ is to give $\mathbf{x}_{T_{zs}}(N) = \mathbf{x}_d$, then it must satisfy the linear equation problem

$$M_{T_c} \mathbf{U}_{T_N} = \mathbf{x}_d \tag{8-65}$$

where $\mathbf{U}_{T_N} \equiv [\mathbf{u}_T^T(N-1), \mathbf{u}_T^T(N-2), \ldots, \mathbf{u}_T^T(0)]^T$

Solvability of this linear equation problem, Eq. (8-65), depends upon some matrix theory concepts yet to be discussed in Chap. 10.

Example 8-8 Consider the following problem: We wish to drive the rocket-sled system from the rest condition to the steady-state value $y_{ss}(t) = 100$ in a minimum number of discrete steps. For the system to maintain $y_{ss}(t) = 100$, a sufficiently large constant input must be applied so that the rocket sled maintains this constant output value. From the classic theory of Chap. 2 it is true that

$$y_{ss}(t) = H(0) u_{ss}$$

where
$$H(s) = \frac{1}{s^3 + 0.4s^2 + 1.14s + 0.22}$$

is the continuous system transfer function. Then $H(0) = 1/0.22$ and we may solve for u_{ss} via

$$u_{ss} = \frac{y_{ss}}{H(0)} = 22$$

This solves part of the problem; but if we simply applied the constant input $u_{ss} = 22$, the system would have its undesirable step-response behavior with many oscillations before finally reaching the steady-state value of $y_{ss} = 100$. Instead, the theory of discrete time controllability may be applied to find the $u_T(0)$, $u_T(1)$, and $u_T(2)$ that bring the system to $y_{ss} = 100$ in only 3 steps. This is termed a *deadbeat* response.

We still do not have all of the pieces necessary to solve this problem. For convenience, we use the discrete time phase-variable canonical form as formed in Example 8-5:

$$\mathbf{x}_T(k) = \begin{bmatrix} 0 & 1 & 0 \\ 0 & 0 & 1 \\ -a_{T_0} & -a_{T_1} & -a_{T_2} \end{bmatrix} \mathbf{x}_T(k-1) + \begin{bmatrix} 0 \\ 0 \\ 1 \end{bmatrix} u_T(k-1)$$

where

$$a_{T_0} = -0.6703200 \qquad a_{T_1} = 1.5636481 \qquad a_{T_2} = -1.7285748$$

Now we wish to bring the system to $y_{ss} = 100$. We already know that the steady-state input required to do this is $u_{ss} = 22$. But what we don't know is the discrete equilibrium state that corresponds to $y_{ss} = 100$ and $u_{ss} = 22$. However, we may use the discrete state update equation to find the appropriate equilibrium state, $\mathbf{x}_{T_{ss}}$. The discrete state is in equilibrium if $\mathbf{x}_T(k) = \mathbf{x}_T(k-1)$ for a constant input (this is analogous to $\dot{\mathbf{x}} = 0$ in the continuous-state differential equation). Using this concept with $u_{ss} = 22$ yields

$$\mathbf{x}_{T_{ss}} = \begin{bmatrix} 0 & 1 & 0 \\ 0 & 0 & 1 \\ -a_{T_0} & -a_{T_1} & -a_{T_2} \end{bmatrix} \mathbf{x}_{T_{ss}} + \begin{bmatrix} 0 \\ 0 \\ 1 \end{bmatrix} (22)$$

Setting this up as a linear equation problem for the unknown $\mathbf{x}_{T_{ss}}$ then yields

$$\begin{bmatrix} 1 & -1 & 0 \\ 0 & 1 & -1 \\ a_{T_0} & a_{T_1} & 1 + a_{T_2} \end{bmatrix} \begin{bmatrix} x_{T_{ss_1}} \\ x_{T_{ss_2}} \\ x_{T_{ss_3}} \end{bmatrix} = \begin{bmatrix} 0 \\ 0 \\ 22 \end{bmatrix}$$

Solving for $\mathbf{x}_{T_{ss}}$ gives

$$\mathbf{x}_{T_{ss}} = \begin{bmatrix} x_{T_{ss_1}} \\ x_{T_{ss_2}} \\ x_{T_{ss_3}} \end{bmatrix} = \begin{bmatrix} 133.5329854 \\ 133.5329854 \\ 133.5329854 \end{bmatrix}$$

This is the desired equilibrium state that will yield $y_{ss} = 100$ when $u_{ss} = 22$. This is the desired state for which we would like to find the three inputs $u_T(0)$, $u_T(1)$, $u_T(2)$ such that $\mathbf{x}_{T_{zs}}(3) = \mathbf{x}_{T_{ss}}$. Setting this up as a linear equation problem with the discrete controllability matrix*

$$M_{T_c} \equiv [B_T, \quad A_T B_T, \quad A_T^2 B_T] = \begin{bmatrix} 0 & 0 & 1 \\ 0 & 1 & -a_{T_2} \\ 1 & -a_{T_2} & -a_{T_1} + a_{T_2}^2 \end{bmatrix}$$

gives

$$\begin{bmatrix} 0 & 0 & 1 \\ 0 & 1 & -a_{T_2} \\ 1 & -a_{T_2} & -a_{T_1} + a_{T_2}^2 \end{bmatrix} \begin{bmatrix} u_T(2) \\ u_T(1) \\ u_T(0) \end{bmatrix} = \begin{bmatrix} 133.5329854 \\ 133.5329854 \\ 133.5329854 \end{bmatrix}$$

Solving this linear equation problem gives the piecewise constant discrete inputs

$$\begin{bmatrix} u_T(2) \\ u_T(1) \\ u_T(0) \end{bmatrix} = \begin{bmatrix} 111.5098308 \\ -97.28876813 \\ 133.5329854 \end{bmatrix}$$

*We choose $N = n = 3$, thus giving a 3×3 controllability matrix for this single-input system.

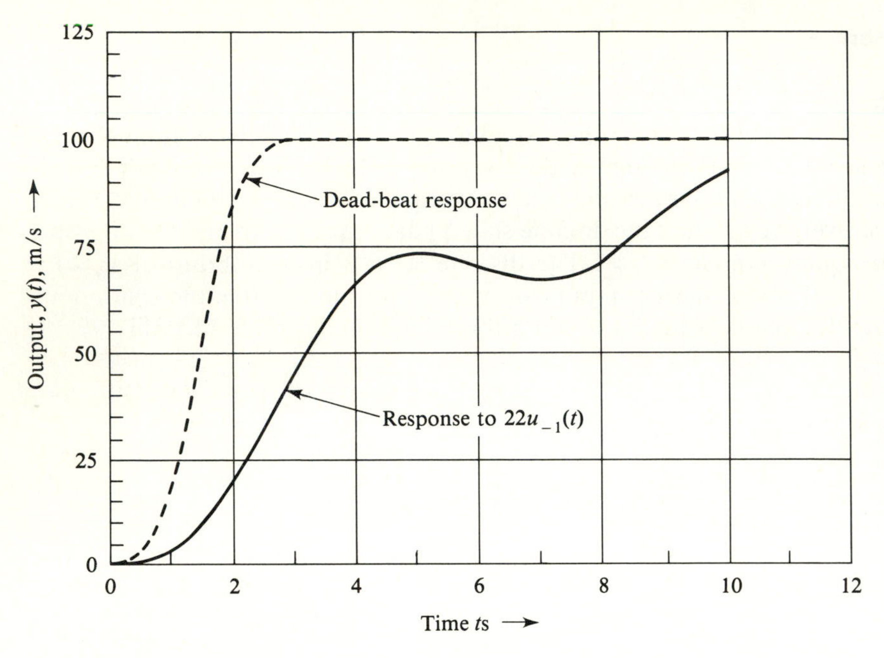

Figure 8-5 Rocket sled "dead-beat" response versus response to $22u_{-1}(t)$.

Naturally it is impossible to apply such a set of discrete pulse inputs with the rocket-sled motor (if for no other reason than that the rocket sled could never produce a negative pulse of force). Ignoring this "minor" practical issue, though, the rocket-sled response with this set of discrete pulse inputs (each held for $T = 1$ second) followed by the application of $u_{ss} = 22$ for $t > 3$, is shown in Fig. 8-5 along with the response to the step input of $u_{ss} = 22$. If the practical issue of producing a negative thrust pulse could be solved, this clearly would be a desirable sequence of control inputs to quickly bring the system to the desired steady-state condition.

To go into control theory topics any further is beyond our scope. However, to be practical a control system must be closed-loop or feedback. In using state-variable methods this implies that the controller constantly estimates the system state from the available output observations. This brings up the important system property of observability.

8-8 INTRODUCTION TO DISCRETE OBSERVABILITY

The counterpart to the system property of state controllability is the system property of state observability; controllability and observability are said to be *dual* properties. State observability is concerned with the question of whether or not a given state in state space, $\mathbf{x}_T(0) \varepsilon R^n$, can be uniquely estimated from the set of N sampled outputs it would produce, $\{\mathbf{y}_{T_{zi}}(0), \mathbf{y}_{T_{zi}}(1), \mathbf{y}_{T_{zi}}(2), \ldots, \mathbf{y}_{T_{zi}}(N-1)\}$.

The zero-input response is used in the formulation of the observability question because the concern is over what observations the state $\mathbf{x}_T(0)$ produces and not over observations that might be produced by the input. Since the zero-input and zero-state response in a linear system are separable, this is not a big concern, but it simplifies the presentation.

The zero-input output response to the initial state $\mathbf{x}_T(0)$ is given by

$$\mathbf{y}_{T_{zi}}(k) = C_T A_T^k \mathbf{x}_T(0) \tag{8-66}$$

for $k = 0, 1, 2, \ldots$. If the system is stable, this decays to zero as $k \to \infty$. Equation (8-66) may be used to formulate a linear equation problem to find $\mathbf{x}_T(0)$ from $\mathbf{y}_{T_{zi}}(k)$, $k = 0, 1, \ldots, n-1$:

$$\begin{aligned} C_T \mathbf{x}_T(0) &= \mathbf{y}_{T_{zi}}(0) = \mathbf{y}_T(0) \\ C_T A_T \mathbf{x}_T(0) &= \mathbf{y}_{T_{zi}}(1) \\ C_T A_T^2 \mathbf{x}_T(0) &= \mathbf{y}_{T_{zi}}(2) \\ &\cdots\cdots\cdots \\ C_T A_T^{n-1} \mathbf{x}_T(0) &= \mathbf{y}_{T_{zi}}(n-1) \end{aligned}$$

In matrix-vector form these linear equations for the unknown $\mathbf{x}_T(0)$ may be written as

$$\begin{bmatrix} C_T \\ C_T A_T \\ \vdots \\ C_T A_T^{n-1} \end{bmatrix}_{mn \times n} \mathbf{x}_T(0) = \begin{bmatrix} \mathbf{y}_{T_{zi}}(0) \\ \mathbf{y}_{T_{zi}}(1) \\ \vdots \\ \mathbf{y}_{T_{zi}}(n-1) \end{bmatrix}_{mn \times 1} \tag{8-67}$$

There are mn equations and n unknowns—the n components of the unknown state vector $\mathbf{x}_T(0)$. If the system is single output, $m = 1$, this represents n simultaneous linear equations in n unknowns.

The $mn \times n$ coefficient matrix in Eq. (8-67) is termed the *discrete observability matrix*

$$M_{T_o} \equiv \begin{bmatrix} C_T \\ C_T A_T \\ \vdots \\ C_T A_T^{n-1} \end{bmatrix}_{mn \times n} \tag{8-68}$$

Like the controllability issue, the question of discrete observability boils down to a question of whether or not Eq. (8-67) produces a unique solution for $\mathbf{x}_T(0)$. If so, the system is said to be *completely observable*. If not, the system is said to be *unobservable.*

Detailed discussion of the above must await matrix theory concepts yet to be developed in Chaps. 9 and 10. However, the following example provides some insight into the special case of the single-output system.

Example 8-9 Consider the hypothetical SISO system

$$\mathbf{x}_T(k) = \begin{bmatrix} 0.4 & 0.6 \\ 0.2 & 0.8 \end{bmatrix} \mathbf{x}_T(k-1) + \begin{bmatrix} 1 \\ 0 \end{bmatrix} u_T(k-1)$$

$$y_T(k) = [1 \quad -1]\mathbf{x}_T(k)$$

The observability matrix is

$$M_{T_o} = \begin{bmatrix} C_T \\ C_T A_T \end{bmatrix} = \begin{bmatrix} 1 & -1 \\ 0.2 & -0.2 \end{bmatrix}$$

The system is unobservable because the 2×2 observability matrix is singular—not invertible. This may be verified by taking the determinant of M_{T_o} det $M_{T_o} = 0$. Hence by Cramer's rule M_{T_o} has no inverse. (See Appendix B.)

To see the effect of unobservability, suppose $\mathbf{x}_T(0) = [1, 1]^T$. Then

$$y_T(0) = y_{T_{zi}}(0) = [1 \quad -1]\begin{bmatrix} 1 \\ 1 \end{bmatrix} = 0$$

and

$$y_{T_{zi}}(1) = [1 \quad -1]\begin{bmatrix} 0.4 & 0.6 \\ 0.2 & 0.8 \end{bmatrix}\begin{bmatrix} 1 \\ 1 \end{bmatrix}$$

$$= 0$$

In fact, $y_{T_{zi}}(k) = C_T A_T^{k-1}\mathbf{x}_T(0) = 0$ for all k for this particular initial state $\mathbf{x}_T(0) = [1, 1]^T$. This initial state is simply unobservable—it produces no observable output response.

Just because the system is unobservable does not imply that the system is uncontrollable. The controllability matrix for the system is

$$M_{T_c} = [B_T, \quad A_T B_T] = \begin{bmatrix} 1 & 0.4 \\ 0 & 0.2 \end{bmatrix}$$

The determinant of M_{T_c} is 0.2; M_{T_c} is invertible and is completely controllable. A set of controls $u_T(0)$ and $u_T(1)$ could be found from the zero state to anywhere in state space by solving the linear equation problem:

$$\begin{bmatrix} 1 & 0.4 \\ 0 & 0.2 \end{bmatrix}\begin{bmatrix} u_T(1) \\ u_T(0) \end{bmatrix} = \begin{bmatrix} x_{T_{zs_1}}(2) \\ x_{T_{zs_2}}(2) \end{bmatrix}$$

8-9 THE SYSTEM HANKEL MATRIX REVISITED

In Sec. 5-1 the curious system Hankel matrix

$$\mathcal{H}_T \equiv \begin{bmatrix} h_T(1) & h_T(2) & \cdots & h_T(n) \\ h_T(2) & h_T(3) & \cdots & h_T(n+1) \\ \cdots & \cdots & \cdots & \cdots \\ h_T(n) & h_T(n+1) & \cdots & h_T(2n-1) \end{bmatrix} \tag{5-18}$$

is encountered. Now using the fact that

$$h_T(k) = C_T A_T^{k-1} B_T \tag{8-55}$$

for the no-feedforward system it may be shown that

$$\mathcal{H}_T = M_{T_o} M_{T_c} = \begin{bmatrix} C_T \\ C_T A_T \\ \vdots \\ C_T A_T^{n-1} \end{bmatrix} [B_T, \quad A_T B_T, \quad \ldots, \quad A_T^{n-1} B_T]$$

$$= \begin{bmatrix} C_T B_T & C_T A_T B_T & \cdots & C_T A_T^{n-1} B_T \\ C_T A_T B_T & C_T A_T^2 B_T & \cdots & C_T A_T^n B_T \\ \cdots & \cdots & \cdots & \cdots \\ C_T A_T^{n-1} B_T & C_T A_T^n B_T & \cdots & C_T A_T^{2n-1} B_T \end{bmatrix} \tag{8-69}$$

In words, this says that the system Hankel matrix is equal to the product of the system observability and controllability matrices. This is regardless of the state coordinate system chosen; this is an input-output property of the system.

In Chap. 5 it is remarked that the system Hankel matrix, which can be determined experimentally from the discrete pulse-response model, can be used to ascertain the state model order n. Equation (8-69) brings us closer to deriving that result, but further matrix theory background is required. Chapter 11 returns to this issue.

8-10 SUMMARY

Computer methods are derived for finding the discrete state model $\{A_T, B_T, C_T, D\}$ from a given continuous time model $\{A, B, C, D\}$. The discrete state model is seen to facilitate digital simulation of the system. Then relationships between the discrete state model and the finite-difference equation and discrete transfer function are examined. Methods analogous to the continuous model results of Chap. 6 are shown for deriving a canonical-form discrete state model from the discrete SISO transfer function. The discrete state model is related to the discrete pulse-response model; this actually provides a convenient method for calculating the discrete pulse-response model. A chain of calculations is then suggested to go from a continuous transfer function, to a continuous state model, to a discrete state model, to the discrete pulse-response model, to the discrete finite-difference equation, and finally to the discrete state model in canonical form.

The last topics presented are an introduction to discrete state controllability and discrete state observability. The curious system Hankel matrix reappears as the product of the discrete observability and controllability matrices. Yet in discussing all of these latter topics, we find ourselves a bit lacking in some fundamental matrix theory concepts. These topics are developed in Chaps. 9 and 10 so that we may return to the important discussion of controllability and observability in Chap. 11.

PROBLEMS

8-1 Find $y(0.3)$ for the system

$$\dot{x}(t) = -2x(t) + 10u(t)$$

$$y(t) = 5x(t)$$

with piecewise constant input sequence

$$u_T(0) = 1 \qquad u_T(1) = -1 \qquad u_T(2) = 1 \qquad T = 0.1$$

and initial state $x(0) = 1$. *Hint:* Discretize the state model at $T = 0.1$, and then propagate forward to find $x(0.3)$ and $y(0.3)$.

8-2 Repeat Prob. 8-1 for the system

$$\dot{x}(t) = 2x(t) + 10u(t)$$

$$y(t) = 5x(t)$$

with the same piecewise constant input sequence and the same initial state.

8-3 Find $y(0.3)$ for the system

$$\begin{bmatrix} \dot{x}_1(t) \\ \dot{x}_2(t) \end{bmatrix} = \begin{bmatrix} -1 & 0 \\ 0 & -3 \end{bmatrix} \begin{bmatrix} x_1(t) \\ x_2(t) \end{bmatrix} + \begin{bmatrix} 1 \\ 1 \end{bmatrix} u(t)$$

$$y(t) = [10 \quad -10]\mathbf{x}(t)$$

for the same piecewise constant imputs as Prob. 8-1 and the initial state $\mathbf{x}(0) = [1, 1]^T$.

8-4 Repeat Prob. 8-3 for the system

$$\begin{bmatrix} \dot{x}_1(t) \\ \dot{x}_2(t) \end{bmatrix} = \begin{bmatrix} 1 & 0 \\ 0 & 3 \end{bmatrix} \begin{bmatrix} x_1(t) \\ x_2(t) \end{bmatrix} + \begin{bmatrix} 1 \\ 1 \end{bmatrix} u(t)$$

$$y(t) = [10 \quad -10]\mathbf{x}(t)$$

with the same input and the same initial state.

8-5 Prove by induction that

$$\mathbf{x}_T(k) = A_T\mathbf{x}_T(k-1) + B_T\mathbf{u}_T(k-1) \tag{8-19}$$

gives $\mathbf{x}_T(k) \equiv \mathbf{x}(kT)$ where $\mathbf{x}(kT)$ is the sampled state solution and

$$\mathbf{u}(t) = \mathbf{u}_T(k) \qquad t \in (kT, kT + T] \tag{8-21}$$

Hint: Start using $\mathbf{x}_T(0) = \mathbf{x}(0)$ and find $\mathbf{x}(T) = \mathbf{x}_T(1)$ using*

$$\mathbf{x}(t) = e^{A(t-t_0)}\mathbf{x}(t_0) + \int_0^{t-t_0} e^{A\xi}B\mathbf{u}(t-\xi)\,d\xi \tag{8-70}$$

Then define $A_T \equiv e^{A \cdot T}$ and $B_T \equiv (\int_0^T e^{A\xi}\,d\xi)B$, and continue proof by induction showing that $\mathbf{x}(kT + T) = \mathbf{x}_T(k + 1)$ assuming $\mathbf{x}(kT) = \mathbf{X}_T(k)$.

8-6 Use the squaring technique to compute e^7.

8-7 Use the squaring technique to compute e^{-7}.

8-8 In using the squaring algorithm, Algorithm 8-1, what would N be for the A matrix

$$A = \begin{bmatrix} 0 & 1 \\ -30 & -200 \end{bmatrix}$$

with $T = 0.8$?

* Let $\xi = t - \tau$ in Eq. (7-11).

8-9 In using the squaring algorithm, Algorithm 8-1, what would N be for the A matrix

$$A = \begin{bmatrix} -10 & 0 & 0 \\ 0 & -20 & 40 \\ 0 & -40 & -20 \end{bmatrix}$$

with $T = 0.6$?

8-10 The eigenvalues (poles) of the A matrix in Prob. 8.8 are $\lambda_1 = -10, \lambda_2 = -20$. What N would you use in the squaring algorithm if you chose N such that

$$\frac{|\lambda_{\max}|T}{2^N} < 1$$

with $T = 0.8$?

8-11 The eigenvalues (poles) of the A matrix in Prob. 8-9 are $\lambda_1 = -10$, $\lambda_2 = -20 + 40j$, $\lambda_3 = -20 - 40j$. What N would you use in the squaring algorithm if you chose N such that

$$\frac{|\lambda_{\max}|T}{2^N} < 1$$

with $T = 0.6$?

8-12 Derivation of Algorithm 8.1

Follow through the following steps to derive Algorithm 8-1 by induction:

(*a*) Show that

$$F_0 = I + AE_0 = \exp\frac{AT}{2^N}$$

and

$$E_0B = \left(\int_0^{T/2^N} e^{A\tau}\,d\tau\right)B$$

are true.

(*b*) For $M = 1$ show that

$$F_1 = F_0F_0 = exp\,\frac{AT}{2^{N-1}}$$

$$E_1B = \left(\int_0^{T/2^{N-1}} e^{A\tau}\,d\tau\right)B$$

where $E_1 = E_0 + F_0E_0$.

Hint:

$$E_1 = \int_0^{T/2^{N-1}} e^{A\tau}\,d\tau = \int_0^{T/2^N} e^{A\tau}\,d\tau + \int_{T/2^N}^{T/2^{N-1}} e^{A\tau}\,d\tau$$

Make the change of variable $\tau = \zeta + T/2^N$ in the second integral to show that $E_1 = E_0 + F_0E_0$.

(*c*) For $M = k$ assume that

$$F_k = F_{k-1}F_{k-1} = \exp\frac{AT}{2^{N-k}}$$

$$E_kB = \left(\int_0^{T/2^{N-k}} e^{A\tau}\,d\tau\right)B$$

where $E_k = E_{k-1} + F_{k-1}E_{k-1}$. Then show that

$$F_{k+1} = F_kF_k = \exp\frac{A\cdot T}{2^{N-k-1}}$$

$$E_{k+1}B = \left(\int_0^{T/2^{N-k-1}} e^{A\tau}\,d\tau\right)B$$

where $E_{k+1} = E_k + F_kE_k$.

Hint: Again let $\tau = \zeta + T/2^{N-k}$ in the integral for E_k.

(*d*) By induction thus show that

$$F_N = F_{N-1}F_{N-1} = \exp A \cdot T \equiv A_T$$

$$E_N B = \left(\int_0^T e^{A\tau}\, d\tau\right) B \equiv B_T$$

where $E_N = E_{N-1} + F_{N-1}E_{N-1}$.

8-13 If the input $\mathbf{u}(t)$ is a continuous function, then the piecewise linear discrete approximation

$$\mathbf{u}(t) \simeq \mathbf{u}_T(k) + \dot{\mathbf{u}}_T(k)(t - kT) \qquad t \in (kT,\quad kT + T] \tag{8-71}$$

is often a better approximation to the actual continuous input $u(t)$ than the piecewise constant, stairstep approximation

$$\mathbf{u}(t) \simeq \mathbf{u}_T(k) \qquad t \in (kT,\quad kT + T] \tag{8-21}$$

Note that the constant value, $\mathbf{u}_T(k)$, in Eq. (8-71) is not necessarily equal to the constant value, $\mathbf{u}_T(k)$, in the stairstep approximation, Eq. (8-21).

Show that the discrete state update equation using the piecewise linear input Eq. (8-71) is given by

$$\mathbf{x}_T(k) = A_T\mathbf{x}_T(k - 1) + B_T\mathbf{u}_T(k - 1) + B'_T\dot{\mathbf{u}}_T(k - 1) \tag{8-72}$$

where A_T and B_T are found the same as previously:

$$A_T = e^{AT} \qquad B_T = \left(\int_0^T e^{A\xi}\, d\xi\right) B$$

and B'_T is found via

$$B'_T = \left(\int_0^T \xi e^{A\xi}\, d\xi\right) B \tag{8-73}$$

Hint: Substitute Eq. (8-71) into Eq. (8-70), Prob. 8-5, starting first at $k = 0$. Then assume true at k, and prove for $k + 1$.

8-14 Show that

$$\int_0^T \xi e^{A\xi}\, d\xi = TA^{-1}e^{AT} - A^{-2}[e^{AT} - I]$$

provided A^{-1} exists. *Hint:* Use integration by parts letting $u = \xi$, $dv = e^{A\xi}\, d\xi$, and the fact that

$$\int_0^T e^{A\xi}\, d\xi = A^{-1}[e^{AT} - I] \tag{7-37}$$

8-15 (*a*) What would be the piecewise linear model

$$u_T(k) + \dot{u}_T(k)(t - kT)$$

for $k = 0, 1, 2$, for $T = 1$, for the continuous input $u(t) = (2t)u_{-1}(t)$?

(*b*) What would be a piecewise constant approximation, $u_T(k)$, to the continuous input of part (*a*), $u(t) = (2t)u_{-1}(t)$?

(*c*) Would the discrete approximation of (*a*) or (*b*) be a better representation? Explain.

8-16 Repeat Prob. 8-15 for the continuous input $u(t) = (2 + t/2)u_{-1}(t)$.

8-17 (*a*) Find the discrete update model of the form Eq. (8-72) for the system of Prob. 8-1

$$\dot{x}(t) = -2x(t) + 10u(t) \qquad x(0) = 0$$

$$y(t) = 5x(t)$$

with $T = 1$.

(*b*) Use the results of part (*a*) to find $y_T(1)$, $y_T(2)$, $y_T(3)$ for the piecewise linear input of Prob. 8-15(*a*). The system starts as the zero initial state.

8-18 (*a*) Find the discrete update model of the form Eq. (8-72) for the system of Prob. 8-2

$$\dot{x}(t) = 2x(t) + 10u(t) \qquad x(0) = 0$$

$$y(t) = 5x(t)$$

with $T = 1$.

(*b*) Use the results of part (*a*) to find $y_T(1)$, $y_T(2)$, $y_T(3)$ for the piecewise linear input of Prob. 8-16(*a*).

8-19 (*a*) Find the discrete update model of the form

$$x_T(k) = a_T x_1(k - 1) + b_T u_T(k - 1)$$

$$y_T(k) = c_T x_T(k)$$

for the system of Prob. 8-17(*a*).

(*b*) Find $y_T(1)$, $y_T(2)$, and $y_T(3)$ for the discrete model of part (*a*) using the discrete inputs from Prob. 8-15(*b*). The system starts at the zero initial state.

(*c*) Will the discrete solution from Prob. 8-17(*b*) or 8-19(*b*) be better? Justify.

8-20 (*a*) Repeat Prob. 8-19(*a*) for the system of Prob. 8-18(*a*).

(*b*) Repeat Prob. 8-19(*b*) for the discrete inputs from Prob. 8-16(*b*).

(*c*) Will the discrete solution from Prob. 8-18(*b*) or 8-20(*b*) be better? Justify.

8-21 Show that for the general linear time-invariant system starting at initial state $\mathbf{x}(0)$ the sampled zero-input response is given by

$$\mathbf{y}_{zi}(kT) = \mathbf{y}_{T_{zi}}(k) = C_T A_T^k \mathbf{x}(0)$$

where $C_T \equiv C$, the continuous output matrix, $A_T \equiv e^{A \cdot T}$, and where A is the continuous plant matrix.

8-22 (*a*) Show that for the general linear time-invariant system with no feedforward ($D = 0$), the sampled zero-state response, $\mathbf{y}_{zs}(kT)$, for the piecewise constant input sequence, $\mathbf{u}_T(i)$, $i = 0, 1, 2, \ldots, k - 1$, is given by the discrete convolution summations

$$\mathbf{y}_{zs}(kT) = \mathbf{y}_{T_{zs}}(k) = \sum_{i=0}^{k-1} C_T A_T^{k-i-1} B_T \mathbf{u}_T(i)$$

$$= \sum_{j=1}^{k} C_T A_T^{j-1} B_T \mathbf{u}_T(k - j)$$

(*b*) Compare the results of part (*a*) with Eqs. (1-20*a*) and (1-20*b*) to show that

$$h_T(k) = C_T A_T^{k-1} B_T$$

for the SISO system.

8-23 Find $h_T(1)$, $h_T(2)$, $h_T(3)$ for the system of Prob. 8-1 via use of Eq. (8-55).

8-24 Repeat Prob. 8-23 for the system of Prob. 8-2.

8-25 Find the phase-variable discrete state model for the system with discrete transfer function

$$H_T(z) = \frac{3z + 2}{z^2 - 0.8z + 0.15}$$

8-26 Find the phase-variable discrete state model for the system with discrete transfer function

$$H_T(z) = \frac{z + 1}{z^2 - 1.7z + 0.72}$$

8-27 Find the discrete state model for the system in Prob. 8-25 in diagonal form.

8-28 Find the discrete state model for the system in Prob. 8-26 in diagonal form.

8-29 Find the discrete state model for the system

$$\dot{\mathbf{x}}(t) = \begin{bmatrix} 0 & 1 \\ -2 & -3 \end{bmatrix}\mathbf{x}(t) + \begin{bmatrix} 0 \\ 1 \end{bmatrix}u(t)$$

$$y(t) = [10 \quad 0]\mathbf{x}(t)$$

in phase-variable form

$$\mathbf{x}_T(k) = \begin{bmatrix} 0 & 1 \\ -a_{T_0} & -a_{T_1} \end{bmatrix}\mathbf{x}_T(k-1) + \begin{bmatrix} 0 \\ 1 \end{bmatrix}u_T(k-1)$$

$$y_T(k) = [b_{T_0} \quad b_{T_1}]\mathbf{x}_T(k)$$

with $T = 0.1$ by going through the following steps:

(*a*) Find A_T and B_T from Eqs. (8-26) and (8-27). Three terms in the power series for E will suffice.

(*b*) Find $h_T(1)$, $h_T(2)$ via Eq. (8-55).

(*c*) Find the finite-difference equation model

$$y_T(k) + a_{T_1}y_T(k-1) + a_{T_0}y_T(k-2) = b_{T_1}u_T(k-1) + b_{T_0}u_T(k-2)$$

by applying Eq. (5-36) and Eqs. (5-11*b*) through (5-14*b*) from Chap. 5. (*Note:* the continuous poles are $p_1 = -1$, $p_2 = -2$.)

(*d*) Put the finite-difference equation model of part (*c*) into phase-variable state-space form.

8-30 Repeat Prob. 8-29 for the continuous system

$$\dot{\mathbf{x}}(t) = \begin{bmatrix} 0 & 1 \\ -5 & -4 \end{bmatrix}\mathbf{x}(t) + \begin{bmatrix} 0 \\ 1 \end{bmatrix}u(t)$$

$$y(t) = [1 \quad 0]\mathbf{x}(t)$$

Again use $T = 0.1$. The continuous poles are $p_1 = -2 + j$, $p_2 = -2 - j$.

8-31 (*a*) Determine whether or not the following discrete system is completely controllable:

$$\mathbf{x}_T(k) = \begin{bmatrix} 0.1 & 0.2 \\ -0.3 & 0.6 \end{bmatrix}\mathbf{x}_T(k-1) + \begin{bmatrix} 1 \\ 1 \end{bmatrix}u_T(k-1)$$

$$y_T(k) = [1, \quad 2]\mathbf{x}_T(k)$$

(*b*) Is the system of (*a*) completely observable?

8-32 (*a*) Determine whether or not the following discrete system is completely controllable:

$$\mathbf{x}_T(k) = \begin{bmatrix} -0.1 & -0.8 \\ 0.1 & 0.5 \end{bmatrix}\mathbf{x}_T(k-1) + \begin{bmatrix} 1 \\ 1 \end{bmatrix}u_T(k-1)$$

$$y_T(k) = [1 \quad 4]\mathbf{x}_T(k)$$

(*b*) Is the system of (*a*) completely observable?

8-33 Find $u_T(0)$, $u_T(1)$, and u_{ss} such that

$$y_T(2) = y_T(3) = y_{ss} = 10 \qquad T = 0.1$$

for the system of Prob. 8-3 assuming that it starts from the zero initial state. *Hint:* see Example 8-8.

8-34 Repeat Prob. 8-33 for the system in Prob. 8-4.

8-35 (*a*) Suppose that the following discrete system

$$\mathbf{x}_T(k) = \begin{bmatrix} 0 & 1 \\ -0.82 & 0.2 \end{bmatrix}\mathbf{x}_T(k-1) + \begin{bmatrix} 0 \\ 1 \end{bmatrix}u_T(k-1)$$

$$y_T(k) = [1, \quad 2]\mathbf{x}_T(k)$$

starts at initial state $\mathbf{x}_T(0) = [0,\ 10]^T$. Find $u_T(0)$, $u_T(1)$ such that $\mathbf{x}_T(2) \equiv 0$. *Hint:* Recall that

$$\mathbf{x}_T(2) = \mathbf{x}_{T_{zi}}(2) + \mathbf{x}_{T_{zs}}(2)$$

and that

$$\mathbf{x}_{T_{zs}}(2) = M_{T_c}\begin{bmatrix} u_T(1) \\ u_T(0) \end{bmatrix}$$

Find $\mathbf{x}_{T_{zi}}(2)$, set $\mathbf{x}_T(2) = 0$, and solve for $u_T(0)$, $u_T(1)$.

(*b*) Verify your answer to (*a*) by propagating the state update expression two steps forward starting from $\mathbf{x}_T(0)$ and using the two computed inputs.

8-36 Repeat Prob. 8-35 for the system

$$\mathbf{x}_T(k) = \begin{bmatrix} 0 & 1 \\ -0.25 & 0.6 \end{bmatrix}\mathbf{x}_T(k-1) + \begin{bmatrix} 0 \\ 1 \end{bmatrix}u_T(k-1)$$

$$y_T(k) = [1,\quad 4]\mathbf{x}_T(k)$$

8-37 (*a*) Suppose the system of Prob. 8-35 has input $u_T(0) = 1$ and the two measurements $y_T(0) = 5$, $y_T(1) = -2$ are taken. Find $\mathbf{x}_T(0)$. *Hint:* Recall that

$$y_T(1) = C_T\mathbf{x}_T(1) = C_T[A_T\mathbf{x}_T(0) + B_Tu_T(0)]$$

(*b*) What is $\mathbf{x}_T(1)$? *Hint:* Use the solution for $\mathbf{x}_T(0)$ to find $\mathbf{x}_T(1)$.

8-38 Repeat Prob. 8-37 for the system of Prob. 8-36.

NOTES AND REFERENCES

Use of a power series to calculate A_T and B_T are well known (e.g., Ref. 1). So too is the concept and importance of using the squaring-up technique to calculate A_T (Ref. 2). The recursive squaring-up Algorithm 8-1 for computing both A_T and B_T simultaneously is original here.

For further background on the deadbeat control problem see Ref. 3. Further results and theory on controllability and observability are presented in Chap. 11, and so further discussion of applicable references is reserved until Chap. 11.

The technique for using higher-order polynomial approximations for the input function can be a powerful technique for accurate digital simulation. This concept is introduced in Probs. 8-13 through 8-22. For further background on this technique see Ref. 4.

1. De Russo, P. M., R. J. Roy, and C. M. Close: *State Variables for Engineers,* Wiley, New York, 1965.
2. Moler, C., and C. Van Loan: "Nineteen Dubious Ways to Compute the Exponential of a Matrix," *SIAM Review,* vol. 20, no. 4, October 1978, pp. 801–836.
3. Cadzow, J. A.: *Discrete Time Systems,* Prentice-Hall, Englewood Cliffs, N.J., 1973.
4. Koenigsberg, W., and L. Lemos: "State Space Simulation of Dynamic Systems," C. S. Draper Laboratory Report. R-1168, Cambridge, Mass., May 1979.

CHAPTER

NINE

FUNDAMENTALS OF LINEAR VECTOR SPACES AND TRANSFORMATION OF STATE COORDINATES

Vectors and linear operations are used extensively in the previous chapters, particularly in regard to state-space concepts. Linear vector space concepts are at the root of linear dynamic-systems theory. It is thus important to comprehend certain basic linear vector space ideas. This chapter defines a linear vector space and subspace, investigates the concept of the linear dependence and independence of vectors, and finally considers the problem of the change of vector coordinate system or basis. This latter topic is directly applied to the state-space model to consider the problem of transformation of the state-space model coordinate system.

This chapter forms a foundation for the linear equation and eigenvector topics of the next chapter. This is a relatively short chapter, but it is essential to have this foundation before proceeding to the more advanced topics of the next chapter.

9-1 THE DEFINITION OF A VECTOR SPACE

As a first matter of business, it is useful to define a vector and a linear vector space.

Definition

A *vector space* V over the field of real or complex numbers is a set of elements, called vectors, for which vector addition and scalar multiplication are defined operations. Three fundamental properties must then be true regarding these operations of vector addition and scalar multiplication:

1. Closure under addition: If $\mathbf{v}_1 \in V$ and $\mathbf{v}_2 \in V$, then their sum $\mathbf{v}_1 + \mathbf{v}_2$ must also be an element of V.
2. Closure under scalar multiplication: If $\mathbf{v} \in \mathbf{V}$, and if α is a scalar from the defined field (a real scalar for our purposes), then the scalar product $\alpha\mathbf{v}$ is also an element of V.
3. Existence of the zero vector: There must be a $\mathbf{0} \in V$ such that $\mathbf{0} + \mathbf{v} = \mathbf{v}$ for all $\mathbf{v} \in V$.

Additionally, vector addition and scalar multiplication must be associative [that is, $(\mathbf{v}_1 + \mathbf{v}_2) + \mathbf{v}_3 = \mathbf{v}_1 + (\mathbf{v}_2 + \mathbf{v}_3)$ and $\alpha(\beta\mathbf{v}) = (\alpha\beta)\mathbf{v}$]; commutative (that is, $\mathbf{v}_1 + \mathbf{v}_2 = \mathbf{v}_2 + \mathbf{v}_1$); and distributive [that is, $\alpha(\mathbf{v}_1 + \mathbf{v}_2) = \alpha\mathbf{v}_1 + \alpha\mathbf{v}_2$ and $(\alpha + \beta)\mathbf{v} = \alpha\mathbf{v} + \beta\mathbf{v}$]. Finally, for each $\mathbf{v} \in V$ there must exist an additive inverse such that $\mathbf{v} + (-\mathbf{v}) = \mathbf{0}$, and it must be true that $1 \cdot \mathbf{v} = \mathbf{v}$.

All of the above conditions are necessary, strictly speaking, to define a vector space. However, proving the first three numbered conditions is usually all that is needed, as the other conditions readily follow except in unusual cases. With this general definition of a linear vector space, there are many familiar entities that qualify as vectors and vector spaces.

Example 9-1 The space of continuous functions defined on the interval [0, 1], and whose values at each end point are zero, is a linear vector space. Letting $f_1(\cdot)$ and $f_2(\cdot)$ be any two vectors (continuous functions in this case) from that space, it is seen that their sum is still a continuous function with zero end conditions, and any scalar multiple of a function from this space is still a continuous function with zero end conditions. Thus the vector space requirements of closure under addition and scalar multiplication are met. The existence of a *zero vector* is met by the zero function (the function which is zero everywhere on the interval [0, 1]).

With such vector spaces of functions, in which the vector space is actually infinite-dimensional, the concepts of linear vector space theory can become somewhat involved. Fortunately, though, most of the endeavors here can be met by restricting attention to what the reader normally might think of as a vector: an n-tuple linear array of the form

$$\mathbf{v} = \begin{bmatrix} v_1 \\ v_2 \\ \vdots \\ v_n \end{bmatrix} \tag{9-1}$$

It may be verified that the space of all such n-tuple arrays satisfies the vector space requirements. Namely, the addition or scalar multiplication of n-dimensional arrays are still n-dimensional arrays, and the n-dimension array of all zeros is the zero vector. If the arrays are restricted to have all real elements, then the vector space is said to be the real n-dimensional vector space and is denoted here as R^n. If the vectors may take on complex elements, then it is termed the complex n-dimensional vector space and denoted as C^n.

For vectors which are n-dimensional arrays, the dimension n is equal to the number of elements in the array. For example,

$$\mathbf{a} = \begin{bmatrix} a_1 \\ a_2 \end{bmatrix}$$

is an element of R^2 (or C^2 if a_1 and a_2 can be complex), while

$$\mathbf{b} = \begin{bmatrix} b_1 \\ b_2 \\ b_3 \end{bmatrix}$$

is an element of R^3 (or C^3). This is written as $\mathbf{a} \in R^2$ and $\mathbf{b} \in R^3$. For convenience in the discussion to follow, real vector spaces are assumed, but the results are easily generalized to C^n.

9-2 VECTOR NORMS AND INNER PRODUCTS

One of the basic concepts associated with vectors is that of the length, or size, or, more properly, the *norm* of the vector. Like the concept of the vector space itself, the norm function may be given a general definition. However, for our purposes in dealing with vectors which are n-dimensional arrays, it suffices to think in terms of the standard or Euclidean norm of a vector $\mathbf{v} \in R^n$:

$$\|\mathbf{v}\| \equiv \left(\sum_{i=1}^{n} v^2(i)\right)^{1/2} \tag{9-2}$$

This is what one normally might think of as the length of a vector—the square root of the sum of the squares of its elements. This is reminiscent of the Froebinius norm of a matrix as encountered in the previous chapter. Indeed, if one thinks of a vector as a single-column matrix, they are one and the same.

As mentioned, the norm of a vector is a general concept, and many possible functions meet the requirements of a valid norm function. Requirements of the norm function over finite-dimensional vector spaces as dealt with here are:

1. $\|\mathbf{v}\| \geq 0$ and $\|\mathbf{v}\| = 0$ if and only if $\mathbf{v} \equiv 0$.
2. $\|\mathbf{v}_1 + \mathbf{v}_2\| \leq \|\mathbf{v}_1\| + \|\mathbf{v}_2\|$ (the triangle inequality).
3. $\|\alpha \mathbf{v}\| = |\alpha| \cdot \|\mathbf{v}\|$ for all scalars α.

These properties may, in fact, be used as the definition of the norm function; any function over a vector space which satisfies these properties is a valid norm function.

Example 9-2 Let $\mathbf{x} \varepsilon R^2$. Then the standard Euclidean norm satisfies the three properties. Namely:

1. $\|\mathbf{x}\| \equiv (x_1^2 + x_2^2)^{1/2} \geq 0$ for all possible elements x_1 and x_2, and it is zero if and only if $x_1 = x_2 = 0$.
2. The sum of $\mathbf{x}_1$ and $\mathbf{x}_2$ vectors is the triangle distance from the tail of $\mathbf{x}_1$ to the tip of $\mathbf{x}_2$. This is shown in Fig. 9-1. Then

$$\|\mathbf{x}_1 + \mathbf{x}_2\| \leq \|\mathbf{x}_1\| + \|\mathbf{x}_2\|$$

   There is equality when $\mathbf{x}_1$ and $\mathbf{x}_2$ point in the same direction.
3. Thirdly,

$$\|\alpha\mathbf{x}\| = (\alpha^2 x_1^2 + \alpha_2^2 x_2^2)^{1/2} = \alpha(x_1^2 + x_2^2)^{1/2}$$
$$= \alpha\|\mathbf{x}\|$$

Example 9-3 Let

$$\mathbf{v} = [2 \quad 1 \quad -2 \quad -4]^T$$

Then $\|\mathbf{v}\| = (4 + 1 + 4 + 16)^{1/2} = 5$

A function of two vectors closely affiliated with the norm function is the *inner product* of two vectors, also known as the *scalar* or *dot* product of two vectors. The inner product that we use for real n-dimensional vectors $\mathbf{v}_1$ and $\mathbf{v}_2$ is

$$\mathbf{v}_1 \bullet \mathbf{v}_2 \equiv \mathbf{v}_1^T \mathbf{v}_2 = \sum_{i=1}^{n} v_1(i) v_2(i) \tag{9-3}$$

By this definition it is seen that the norm squared of a vector is just the inner product of $\mathbf{v}$ with itself:

$$\|\mathbf{v}\|^2 = \mathbf{v} \bullet \mathbf{v} = \mathbf{v}^T \mathbf{v} = \sum_{i=1}^{n} v(i) v(i) \tag{9-4}$$

Example 9-4 Consider the inner product of the vector

$$\mathbf{v} = [2 \quad 1 \quad -2 \quad -4]^T$$

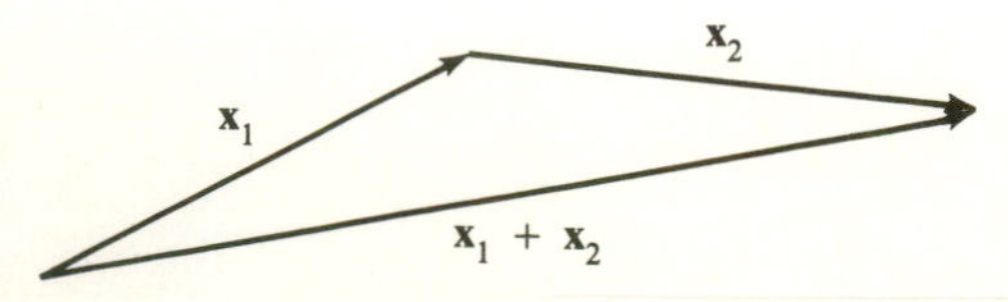

Figure 9-1 Illustration of triangle inequality, $\|x_1 + x_2\| \leq \|x_1\| + \|x_2\|$.

with each of the three vectors

$$\mathbf{v}_1 = \begin{bmatrix} 0 \\ 0 \\ 1 \\ 0 \end{bmatrix} \qquad \mathbf{v}_2 = \begin{bmatrix} 2/5 \\ 1/5 \\ -2/5 \\ -4/5 \end{bmatrix} \qquad \mathbf{v}_3 = \begin{bmatrix} 2 \\ 0 \\ 0 \\ 1 \end{bmatrix}$$

These inner products yield

$$\mathbf{v}^T\mathbf{v}_1 = -2 \qquad \mathbf{v}^T\mathbf{v}_2 = 5 \qquad \mathbf{v}^T\mathbf{v}_3 = 0$$

The inner product of $\mathbf{v}$ with $\mathbf{v}_1$, $\mathbf{v}_2$, and $\mathbf{v}_3$ illustrates several interesting points. The vectors $\mathbf{v}_1$ and $\mathbf{v}_2$ both have unit norms. Therefore, the inner product of $\mathbf{v}$ with each of these two vectors indicates how much of $\mathbf{v}$ lies in the $\mathbf{v}_1$ and $\mathbf{v}_2$ directions, respectively. The vectors $\mathbf{v}$ and $\mathbf{v}_2$ are in the same direction,

$$\mathbf{v} = 5\mathbf{v}_2 = \|\mathbf{v}\|\mathbf{v}_2$$

Then

$$\mathbf{v}^T\mathbf{v}_2 = \|\mathbf{v}\| \, \|\mathbf{v}_2\| = \|\mathbf{v}\| = 5$$

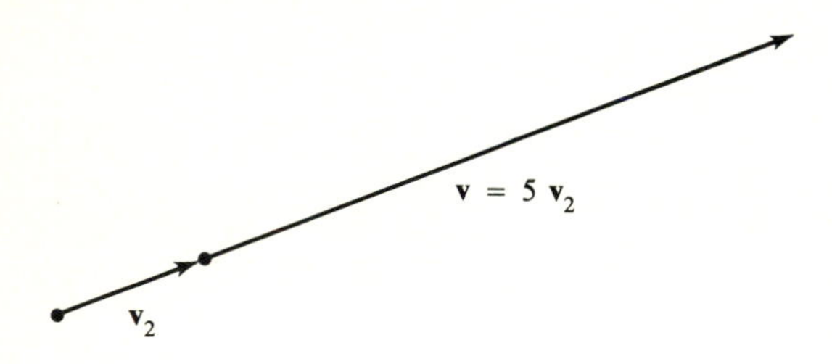

Figure 9-2 Illustration of v and v_2 in same direction.

Figure 9-3 Illustration of orthogonality of v and v_3.

The inner product of $\mathbf{v}$ and $\mathbf{v}_3$ is zero. The vectors $\mathbf{v}$ and $\mathbf{v}_3$ are thus said to be *orthogonal,* or perpendicular. The property of orthogonality plays a key role in later discussions. Figures 9-2 and 9-3 illustrate these points regarding $\mathbf{v}$, $\mathbf{v}_2$ and $\mathbf{v}_3$.

Both the norm and the inner product may be given general definitions, but the restricted definitions of the norm and inner product as given here will suffice for our purposes.

9-3 LINEAR COMBINATIONS OF VECTORS AND VECTOR SUBSPACES

Let $\mathbf{a}_1$ and $\mathbf{a}_2$ be two given vectors in real n-dimensional vector space, R^n. Then all possible linear combinations of $\mathbf{a}_1$ and $\mathbf{a}_2$,

$$\mathbf{v} = \alpha_1\mathbf{a}_1 + \alpha_2\mathbf{a}_2 \tag{9-5}$$

for all real α_1 and α_2 scalars, form what is termed a *subspace* of R^n. In this case (since it is linear combinations of only *two* vectors, $\mathbf{a}_1$ and $\mathbf{a}_2$) the subspace is a plane, or perhaps a line if $\mathbf{a}_1$ and $\mathbf{a}_2$ happen to be colinear (that is, $\mathbf{a}_1 = k\mathbf{a}_2$ for a given scalar k). This subspace is denoted as

$$\text{span}\ \{\mathbf{a}_1, \mathbf{a}_2\} \equiv \text{all linear combinations of } \mathbf{a}_1 \text{ and } \mathbf{a}_2$$

The subspace $V = \text{span}\ \{\mathbf{a}_1, \mathbf{a}_2\}$ is seen to satisfy the requirements for a vector space itself. Namely, if $\mathbf{v}_1 \in \text{span}\ \{\mathbf{a}_1, \mathbf{a}_2\}$ and $\mathbf{v}_2 \in \text{span}\ \{\mathbf{a}_1, \mathbf{a}_2\}$ then, by definition, there are some constants (α_1, α_2) and (β_1, β_2) such that

$$\mathbf{v}_1 = \alpha_1\mathbf{a}_1 + \alpha_2\mathbf{a}_2$$

$$\mathbf{v}_2 = \beta_1\mathbf{a}_1 + \beta_2\mathbf{a}_2$$

It follows that

$$(k\mathbf{v}_1) = (k\alpha_1)\mathbf{a}_1 + (k\alpha_2)\mathbf{a}_2$$

and

$$(\mathbf{v}_1 + \mathbf{v}_2) = (\alpha_1 + \beta_1)\mathbf{a}_1 + (\alpha_2 + \beta_2)\mathbf{a}_2$$

are also elements of span $\{\mathbf{a}_1, \mathbf{a}_2\}$. Hence the vector subspace span $\{\mathbf{a}_1, \mathbf{a}_2\}$ is closed under scalar multiplication and addition. Finally the zero vector is an element of the subspace V since $\mathbf{0} = 0 \cdot \mathbf{a}_1 + 0 \cdot \mathbf{a}_2$.

To illustrate, let

$$\mathbf{a}_1 = \begin{bmatrix} 1 \\ 1 \\ 0 \end{bmatrix} \quad \text{and} \quad \mathbf{a}_2 = \begin{bmatrix} 0 \\ 1 \\ 1/2 \end{bmatrix}$$

Then span $\{\mathbf{a}_1, \mathbf{a}_2\}$ forms the two-dimensional subspace as shown in Fig. 9-4. Any vector that lies in this plane is contained in span $\{\mathbf{a}_1, \mathbf{a}_2\}$ and may be written as the linear combination of the two vectors $\mathbf{a}_1$ and $\mathbf{a}_2$. Notice that the plane passes through the origin, and so the zero vector is always contained in the subspace. Also notice that the addition or scalar multiplication of any number of vectors

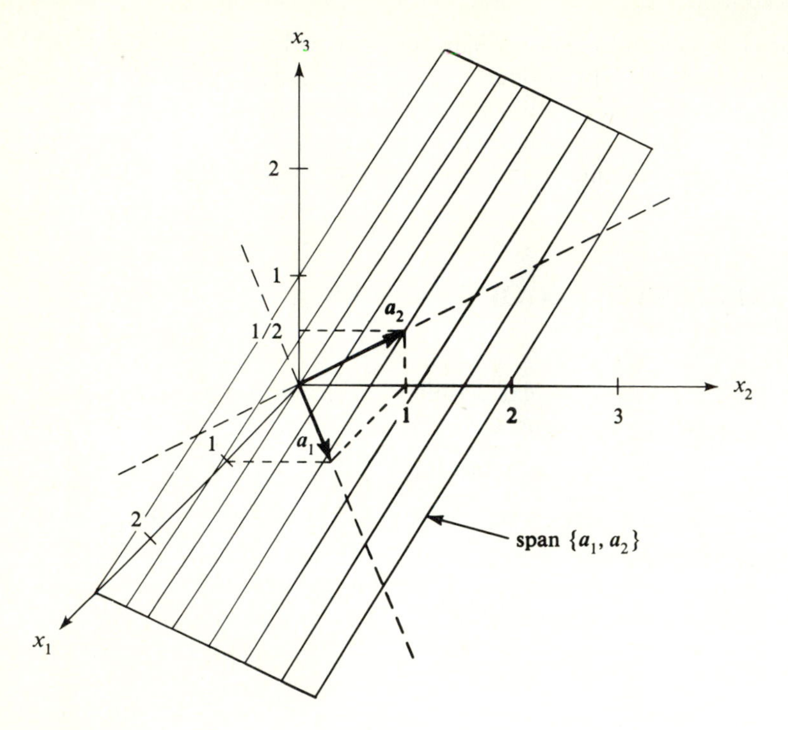

Figure 9-4 Illustration of "plane" span $\{a_1, a_2\}$.

that lie in the plane always remains in the plane. This is the concept of closure under scalar multiplication and addition.

Also note that any two noncolinear vectors in the plane may be used to form the subspace. For instance,

$$\mathbf{b}_1 = (1)\begin{bmatrix}1\\1\\0\end{bmatrix} + (2)\begin{bmatrix}0\\1\\1/2\end{bmatrix} = \begin{bmatrix}1\\3\\1\end{bmatrix}$$

$$\mathbf{b}_2 = (-1)\begin{bmatrix}1\\1\\0\end{bmatrix} + (2)\begin{bmatrix}0\\1\\1/2\end{bmatrix} = \begin{bmatrix}-1\\1\\1\end{bmatrix}$$

are also in the plane (since they are linear combinations of $\mathbf{a}_1$ and $\mathbf{a}_2$), and they are not colinear. Then span $\{\mathbf{b}_1, \mathbf{b}_2\}$ is exactly this same plane, and so

$$\text{span }\{\mathbf{a}_1, \mathbf{a}_2\} = \text{span }\{\mathbf{b}_1, \mathbf{b}_2\}$$

That is, the span of $\mathbf{a}_1$ and $\mathbf{a}_2$ and the span of $\mathbf{b}_1$ and $\mathbf{b}_2$ form exactly the same subspace (a plane in this case).

How about span $\{\mathbf{a}_1, \mathbf{a}_2, \mathbf{b}_1, \mathbf{b}_2\} \equiv$ all possible linear combinations of $\mathbf{a}_1$, $\mathbf{a}_2$, $\mathbf{b}_1$, $\mathbf{b}_2$? Since all of these vectors lie in the same two-dimensional plane, linear combinations of these vectors again never produce a vector outside the plane.

Hence

$$\text{span}\{\mathbf{a}_1, \mathbf{a}_2\} = \text{span}\{\mathbf{b}_1, \mathbf{b}_2\} = \text{span}\{\mathbf{a}_1, \mathbf{a}_2, \mathbf{b}_1, \mathbf{b}_2\}$$

But now suppose that $\mathbf{b}_3$ is formed from

$$\mathbf{b}_3 = (-1)\begin{bmatrix} 1 \\ 1 \\ 0 \end{bmatrix} + (-2)\begin{bmatrix} 0 \\ 1 \\ 1/2 \end{bmatrix} = \begin{bmatrix} -1 \\ -3 \\ -1 \end{bmatrix}$$

Again $\mathbf{b}_3$ is in the subspace since it is a linear combination of $\mathbf{a}_1$ and $\mathbf{a}_2$. However, now

$$\text{span}\{\mathbf{a}_1, \mathbf{a}_2\} \neq \text{span}\{\mathbf{b}_1, \mathbf{b}_3\}$$

since $\mathbf{b}_1 = -\mathbf{b}_3$ are colinear. Thus

$$\text{span}\{\mathbf{b}_1, \mathbf{b}_3\} = \text{span}\{\mathbf{b}_1\}$$

is the one-dimensional line subspace and no longer forms the two-dimensional planar subspace; it takes two noncolinear vectors to form a plane.

Definition The *dimension of a given subspace* is equal to the minimal number of vectors required to span that subspace.

By this definition it is seen that, for the previous example, span $\{\mathbf{a}_1, \mathbf{a}_2, \mathbf{b}_1, \mathbf{b}_2\}$ has dimension 2 since this subspace may be spanned by a minimum of two vectors, while span $\{\mathbf{b}_1, \mathbf{b}_3\}$ has dimension 1 since this subspace may be spanned by a minimum of one vector. Thus the dimension of a subspace has nothing to do with the number of vectors in the subspace; there are always an infinite number of vectors in a subspace. Likewise, it has nothing to do with the dimension of the vectors themselves, except that the dimension of the subspace can never exceed the dimension of the original vector space itself—the largest dimension subspace in R^3 is three, or R^3 itself.

How, then, is the minimal number of vectors required to span some given subspace $V = \text{span}\{\mathbf{v}_1, \mathbf{v}_2, \ldots, \mathbf{v}_m\}$ determined? It seems that the previous definition of the dimension of a subspace may be true, but it really does not help to *determine* that dimension except for obvious cases like a line or a plane. And even in these cases it would not have been obvious at first glance that $\mathbf{a}_1$, $\mathbf{a}_2$, $\mathbf{b}_1$, and $\mathbf{b}_2$ all lie in the same plane, or perhaps even that $\mathbf{b}_1$ and $\mathbf{b}_3$ are colinear. The answer to this dilemma is found in the concept of linear dependence and independence of vectors.

9-4 LINEAR DEPENDENCE, INDEPENDENCE, AND BASIS

A set of vectors $\{\mathbf{v}_1, \mathbf{v}_2, \ldots, \mathbf{v}_m\}$ are said to be *linearly independent* if the only set of scalars $\{\alpha_1, \alpha_2, \ldots, \alpha_m\}$ such that

$$\alpha_1\mathbf{v}_1 + \alpha_2\mathbf{v}_2 + \cdots + \alpha_m\mathbf{v}_m = \mathbf{0} \tag{9-6}$$

is the set of scalars which are all identically zero. The set is *linearly dependent* if there is a set of scalars $\{\alpha_1, \alpha_2, \ldots, \alpha_m\}$, not all zero, such that Eq. (9-6) is true. This is equivalent to saying that any one of the set, $\mathbf{v}_i$, may be written as a linear combination of the others in the set; that is, picking any $\mathbf{v}_i$ from the set, it is possible to find a set of scalars $\beta_{i,j}$, $j = 1, 2, \ldots, m$, $j \neq i$, such that

$$\mathbf{v}_i = \sum_{\substack{j=1 \\ j \neq i}}^{m} \beta_{ij} \mathbf{v}_j \tag{9-7}$$

This set of scalars may not be unique, but it is always possible to find a set of $\beta_{i,j}$ scalars such that Eq. (9-7) is true when the set of vectors $\{\mathbf{v}_1, \mathbf{v}_2, \ldots, \mathbf{v}_m\}$ are linearly dependent.

Using the example vectors of the previous section

$$\mathbf{a}_1 = \begin{bmatrix} 1 \\ 1 \\ 0 \end{bmatrix} \quad \mathbf{a}_2 = \begin{bmatrix} 0 \\ 1 \\ 1/2 \end{bmatrix} \quad \mathbf{b}_1 = \begin{bmatrix} 1 \\ 3 \\ 1 \end{bmatrix} \quad \mathbf{b}_2 = \begin{bmatrix} -1 \\ 1 \\ 1 \end{bmatrix}$$

it is seen that a set of any two of these vectors is linearly independent; for example, the set $\{\mathbf{a}_1, \mathbf{b}_1\}$ is linearly independent, since the only way that

$$\alpha_1 \begin{bmatrix} 1 \\ 1 \\ 0 \end{bmatrix} + \alpha_2 \begin{bmatrix} 1 \\ 3 \\ 1 \end{bmatrix} = \begin{bmatrix} 0 \\ 0 \\ 0 \end{bmatrix}$$

is for $\alpha_1 = \alpha_2 = 0$. However, a set of any three of these vectors are linearly dependent. For example,

$$\mathbf{b}_2 = -2\mathbf{a}_1 + \mathbf{b}_1$$

This is a unique relation. But if the set of four is considered, there are an infinity of linear relations between the vectors; for example,

$$\mathbf{b}_2 = -2\mathbf{a}_1 + \mathbf{b}_1 + \alpha(\mathbf{a}_1 + 2\mathbf{a}_2 - \mathbf{b}_1)$$

for any scalar α. Note that this is true since the factor $(\mathbf{a}_1 + 2\mathbf{a}_2 - \mathbf{b}_1)$ is the zero vector, and so it adds nothing to the particular relation $\mathbf{b}_2 = -2\mathbf{a}_1 + \mathbf{b}_1$.

It is not a coincidence that any two vectors in the set $\mathbf{a}_1$, $\mathbf{a}_2$, $\mathbf{b}_1$, and $\mathbf{b}_2$ are linearly independent, while a set of three or more of these vectors are linearly dependent; each of these four vectors lies in the two-dimensional subspace

$$\text{span}\,\{\mathbf{a}_1, \mathbf{a}_2\} = \text{span}\,\{\mathbf{b}_1, \mathbf{b}_2\} = \text{span}\,\{\mathbf{a}_1, \mathbf{a}_2, \mathbf{b}_1, \mathbf{b}_2\}$$

The dimension of the subspace can then be related to the maximum number of linearly independent vectors that can be found in that subspace.

Definition The *dimension of a subspace* is equal to the maximum number of linearly independent vectors that can be selected from that subspace.

This is a somewhat more useful definition of the dimension of a subspace—

provided, of course, that a way can be found to conveniently determine whether or not a given set of vectors are linearly independent or linearly dependent. Fortunately, there are straightforward means to do this that are described in the next chapter using matrix operation techniques.

It is also not a coincidence that two independent vectors selected from the two-dimensional subspace can be used to *uniquely* represent any other vector in the subspace, while a set of three linearly dependent vectors (for example, $\mathbf{a}_1$, $\mathbf{a}_2$, and $\mathbf{b}_1$) gave an infinity of solutions. Because of the *uniqueness* of the representation in the former case, that set of linearly independent vectors, $\{\mathbf{a}_1, \mathbf{b}_1\}$, from the two-dimensional subspace (in fact, *any* two linearly independent vectors from the two-dimensional subspace) form what is termed a *basis* for the subspace.

Definition A *basis* for an m-dimensional subspace, V, is formed by a set of *any* m linearly independent vectors from that subspace, $\mathbf{v}_1 \varepsilon V, \mathbf{v}_2 \varepsilon V, \ldots, \mathbf{v}_m \varepsilon V$. Furthermore, for any given $\mathbf{v} \varepsilon V$ there is a *unique* representation

$$\mathbf{v} = \alpha_1 \mathbf{v}_1 + \alpha_2 \mathbf{v}_2 + \cdots + \alpha_m \mathbf{v}_m$$

in terms of the $\{\mathbf{v}_1, \mathbf{v}_2, \ldots, \mathbf{v}_m\}$ basis vectors.

Example 9-4 Again consider

$$\mathbf{a}_1 = \begin{bmatrix} 1 \\ 1 \\ 0 \end{bmatrix} \qquad \mathbf{a}_2 = \begin{bmatrix} 0 \\ 1 \\ 1/2 \end{bmatrix} \qquad \mathbf{b}_1 = \begin{bmatrix} 1 \\ 3 \\ 1 \end{bmatrix} \qquad \mathbf{b}_2 = \begin{bmatrix} -1 \\ 1 \\ 1 \end{bmatrix}$$

The span $\{\mathbf{a}_1, \mathbf{a}_2, \mathbf{b}_1, \mathbf{b}_2\} \equiv V$ was shown to be a planar subspace and to have dimension 2. Therefore, any two linearly independent vectors from the plane V may be used to define a basis for V. For instance, the set $\mathcal{A} = \{\mathbf{a}_1, \mathbf{a}_2\}$ may be used to define the $\mathcal{A}$ basis for V. Then any vector in the plane has a *unique* representation in terms of the $\mathcal{A}$ basis. For instance,

$$\mathbf{b}_1 \Big|_{\mathcal{A}} = 1 \cdot \mathbf{a}_1 + 2 \cdot \mathbf{a}_2$$

In the $\mathcal{A}$-basis we may thus write

$$\mathbf{b}_1 \Big|_{\mathcal{A}} = \begin{bmatrix} 1 \\ 2 \end{bmatrix}\Bigg|_{\mathcal{A}} \equiv 1 \cdot \mathbf{a}_1 + 2 \cdot \mathbf{a}_2$$

We may also define, for example, the $\mathcal{B}$ basis by the set $\{\mathbf{b}_1, \mathbf{b}_2\}$. In terms of the $\mathcal{B}$ basis,

$$\mathbf{a}_1 \Big|_{\mathcal{B}} = \begin{bmatrix} 1/2 \\ 1/2 \end{bmatrix}\Bigg|_{\mathcal{B}} \equiv 1/2\mathbf{b}_1 + 1/2\mathbf{b}_2$$

Also note that we may represent $\mathbf{b}_1$ and $\mathbf{b}_2$ themselves with respect to the $\mathcal{B}$

basis. These representations are

$$\mathbf{b}_1\Big|_{\mathcal{B}} = \begin{bmatrix}1\\0\end{bmatrix}\Bigg|_{\mathcal{B}} \equiv 1\cdot\mathbf{b}_1 + 0\cdot\mathbf{b}_2$$

$$\mathbf{b}_2\Big|_{\mathcal{B}} = \begin{bmatrix}0\\1\end{bmatrix}\Bigg|_{\mathcal{B}} \equiv 0\cdot\mathbf{b}_1 + 1\cdot\mathbf{b}_2$$

Note that in the definition for basis there is no requirement for orthoganality among the basis vectors; the m-basis vectors for the given m-dimensional space merely have to be linearly independent. If each of the basis vectors also happens to be orthogonal to one another, it is termed an *orthogonal basis*. If the basis is orthogonal and each of the basis vectors also happens to have unit norm, it is termed an *orthonormal basis*.

Example 9-5 The standard euclidean basis

$$\begin{bmatrix}1\\0\\0\end{bmatrix} \quad \begin{bmatrix}0\\1\\0\end{bmatrix} \quad \begin{bmatrix}0\\0\\1\end{bmatrix}$$

represents an orthonormal basis for R^3. The vectors

$$\begin{bmatrix}\sqrt{3}\\0\\1\end{bmatrix} \quad \begin{bmatrix}0\\1\\0\end{bmatrix} \quad \begin{bmatrix}-1\\0\\\sqrt{3}\end{bmatrix}$$

are all orthogonal and hence linearly independent, and so would form an orthogonal basis for R^3. If we normalize each of these vectors by dividing each by their respective norm we obtain an orthonormal basis for R^3:

$$\begin{bmatrix}\sqrt{3}/2\\0\\1/2\end{bmatrix} \quad \begin{bmatrix}0\\1\\0\end{bmatrix} \quad \begin{bmatrix}-1/2\\0\\\sqrt{3}/2\end{bmatrix}$$

By the previous examples and discussions it is seen that there are an infinite number of possible bases for a given space or subspace. The problem remaining is how to tranform from one basis representation (or coordinate system) to another. This problem is addressed in the next section.

9-5 TRANSFORMATION OF BASIS

It follows from the discussion in the previous section that *any* set of n linearly independent vectors in an n-dimensional vector space forms a basis for that space.

For example, any three noncoplanar vectors in R^3 form a basis for R^3 and can be used to uniquely represent any other vector in R^3. Notice that there is no requirement for the vectors to be perpendicular in this concept of basis, as is the case, for example, in the standard euclidean basis

$$\begin{bmatrix} 1 \\ 0 \\ 0 \end{bmatrix} \quad \begin{bmatrix} 0 \\ 1 \\ 0 \end{bmatrix} \quad \begin{bmatrix} 0 \\ 0 \\ 1 \end{bmatrix}$$

The basis vectors merely have to be linearly independent.

This raises an interesting question: since any basis provides a unique representation of all vectors in the space, can one transform one vector representation to another by changing the basis coordinate system? Indeed, in Chap. 6, several different state-space models are derived for the same system: the phase-variable states, the states derived from the simulation diagram, and the diagonal-form states. And these are just three of an infinite number of possible state vector representations. It is remarked that these different state representations are state models in different state coordinate systems or different bases. If a way can be found to conveniently transform a vector representation from one basis to another, then maybe these same procedures can be used to mathematically transform from one state vector representation to another without going back to the transfer function.

To develop this notion consider a given vector in R^3:

$$\mathbf{x} = \begin{bmatrix} x_1 \\ x_2 \\ x_3 \end{bmatrix} = x_1 \begin{bmatrix} 1 \\ 0 \\ 0 \end{bmatrix} + x_2 \begin{bmatrix} 0 \\ 1 \\ 0 \end{bmatrix} + x_3 \begin{bmatrix} 0 \\ 0 \\ 1 \end{bmatrix}$$

This is the 3-tuple representation of $\mathbf{x}$ with respect to the three linearly independent orthonormal basis vectors

$$\begin{bmatrix} 1 \\ 0 \\ 0 \end{bmatrix} \quad \begin{bmatrix} 0 \\ 1 \\ 0 \end{bmatrix} \quad \begin{bmatrix} 0 \\ 0 \\ 1 \end{bmatrix}$$

Now suppose that one wishes to find the representation of $\mathbf{x}$ in a new basis $\mathcal{A} = \{\mathbf{a}_1, \mathbf{a}_2, \mathbf{a}_3\}$, where $\mathbf{a}_1$, $\mathbf{a}_2$, and $\mathbf{a}_3$ have the representation in the old euclidean basis

$$\mathbf{a}_1 = \begin{bmatrix} 1 \\ 1 \\ 0 \end{bmatrix} \qquad \mathbf{a}_2 = \begin{bmatrix} 0 \\ 1 \\ 1/2 \end{bmatrix} \qquad \mathbf{a}_3 = \begin{bmatrix} 1 \\ 0 \\ 1 \end{bmatrix}$$

That is, it is desired to find α_1, α_2, and α_3 such that

$$\mathbf{x} = \begin{bmatrix} x_1 \\ x_2 \\ x_3 \end{bmatrix} = \alpha_1 \begin{bmatrix} 1 \\ 1 \\ 0 \end{bmatrix} + \alpha_2 \begin{bmatrix} 0 \\ 1 \\ 1/2 \end{bmatrix} + \alpha_3 \begin{bmatrix} 1 \\ 0 \\ 1 \end{bmatrix}$$

This equation can then be formulated as a linear equation problem:

$$\begin{bmatrix} 1 & 0 & 1 \\ 1 & 1 & 0 \\ 0 & 1/2 & 1 \end{bmatrix} \begin{bmatrix} \alpha_1 \\ \alpha_2 \\ \alpha_3 \end{bmatrix} = \begin{bmatrix} x_1 \\ x_2 \\ x_3 \end{bmatrix}$$

or

$$\begin{bmatrix} \alpha_1 \\ \alpha_2 \\ \alpha_3 \end{bmatrix} = \begin{bmatrix} 1 & 0 & 1 \\ 1 & 1 & 0 \\ 0 & 1/2 & 1 \end{bmatrix}^{-1} \begin{bmatrix} x_1 \\ x_2 \\ x_3 \end{bmatrix}$$

But

$$\begin{bmatrix} \alpha_1 \\ \alpha_2 \\ \alpha_3 \end{bmatrix}$$

is the representation of **x** with respect to the new $\mathcal{A}$ basis:

$$\mathbf{x}|_{\mathcal{A}} = \begin{bmatrix} \alpha_1 \\ \alpha_2 \\ \alpha_3 \end{bmatrix} = \begin{bmatrix} 1 & 0 & 1 \\ 1 & 1 & 0 \\ 0 & 1/2 & 1 \end{bmatrix}^{-1} \mathbf{x}|_{\text{old basis}}$$

or, going the other way,

$$\mathbf{x}|_{\text{old basis}} = \begin{bmatrix} 1 & 0 & 1 \\ 1 & 1 & 0 \\ 0 & 1/2 & 1 \end{bmatrix} \mathbf{x}|_{\mathcal{A}}$$

To be specific, let **x** in the original basis be given by

$$\mathbf{x} = \begin{bmatrix} 1 \\ 2 \\ 3 \end{bmatrix}$$

Then **x** in the new basis is given by

$$\begin{aligned} \mathbf{x}|_{\mathcal{A}} &= \begin{bmatrix} 1 & 0 & 1 \\ 1 & 1 & 0 \\ 0 & 1/2 & 1 \end{bmatrix}^{-1} \begin{bmatrix} 1 \\ 2 \\ 3 \end{bmatrix} \\ &= \begin{bmatrix} 2/3 & 1/3 & -2/3 \\ -2/3 & 2/3 & 2/3 \\ 1/3 & -1/3 & 2/3 \end{bmatrix} \begin{bmatrix} 1 \\ 2 \\ 3 \end{bmatrix} \\ &= \begin{bmatrix} -2/3 \\ 8/3 \\ 5/3 \end{bmatrix} \end{aligned}$$

To generalize this discussion, *let P be the matrix whose column vectors are the representations of the new basis vectors with respect to old basis.* Then a given n-dimensional vector written with respect to the old basis is represented in the new basis via

$$\boxed{\mathbf{x}|_{\text{new basis}} = P^{-1}\mathbf{x}|_{\text{old basis}}} \tag{9-8}$$

or

$$\boxed{\mathbf{x}|_{\text{old basis}} = P\mathbf{x}|_{\text{new basis}}} \tag{9-9}$$

Example 9-6 Again consider

$$\mathbf{a}_1 = \begin{bmatrix} 1 \\ 1 \\ 0 \end{bmatrix} \qquad \mathbf{a}_2 = \begin{bmatrix} 0 \\ 1 \\ 1/2 \end{bmatrix} \qquad \mathbf{b}_1 = \begin{bmatrix} 1 \\ 3 \\ 1 \end{bmatrix} \qquad \mathbf{b}_2 = \begin{bmatrix} -1 \\ 1 \\ 1 \end{bmatrix}$$

Define the $\mathcal{A}$ basis as $\{\mathbf{a}_1, \mathbf{a}_2\}$ and the $\mathcal{B}$ basis as $\{\mathbf{b}_1, \mathbf{b}_2\}$. Suppose that a given $\mathbf{x} \in V = \text{span}\{\mathbf{a}_1, \mathbf{a}_2\} = \text{span}\{\mathbf{b}_1, \mathbf{b}_2\}$ has the representation with respect to the $\mathcal{A}$ basis as

$$\mathbf{x}|_{\mathcal{A}} = \begin{bmatrix} 3 \\ -5 \end{bmatrix}\Bigg|_{\mathcal{A}} \equiv 3 \cdot \mathbf{a}_1 - 5 \cdot \mathbf{a}_2$$

Suppose that we wish to find $\mathbf{x}$ with respect to the $\mathcal{B}$ basis. From Eq. (9-8) it is true that

$$\mathbf{x}|_{\mathcal{B}} = P^{-1}\mathbf{x}|_{\mathcal{A}}$$

where

$$P \equiv \left[\mathbf{b}_1\big|_{\mathcal{A}}, \mathbf{b}_2\big|_{\mathcal{A}}\right]$$

We thus need to find $\mathbf{b}_1$ and $\mathbf{b}_2$ represented in the $\mathcal{A}$ basis.
Now it is true that

$$\mathbf{b}_1 = \begin{bmatrix} 1 \\ 3 \\ 1 \end{bmatrix} = 1 \cdot \begin{bmatrix} 1 \\ 1 \\ 0 \end{bmatrix} + 2 \cdot \begin{bmatrix} 0 \\ 1 \\ 1/2 \end{bmatrix}$$

$$\mathbf{b}_2 = \begin{bmatrix} -1 \\ 1 \\ 1 \end{bmatrix} = -1 \cdot \begin{bmatrix} 1 \\ 1 \\ 0 \end{bmatrix} + 2 \cdot \begin{bmatrix} 0 \\ 1 \\ 1/2 \end{bmatrix}$$

Therefore

$$\mathbf{b}_1\big|_{\mathcal{A}} = \begin{bmatrix} 1 \\ 2 \end{bmatrix} = 1 \cdot \mathbf{a}_1 + 2 \cdot \mathbf{a}_2$$

$$\mathbf{b}_2\big|_{\mathcal{A}} = \begin{bmatrix} -1 \\ 2 \end{bmatrix} = -1 \cdot \mathbf{a}_1 + 2 \cdot \mathbf{a}_2$$

and

$$P = \begin{bmatrix} 1 & -1 \\ 2 & 2 \end{bmatrix} \qquad P^{-1} = \begin{bmatrix} 1/2 & 1/4 \\ -1/2 & 1/4 \end{bmatrix}$$

Thus

$$\mathbf{x}|_{\mathcal{B}} = \begin{bmatrix} 1/2 & 1/4 \\ -1/2 & 1/4 \end{bmatrix} \begin{bmatrix} 3 \\ -5 \end{bmatrix}\Bigg|_{\mathcal{A}}$$

$$= \begin{bmatrix} 1/4 \\ -11/4 \end{bmatrix}\Bigg|_{\mathcal{B}} = 1/4 \cdot \mathbf{b}_1 - 11/4 \cdot \mathbf{b}_2$$

is the representation of $\mathbf{x}$ with respect to the $\mathcal{B}$ basis.

9-6 TRANSFORMATION OF STATE COORDINATES

In Chap. 6 it is remarked that a change in state-space model (say, from the phase-variable to the diagonal form) is simply a change in state-space coordinates. Now with the results of the previous section this fact may be demonstrated. Furthermore it may be shown how to transform from one model to another by matrix-vector transformation techniques rather than going back to the transfer function to rederive a state model in a different form.

The trick for transformation is to find a matrix, P, which relates the new state coordinates, $\tilde{\mathbf{x}}(t)$, to the old state coordinates, $\mathbf{x}(t)$, via

$$P\tilde{\mathbf{x}}(t) = \mathbf{x}(t) \tag{9-10}$$

Such a matrix, P, is called a *matrix of transition*. Then if

$$\dot{\mathbf{x}}(t) = A\mathbf{x}(t) + B\mathbf{u}(t) \tag{9-11}$$

$$\mathbf{y}(t) = C\mathbf{x}(t) + Du(t) \tag{9-12}$$

substitution of Eq. (9-10) into the state model Eqs. (9-11) and (9-12) gives

$$\frac{d}{dt}[P\tilde{\mathbf{x}}(t)] = A[P\tilde{\mathbf{x}}(t)] + B\mathbf{u}(t)$$

$$\mathbf{y}(t) = C[P\tilde{\mathbf{x}}(t)] + D\mathbf{u}(t)$$

Assuming P is constant and invertible, this model may be reformulated as

$$\dot{\tilde{\mathbf{x}}}(t) = (P^{-1}AP)\tilde{\mathbf{x}}(t) + (P^{-1}B)\mathbf{u}(t) \tag{9-13}$$

$$\mathbf{y}(t) = (CP)\tilde{\mathbf{x}}(t) + D\mathbf{u}(t) \tag{9-14}$$

In the $\tilde{\mathbf{x}}(t)$ coordinate system, the $\{\tilde{A}, \tilde{B}, \tilde{C}, \tilde{D}\}$ model is

$$\tilde{A} = P^{-1}AP \qquad \tilde{B} = P^{-1}B$$

$$\tilde{C} = CP \qquad \tilde{D} = D$$

Notice that the D matrix is invariant to the state-space, internal-model coordinates. This confirms what was found in Chap. 6.

The remaining step is to indeed verify that two different state models may be related by some matrix of transition, P. We use the example from Chap. 6.

Example 9-7 In Chap. 6 the transfer function

$$H(s) = \frac{K(s^2 + 2s - 15)}{s^3 + 4s^2 + s - 6}$$

is used to illustrate the derivation of a state-space model in several different state coordinate systems: phase-variable (controllability), simulation-diagram (observability), and diagonal. In this example we pick the phase-variable and diagonal forms and show that they can be related by a matrix of transition, P.

The phase variable form was found directly from $H(s)$ to be

$$\begin{bmatrix} \dot{x}_1 \\ \dot{x}_2 \\ \dot{x}_3 \end{bmatrix} = \begin{bmatrix} 0 & 1 & 0 \\ 0 & 0 & 1 \\ 6 & -1 & -4 \end{bmatrix} \begin{bmatrix} x_1 \\ x_2 \\ x_3 \end{bmatrix} + \begin{bmatrix} 0 \\ 0 \\ K \end{bmatrix} u$$

$$y = [-15 \quad 2 \quad 1] \begin{bmatrix} x_1 \\ x_2 \\ x_3 \end{bmatrix}$$

where the dependence on t has been deleted for convenience. The diagonal form was found by partial fraction expansion of $H(s)$ to be

$$\begin{bmatrix} \dot{\hat{x}}_1 \\ \dot{\hat{x}}_2 \\ \dot{\hat{x}}_3 \end{bmatrix} = \begin{bmatrix} 1 & 0 & 0 \\ 0 & -2 & 0 \\ 0 & 0 & -3 \end{bmatrix} \begin{bmatrix} \hat{x}_1 \\ \hat{x}_2 \\ \hat{x}_3 \end{bmatrix} + \begin{bmatrix} 1 \\ 1 \\ 1 \end{bmatrix} u$$

$$y = [-K \quad 5K \quad -3K] \begin{bmatrix} \hat{x}_1 \\ \hat{x}_2 \\ \hat{x}_3 \end{bmatrix}$$

The problem is how to relate $\mathbf{x}$ in phase variable form to $\hat{\mathbf{x}}$ in diagonal form. A hint may be in the curious result in solving for the initial state vector for all three of the models looked at in Chap. 6, and for these two models now in particular. Starting with the output relations

$$[-15 \quad 2 \quad 1]\mathbf{x} = y$$

$$[-K \quad 5K \quad -3K]\hat{\mathbf{x}} = y$$

and differentiating these twice, while each time substituting for $\dot{\mathbf{x}}$ (or $\dot{\hat{\mathbf{x}}}$), gave the linear equation problems

$$\begin{bmatrix} -15 & 2 & 1 \\ 6 & -16 & -2 \\ -12 & 8 & -8 \end{bmatrix} \begin{bmatrix} x_1 \\ x_2 \\ x_3 \end{bmatrix} = \begin{bmatrix} y \\ \dot{y} - Ku \\ \ddot{y} + 2Ku - K\dot{u} \end{bmatrix}$$

$$\begin{bmatrix} -K & 5K & -3K \\ -K & -10K & 9K \\ -K & 20K & -27K \end{bmatrix} \begin{bmatrix} \hat{x}_1 \\ \hat{x}_2 \\ \hat{x}_3 \end{bmatrix} = \begin{bmatrix} y \\ \dot{y} - Ku \\ \ddot{y} + 2Ku - K\dot{u} \end{bmatrix}$$

These equations were set up for an arbitrary initial time t_0. But it was found that the right-hand sides are everywhere continuous. Furthermore, these relations are true for all t. Since these relations are true for all t and since the right-hand sides are the same, we may equate them to relate $\mathbf{x}(t)$ and $\hat{\mathbf{x}}(t)$ for all t. In particular

$$P\hat{\mathbf{x}}(t) = \mathbf{x}(t)$$

where $$P = \begin{bmatrix} -15 & 2 & 1 \\ 6 & -16 & -2 \\ -12 & 8 & -8 \end{bmatrix}^{-1} \begin{bmatrix} -K & 5K & -3K \\ -K & -10K & 9K \\ -K & 20K & -27K \end{bmatrix}$$

Starting with the state model for $\mathbf{x}(t)$ in phase-variable form we could then derive the diagonal $\hat{\mathbf{x}}(t)$ state model from Eqs. (9-13) and (9-14).

This example illustrates that the two different state-space models may be related by a state coordinate transformation, P. But it is not a useful method for finding the matrix of transition, P, because we had to know the complete form of the two models to find that P. Once we already know the new state model, there is no reason to find P. There are certainly situations, though, where the matrix of transition from old state coordinates to new state coordinates is known, and it remains to find the new state model.

Example 9-8 Consider the phase-variable state model from Example 9-7:

$$\dot{\mathbf{x}} = \begin{bmatrix} 0 & 1 & 0 \\ 0 & 0 & 1 \\ 6 & -1 & -4 \end{bmatrix} \mathbf{x} + \begin{bmatrix} 0 \\ 0 \\ K \end{bmatrix} u$$

$$y = [-15 \quad 2 \quad 1]\mathbf{x}$$

and suppose that we would like to transform to the new state model with states defined by

$$\mathbf{x}' = \begin{bmatrix} x_1' \\ x_2' \\ x_3' \end{bmatrix} \equiv \begin{bmatrix} y \\ \dot{y} - Ku \\ \ddot{y} + 2Ku - K\dot{u} \end{bmatrix}$$

This state is the right-hand side of the initial state relation

$$\begin{bmatrix} -15 & 2 & 1 \\ 6 & -16 & -2 \\ -12 & 8 & -8 \end{bmatrix} \begin{bmatrix} x_1 \\ x_2 \\ x_3 \end{bmatrix} = \begin{bmatrix} y \\ \dot{y} - Ku \\ \ddot{y} + 2Ku - K\dot{u} \end{bmatrix}$$

Then the P such that

$$P\mathbf{x}' = \mathbf{x}$$

is given by

$$P = \begin{bmatrix} -15 & 2 & 1 \\ 6 & -16 & -2 \\ -12 & 8 & -8 \end{bmatrix}^{-1}$$

$$= \begin{bmatrix} -1/15 & -1/90 & -1/180 \\ -1/30 & -11/180 & 1/90 \\ 1/15 & -4/90 & -19/180 \end{bmatrix}$$

The $\mathbf{x}'$ state equation is then given by

$$\dot{\mathbf{x}}' = (P^{-1}AP)\mathbf{x}' + (P^{-1}B)u$$

$$y = (CP)\mathbf{x}'$$

where

$$A' = P^{-1}AP = \begin{bmatrix} 0 & 1 & 0 \\ 0 & 0 & 1 \\ 6 & -1 & -4 \end{bmatrix}$$

$$B' = P^{-1}B = \begin{bmatrix} K \\ -2K \\ -8K \end{bmatrix}$$

$$C' = CP = [1 \quad 0 \quad 0]$$

This is a curious result. The plant A' matrix is the same as the original phase-variable matrix, and the output C' matrix is similar to the observability canonical form (the states derived from the simulation diagram), but the B' matrix is different from either of these. A simulation diagram along with the simulation diagram of the observability canonical form

$$\begin{bmatrix} \dot{\tilde{x}}_1 \\ \dot{\tilde{x}}_2 \\ \dot{\tilde{x}}_3 \end{bmatrix} = \begin{bmatrix} 0 & 0 & 6 \\ 1 & 0 & -1 \\ 0 & 1 & -4 \end{bmatrix} \begin{bmatrix} \tilde{x}_1 \\ \tilde{x}_2 \\ \tilde{x}_3 \end{bmatrix} + \begin{bmatrix} -15K \\ 2K \\ K \end{bmatrix} u$$

$$y = [0 \quad 0 \quad 1]\tilde{\mathbf{x}}$$

is shown in Fig. 9-5.

The observability matrix of the $\mathbf{x}'$ coordinate frame also has a curious form:

$$M_o' = \begin{bmatrix} C' \\ C'A' \\ C'A' \end{bmatrix} = \begin{bmatrix} 1 & 0 & 0 \\ 0 & 1 & 0 \\ 0 & 0 & 1 \end{bmatrix} \tag{9-15}$$

It is the identity matrix. This fact could be used to find the transformation matrix to transform to the $\mathbf{x}'$ state coordinate system. Indeed, using the observability or controllability matrix provides a general method to transform to a desired coordinate frame.

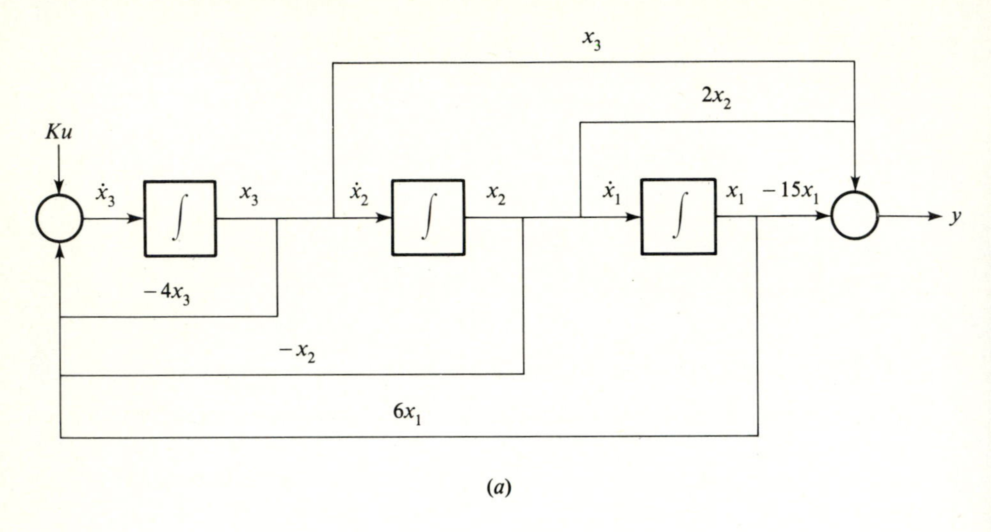

(a)

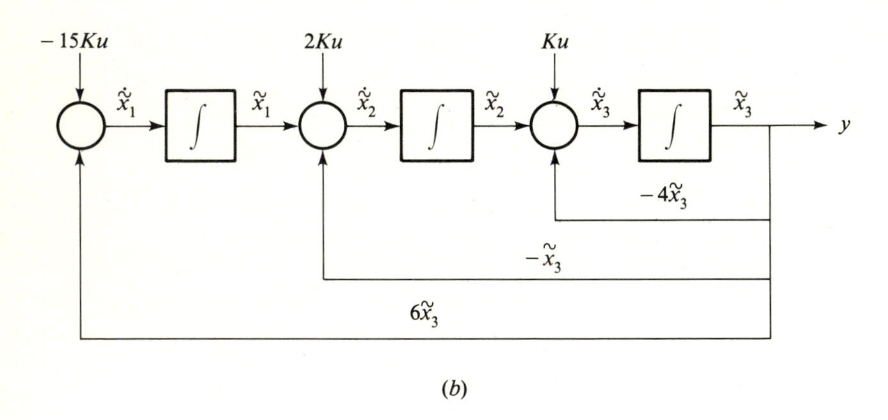

(b)

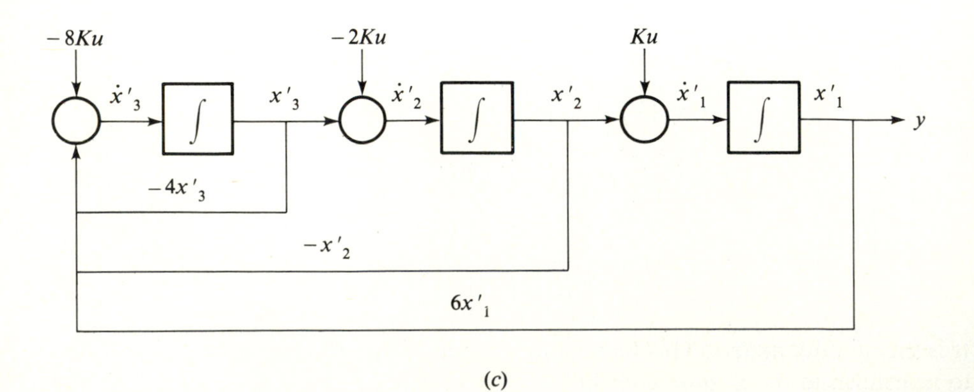

(c)

Figure 9-5 Simulation diagrams for alternate-state coordinate, Example 9-8. (a) Phase variable. (b) Observability. (c) State coordinates with $M_o = I$.

9-7 USE OF CONTROLLABILITY AND OBSERVABILITY MATRICES TO TRANSFORM STATE COORDINATES

The controllability and observability matrices are introduced in Chap. 8 in conjunction with the discrete state model. But in Chap. 11 it is seen that the matrices

$$M_c \equiv [B, AB, \ldots, A^{n-1}B] \tag{9-16}$$

$$M_o \equiv \begin{bmatrix} C \\ CA \\ \vdots \\ CA^{n-1} \end{bmatrix} \tag{9-17}$$

are also applicable to the controllability and observability properties for the continuous state model.

Here we use the matrices M_c and M_o to find a useful technique for transformation to a desired state coordinate system. In particular, suppose we wish to transform to some desired state coordinate system, $\tilde{\mathbf{x}}$ such that $P\tilde{\mathbf{x}} = \mathbf{x}$ for all t. Then $\tilde{A} = P^{-1}AP$, $\tilde{B} = P^{-1}B$, $\tilde{C} = CP$, and we may find $\tilde{M}_c$ and $\tilde{M}_o$ by

$$\begin{aligned} \tilde{M}_c &\equiv [\tilde{B}, \tilde{A}\tilde{B}, \ldots, \tilde{A}^{n-1}\tilde{B}] \\ &= [(P^{-1}B), (P^{-1}AP)(P^{-1}B), \ldots, (P^{-1}AP)^{n-1}(P^{-1}B)] \end{aligned}$$

$$\tilde{M}_o \equiv \begin{bmatrix} \tilde{C} \\ \tilde{C}\tilde{A} \\ \vdots \\ \tilde{C}\tilde{A}^{n-1} \end{bmatrix} = \begin{bmatrix} (CP) \\ (CP)(P^{-1}AP) \\ \vdots \\ (CP)(P^{-1}AP)^{n-1} \end{bmatrix}$$

It may be shown that $(P^{-1}AP)^k = P^{-1}A^kP$, and so the expressions for $\tilde{M}_c$ and $\tilde{M}_o$ become

$$\tilde{M}_c = [P^{-1}B, \quad P^{-1}AB, \quad \ldots, \quad P^{-1}A^{n-1}B]$$

$$\tilde{M}_o = \begin{bmatrix} CP \\ CAP \\ \vdots \\ CA^{n-1}P \end{bmatrix}$$

Then P^{-1} can be factored from $\tilde{M}_c$ and P from $\tilde{M}_o$ to give the final result

$$\tilde{M}_c = P^{-1}M_C \tag{9-18}$$

$$\tilde{M}_o = M_oP \tag{9-19}$$

These expressions may be used to find P. In particular, if we know *either* the desired $\tilde{M}_c$ or $\tilde{M}_o$ and the original M_c or M_o, then Eq. (9-18) or Eq. (9-19) may be used to find P:

$$P = M_c\tilde{M}_c^{-1} \tag{9-20}$$

or

$$\boxed{P = M_o^{-1}\widetilde{M}_o} \tag{9-21}$$

provided $\widetilde{M}_c$ and M_o are nonsingular.

Example 9-9 Reconsider the phase-variable state model from Example 9-8:

$$\dot{\mathbf{x}} = \begin{bmatrix} 0 & 1 & 0 \\ 0 & 0 & 1 \\ 6 & -1 & -4 \end{bmatrix}\mathbf{x} + \begin{bmatrix} 0 \\ 0 \\ K \end{bmatrix}u$$

$$y = [-15 \quad 2 \quad 1]\mathbf{x}$$

Suppose we would like to transform to the $\mathbf{x}'$ state found in Example 9-8. The observability matrix for $\mathbf{x}'$ was found to be

$$M_o' = \begin{bmatrix} 1 & 0 & 0 \\ 0 & 1 & 0 \\ 0 & 0 & 1 \end{bmatrix} \tag{9-15}$$

the identity matrix. Then from Eq. (9-21)

$$P = M_o^{-1}M_o' = M_o^{-1}$$

where M_o is the phase-variable observability matrix. The phase-variable observability matrix is

$$M_o = \begin{bmatrix} -15 & 2 & 1 \\ 6 & -16 & -2 \\ -12 & 8 & -8 \end{bmatrix}$$

Then $P = M_o^{-1}$ is seen to be exactly the same as determined in Example 9-8.

Example 9-10 Again consider the phase-variable form of Example 9-9. Suppose that we know the system eigenvalues (poles) $\lambda_1 = 1, \lambda_2 = -2, \lambda_3 = -3$, and we would like to transform to diagonal form

$$\dot{\hat{\mathbf{x}}} = \begin{bmatrix} 1 & 0 & 0 \\ 0 & -2 & 0 \\ 0 & 0 & -3 \end{bmatrix}\hat{\mathbf{x}} + \begin{bmatrix} 1 \\ 1 \\ 1 \end{bmatrix}u$$

but we do not know the $\widehat{C} = CP$ matrix.

However, since we do know $\widehat{A}$ and $\widehat{B}$ we may find $\widehat{M}_c$:

$$\widehat{M}_c = [\widehat{B}, \widehat{A}\widehat{B}, \widehat{A}^2\widehat{B}]$$

$$= \begin{bmatrix} 1 & 1 & 1 \\ 1 & -2 & 4 \\ 1 & -3 & 9 \end{bmatrix}$$

We also can find M_c:

$$M_c = [B, AB, A^2B]$$
$$= \begin{bmatrix} 0 & 0 & K \\ 0 & K & -4K \\ K & -4K & 15K \end{bmatrix}$$

Then from Eq. (9-20)

$$P = M_c \widehat{M}_c^{-1}$$
$$= K \begin{bmatrix} 0 & 0 & 1 \\ 0 & 1 & -4 \\ 1 & -4 & 15 \end{bmatrix} \begin{bmatrix} 1 & 1 & 1 \\ 1 & -2 & 4 \\ 1 & -3 & 9 \end{bmatrix}^{-1}$$

Carrying out the calculations gives

$$P = K \begin{bmatrix} 1/12 & -1/3 & 1/4 \\ 1/12 & 2/3 & -3/4 \\ 1/12 & -4/3 & 5/4 \end{bmatrix}$$

Then $$\widehat{C} = CP = [-K \quad 5K \quad -3K]$$

which is the same as $\widehat{C}$ from Example 9-7, where $\widehat{C}$ is found by partial fraction expansions.

These two examples have shown how the controllability and observability matrices can be used to transform to some desired canonical form when one happens to know the desired controllability and observability matrices. Chapter 10 next shows how to transform a given A to the diagonal form using the system eigenvectors.

9-8 SUMMARY

This chapter has introduced some important vector space concepts. The notion of linear dependence and independence of vectors is particularly relevant to topics of the later chapters. Also discussed is the topic of transforming from one state coordinate system to another. The controllability and observability matrices introduced in Chap. 8 are seen to provide an interesting technique for finding the desired transformation. Numerically this technique may not be suitable, though, because of possible ill-conditioning of M_o or M_c. See Appendix E.

The next chapter continues with this somewhat mathematical discussion by considering the matrix eigenvalues and eigenvectors. The eigenvalues and eigenvectors, in turn, allow a much deeper analysis of the matrix exponential function, e^{At}, than was feasible in Chap. 7.

PROBLEMS

9-1 Let $\mathbf{v}$ be an arbitrary element of R^4.

(*a*) What is the general form of $\mathbf{v}$?

(*b*) What is the general form of $\alpha\mathbf{v}$ for an arbitrary scalar α? Is $(\alpha\mathbf{v})$ still a member of R^4?

(*c*) Let $\mathbf{v}_a$ and $\mathbf{v}_b$ be two arbitrary vectors from R^4. What is the general form of $(\mathbf{v}_a + \mathbf{v}_b)$? Is their sum still a vector in R^4?

(*d*) What is the zero vector in R^4?

9-2 Repeat Prob. 9-1 for an arbitrary vector in R^5.

9-3 (*a*) What are the norms of the following vectors:

$$\mathbf{v}_1 = \begin{bmatrix} 1 \\ 0 \\ 2 \\ 0 \end{bmatrix} \quad \mathbf{v}_2 = \begin{bmatrix} -2 \\ -1 \\ 1 \\ 2 \end{bmatrix} \quad \mathbf{v}_3 = \begin{bmatrix} 1 \\ 1/2 \\ -1/2 \\ -1 \end{bmatrix} \quad \mathbf{v}_4 = \begin{bmatrix} -3 \\ -2 \\ 4 \\ 4 \end{bmatrix}$$

(*b*) Normalize each of the vectors in part *a*.

9-4 (*a*) What are the norms of the following vectors:

$$\mathbf{v}_1 = \begin{bmatrix} 3 \\ 1 \\ -2 \\ 0 \\ 1 \end{bmatrix} \quad \mathbf{v}_2 = \begin{bmatrix} 0 \\ 1 \\ 1 \\ -4 \\ 1 \end{bmatrix} \quad \mathbf{v}_3 = \begin{bmatrix} 6 \\ 4 \\ -2 \\ -8 \\ 4 \end{bmatrix} \quad \mathbf{v}_4 = \begin{bmatrix} -3 \\ -2 \\ 1 \\ 4 \\ -2 \end{bmatrix}$$

(*b*) Normalize each of the vectors in part *a*.

9-5 (*a*) Find the inner product of the vector

$$\mathbf{v} = [4, \quad 1, \quad -2, \quad -1]^T$$

with each of the vectors in Prob. 9-3*a*.

(*b*) To which vectors is $\mathbf{v}$ orthogonal?

9-6 Repeat Prob. 9-5 for the vector

$$\mathbf{v} = [3, \quad 2, \quad 2, \quad 1, \quad 0]^T$$

using the vectors of Prob. 9-4.

9-7 For the four vectors in Prob. 9-3 verify that

$$\mathbf{v}_1 \cdot \mathbf{v}_2 = 0 \qquad \mathbf{v}_1 \cdot \mathbf{v}_3 = 0 \qquad \mathbf{v}_3 = \frac{-\mathbf{v}_2}{2} \qquad \mathbf{v}_4 = \mathbf{v}_1 + 2\mathbf{v}_2$$

9-8 For the four vectors in Prob. 9-4 verify that

$$\mathbf{v}_1 \cdot \mathbf{v}_2 = 0 \qquad \mathbf{v}_3 = 2\mathbf{v}_1 + 2\mathbf{v}_2 \qquad \mathbf{v}_4 = \frac{-\mathbf{v}_3}{2}$$

9-9 For the vectors in Prob. 9-3, what are the dimensions of the following subspaces:

(*a*) span $\{\mathbf{v}_1, \mathbf{v}_2\}$ (*b*) span $\{\mathbf{v}_1, \mathbf{v}_3\}$
(*c*) span $\{\mathbf{v}_1, \mathbf{v}_4\}$ (*d*) span $\{\mathbf{v}_2, \mathbf{v}_3\}$
(*e*) span $\{\mathbf{v}_2, \mathbf{v}_4\}$ (*f*) span $\{\mathbf{v}_3, \mathbf{v}_4\}$
(*g*) span $\{\mathbf{v}_1, \mathbf{v}_2, \mathbf{v}_3\}$ (*h*) span $\{\mathbf{v}_1, \mathbf{v}_2, \mathbf{v}_4\}$
(*i*) span $\{\mathbf{v}_2, \mathbf{v}_3, \mathbf{v}_4\}$ (*j*) span $\{\mathbf{v}_1, \mathbf{v}_2, \mathbf{v}_3, \mathbf{v}_4\}$

Hint: Use the results of Prob. 9-7.

9-10 Repeat Prob. 9-9 for the four vectors in Prob. 9-4. *Hint:* Use the results of Prob. 9-8.

9-11 (*a*) For the combinations of vectors in Prob. 9-9, what combinations would form a basis for span $\{\mathbf{v}_1, \mathbf{v}_2, \mathbf{v}_3, \mathbf{v}_4\}$?

(*b*) Which would form an orthogonal basis for span $\{\mathbf{v}_1, \mathbf{v}_2, \mathbf{v}_3, \mathbf{v}_4\}$?

(*c*) What would be an orthonormal basis for span $\{\mathbf{v}_1, \mathbf{v}_2, \mathbf{v}_3, \mathbf{v}_4\}$? Use the results of Prob. 9-3(*b*).

9-12 Repeat Prob. 9-11 using the four vectors of Prob. 9-4.

9-13 Using the four vectors defined in Prob. 9-3, define the set $\{\mathbf{v}_1, \mathbf{v}_2\}$ as the $\mathcal{A}$ basis and define the set $\{\mathbf{v}_3, \mathbf{v}_4\}$ as the $\mathcal{B}$ basis for the subspace span $\{\mathbf{v}_1, \mathbf{v}_2, \mathbf{v}_3, \mathbf{v}_4\}$.

(*a*) Find the coordinate transformation matrix, P, which transforms vectors represented in the $\mathcal{A}$ basis into vectors represented in the $\mathcal{B}$ basis.

(*b*) Let

$$\mathbf{v} = [1, \quad -1, \quad 7, \quad 2]^T$$

Verify that $\mathbf{v}$ represented in the $\mathcal{A}$ basis is

$$\mathbf{v}|_{\mathcal{A}} = \begin{bmatrix} 3 \\ 1 \end{bmatrix}$$

(*c*) Use the results of *a* to find the representation of $\mathbf{v}$ with respect to the $\mathcal{B}$ basis.

(*d*) Verify that your answer to part *c* is correct.

9-14 Using the four vectors defined in Prob. 9-4, define $\{\mathbf{v}_1, \mathbf{v}_2\}$ as the $\mathcal{A}$ basis and define $\{\mathbf{v}_2, \mathbf{v}_3\}$ as the $\mathcal{B}$ basis for the subspace span $\{\mathbf{v}_1, \mathbf{v}_2, \mathbf{v}_3, \mathbf{v}_4\}$.

(*a*) Find the coordinate transformation matrix, P, that transforms vectors represented in the $\mathcal{A}$ basis into vectors represented in the $\mathcal{B}$ basis.

(*b*) Let

$$\mathbf{v} = [3, \quad 4, \quad 1, \quad -12, \quad 4]^T$$

Verify that $\mathbf{v}$ represented in the $\mathcal{A}$ basis is

$$\mathbf{v}|_{\mathcal{A}} = \begin{bmatrix} 1 \\ 3 \end{bmatrix}$$

(*c*) Use the results of *a* to find the representation of $\mathbf{v}$ with respect to the $\mathcal{B}$ basis.

(*d*) Verify that your answer to part *c* is correct.

9-15 Suppose the states $\mathbf{x}(t)$ and $\tilde{\mathbf{x}}(t)$ are related by

$$\mathbf{x}(t) = \begin{bmatrix} 5 & 3 \\ 2 & 1 \end{bmatrix} \tilde{\mathbf{x}}(t)$$

Also suppose the state-space model for $\mathbf{x}(t)$ is

$$\dot{\mathbf{x}}(t) = \begin{bmatrix} 0 & 1 \\ -2 & -3 \end{bmatrix} \mathbf{x}(t) + \begin{bmatrix} 0 \\ 10 \end{bmatrix} u(t)$$

$$y(t) = [2, \quad 1]\mathbf{x}(t)$$

Find the state-space model for $\tilde{\mathbf{x}}(t)$.

9-16 Suppose that the states $\mathbf{x}(t)$ and $\tilde{\mathbf{x}}(t)$ are related by

$$\tilde{\mathbf{x}}(t) = \begin{bmatrix} 1 & 2 \\ 1 & 3 \end{bmatrix} \mathbf{x}(t)$$

Suppose also that the state-space model for $\mathbf{x}(t)$ is

$$\dot{\mathbf{x}}(t) = \begin{bmatrix} 0 & 1 \\ -5 & -4 \end{bmatrix} \mathbf{x}(t) + \begin{bmatrix} 0 \\ 1 \end{bmatrix} u(t)$$

$$y(t) = [10, \quad 2]\mathbf{x}(t)$$

Find the state-space model for $\tilde{\mathbf{x}}(t)$.

9-17 Using the methods of Sec. 9-6 find the matrix of transition, P, that relates states in the phase-variable coordinate system, $\mathbf{x}(t)$, to states in the diagonal coordinate system, $\widehat{\mathbf{x}}(t)$, via $\mathbf{x}(t) = P\widehat{\mathbf{x}}(t)$, for the following system:

$$H(s) = \frac{s + 5}{s^2 + 3s + 2}$$

9-18 Repeat Prob. 9-17 for the system:

$$H(s) = \frac{s - 2}{s^2 + 6s + 5}$$

9-19 Repeat Prob. 9-17 by relating the controllability matrix of the phase-variable states to the controllability matrix of the diagonal state coordinates.

9-20 Repeat Prob. 9-19 for the systems in Prob. 9-18.

9-21 For the system in Prob. 9-17 find the matrix of transition, P, such that $\mathbf{x}(t) = P\widetilde{\mathbf{x}}(t)$, where $\mathbf{x}(t)$ is the phase-variable state coordinates and $\widetilde{\mathbf{x}}(t)$ is the observability canonical-form states. Do this by relating the observability matrices for $\mathbf{x}(t)$ and $\widetilde{\mathbf{x}}(t)$.

9-22 Repeat Prob. 9-21 for the system in Prob. 9-18.

9-23 (*a*) For the system in Prob. 9-17 find the state model for the $\mathbf{x}'(t)$ coordinate system in which the observability matrix of $\mathbf{x}'(t)$ is the identity matrix. (*Hint:* Work part *b* and use this result to find the $\mathbf{x}'(t)$ state model.)

(*b*) Find the matrix of transition P such that $\mathbf{x}(t) = P\mathbf{x}'(t)$ and $\mathbf{x}(t)$ is in phase-variable form.

(*c*) Find the matrix of transition $\widehat{P}$ such that $\widehat{\mathbf{x}}(t) = \widehat{P}\mathbf{x}'(t)$ and $\widehat{\mathbf{x}}(t)$ is in diagonal form.

(*d*) Find the matrix of transition $\widetilde{P}$ such that $\widetilde{\mathbf{x}}(t) = \widetilde{P}\mathbf{x}'(t)$ and $\widetilde{\mathbf{x}}(t)$ is in observability canonical form.

9-24 Repeat Prob. 9-23 for the system in Prob. 9-18.

NOTES AND REFERENCES

The material in this chapter is basically a review of important vector space ideas necessary for later chapters. For further mathematical background the reader is referred to, for example, Refs. 1 and 2.

1. Bellman, R.: *Introduction to Matrix Analysis,* McGraw-Hill, New York, 2d ed., 1970.
2. Strang, G.: *Linear Algebra and Its Application,* Academic, New York, 1976.

CHAPTER

TEN

EIGENVALUES, EIGENVECTORS, AND e^{At}

Chapters 6 and 7 introduce the continuous-time linear system state-space model and its solution. The continuous-time solution fundamentally involves the matrix exponential function, e^{At}. This chapter examines e^{At} in depth. Examination of e^{At} requires, though, an understanding of the system eigenvalues and eigenvectors. These are the topics of this chapter.

10-1 THE EIGENVALUE AND EIGENVECTOR LINEAR-EQUATION PROBLEM

The eigenvalue and eigenvector linear-equation problem has the following form: find the scalars λ and the n-dimensional vectors $\boldsymbol{\xi}$ such that

$$A\boldsymbol{\xi} = \lambda\boldsymbol{\xi} \tag{10-1}$$

where A is a given square $n \times n$ matrix. Any scalar λ and vector $\boldsymbol{\xi}$ which satisfy Eq. (10-1) are termed an *eigenvalue* and *eigenvector* of the matrix A, respectively.

Equation (10-1) is an interesting relation. It says that the $n \times n$ matrix A times its eigenvector $\boldsymbol{\xi}$ yields the eigenvector $\boldsymbol{\xi}$ back, only scaled by the amount λ. The eigenvector is therefore often referred to as an *invariant direction* of the matrix A.

Equation (10-1) may be reformed as

$$[\lambda I - A]\boldsymbol{\xi} = \mathbf{0} \tag{10-2}$$

where both λ and $\boldsymbol{\xi}$ are still unknown. Considering the scalar λ first, it is seen that λ must be a value, real or complex, such that the coefficient matrix $[\lambda I - A]$ is singular (not invertible); if $[\lambda I - A]$ were invertible, then the solution ξ would simply be the trivial zero vector. The matrix $[\lambda I - A]$ is termed the *resolvent* matrix.

One condition for the coefficient matrix $[\lambda I - A]$ to be singular is for its determinant to be zero. The determinant of $[\lambda I - A]$ yields an nth-order polynomial in λ, termed the *characteristic polynomial,* $\Delta(\lambda)$,

$$\begin{aligned}\Delta(\lambda) &\equiv \det[\lambda I - A]\\ &= \lambda^n + a_{n-1}\lambda^{n-1} + \cdots + a_1\lambda + a_0\\ &= \prod_{i=1}^{n^*} (\lambda - \lambda_i)^{n_i}\end{aligned} \tag{10-3}$$

The eigenvalues, λ_i, are the distinct roots of $\Delta(\lambda)$; n^* is the number of distinct eigenvalues; and n_i is the multiplicity of λ_i. An nth-order polynomial has n roots (real or complex), and so it is true that the multiplicities all add up to n:

$$\sum_{i=1}^{n^*} n_i = n \tag{10-4}$$

If A is a real matrix, all the coefficients in the characteristic polynomial [Eq. (10-3)] are real, and all complex roots occur in complex conjugate pairs.

Example 10-1 Let

$$A = \begin{bmatrix} 1 & 2 \\ -1 & 4 \end{bmatrix}$$

Then

$$\begin{aligned}\det[\lambda I - A] &= \det\begin{bmatrix} \lambda - 1 & -2 \\ 1 & \lambda - 4 \end{bmatrix}\\ &= \lambda^2 - 5\lambda + 6\\ &= (\lambda - 2)(\lambda - 3)\end{aligned}$$

Hence the two eigenvalues are $\lambda_1 = 2$ and $\lambda_2 = 3$. The eigenvector linear equation problems are

$$[\lambda_1 I - A]\xi_1 = \begin{bmatrix} 1 & -2 \\ 1 & -2 \end{bmatrix}\begin{bmatrix} \xi_1(1) \\ \xi_1(2) \end{bmatrix} = \begin{bmatrix} 0 \\ 0 \end{bmatrix}$$

and

$$[\lambda_2 I - A]\xi_2 = \begin{bmatrix} 2 & -2 \\ 1 & -1 \end{bmatrix}\begin{bmatrix} \xi_2(1) \\ \xi_2(2) \end{bmatrix} = \begin{bmatrix} 0 \\ 0 \end{bmatrix}$$

Solutions to these eigenvector linear equation problems are

$$\xi_1 = \alpha_1\begin{bmatrix} 2 \\ 1 \end{bmatrix} \quad \text{and} \quad \xi_2 = \alpha_2\begin{bmatrix} 1 \\ 1 \end{bmatrix}$$

for any scalars α_1 and α_2. Further discussion of the solution of these eigenvector linear equation problems is deferred until after discussion of the linear equation problem in general in the next section.

Example 10-2 Let

$$A = \begin{bmatrix} 1 & -4 \\ 1 & 1 \end{bmatrix}$$

Then

$$\Delta(\lambda) = \det \begin{bmatrix} \lambda - 1 & 4 \\ -1 & \lambda - 1 \end{bmatrix} = \lambda^2 + 2\lambda + 5$$

$$= (\lambda - 1 - 2j)(\lambda - 1 + 2j)$$

Hence $\lambda_1 = 1 + 2j$ and $\lambda_2 = 1 - 2j$.

Example 10-3 Let

$$A = \begin{bmatrix} 4 & 1 & 2 \\ 0 & 4 & 3 \\ 0 & 0 & 3 \end{bmatrix}$$

Because of the upper triangular form of A, the characteristic polynomial of A is easily seen to be $\Delta(\lambda) = (\lambda - 4)^2(\lambda - 3)$. The eigenvalues are $\lambda_1 = 4$ and $\lambda_2 = 3$ with multiplicities $n_1 = 2$ and $n_2 = 1$, respectively.

Relation of the Eigenvalues and the System Poles

From the foregoing discussion it should be apparent that the eigenvalues of A and the system poles for the state-space model with state plant matrix A exactly coincide. Namely, the transfer function for the state-space model is

$$H(s) = C[sI - A]^{-1}B + D \tag{10-5}$$

with characteristic polynomial

$$\Delta(s) = \det [sI - A] \tag{10-6}$$

This is the same as the characteristic polynomial Eq. (10-3). Hence the system poles and the A matrix eigenvalues exactly coincide.

Eigenvalues Under a Change of State Coordinates

In Sec. 9-4 it is shown that when a new state $\tilde{\mathbf{x}}(t)$ is related to the old state $\mathbf{x}(t)$ by the matrix of transition P such that $\mathbf{x}(t) = P\tilde{\mathbf{x}}(t)$, the new state matrix $\tilde{A}$ equals

$$\tilde{A} = P^{-1}AP \tag{10-7}$$

The relation [Eq. (10-7)] between $\tilde{A}$ and A is termed a *similarity transformation*.

If the new state-space model represents the same system, it has the same transfer function and hence the same characteristic polynomial and eigenvalues. Indeed, it is true that all matrices which are related by a similarity transformation of the form of Eq. (10-7) have identical eigenvalues (see Prob. 10-3).

With this introduction to the system eigenvalues, the discussion turns to consideration of the eigenvectors. Analysis is first given to the linear equation problem in its most general form. The eigenvector linear equation then becomes a special case.

10-2 THE GENERAL LINEAR-EQUATION PROBLEM

This section examines the solution of the general linear-equation problem

$$M\mathbf{x} = \mathbf{d} \tag{10-8}$$

where M is a given $m \times p$-dimension coefficient matrix, $\mathbf{x}$ is a p-dimensional vector of unknowns to find, and $\mathbf{d}$ is a given m-dimensional set of data. In general, the dimension m may be equal to, greater than, or less than p. Regardless, there are three basic situations that may be encountered:

1. There is one unique solution.
2. There are an infinity of solutions.
3. There is no solution.

The eigenvector problem is a special case of number 2. The coefficient matrix M is the square, singular, resolvent matrix $[\lambda_i I - A]$; the unknown is the eigenvector $\boldsymbol{\xi}_i$; and the set of data is the zero vector.

This section considers, however, the theoretical aspects of the most general case. These results are specialized to the eigenvector problem in the section to follow. A convenient hand-analysis technique, employing elementary row operations, is used to investigate the solution properties. This procedure is quite similar to Gauss-Jordan elimination as studied in Sec. 2-7. However, while this technique serves our purposes for hand analysis and initial concept understanding, it should not be misinterpreted as an efficient or necessarily stable numerical algorithm to apply to the digital computer. The reader is referred to a number of excellent sources for further information on the computational aspects of the linear equation problem (e.g., Stewart [1]). Expertly written software packages are available at nearly all major computational centers to solve the general linear-equation problem (e.g., LINPACK [2] or IMSL [3]).

The General Form of the Solution

Writing M in terms of its p column vectors, $\mathbf{m}_i$, the linear-equation problem Eq. (10-8) takes the form

$$[\mathbf{m}_1, \quad \mathbf{m}_2, \quad \ldots, \quad \mathbf{m}_p]\begin{bmatrix} x_1 \\ x_2 \\ \vdots \\ x_p \end{bmatrix} = \mathbf{d}$$

This may be rearranged to

$$x_1\mathbf{m}_1 + x_2\mathbf{m}_2 + \cdots + x_p\mathbf{m}_p = \mathbf{d} \qquad (10\text{-}8a)$$

Each x_i is a scalar unknown, and so Eq. (10-8*a*) reads: **d** is equal to a linear combination of the columns of M: $\mathbf{m}_1, \mathbf{m}_2, \ldots, \mathbf{m}_p$. Equivalently, Eq. (10-8*a*) says that **d** is in the span of $\{\mathbf{m}_1, \mathbf{m}_2, \ldots, \mathbf{m}_p\}$. Or more precisely, it should be said that *perhaps* $\mathbf{d} \in \text{span}\{\mathbf{m}_1, \mathbf{m}_2, \ldots, \mathbf{m}_p\}$, for it is not yet known whether or not Eq. (10-8) has a solution. It may not, in which case **d** would not be contained in the span of the column vectors of M.

Example 10-4

$$\begin{bmatrix} 1 & 0 \\ 0 & 1 \\ 1 & 0 \end{bmatrix}\begin{bmatrix} x_1 \\ x_2 \end{bmatrix} \stackrel{?}{=} \begin{bmatrix} 1 \\ 2 \\ 3 \end{bmatrix}$$

or

$$x_1\begin{bmatrix} 1 \\ 0 \\ 1 \end{bmatrix} + x_2\begin{bmatrix} 0 \\ 1 \\ 0 \end{bmatrix} \stackrel{?}{=} \begin{bmatrix} 1 \\ 2 \\ 3 \end{bmatrix}$$

has no solution because the data vector

$$\begin{bmatrix} 1 \\ 2 \\ 3 \end{bmatrix}$$

is not contained in the plane

$$\text{span}\left\{\begin{bmatrix} 1 \\ 0 \\ 1 \end{bmatrix}\begin{bmatrix} 0 \\ 1 \\ 0 \end{bmatrix}\right\}$$

The results above can be summarized in the following form:

Proposition 10-1 The linear equation problem $M\mathbf{x} = \mathbf{d}$ has a solution (it may or may not be unique) if and only if $\mathbf{d} \in \text{span}\{\mathbf{m}_1, \mathbf{m}_2, \ldots, \mathbf{m}_p\}$, where the $\mathbf{m}_i$ are the column vectors of M.

The span of the column vectors of M is a subspace called the *range space* (or sometimes the *column space* or *image* of M) and given the symbol

$$\mathcal{R}(M) \equiv \text{span}\{\mathbf{m}_1, \mathbf{m}_2, \ldots, \mathbf{m}_p\} \qquad (10\text{-}9)$$

Since the columns of M are assumed to be m-dimensional (in the example above they were three-dimensional) the range space of M is a subspace of R^m. If the range space of M is the entire space R^m, then Eq. (10-8) always has a solution. Techniques are discussed shortly to determine the range space and to determine whether or not $\mathbf{d}$ is contained in that range space.

Assuming Eq. (10-8) has a solution, the general form of the solution is

$$\mathbf{x} = \mathbf{x}_p + \mathbf{x}_h \tag{10-10}$$

where $\mathbf{x}_p$ is the particular solution and $\mathbf{x}_h$ is the homogeneous solution. The homogeneous solution must satisfy the homogeneous problem

$$M\mathbf{x}_h = \mathbf{0} \tag{10-11}$$

Note the similarity here to the classic solution of the linear ordinary differential equation. However, unlike the solution of the linear differential equation, where the n boundary conditions provided a unique solution by giving n conditions for the n arbitrary constants in the homogeneous response, there is nothing here which provides such boundary conditions. Thus the arbitrary constants in the homogeneous solution remain, and Eq. (10-10) represents an infinity of solutions.

To examine the solution technique, consider the following hypothetical linear equation problem:

$$\begin{bmatrix} 1 & -2 & 0 & 1 \\ 0 & 0 & 1 & 2 \end{bmatrix} \begin{bmatrix} x_1 \\ x_2 \\ x_3 \\ x_4 \end{bmatrix} = \begin{bmatrix} 5 \\ 6 \end{bmatrix} \tag{10-12}$$

Inserting zero rows into the coefficient matrix does not alter the solution of the linear equation, but it provides a convenient form by which to discern the solution:

$$\begin{bmatrix} 1 & -2 & 0 & 1 \\ 0 & 0 & 0 & 0 \\ 0 & 0 & 1 & 2 \\ 0 & 0 & 0 & 0 \end{bmatrix} \begin{bmatrix} x_1 \\ x_2 \\ x_3 \\ x_4 \end{bmatrix} = \begin{bmatrix} 5 \\ 0 \\ 6 \\ 0 \end{bmatrix} \tag{10-13}$$

This is the form of the linear equation problem that is sought later using the method of elementary row operations. It is termed the *modified Hermite normal form,* and it has the following properties:

1. It has at least as many rows as columns (zero rows are added if $m < p$ originally).
2. It is upper-triangular (all elements below the main diagonal are zero).
3. Either ones or zeros are on the diagonal.
4. Any column with unity on the diagonal is otherwise zero.

Because the coefficient matrix [Eq. (10-13)] is in this special form, the solution is quite easy to see. A particular solution is

$$\mathbf{x}_p = \begin{bmatrix} 5 \\ 0 \\ 6 \\ 0 \end{bmatrix}$$

(note that

$$M\mathbf{x}_p = \begin{bmatrix} 5 \\ 6 \end{bmatrix}$$

as required). There are two independent homogeneous solutions:

$$\mathbf{x}_{h_1} = \begin{bmatrix} -2 \\ -1 \\ 0 \\ 0 \end{bmatrix} \qquad \mathbf{x}_{h_2} = \begin{bmatrix} 1 \\ 0 \\ 2 \\ -1 \end{bmatrix}$$

(note that $M\mathbf{x}_{h_1} = M\mathbf{x}_{h_2} = \mathbf{0}$). These are found by inspection from the modified problem of Eq. (10-13); $\mathbf{x}_p$ is the right-hand side of Eq. (10-13), and $\mathbf{x}_{h_1}$ and $\mathbf{x}_{h_2}$ are the second- and fourth-column vectors of the modified coefficient matrix M' with a -1 added to the row corresponding to the column number (the second and fourth rows, respectively). The second and fourth columns of M' are the columns with a zero on the diagonal.

The complete solution to Eq. (10-12) is then given by

$$\mathbf{x} = \begin{bmatrix} 5 \\ 0 \\ 6 \\ 0 \end{bmatrix} + \alpha_1 \begin{bmatrix} -2 \\ -1 \\ 0 \\ 0 \end{bmatrix} + \alpha_2 \begin{bmatrix} 1 \\ 0 \\ 2 \\ -1 \end{bmatrix}$$

or equivalently by

$$\mathbf{x} = \begin{bmatrix} 5 \\ 0 \\ 6 \\ 0 \end{bmatrix} + \text{span} \left\{ \begin{bmatrix} -2 \\ -1 \\ 0 \\ 0 \end{bmatrix} \begin{bmatrix} 1 \\ 0 \\ 2 \\ -1 \end{bmatrix} \right\}$$

Hence there are an infinity of solutions.

The homogeneous solution forms a subspace (for this example the subspace is span $\{\mathbf{x}_{h_1}, \mathbf{x}_{h_2}\}$). This subspace is termed the *null space of M* (or sometimes the *kernel of M*) and given the symbol

$$\mathfrak{N}(M) \equiv \text{span}\,\{\mathbf{x}_{h_1}, \mathbf{x}_{h_2}, \ldots, \mathbf{x}_{h_\nu}\} \tag{10-14}$$

where $\mathbf{x}_{h_1}, \mathbf{x}_{h_2}, \ldots, \mathbf{x}_{h_\nu}$ are linearly independent solutions to the homogeneous problem

$$M\mathbf{x}_{h_i} = \mathbf{0}$$

The number of these linearly independent solutions, or equivalently the dimension of the null space, is termed the *nullity of M* and given the symbol $\nu \equiv$ dimension of the subspace $\mathfrak{N}(M)$.

Note that the null space of M is formed from linear combinations of homogeneous solution vectors, $\mathbf{x}_h$. Hence the null space of M is a subspace of R^p where p is the column dimension of M. The nullity of M thus satisfies the inequality

$$0 \le \nu \le p$$

where $\nu = p$ if and only if M is the zero matrix.

The relation of the nullity ν and the column dimension p may be made more precise by introducing the notion of the rank of M, given the symbol $\gamma \equiv$ rank (M). The *rank of M* is defined as the number of linearly independent columns of M; equivalently, the rank of M is equal to the dimension of the range space of M:

$$\gamma \equiv \text{rank } (M) = \text{dimension } [\mathcal{R}(M)]$$

Then it is true that the rank of M plus the nullity of M equals the column dimension p:

$$\boxed{\gamma + \nu = p}$$

That this is true may be observed from the modified Hermite normal form as illustrated by the coefficient matrix in Eq. (10-13):

$$M' = \begin{bmatrix} 1 & -2 & 0 & 1 \\ 0 & 0 & 0 & 0 \\ 0 & 0 & 1 & 2 \\ 0 & 0 & 0 & 0 \end{bmatrix}$$

The nullity of M' is equal to the number of columns with zeros on the diagonal—columns 2 and 4 in this case. Hence $\nu = 2$. Then, because of the upper-triangular form of M', the columns with ones on the diagonal become a basis for the range space of M'. Namely,

$$\mathcal{R}(M') = \text{span} \left\{ \begin{bmatrix} 1 \\ 0 \\ 0 \\ 0 \end{bmatrix} \begin{bmatrix} -2 \\ 0 \\ 0 \\ 0 \end{bmatrix} \begin{bmatrix} 0 \\ 0 \\ 1 \\ 0 \end{bmatrix} \begin{bmatrix} 1 \\ 0 \\ 2 \\ 0 \end{bmatrix} \right\}$$

$$= \text{span} \left\{ \begin{bmatrix} 1 \\ 0 \\ 0 \\ 0 \end{bmatrix} \begin{bmatrix} 0 \\ 0 \\ 1 \\ 0 \end{bmatrix} \right\}$$

Hence the rank $(M') = \gamma = 2$. Thus $\nu + \gamma = 4$, which is the column dimension of M'.

This shows the relation to be true for the modified matrix M'. But what about for the original matrix

$$M = \begin{bmatrix} 1 & -2 & 0 & 1 \\ 0 & 0 & 1 & 2 \end{bmatrix}$$

It has already been seen that

$$\mathfrak{N}(M) = \mathfrak{N}(M')$$

and hence the nullity of M equals the nullity of M'. Moreover, the addition of zero rows in M' does not alter the linear independence of the columns of M and hence the *rank* of M.

Other interesting properties regarding rank are the following.

Proposition 10-2 Let M be a given $m \times p$ matrix. Then the rank of M satisfies:

(*a*) rank $M \equiv$ the number of linearly independent columns
(*b*) rank $M \equiv$ the number of linearly independent rows
(*c*) rank M = rank M^T
(*d*) rank $M \leq$ minimum $\{m \text{ or } p\}$
(*e*) If M is square $m \times m$, then M^{-1} exists if and only if rank $M = m$.
(*f*) rank M is not changed if M is either premultiplied or postmultiplied by a square, nonsingular matrix.
(*g*) If rank $M_1 = \gamma_1$ and rank $M_2 = \gamma_2$, then rank $M_1M_2 \leq$ minimum $\{\gamma_1 \text{ or } \gamma_2\}$

PROOF We may take properties *a* and *b* as the definition of rank. Properties *c* and *d* then follow quite readily. Property *e* follows from the fact that a square matrix is invertible (nonsingular) if and only if all of its columns (or rows) are linearly independent; its determinant is zero if it has a linearly dependent row or column (see Appendix B). Premultiplication or postmultiplication of M by a square nonsingular matrix is equivalent to forming linearly independent combinations of its rows or columns; linearly independent combinations of linearly independent vectors are still linearly independent. Hence the rank of M is not changed. Lastly, property *g* follows by similar arguments.

The next section introduces an elementary row operation technique to implement the theoretical results of this section. These theoretical results regarding the solution to the linear equation problem [Eq. (10-8)] are summarized in the following theorem:

Theorem 10-1 The linear equation problem [Eq. (10-8)] has a solution if and only if $\mathbf{d} \in \mathfrak{R}(M)$. If it exists, this solution has the general form

$$\mathbf{x} = \mathbf{x}_p + \mathfrak{N}(M) \tag{10-15}$$

The solution is unique if and only if the null space of M is empty. Equivalently, the solution is unique if and only if the rank of M equals p, the column dimension of M.

10-3 ELEMENTARY ROW OPERATIONS AND THE COMPLETE SOLUTION

The modified Hermite normal form provides a convenient form to discern the complete solution of the general linear equation problem

$$M\mathbf{x} = \mathbf{d} \tag{10-8a}$$

where M has dimension $m \times p$. As introduced in Sec. 2-7, elementary row operations may be used to simultaneously operate on both sides of Eq. (10-8) to obtain

$$(EM)\mathbf{x} = (E\mathbf{d}) \tag{10-8b}$$

where (EM) has the desired modified Hermite normal form. The procedure is to form the augmented matrix $[M, \mathbf{d}]$ and use elementary row operations until it is in the desired form $[(EM), (E\mathbf{d})]$.

Two changes from the methods presented in Sec. 2-7 are now required. First, the coefficient matrix M and data vector $\mathbf{d}$ are modified by adding rows of zeros until the point that M' has at least as many rows as columns (nothing needs to be done if M is square or has more rows than columns). Second, row operations are carried out on $[M', \mathbf{d}']$ until (EM') is in the modified Hermite normal form rather than being the diagonal identity matrix. If M is square and nonsingular, then the modified Hermite normal form *is* the indentity matrix, and so Sec. 2-7 is a special case of that considered here. The complete solution (or lack of solution) is then determined from

$$(EM')\mathbf{x} = (E\mathbf{d}')$$

as described in the previous section.

Example 10-5

$$\begin{bmatrix} 0 & 2 & 4 & -2 \\ 1 & 1 & 3 & 0 \\ 2 & 5 & 12 & -3 \end{bmatrix} \begin{bmatrix} x_1 \\ x_2 \\ x_3 \\ x_4 \end{bmatrix} = \begin{bmatrix} 1 \\ 2 \\ 3 \end{bmatrix}$$

Forming $[M', \mathbf{d}']$ and carrying out elementary row operations yields

$$\begin{bmatrix} 0 & 2 & 4 & -2 & 1 \\ 1 & 1 & 3 & 0 & 2 \\ 2 & 5 & 12 & -3 & 3 \\ 0 & 0 & 0 & 0 & 0 \end{bmatrix} \sim \begin{bmatrix} 1 & 1 & 3 & 0 & 2 \\ 0 & 2 & 4 & -2 & 1 \\ 0 & 3 & 6 & -3 & -1 \\ 0 & 0 & 0 & 0 & 0 \end{bmatrix} \sim$$

$$\begin{bmatrix} 1 & 1 & 3 & 0 & 2 \\ 0 & 1 & 2 & -1 & 1/2 \\ 0 & 0 & 0 & 0 & -5/2 \\ 0 & 0 & 0 & 0 & 0 \end{bmatrix} \sim \begin{bmatrix} 1 & 0 & 1 & 1 & 3/2 \\ 0 & 1 & 2 & -1 & 1/2 \\ 0 & 0 & 0 & 0 & -5/2 \\ 0 & 0 & 0 & 0 & 0 \end{bmatrix}$$

and so the revised linear equation problem is

$$\begin{bmatrix} 1 & 0 & 1 & 1 \\ 0 & 1 & 2 & -1 \\ 0 & 0 & 0 & 0 \\ 0 & 0 & 0 & 0 \end{bmatrix} \begin{bmatrix} x_1 \\ x_2 \\ x_3 \\ x_4 \end{bmatrix} = \begin{bmatrix} 3/2 \\ 1/2 \\ -5/2 \\ 0 \end{bmatrix}$$

which obviously has no solution since the revised data vector $(E\mathbf{d}')$ is not in the span of the columns of (EM'). This could have been discerned at the next to last step, but the last step is taken in order to find the null space of M. Namely, going to the third and fourth columns of (EM') (the columns with zeros on the diagonal instead of unity) and adding a -1 at the corresponding row number gives

$$\mathcal{N}(M) = \text{span}\left\{ \begin{bmatrix} 1 \\ 2 \\ -1 \\ 0 \end{bmatrix} \begin{bmatrix} 1 \\ -1 \\ 0 \\ -1 \end{bmatrix} \right\}$$

This is determined by inspection from the modified Hermite normal form. This is one of the main reasons to go to this particular form.

The result regarding the fact that this particular example had no solution is easily generalized. Namely, it is seen that the rank of (EM') is 2, while the rank of $[(EM'), (E\mathbf{d}')]$ is 3. But $\text{rank } M = \text{rank }(EM') = 2$, and $\text{rank }[M, \mathbf{d}] = \text{rank }[(EM'), (E\mathbf{d}')] = 3$. Therefore, in general it is true that:

Proposition 10-3 The linear equation problem

$$M\mathbf{x} = \mathbf{d} \tag{10-8}$$

has a solution (it may or may not be unique) if and only if $\text{rank } M = \text{rank }[M, \mathbf{d}]$.

Note that this proposition is equivalent to saying that $M\mathbf{x} = \mathbf{d}$ has no solution when $\mathbf{d}$ is not in the range space of M. However, the advantage of this proposition is that it is not necessary to find the range space of M.

Example 10-6

$$\begin{bmatrix} 1 & 0 & 1 \\ 2 & 1 & 1 \\ 0 & -2 & 2 \\ 3 & 2 & 1 \end{bmatrix} \begin{bmatrix} x_1 \\ x_2 \\ x_3 \end{bmatrix} = \begin{bmatrix} -1 \\ -3 \\ 2 \\ -5 \end{bmatrix}$$

This problem has more rows than columns, and so no row of zeros is added to the problem. The augmented matrix is

$$[M, \quad \mathbf{d}] = \begin{bmatrix} 1 & 0 & 1 & -1 \\ 2 & 1 & 1 & -3 \\ 0 & -2 & 2 & 2 \\ 3 & 2 & 1 & -5 \end{bmatrix}$$

Performing elementary row operations to zero the first column, and then progressing to the final Hermite normal form, gives

$$\begin{bmatrix} 1 & 0 & 1 & -1 \\ 0 & 1 & -1 & -1 \\ 0 & -2 & 2 & 2 \\ 0 & 2 & -2 & -2 \end{bmatrix} \sim \begin{bmatrix} 1 & 0 & 1 & -1 \\ 0 & 1 & -1 & -1 \\ 0 & 0 & 0 & 0 \\ 0 & 0 & 0 & 0 \end{bmatrix}$$

From this the solution is found by inspection to be

$$\mathbf{x} = \begin{bmatrix} -1 \\ -1 \\ 0 \end{bmatrix} + \text{span} \left\{ \begin{bmatrix} 1 \\ -1 \\ -1 \end{bmatrix} \right\}$$

Note that the solution has dimension equal to the column dimension of M and is unrelated to the row dimension of M. To see that $\mathbf{x}$ is the solution note that the revised linear equation problem may be written as

$$x_1 \begin{bmatrix} 1 \\ 0 \\ 0 \\ 0 \end{bmatrix} + x_2 \begin{bmatrix} 0 \\ 1 \\ 0 \\ 0 \end{bmatrix} + x_3 \begin{bmatrix} 1 \\ -1 \\ 0 \\ 0 \end{bmatrix} = \begin{bmatrix} -1 \\ -1 \\ 0 \\ 0 \end{bmatrix}$$

Thus picking $x_1 = -1$, $x_2 = -1$, and $x_3 = 0$ gives a particular solution, and picking $x_1 = 1$, $x_2 = -1$, and $x_3 = -1$ gives a homogeneous solution.

Based on the final Hermite normal form of the augmented $[M, \mathbf{d}]$, the following conclusions regarding the example linear equation problem may be made:

1. $\mathbf{d} \in \mathcal{R}(M) = \text{span} \left\{ \begin{bmatrix} 1 \\ 2 \\ 0 \\ 3 \end{bmatrix} \begin{bmatrix} 0 \\ 1 \\ -2 \\ 2 \end{bmatrix} \begin{bmatrix} 1 \\ 1 \\ 2 \\ 1 \end{bmatrix} \right\} = \text{span} \left\{ \begin{bmatrix} 1 \\ 2 \\ 0 \\ 3 \end{bmatrix} \begin{bmatrix} 0 \\ 1 \\ -2 \\ 2 \end{bmatrix} \right\}$
2. rank M = rank $[M, \mathbf{d}] = 2$.
3. The nullity of M is one and

$$\mathfrak{N}(M) = \text{span} \left\{ \begin{bmatrix} 1 \\ -1 \\ -1 \end{bmatrix} \right\}$$

These last two examples may have been deceptive. Normally a problem with only three equations to satisfy, but four unknowns, would be expected to not only have a solution but to have an infinity of solutions. Example 10-5 had no solution, even though the associated coefficient matrix did indeed have a null space. Likewise, a problem with four equations and three unknowns is overdetermined and normally would be expected to have no solution. But not only did Example 10-6 have a solution, it had an infinity of solutions. Fortunately, both of these unusual situations are handled readily by the procedure of elementary row operations and the modified Hermite normal form. However, as mentioned previously, there are better digital computation methods for determination of these same things, and these are discussed in Appendix E. Hopefully, though, the analysis here has provided some new insight into the potentially insidious nature of the innocent-looking linear equation problem. In any event, we now have the tools to solve the eigenvector linear equation problem.

10-4 THE EIGENVECTOR SOLUTION

Using the methods of the previous two sections, solution of the eigenvector linear equation problem is now straightforward.

Example 10-1, cont'd. Let

$$A = \begin{bmatrix} 1 & 2 \\ -1 & 4 \end{bmatrix}$$

with eigenvalues $\lambda_1 = 2$ and $\lambda_2 = 3$ (see Example 10-1, Sec. 10.1). The eigenvector problem for $\boldsymbol{\xi}_1$ is

$$[\lambda_1 I - A]\boldsymbol{\xi}_1 = \begin{bmatrix} 1 & -2 \\ 1 & -2 \end{bmatrix}\boldsymbol{\xi}_1 = \begin{bmatrix} 0 \\ 0 \end{bmatrix}$$

Using row operations on the coefficient matrix gives

$$\begin{bmatrix} 1 & -2 \\ 1 & -2 \end{bmatrix} \sim \begin{bmatrix} 1 & -2 \\ 0 & 0 \end{bmatrix}$$

The null space of $[\lambda_1 I - A]$ is given by

$$\boldsymbol{\xi}_1 = \text{span}\left\{\begin{bmatrix} -2 \\ -1 \end{bmatrix}\right\} = \text{span}\left\{\begin{bmatrix} 2 \\ 1 \end{bmatrix}\right\}$$

The eigenvector problem for $\boldsymbol{\xi}_2$ is

$$[\lambda_2 I - A]\boldsymbol{\xi}_2 = \begin{bmatrix} 2 & -2 \\ 1 & -1 \end{bmatrix}\boldsymbol{\xi}_2 = \begin{bmatrix} 0 \\ 0 \end{bmatrix}$$

Applying row operations to the coefficient matrix gives

$$\begin{bmatrix} 2 & -2 \\ 1 & -1 \end{bmatrix} \sim \begin{bmatrix} 1 & -1 \\ 0 & 0 \end{bmatrix}$$

Thus $$\xi_2 = \text{span}\left\{\begin{bmatrix}-1\\-1\end{bmatrix}\right\} = \text{span}\left\{\begin{bmatrix}1\\1\end{bmatrix}\right\}$$

Example 10-2, cont'd. Let

$$A = \begin{bmatrix}1 & -4\\1 & 1\end{bmatrix}$$

with eigenvalues $\lambda_1 = 1 + 2j$ and $\lambda_2 = 1 - 2j$. (See Example 10-2, Sec. 10-1.) The eigenvector problem for ξ_1 is

$$[\lambda_1 I - A]\xi_1 = \begin{bmatrix}2j & 4\\-1 & 2j\end{bmatrix}\xi_1 = \begin{bmatrix}0\\0\end{bmatrix}$$

Using row operations on the coefficient matrix gives

$$\begin{bmatrix}2j & 4\\-1 & 2j\end{bmatrix} \sim \begin{bmatrix}1 & -2j\\2j & 4\end{bmatrix} \sim \begin{bmatrix}1 & -2j\\0 & 0\end{bmatrix}$$

Notice that the elementary row operations are exactly the same, except now complex scalar multiples of rows must be added to other rows. The final form is still the same with either ones or zeros on the diagonal. The same procedure is still applied, also, for finding the null space: select the column with a zero on the diagonal and place a -1 in the corresponding row. Hence

$$\xi_1 = \text{span}\left\{\begin{bmatrix}-2j\\-1\end{bmatrix}\right\} = \text{span}\left\{\begin{bmatrix}2j\\1\end{bmatrix}\right\}$$

The eigenvector problem for ξ_2 is

$$[\lambda_2 I - A]\xi_2 = \begin{bmatrix}-2j & 4\\-1 & -2j\end{bmatrix}\xi_2 = \begin{bmatrix}0\\0\end{bmatrix}$$

Elementary row operations on the coefficient matrix gives

$$\begin{bmatrix}-2j & 4\\-1 & -2j\end{bmatrix} \sim \begin{bmatrix}-1 & -2j\\-2j & 4\end{bmatrix} \sim \begin{bmatrix}1 & 2j\\0 & 0\end{bmatrix}$$

Thus $$\xi_2 = \text{span}\left\{\begin{bmatrix}2j\\-1\end{bmatrix}\right\} = \text{span}\left\{\begin{bmatrix}-2j\\1\end{bmatrix}\right\}$$

with the restriction that ξ_1 and ξ_2 must always be selected as complex conjugate pairs. Thus once the problem for ξ_1 is solved, it really is not necessary to form the eigenvector problem for ξ_2; ξ_2 is simply the complex conjugate of ξ_1.

Example 10-7 Let

$$A = \begin{bmatrix}2 & 3 & -1\\0 & 2 & 0\\0 & 0 & 2\end{bmatrix}$$

with eigenvalue $\lambda_1 = 2$ repeated three times. The eigenvector problem is

$$[\lambda_1 I - A]\xi_1 = \begin{bmatrix} 0 & -3 & 1 \\ 0 & 0 & 0 \\ 0 & 0 & 0 \end{bmatrix} \xi_1 = \begin{bmatrix} 0 \\ 0 \\ 0 \end{bmatrix}$$

Performing row operations gives

$$\begin{bmatrix} 0 & -3 & 1 \\ 0 & 0 & 0 \\ 0 & 0 & 0 \end{bmatrix} \sim \begin{bmatrix} 0 & 0 & 0 \\ 0 & 1 & -1/3 \\ 0 & 0 & 0 \end{bmatrix}$$

Thus

$$\xi_1 = \text{span} \left\{ \begin{bmatrix} -1 \\ 0 \\ 0 \end{bmatrix} \begin{bmatrix} 0 \\ -1/3 \\ -1 \end{bmatrix} \right\} = \text{span} \left\{ \begin{bmatrix} 1 \\ 0 \\ 0 \end{bmatrix} \begin{bmatrix} 0 \\ 1/3 \\ 1 \end{bmatrix} \right\}$$

The eigenvector space corresponding to $\lambda_1 = 2$ (which has multiplicity 3) is now two-dimensional. It thus might be said that there are two independent eigenvectors corresponding to $\lambda_1 = 2$.

Example 10-8 Let

$$A = \begin{bmatrix} 2 & 3 & 0 \\ 0 & 2 & -1 \\ 0 & 0 & 2 \end{bmatrix}$$

The eigenvalues of A are again $\lambda_1 = 2$ with multiplicity 3. The eigenvector problem is

$$[\lambda_1 I - A]\xi_1 = \begin{bmatrix} 0 & 3 & 0 \\ 0 & 0 & -1 \\ 0 & 0 & 0 \end{bmatrix} \xi_1 = \begin{bmatrix} 0 \\ 0 \\ 0 \end{bmatrix}$$

Performing row operations

$$\begin{bmatrix} 0 & 3 & 0 \\ 0 & 0 & -1 \\ 0 & 0 & 0 \end{bmatrix} \sim \begin{bmatrix} 0 & 0 & 0 \\ 0 & 1 & 0 \\ 0 & 0 & 1 \end{bmatrix}$$

Hence

$$\xi_1 = \text{span} \left\{ \begin{bmatrix} -1 \\ 0 \\ 0 \end{bmatrix} \right\} = \text{span} \left\{ \begin{bmatrix} 1 \\ 0 \\ 0 \end{bmatrix} \right\}$$

and so this time there is only one eigenvector for $\lambda_1 = 2$ with multiplicity 3.

The Phase-Variable Special Case

The previous examples have shown that the number of independent eigenvectors is somewhat problematic when there are repeated eigenvalues. The number of eigenvectors will never exceed the multiplicity of the eigenvalue, but it seems that the number of independent eigenvectors is otherwise problem-dependent. This is basically true, but there are two special cases to mention in which there is certainty. In particular, if the matrix is in the phase-variable form and has repeated eigenvalues, there is only one independent eigenvector for each distinct eigenvalue. Conversely, if the matrix is real and symmetric, then the number of independent eigenvectors is always n, the dimension of A, regardless of the multiplicities. The former phase-variable case is demonstrated in the next example, but the symmetric-matrix case is such an important situation that it is presented separately in Sec. 10-6.

Example 10-9 Let

$$A = \begin{bmatrix} 0 & 1 & 0 \\ 0 & 0 & 1 \\ 8 & -12 & 6 \end{bmatrix}$$

which is in phase-variable form and has eigenvalue $\lambda_1 = 2$ with multiplicity 3. Then

$$[\lambda_1 I - A]\xi_1 = \begin{bmatrix} 2 & -1 & 0 \\ 0 & 2 & -1 \\ -8 & 12 & -4 \end{bmatrix} \xi_1 = \begin{bmatrix} 0 \\ 0 \\ 0 \end{bmatrix}$$

Performing row operations gives

$$\begin{bmatrix} 2 & -1 & 0 \\ 0 & 2 & -1 \\ -8 & 12 & -4 \end{bmatrix} \sim \begin{bmatrix} 1 & -1/2 & 0 \\ 0 & 1 & -1/2 \\ 0 & 8 & -4 \end{bmatrix} \sim \begin{bmatrix} 1 & 0 & -1/4 \\ 0 & 1 & -1/2 \\ 0 & 0 & 0 \end{bmatrix}$$

Thus

$$\xi_1 = \text{span}\left\{\begin{bmatrix} -1/4 \\ -1/2 \\ -1 \end{bmatrix}\right\} = \text{span}\left\{\begin{bmatrix} 1 \\ 2 \\ 4 \end{bmatrix}\right\}$$

and so, as claimed, there is only one eigenvector direction for $\lambda_1 = 2$ even though the multiplicity is 3.

Another useful observation concerning the eigenvectors for the phase-variable form is that the eigenvector corresponding to a given λ_i is given by

$$\xi_i = \text{span}\left\{\begin{bmatrix} 1 \\ \lambda_i \\ \lambda_i^2 \\ \vdots \\ \lambda_i^{n-1} \end{bmatrix}\right\} \tag{10-16}$$

This is only true for the phase-variable matrix, but it is a useful fact. The eigenvectors can be immediately determined from the eigenvalues without having to form and solve the eigenvector linear equation problems.

Consideration of Digital Computation

It is beyond our scope here to consider the detailed computational aspects of the eigenvalue and eigenvector problems. But it must be emphasized that the hand-analysis techniques presented in this section for solving the eigenvector problem and in Sec. 10-1 for solving the eigenvalue problem *do not* represent practical digital computational algorithms or even simplified approximations of the practical techniques. It is generally acknowledged by the numerical analyst specialists that the iterative QR technique developed extensively and documented in exhustive detail by Wilkinson [4] is the most accurate and numerically suitable method for determining both the system eigenvalues and eigenvectors. Fortunately, expert computational software is readily available for application (e.g., Refs. 3, 5, and 6) without requiring one to be an expert in its derivation.

10-5 TRANSFORMING TO THE DIAGONAL AND BLOCK-DIAGONAL CANONICAL FORMS

The means have now been shown to find the eigenvalues and eigenvectors of a given $n \times n$ matrix A, but really a more pressing concern is exactly how to *use* the eigenvalues and the eigenvectors. Their main significance to the linear dynamic system comes in their application to e^{At}. A stepping stone to this application comes, though, in their use to diagonalize a given A matrix. Chapter 7 shows that the diagonal state coordinate system is particularly convenient to use because it decouples the states. Also, for a diagonal Λ matrix, $\Lambda = \text{diagonal}\,[\lambda_1, \lambda_2, \ldots, \lambda_n]$, it is true that

$$e^{\Lambda t} = \text{diagonal}\,[e^{\lambda_1 t}, \quad e^{\lambda_2 t}, \quad \ldots, \quad e^{\lambda_n t}] \tag{10-17}$$

Diagonalizing A when A has n independent eigenvectors is considered in this section. The next section briefly examines the more complicated case when A has less than n independent eigenvectors, and the almost-diagonal Jordan canonical form is used.

Let A have eigenvalues $\lambda_1, \lambda_2, \ldots, \lambda_n$, real or complex, distinct or repeated, and let $\boldsymbol{\xi}_1, \boldsymbol{\xi}_2, \ldots, \boldsymbol{\xi}_n$ be n corresponding linearly independent eigenvectors. Forming what is called the *modal matrix*

$$\boxed{P = [\boldsymbol{\xi}_1, \quad \boldsymbol{\xi}_2, \quad \ldots, \quad \boldsymbol{\xi}_n]_{n \times n}} \tag{10-18}$$

it is then true that

$$\boxed{P^{-1}AP = \text{diagonal}\,[\lambda_1, \quad \lambda_2, \quad \ldots, \quad \lambda_n]} \tag{10-19}$$

PROOF From the n relations

$$A\boldsymbol{\xi}_i = \lambda_i \boldsymbol{\xi}_i \qquad i = 1, 2, \ldots, n$$

it is seen that

$$AP = P\Lambda \tag{10-20}$$

where
$$P = [\boldsymbol{\xi}_1, \quad \boldsymbol{\xi}_2, \quad \ldots, \quad \boldsymbol{\xi}_n]_{n\times n}$$
$$\Lambda = \text{diagonal}\,[\lambda_1, \quad \lambda_2, \quad \ldots, \quad \lambda_n]$$

(see Prob. 10-24). Then, since P is assumed to have n linearly independent columns, it has rank n and is thus invertible. Premultiplication of both sides of Eq. (10-20) by P^{-1} then yields the desired result, Eq. (10-19).

The columns of P are set equal to the n eigenvectors, $\boldsymbol{\xi}_i$, $i = 1, 2, \ldots, n$. The *row* vectors of P^{-1} also have special properties. We denote these row vectors of P^{-1} as $\boldsymbol{\eta}_i^T$, $i = 1, 2, \ldots, n$. These $\boldsymbol{\eta}_i$ are termed the *left* or *reciprocal* eigenvectors of A. It may be shown that the $\boldsymbol{\eta}_i$ satisfy

$$[\lambda_i I - A^T]\boldsymbol{\eta}_i = 0 \tag{10-21a}$$

or, equivalently, by taking the transpose of Eq. (10-21a),

$$\boldsymbol{\eta}_i^T[\lambda_i I - A] = 0 \tag{10-21b}$$

(see Prob. 10-24). Also it may be shown that

$$\boldsymbol{\eta}_i^T\boldsymbol{\xi}_j = \begin{cases} 1 & \text{for } i = j \\ 0 & \text{for } i \neq j \end{cases}$$

(see Prob. 10-23).

Example 10-10 Given a state-space model in phase-variable form

$$\begin{bmatrix} \dot{x}_1 \\ \dot{x}_2 \end{bmatrix} = \begin{bmatrix} 0 & 1 \\ -2 & -3 \end{bmatrix} \begin{bmatrix} x_1 \\ x_2 \end{bmatrix} + \begin{bmatrix} 0 \\ K \end{bmatrix} u$$

$$y = [1 \quad 0] \begin{bmatrix} x_1 \\ x_2 \end{bmatrix}$$

It is desired to find the equivalent state-space model in diagonal form. One approach is to form $H(s) = C[sI - A]^{-1}B$ and then form the diagonal model by partial fraction expansion. A second approach is via the controllability matrix as discussed in Sec. 9-7.

A third approach is now available. Namely, the matrix of transition is the modal matrix P formed from the matrix of eigenvectors, and P may be used to transform to the diagonal form.

The eigenvalues of the A matrix are $\lambda_1 = -1$ and $\lambda_2 = -2$. Since A is in phase-variable form, the modal matrix and reciprocal modal matrix are

$$P = \begin{bmatrix} 1 & 1 \\ -1 & -2 \end{bmatrix} \qquad P^{-1} = \begin{bmatrix} 2 & 1 \\ -1 & -1 \end{bmatrix}$$

(Recall that $\boldsymbol{\xi}_i = \begin{bmatrix} 1 \\ \lambda_i \end{bmatrix}$ for a matrix in phase-variable form.) Thus

$$P^{-1}AP = \begin{bmatrix} -1 & 0 \\ 0 & -2 \end{bmatrix}$$

The diagonal state-space model is then

$$\begin{bmatrix} \dot{x}'_1 \\ \dot{x}'_2 \end{bmatrix} = \begin{bmatrix} -1 & 0 \\ 0 & -2 \end{bmatrix} \begin{bmatrix} x'_1 \\ x'_2 \end{bmatrix} + \begin{bmatrix} K \\ -K \end{bmatrix} u$$

$$y = [1 \quad 1] \begin{bmatrix} x'_1 \\ x'_2 \end{bmatrix}$$

This is not the same form of the diagonal state model as would be derived with either the partial fraction expansion method or the controllability method. With these methods one would obtain

$$\hat{A} = A' \qquad \hat{B} = \begin{bmatrix} 1 \\ 1 \end{bmatrix} \qquad \hat{C} = [K \quad -K]$$

This raises an important point, though. Like the eigenvectors, the modal matrix that transforms to the diagonal form is not unique. Hence there are an infinite number of possible state-space models, all having a diagonal A matrix.

Example 10-11

$$\begin{bmatrix} \dot{x}_1 \\ \dot{x}_2 \end{bmatrix} = \begin{bmatrix} -1 & 2 \\ -2 & -1 \end{bmatrix} \begin{bmatrix} x_1 \\ x_2 \end{bmatrix} + \begin{bmatrix} 1 \\ 1 \end{bmatrix} u$$

$$y = [3 \quad 4] \begin{bmatrix} x_1 \\ x_2 \end{bmatrix}$$

The eigenvalues and corresponding eigenvectors of A are

$$\lambda_1 = -1 + 2j \qquad \boldsymbol{\xi}_1 = \text{span}\left\{\begin{bmatrix} j \\ -1 \end{bmatrix}\right\} = \text{span}\left\{\begin{bmatrix} 1 \\ j \end{bmatrix}\right\}$$

$$\lambda_2 = -1 - 2j \qquad \boldsymbol{\xi}_2 = \text{span}\left\{\begin{bmatrix} -j \\ -1 \end{bmatrix}\right\} = \text{span}\left\{\begin{bmatrix} 1 \\ -j \end{bmatrix}\right\}$$

Then $$P = \begin{bmatrix} 1 & 1 \\ j & -j \end{bmatrix} \qquad P^{-1} = \begin{bmatrix} 1/2 & -j/2 \\ 1/2 & j/2 \end{bmatrix}$$

and the diagonal state-space model is

$$\begin{bmatrix} \dot{x}'_1 \\ \dot{x}'_2 \end{bmatrix} = \begin{bmatrix} -1 + 2j & 0 \\ 0 & -1 - 2j \end{bmatrix} \begin{bmatrix} x'_1 \\ x'_2 \end{bmatrix} + \begin{bmatrix} 1/2 - j/2 \\ 1/2 + j/2 \end{bmatrix} u$$

$$y = [3 + 4j \quad 3 - 4j] \begin{bmatrix} x'_1 \\ x'_2 \end{bmatrix}$$

The problem with the diagonal state model is that the diagonal model is complex-valued. The input, output, and state are still real-valued, but the complex-valued state model is difficult to work with. Therefore, the original form in this example is actually preferred. The original form is called the *block diagonal form.* It has nearly all the same attributes as the diagonal model, only the model is real-valued.

Transformation to the Block-Diagonal Form

The results of the previous example may be generalized to provide a technique to transform to the block-diagonal form for cases in which there are distinct, complex-conjugate-pair eigenvalues.

Proposition 10-4 Let A have a set of $\mu + 1$ distinct complex-conjugate-pair eigenvalues $\lambda_{1,2} = \sigma_1 \pm j\omega_1, \lambda_{3,4} = \sigma_3 \pm j\omega_3, \ldots, \lambda_{\mu,\mu+1} = \sigma_\mu \pm j\omega_\mu$, and the remaining eigenvalues real and distinct. Then the real $n \times n$ transformation matrix

$$\begin{aligned} P_b = [&\text{Re}\,\{\xi_1\},\ \text{Im}\,\{\xi_1\},\ \text{Re}\,\{\xi_3\},\ \text{Im}\,\{\xi_3\},\ \ldots, \\ &\text{Re}\,\{\xi_\mu\},\ \text{Im}\,\{\xi_\mu\},\ \xi_{\mu+2},\ \ldots,\ \xi_n] \end{aligned} \tag{10-22}$$

transforms A to the block-diagonal form

$$\begin{aligned} \Lambda_b \equiv P_b^{-1} A P_b = \text{diagonal} &\left\{ \begin{bmatrix} \sigma_1 & \omega_1 \\ -\omega_1 & \sigma_1 \end{bmatrix}, \begin{bmatrix} \sigma_3 & \omega_3 \\ -\omega_3 & \sigma_3 \end{bmatrix}, \ldots, \right. \\ &\left. \begin{bmatrix} \sigma_\mu & \omega_\mu \\ -\omega_\mu & \sigma_\mu \end{bmatrix},\ \lambda_{m+2},\ \ldots,\ \lambda_n \right\} \end{aligned} \tag{10-23}$$

PROOF Proof is notationally cumbersome for the general case. We demonstrate by example.

Example 10-12

$$\dot{\mathbf{x}} = \begin{bmatrix} 0 & 1 & 0 \\ 0 & 0 & 1 \\ -13 & -17 & -5 \end{bmatrix} \mathbf{x} + \begin{bmatrix} 0 \\ 0 \\ K \end{bmatrix} u$$

$$y = [1 \quad 0 \quad 0]\mathbf{x}$$

The eigenvalues of A are $\lambda_1 = -1$, $\lambda_{2,3} = -2 \pm 3j$. The modal matrix to transform to diagonal form is

$$P = \begin{bmatrix} 1 & 1 & 1 \\ -1 & -2 + 3j & -2 - 3j \\ 1 & -5 - 12j & -5 + 12j \end{bmatrix}$$

and the transformation matrix to block-diagonal form is

$$P_b = \begin{bmatrix} 1 & 1 & 0 \\ -1 & -2 & 3 \\ 1 & -5 & -12 \end{bmatrix}$$

Inverting P_b gives

$$P_b^{-1} = \begin{bmatrix} 13/10 & 2/5 & 1/10 \\ -3/10 & -2/5 & -1/10 \\ 7/30 & 1/5 & -1/30 \end{bmatrix}$$

Then the state-space model in block-diagonal form becomes

$$\dot{\mathbf{x}}'' = P_b^{-1}AP_b\mathbf{x}'' + P_b^{-1}Bu$$

$$y = CP_b\mathbf{x}''$$

or

$$\dot{\mathbf{x}}'' = \begin{bmatrix} -1 & 0 & 0 \\ 0 & -2 & 3 \\ 0 & -3 & -2 \end{bmatrix}\mathbf{x}'' + \begin{bmatrix} K/10 \\ -K/10 \\ -K/30 \end{bmatrix}u$$

$$y = [1 \quad 1 \quad 0]\mathbf{x}''$$

10-6 THE JORDAN CANONICAL FORM

From the examples of Sec. 10-4 it is seen that there are cases when A has repeated eigenvalues and it does not have n linearly independent eigenvectors. In this case it is not possible to find a modal matrix, P, which diagonalizes the A matrix. However, a form is possible which is almost diagonal. This almost-diagonal form is called the *Jordan canonical form*. The form has the eigenvalues down the diagonal, repeated as many times as the multiplicity in the characteristic polynomial, and a unity element above the repeated eigenvalues for each missing eigenvector. For example,

$$J = \begin{bmatrix} 2 & 1 & 0 & 0 & 0 \\ 0 & 2 & 1 & 0 & 0 \\ 0 & 0 & 2 & 0 & 0 \\ 0 & 0 & 0 & 3 & 1 \\ 0 & 0 & 0 & 0 & 3 \end{bmatrix}$$

is the Jordan canonical form for the A matrix with characteristic polynomial

$$\Delta(\lambda) = (\lambda_1 - 2)^3(\lambda_2 - 3)^2$$

and with only one eigenvector ξ_1 for λ_1 and one eigenvector ξ_2 for λ_2.

The Jordan canonical form can get rather complicated for cases where, say, λ_i has multiplicity $n_i = 5$ and has two independent eigenvectors. Such cases are rare,

*The methods of this section are mathematically correct, but for digital calculation very poor. See [8].

however, so we cover only the one case where each eigenvalue has but one eigenvector regardless of the multiplicity. As stated earlier, this is the case for the phase-variable form and hence is the case for the state model derived from any single-input, single-output transfer function.

As a hypothetical example to develop the technique, consider the phase-variable A matrix

$$A = \begin{bmatrix} 0 & 1 & 0 \\ 0 & 0 & 1 \\ 8 & -12 & 6 \end{bmatrix}$$

This A matrix has characteristic polynomial $\Delta(\lambda) = (\lambda_1 - 2)^3$ and only one eigenvector

$$\boldsymbol{\xi}_1 = \begin{bmatrix} 1 \\ \lambda_1 \\ \lambda_1^2 \end{bmatrix} = \begin{bmatrix} 1 \\ 2 \\ 4 \end{bmatrix}$$

However, three independent vectors are required to form a nonsingular matrix of transition, P. The technique for finding these two additional independent columns is straightforward. A chain of linear equation problems is set up from

$$[A - \lambda_1 I]\mathbf{g}_{1,1} = \boldsymbol{\xi}_1 \tag{10-24}$$

$$[A - \lambda_1 I]\mathbf{g}_{1,2} = \mathbf{g}_{1,1} \tag{10-25}$$

The $\mathbf{g}_{1,1}$ and $\mathbf{g}_{1,2}$ vectors are termed *generalized eigenvectors*. Here two generalized eigenvectors are required because the multiplicity of λ_1 is 3 and there is only one eigenvector. The chain Eqs. (10-24) and (10-25) would simply be continued if λ had a higher multiplicity and more generalized eigenvectors were required.

For this example the linear equation problem Eq. (10-24) becomes

$$\begin{bmatrix} -2 & 1 & 0 \\ 0 & -2 & 1 \\ 8 & -12 & 4 \end{bmatrix} \mathbf{g}_{1,1} = \begin{bmatrix} 1 \\ 2 \\ 4 \end{bmatrix}$$

Elementary row operations provide the solution

$$\begin{bmatrix} -2 & 1 & 0 & 1 \\ 0 & -2 & 1 & 2 \\ 8 & -12 & 4 & 4 \end{bmatrix} \sim \begin{bmatrix} 1 & -1/2 & 0 & -1/2 \\ 0 & 1 & -1/2 & -1 \\ 0 & -8 & 4 & 8 \end{bmatrix}$$

$$\sim \begin{bmatrix} 1 & 0 & -1/4 & -1 \\ 0 & 1 & -1/2 & -1 \\ 0 & 0 & 0 & 0 \end{bmatrix}$$

For simplicity the particular solution is selected

$$\mathbf{g}_{1,1} = \begin{bmatrix} -1 \\ -1 \\ 0 \end{bmatrix}$$

The linear equation problem Eq. (10-25) is then

$$\begin{bmatrix} -2 & 1 & 0 \\ 0 & -2 & 1 \\ 8 & -12 & 4 \end{bmatrix} \mathbf{g}_{2,1} = \begin{bmatrix} -1 \\ -1 \\ 0 \end{bmatrix}$$

with the particular solution

$$\mathbf{g}_{2,1} = \begin{bmatrix} 3/4 \\ 1/2 \\ 0 \end{bmatrix}$$

The matrix of transition is

$$P = [\boldsymbol{\xi}_1, \ \mathbf{g}_{1,1}, \ \mathbf{g}_{1,2}] = \begin{bmatrix} 1 & -1 & 3/4 \\ 2 & -1 & 1/2 \\ 4 & 0 & 0 \end{bmatrix}$$

$$P^{-1} = \begin{bmatrix} 0 & 0 & 1/4 \\ 2 & -3 & 1 \\ 4 & -4 & 1 \end{bmatrix}$$

and

$$P^{-1}AP = \begin{bmatrix} 2 & 1 & 0 \\ 0 & 2 & 1 \\ 0 & 0 & 2 \end{bmatrix}$$

is the desired Jordan canonical form of A.

*10-7 EIGENVALUES AND EIGENVECTORS OF REAL SYMMETRIC MATRICES

One of the most important classes of matrices in the theory and application of linear systems is real symmetric matrices; that is, matrices for which $A = A^T$. Symmetric matrices occur in many different applications of system estimation and control theory (the covariance matrix, for example, in an estimation problem is always symmetric). Furthermore, the concepts behind symmetric matrices form the cornerstone of many computational approaches.

The eigenvalues and eigenvectors of a real symmetric matrix have special properties which single them out from other general square matrices.

Theorem 10-2 Let A be $n \times n$, real, and symmetric. Then

1. Each eigenvalue λ_i is real.
2. Each eigenvalue has n_i independent eigenvectors where n_i is the multiplicity of λ_i.
3. The eigenvectors for distinct eigenvalues are orthogonal.

4. There exists an orthogonal matrix Q (i.e., $Q^T Q = I$) such that $Q^T A Q =$ diagonal $[\lambda_1, \lambda_2, \ldots, \lambda_n]$ even if any of the eigenvalues are repeated (that eigenvalue is simply repeated n_i times on the diagonal).

PROOF For property 1, assume that λ_i is complex and we show that there is a contradiction. If λ_i is complex, then another eigenvalue must be its complex conjugate—say, $\lambda_{i+1} = \bar{\lambda}_i$. It must also be true that the eigenvectors satisfy $\boldsymbol{\xi}_{i+1} = \bar{\boldsymbol{\xi}}_i$. Hence, the eigenvector equation for $\boldsymbol{\xi}_{i+1}$ is $A\bar{\boldsymbol{\xi}}_i = \bar{\lambda}_i \bar{\boldsymbol{\xi}}_i$.

Taking the dot product* of $(A\boldsymbol{\xi}_i)$ into $\bar{\boldsymbol{\xi}}_i$ gives

$$(A\boldsymbol{\xi}_i) \cdot \bar{\boldsymbol{\xi}}_i \equiv (\bar{\boldsymbol{\xi}}_i^T A^T)\bar{\boldsymbol{\xi}}_i = (\lambda_i \bar{\boldsymbol{\xi}}_i^T)\bar{\boldsymbol{\xi}}_i \tag{10-26}$$

where the last equality follows from the fact that $A\boldsymbol{\xi}_i = \lambda_i \boldsymbol{\xi}_i$. Next, taking the dot product of $\boldsymbol{\xi}_i$ into $(A\bar{\boldsymbol{\xi}}_i) = (\bar{\lambda}_i \bar{\boldsymbol{\xi}}_i)$ gives

$$\boldsymbol{\xi}_i \cdot (A\bar{\boldsymbol{\xi}}_i) \equiv \bar{\boldsymbol{\xi}}_i^T (A\bar{\boldsymbol{\xi}}_i) = \bar{\boldsymbol{\xi}}_i^T (\bar{\lambda}_i \bar{\boldsymbol{\xi}}_i) \tag{10-27}$$

But since $A^T = A$, it follows that Eqs. (10-26) and (10-27) are equal, and so $\lambda_i = \bar{\lambda}_i$, requiring λ_i to be real.

Property 2 is difficult to prove, in general, and so it is demonstrated by an example to follow. For proof see Ref. 7.

For property 3, assume that $\lambda_i \neq \lambda_j$. Then

$$(\lambda_i \boldsymbol{\xi}_i) \cdot \boldsymbol{\xi}_j = (A\boldsymbol{\xi}_i) \cdot \boldsymbol{\xi}_j = \boldsymbol{\xi}_i^T A^T \boldsymbol{\xi}_j = \boldsymbol{\xi}_i^T A \boldsymbol{\xi}_j$$

when the last equality follows from the fact that A is symmetric, that is, $A = A^T$. Likewise

$$\boldsymbol{\xi}_i \cdot (\lambda_j \boldsymbol{\xi}_j) = \boldsymbol{\xi}_i^T A \boldsymbol{\xi}_j$$

The difference between these two equations is zero because they both equal $\boldsymbol{\xi}_i^T A \boldsymbol{\xi}_j$:

$$(\lambda_i \boldsymbol{\xi}_i) \cdot \boldsymbol{\xi}_j - \boldsymbol{\xi}_i \cdot (\lambda_j \boldsymbol{\xi}_j) = 0$$

or

$$(\lambda_i - \lambda_j)(\boldsymbol{\xi}_i \cdot \boldsymbol{\xi}_j) = 0$$

Since $\lambda_i \neq \lambda_j$, $\boldsymbol{\xi}_i$ and $\boldsymbol{\xi}_j$ must be orthogonal.

Property 4 follows from properties 2 and 3 in that an orthogonal basis can always be selected for each eigenvector space of dimension n_i (say, by using Gram-Schmidt orthogonalization [1]), and each orthogonal eigenvector can be normalized to have unit norm. The resulting modal matrix of eigenvectors, Q, then has orthogonal columns each with unit norm. Such a matrix is called an *orthogonal matrix* and has the property that $Q^T = Q^{-1}$ (see Prob. 10-31). Hence property 4 follows directly from the results of the previous section.

*For complex vectors **a** and **b** the dot or inner product of **a** and **b** is defined by $\mathbf{a} \cdot \mathbf{b} \equiv \bar{\mathbf{a}}^T \mathbf{b}$ where $\bar{\mathbf{a}}$ is the complex conjugate of **a**.

Example 10-13 Let

$$A = \begin{bmatrix} 2 & \sqrt{2} & 0 \\ \sqrt{2} & 3 & 0 \\ 0 & 0 & 1 \end{bmatrix}$$

It is seen that $\lambda_1 = 1$ with multiplicity 2, and $\lambda_2 = 4$. The eigenvector problem for $\boldsymbol{\xi}_1$ is

$$[\lambda_1 I - A]\boldsymbol{\xi}_1 = \begin{bmatrix} -1 & -\sqrt{2} & 0 \\ -\sqrt{2} & -2 & 0 \\ 0 & 0 & 0 \end{bmatrix} \boldsymbol{\xi}_1 = \begin{bmatrix} 0 \\ 0 \end{bmatrix}$$

Using row operations gives

$$\begin{bmatrix} -1 & -\sqrt{2} & 0 \\ -\sqrt{2} & -2 & 0 \\ 0 & 0 & 0 \end{bmatrix} \sim \begin{bmatrix} 1 & \sqrt{2} & 0 \\ 0 & 0 & 0 \\ 0 & 0 & 0 \end{bmatrix}$$

Thus $$\boldsymbol{\xi}_1 = \text{span} \left\{ \begin{bmatrix} \sqrt{2} \\ -1 \\ 0 \end{bmatrix} \begin{bmatrix} 0 \\ 0 \\ -1 \end{bmatrix} \right\} = \text{span} \left\{ \begin{bmatrix} \sqrt{2}/\sqrt{3} \\ -1/\sqrt{3} \\ 0 \end{bmatrix} \begin{bmatrix} 0 \\ 0 \\ 1 \end{bmatrix} \right\}$$

and there are two orthogonal, unit-norm eigenvectors corresponding to $\lambda_1 = 1$ with multiplicity 2.

For $\boldsymbol{\xi}_2$ it is readily found that

$$\boldsymbol{\xi}_2 = \text{span} \left\{ \begin{bmatrix} -\sqrt{2}/2 \\ -1 \\ 0 \end{bmatrix} \right\} = \text{span} \left\{ \begin{bmatrix} 1/\sqrt{3} \\ \sqrt{2}/\sqrt{3} \\ 0 \end{bmatrix} \right\}$$

The orthogonal Q matrix is

$$Q = \begin{bmatrix} \sqrt{2}/\sqrt{3} & 0 & 1/\sqrt{3} \\ -1/\sqrt{3} & 0 & \sqrt{2}/\sqrt{3} \\ 0 & 1 & 0 \end{bmatrix}$$

and $$Q^T A Q = \begin{bmatrix} 1 & 0 & 0 \\ 0 & 1 & 0 \\ 0 & 0 & 4 \end{bmatrix}$$

10-8 e^{At} AND THE EIGENVALUES AND EIGENVECTORS

In Chap. 7 a relation of e^{At} is the matrix power series

$$e^{At} = I + At + A^2 \frac{t^2}{2!} + A^3 \frac{t^3}{3!} + \dots \tag{7-33}$$

Armed now with the tools of the system eigenvalues and eigenvectors, considerably more insight can be gained concerning the nature of this important matrix exponential function.

If A has n independent eigenvectors, then it is known that $A = P\Lambda P^{-1}$ where P is the modal matrix of eigenvectors and Λ is the diagonal matrix of eigenvalues, some of which may be repeated. This substitution may be made into the power series Eq. (7-33) to yield

$$e^{At} = I + (P\Lambda P^{-1})t + (P\Lambda P^{-1})^2 \frac{t^2}{2!} + (P\Lambda P^{-1})^3 \frac{t^3}{3!} + \cdots \tag{10-28}$$

But $I = P^{-1}P$, and it is seen that

$$(P\Lambda P^{-1})^2 = (P\Lambda P^{-1})(P\Lambda P^{-1}) = P\Lambda^2 P^{-1}$$

$$(P\Lambda P^{-1})^3 = (P\Lambda P^{-1})(P\Lambda P^{-1})^2 = P\Lambda^3 P^{-1}$$

and so on for every term

$$(P\Lambda P^{-1})^n = P\Lambda^n P^{-1}$$

Thus Eq. (10-28) becomes

$$e^{At} = P\left[I + \Lambda t + \Lambda^2 \frac{t^2}{2!} + \Lambda^3 \frac{t^3}{3!} + \cdots\right]P^{-1}$$

The power series inside the brackets is, by definition, $e^{\Lambda t}$. Hence the final result is

$$\boxed{e^{At} = Pe^{\Lambda t}P^{-1}} \tag{10-28a}$$

Because of the ease in computing $e^{\Lambda t}$

$$e^{\Lambda t} = \text{diagonal}\,[e^{\lambda_1 t}, \quad e^{\lambda_2 t}, \quad \ldots, \quad e^{\lambda_n t}]$$

Equation (10-28*a*) is a valuable computational form for finding e^{At}. However, substituting the expressions for P and P^{-1} into Eq. (10-28*a*) gives an equivalent form providing considerable analysis insight:

$$e^{At} = [\boldsymbol{\xi}_1, \quad \boldsymbol{\xi}_2, \quad \ldots, \quad \boldsymbol{\xi}_n]\begin{bmatrix} e^{\lambda_1 t} & 0 & \cdots & 0 \\ 0 & e^{\lambda_2 t} & & \\ \cdots & \cdots & \cdots & \cdots \\ 0 & & & e^{\lambda_n t} \end{bmatrix}\begin{bmatrix} \boldsymbol{\eta}_1^T \\ \boldsymbol{\eta}_2^T \\ \vdots \\ \boldsymbol{\eta}_n^T \end{bmatrix}$$

Letting the diagonal matrix of $e^{\lambda_i t}$ multiply through P^{-1} gives

$$e^{At} = [\boldsymbol{\xi}_1, \quad \boldsymbol{\xi}_2, \quad \ldots, \quad \boldsymbol{\xi}_n]\begin{bmatrix} e^{\lambda_1 t}\boldsymbol{\eta}_1^T \\ e^{\lambda_2 t}\boldsymbol{\eta}_2^T \\ \vdots \\ e^{\lambda_n t}\boldsymbol{\eta}_n^T \end{bmatrix}$$

Finally, taking the product of the partitioned matrix P with the partitioned matrix

$(e^{\Lambda t}P^{-1})$ gives the final *dyadic sum*

$$\boxed{e^{At} = \sum_{i=1}^{n} e^{\lambda_i t}\boldsymbol{\xi}_i \boldsymbol{\eta}_i^T} \tag{10-28b}$$

Example 10-14 This example considers the use of the dyadic sum, Eq. (10-28*b*) for finding e^{At}. For the continuous-time case this is an alternative to

$$e^{At} = \mathcal{L}^{-1}\{[sI - A]^{-1}\} \tag{7-31}$$

as presented in Sec. 7-3. We choose

$$A = \begin{bmatrix} 0 & 1 & 0 \\ 0 & 0 & 1 \\ -13 & -17 & -5 \end{bmatrix}$$

This is the same A as considered in Example 7-3, where e^{At} is found by use of Eq. (7-31). This method proved to involve a great deal of labor even for this sample 3×3 case.

The eigenvalues of A are $\lambda_1 = -1, \lambda_{2,3} = -2 \pm 3j$. Since A is in phase-variable form,

$$P = \begin{bmatrix} 1 & 1 & 1 \\ \lambda_1 & \lambda_2 & \lambda_3 \\ \lambda_1^2 & \lambda_2^2 & \lambda_3^2 \end{bmatrix} = [\boldsymbol{\xi}_1, \ \boldsymbol{\xi}_2, \ \boldsymbol{\xi}_3]$$

$$= \begin{bmatrix} 1 & 1 & 1 \\ -1 & -2+3j & -2-3j \\ 1 & -5-12j & -5+12j \end{bmatrix}$$

Inverting P gives the left or reciprocal eigenvectors

$$P^{-1} = \begin{bmatrix} \boldsymbol{\eta}_1^T \\ \boldsymbol{\eta}_2^T \\ \boldsymbol{\eta}_3^T \end{bmatrix} = \begin{bmatrix} \dfrac{13}{10} & \dfrac{2}{5} & \dfrac{1}{10} \\ -\dfrac{3}{20} - \dfrac{7j}{60} & -\dfrac{1}{5} - \dfrac{j}{10} & -\dfrac{1}{20} + \dfrac{j}{60} \\ -\dfrac{3}{20} + \dfrac{7j}{60} & -\dfrac{1}{5} + \dfrac{j}{10} & -\dfrac{1}{20} - \dfrac{j}{60} \end{bmatrix}$$

To invert the complex P is a major undertaking, particularly by hand. Shortly we see how to use the all-real block-diagonal form as considered in Example 10-12 to considerably simplify this calculation. Indeed, the author must confess that this is the method he used to find P^{-1} here.

Now, using Eq. (10-28*b*) the remaining work to find e^{At} is straightforward (but still tedious):

$$e^{At} = \sum_{i=1}^{3} e^{\lambda_i t} \boldsymbol{\xi}_i \boldsymbol{\eta}_i^T$$

$$= e^{-t} \begin{bmatrix} 1 \\ -1 \\ 1 \end{bmatrix} [13/10, \quad 2/5, \quad 1/10]$$

$$+ e^{(-2+3j)t} \begin{bmatrix} 1 \\ -2+3j \\ -5-12j \end{bmatrix} \left[-\frac{3}{20} - \frac{7j}{60}, -\frac{1}{5} - \frac{j}{10}, -\frac{1}{20} + \frac{j}{60} \right]$$

$$+ e^{(-2-3j)t} \begin{bmatrix} 1 \\ -2-3j \\ -5+12j \end{bmatrix} \left[-\frac{3}{20} + \frac{7j}{60}, -\frac{1}{5} + \frac{j}{10}, -\frac{1}{20} - \frac{j}{60} \right]$$

This dyadic sum form for e^{At} may be multiplied out to form e^{At} as the sum of three 3×3 complex *component matrices* corresponding to each *mode*, $e^{\lambda_i t}$. Notice that the matrices multiplying the two terms $e^{(-2+3j)t}$ and $e^{(-2-3j)t}$ are complex conjugates of each other.

System Modal Representation

Equation (10-28*b*) demonstrates the explicit dependence of e^{At} on its eigenvalues, λ_i, right eigenvectors, $\boldsymbol{\xi}_i$, and left eigenvectors, $\boldsymbol{\eta}_i$. Equation (10-28*b*) also provides useful insights into the input-output state relationship. The output relation is found in Chap. 7 to be

$$y(t) = Ce^{A(t-t_0)}\mathbf{x}(t_0) + \int_{t_0}^{t} Ce^{A(t-\tau)}B\mathbf{u}(\tau)\,d\tau \tag{10-29}$$

For simplicity this assumes $D \equiv 0$ (the system has no direct feedforward). Applying Eq. (10-28*b*) to Eq. (10-29) then yields

$$\mathbf{y}(t) = \sum_{i=1}^{n} [C\boldsymbol{\xi}_i][\boldsymbol{\eta}_i^T \mathbf{x}(t_0)] e^{\lambda_i (t-t_0)} + \sum_{i=1}^{n} [C\boldsymbol{\xi}_i][\boldsymbol{\eta}_i^T B] \int_{t_0}^{t} e^{\lambda_i (t-\tau)} \mathbf{u}(\tau)\,d\tau \tag{10-29a}$$

This is termed the *modal* representation of the system output response, as it reveals how each mode $e^{\lambda_i t}$ enters into the output solution. This modal form of the system response plays a key role in Chap. 11 for analysis of controllability and observability.

Calculation of the Transfer Function

If the system is single-input, single-output, then C and B are row and column vectors, respectively. Then the factors $(C\boldsymbol{\xi}_i)$ and $(\boldsymbol{\eta}_i^T B)$ are both scalars. In this case the zero-state response takes the form

$$y_{zs}(t) = \sum_{i=1}^{n} r_i \int_{t_0}^{t} e^{\lambda_i(t-\tau)} u(\tau)\, d\tau \tag{10-30}$$

where the r_i scalars are

$$r_i = (C\boldsymbol{\xi}_i)(\boldsymbol{\eta}_i^T B) \tag{10-31}$$

For distinct eigenvalues these r_i are, in fact, the residues in the partial fraction expansion

$$H(s) = \frac{r_1}{s-\lambda_1} + \frac{r_2}{s-\lambda_2} + \cdots + \frac{r_n}{s-\lambda_n} \tag{10-32}$$

Indeed, for the multi-input, multi-output case, letting

$$R_i \equiv (C\boldsymbol{\xi}_i)(\boldsymbol{\eta}_i^T B) \tag{10-33}$$

be an $m \times \ell$ matrix (m outputs, ℓ inputs), it is true that the matrix transfer function has the partial fraction form

$$H(s) = \frac{R_1}{s-\lambda_1} + \frac{R_2}{s-\lambda_2} + \cdots + \frac{R_n}{s-\lambda_n} \tag{10-34a}$$

This becomes, then, a way to find the transfer function

$$H(s) = C[sI - A]^{-1}B \tag{10-34b}$$

without having to find $[sI - A]^{-1}$; computer tools are readily available to find the eigenvalue and eigenvectors of a given matrix A. (The eigenvalue/eigenvector problem can often be ill-conditioned, though. See Ref. 11 for numerically stable alternatives.)

Example 10-15 Consider the phase-variable SISO system

$$\dot{\mathbf{x}} = \begin{bmatrix} 0 & 1 & 0 \\ 0 & 0 & 1 \\ 6 & -1 & 4 \end{bmatrix} \mathbf{x} + \begin{bmatrix} 0 \\ 0 \\ K \end{bmatrix} u$$

$$y = [-15 \quad 2 \quad 1]\mathbf{x}$$

This is the example considered in Sec. 6-1 and then again in Sec. 9-4. The eigenvalues are $\lambda_1 = 1$, $\lambda_2 = -2$, $\lambda_3 = -3$. Since the system is in phase-variable form the eigenvectors are written down by inspection as

$$\xi_1 = \begin{bmatrix} 1 \\ 1 \\ 1 \end{bmatrix} \qquad \xi_2 = \begin{bmatrix} 1 \\ -2 \\ 4 \end{bmatrix} \qquad \xi_3 = \begin{bmatrix} 1 \\ -3 \\ 9 \end{bmatrix}$$

The reciprocal or left eigenvectors are found from the rows of

$$\begin{bmatrix} \eta_1^T \\ \eta_2^T \\ \eta_3^T \end{bmatrix} = \begin{bmatrix} 1 & 1 & 1 \\ 1 & -2 & -3 \\ 1 & 4 & 9 \end{bmatrix}^{-1} = \begin{bmatrix} 1/2 & 5/12 & 1/12 \\ 1 & -2/3 & -1/3 \\ -1/2 & 1/4 & 1/4 \end{bmatrix}$$

Using Eqs. (10-31) and (10-32), the transfer function is

$$H(s) = \frac{r_1}{s-1} + \frac{r_2}{s+2} + \frac{r_3}{s+3}$$

where

$$r_1 = (C\xi_1)(\eta_1^T B) = (-12)(K/12) = -K$$

$$r_2 = (C\xi_2)(\eta_2^T B) = (-15)(-K/3) = 5K$$

$$r_3 = (C\xi_3)(\eta_3^T B) = (-12)(K/4) = -3K$$

These are the same residues as found in Sec. 6-3 by partial fraction expansions.

10-9 FINDING e^{At} USING THE BLOCK-DIAGONAL FORM

In Sec. 10-5 it is seen that the block-diagonal canonical form may be used to avoid complex variable operations with a complex P, modal matrix (see Proposition 10-4). Therefore, it is natural to consider what $\exp\{\Lambda_b t\}$ equals, where Λ_b is the block-diagonal canonical form, Eq. (10-23).

Since each block of Λ_b is decoupled from the other blocks, it is only necessary to consider one block of the matrix exponential for Λ_b. Namely, consider

$$\exp \begin{bmatrix} \sigma & \omega \\ -\omega & \sigma \end{bmatrix} t$$

for the complex eigenvalue pair $\lambda = \sigma \pm \omega j$. The subscript i is delected for notational convenience.

The block matrix may be transformed to diagonal form via (see Example 10-11)

$$\begin{bmatrix} \sigma + \omega j & 0 \\ 0 & \sigma - \omega j \end{bmatrix} = \begin{bmatrix} 1/2 & -j/2 \\ 1/2 & j/2 \end{bmatrix} \begin{bmatrix} \sigma & \omega \\ -\omega & \sigma \end{bmatrix} \begin{bmatrix} 1 & 1 \\ j & -j \end{bmatrix}$$

Therefore,

$$\exp \begin{bmatrix} \sigma & \omega \\ -\omega & \sigma \end{bmatrix} t = \begin{bmatrix} 1 & 1 \\ j & -j \end{bmatrix} \begin{bmatrix} e^{(\sigma+\omega j)t} & 0 \\ 0 & e^{(\sigma-\omega j)t} \end{bmatrix} \begin{bmatrix} 1/2 & -j/2 \\ 1/2 & j/2 \end{bmatrix}$$

Performing all of the matrix multiplications and using Euler's identities

$$\cos(\omega t) = \frac{e^{j\omega t} + e^{-j\omega t}}{2}$$

$$\sin(\omega t) = \frac{e^{j\omega t} - e^{-j\omega t}}{2j}$$

yields

$$\boxed{\exp\begin{bmatrix} \sigma & \omega \\ -\omega & \sigma \end{bmatrix} t = \begin{bmatrix} \cos\omega t & \sin\omega t \\ -\sin\omega t & \cos\omega t \end{bmatrix} e^{\sigma t}} \qquad (10\text{-}35)$$

Combining Eq. (10-13) with Proposition 10-4 then gives the useful result

$$\boxed{e^{At} = P_b e^{\Lambda_b t} P_b^{-1}} \qquad (10\text{-}28c)$$

where the real-valued P_b is given by Eq. (10-22) and the real-valued $\exp(\Lambda_b t)$ is given by

$$\boxed{\begin{aligned} e^{\Lambda_b t} = \text{block diagonal}\Bigg\{ &\begin{bmatrix} \cos\omega_1 t & \sin\omega_1 t \\ -\sin\omega_1 t & \cos\omega_1 t \end{bmatrix} e^{\sigma_1 t}, \quad \ldots, \\ &\begin{bmatrix} \cos\omega_\mu t & \sin\omega_\mu t \\ -\sin\omega_\mu t & \cos\omega_\mu t \end{bmatrix} e^{\sigma_\mu t}, \quad e^{\lambda_{\mu+2} t}, \quad \ldots, \quad e^{\lambda_n t} \Bigg\} \end{aligned}} \qquad (10\text{-}36)$$

The A matrix is assumed to have $\mu + 1$ distinct complex-conjugate-pair eigenvalues and $(n - \mu - 1)$ distinct real eigenvalues.

Example 10-16 Again choose

$$A = \begin{bmatrix} 0 & 1 & 0 \\ 0 & 0 & 1 \\ -13 & -17 & -5 \end{bmatrix}$$

the same as Examples 7-3, 10-12, and 10-14. The eigenvalues of A are

$$\lambda_1 = -1 \qquad \lambda_{2,3} = -2 \pm 3j$$

Example 10-12 defines the transformation matrices P_b and P_b^{-1} to put A in the block-diagonal form

$$\Lambda_b = \begin{bmatrix} -1 & 0 & 0 \\ 0 & -2 & 3 \\ 0 & -3 & -2 \end{bmatrix}$$

Defining $\boldsymbol{\xi}_{b_1}, \boldsymbol{\xi}_{b_2}, \boldsymbol{\xi}_{b_3}$ and $\boldsymbol{\eta}_{b_1}^T, \boldsymbol{\eta}_{b_2}^T, \boldsymbol{\eta}_{b_3}^T$ as the columns of rows of P_b and

P_b^{-1}, respectively, gives a dyadic sum representation of Eq. (10-28c) for this example:

$$
\begin{aligned}
e^{At} &= e^{-t}[\boldsymbol{\xi}_{b_1}\boldsymbol{\eta}_{b_1}^T] \\
&\quad + e^{-2t}\cos(3t)[\boldsymbol{\xi}_{b_2}\boldsymbol{\eta}_{b_2}^T + \boldsymbol{\xi}_{b_3}\boldsymbol{\eta}_{b_3}^T] \\
&\quad + e^{-2t}\sin(3t)[\boldsymbol{\xi}_{b_2}\boldsymbol{\eta}_{b_3}^T - \boldsymbol{\xi}_{b_3}\boldsymbol{\eta}_{b_2}^T]
\end{aligned}
$$

This is still not a particularly desirable form for calculation of e^{At} by hand, but it does avoid complex-variable calculations for digital computations. The operations of Eq. (10-28c) are carried out readily on the computer. There is no good way to determine the continuous e^{At}, but the use of Eq. (10-28c) is perhaps one of the best for distinct eigenvalues.* The situation for finding e^{At} for A with repeated eigenvalues is more of a problem still.

*10-10 COMPUTING e^{At} WITH THE JORDAN CANONICAL FORM

If A has fewer than n independent eigenvectors, it may be put in the Jordan canonical form as discussed in Sec. 10-6. Hence it is true that

$$A = PJP^{-1} \tag{10-37}$$

where the columns of the matrix P now contain eigenvectors and generalized eigenvectors, and J is a block diagonal with blocks of the form such as

$$\begin{bmatrix} \lambda_i & 1 & 0 \\ 0 & \lambda_i & 1 \\ 0 & 0 & \lambda_i \end{bmatrix}$$

for, say, an eigenvalue λ_i with multiplicity 3 but only one independent eigenvector.

Again,

$$e^{At} = Pe^{Jt}P^{-1} \tag{10-38}$$

and so it is of interest to examine e^{Jt}.

The matrix e^{Jt} may be computed by several methods, including inverse Laplace transform of $[sI - J]^{-1}$, infinite power series, or the Cayley-Hamilton theorem method of the next section. But, since J is block diagonal, e^{Jt} will also be block diagonal. It is thus of value to consider several standard blocks. For instance,

$$\exp\begin{bmatrix} \lambda & 1 \\ 0 & \lambda \end{bmatrix} t = \begin{bmatrix} e^{\lambda t} & te^{\lambda t} \\ 0 & e^{\lambda t} \end{bmatrix} \tag{10-39}$$

*The interested reader is referred to the informative article, "Nineteen Dubious Ways to Compute e^{At}," Ref. 9. This article discusses some of the basic computational problems associated with finding e^{At}, continuous t.

$$\exp\begin{bmatrix}\lambda & 1 & 0\\ 0 & \lambda & 1\\ 0 & 0 & \lambda\end{bmatrix}t = \begin{bmatrix}e^{\lambda t} & te^{\lambda t} & (t^2/2!)e^{\lambda t}\\ 0 & e^{\lambda t} & te^{\lambda t}\\ 0 & 0 & e^{\lambda t}\end{bmatrix} \tag{10-40}$$

and so forth. Blocks such as Eqs. (10-39) and (10-40) may then be used in conjunction with Eq. (10-38) to compute e^{At} in the general case.

10-11 e^{At} AND THE CAYLEY-HAMILTON THEOREM

A well-known and often very useful result in the theory of matrices is the Cayley-Hamilton theorem. This theorem states that every matrix satisfies its own characteristic polynomial; that is, if A has characteristic polynomial

$$\Delta(\lambda) = \lambda^n + a_{n-1}\lambda^{n-1} + \cdots + a_1\lambda + a_0 \tag{10.41}$$

then it is true that

$$\Delta(A) = A^n + a_{n-1}A^{n-1} + \cdots + a_1 A + a_0 I = 0 \tag{10-42}$$

Since each eigenvalue λ_i satisifies the characteristic polynomial (that is, $\Delta(\lambda_i) = 0$), this theorem is easily proven for the case of n independent eigenvectors (see Prob. 10-41). For the general case, it is more difficult to prove, but it is still true.

The Cayley-Hamilton theorem can be used in a variety of ways, but one important use is to turn an infinite sum on A (such as the infinite power series for e^{At}) into a finite sum with no more than n terms. To be specific, from the Cayley-Hamilton theorem it is known that

$$A^n = -a_{n-1}A^{n-1} - a_{n-2}A^{n-2} - \cdots - a_1 A - a_0 I \tag{10-43}$$

$$\begin{aligned} A \cdot A^n &= -a_{n-1}A^n - a_{n-2}A^{n-1} - \cdots - a_1 A^2 - a_0 A \\ &= -a_{n-1}(-a_{n-1}A^{n-1} - \cdots - a_1 A - a_0 I) \\ &\quad - a_{n-2}A^{n-1} - \cdots - a_1 A^2 - a_0 A \end{aligned}$$

or

$$\begin{aligned} A^{n+1} &= (a_{n-1}^2 - a_{n-2})A^{n-1} + (a_{n-1}a_{n-2} - a_{n-3})A^{n-2} \\ &\quad + \cdots + (a_{n-1}a_1 - a_0)A + (a_{n-1}a_0)I \end{aligned} \tag{10-44}$$

and so on for all higher-order powers of A. Thus *all powers* of A may be found as the linear combination of the n *basis matrices* $\{I, A, A^2, \ldots, A^{n-1}\}$.

Since *all powers* of A are linear combinations of these n basis matrices, it is thus true that an infinite power series of A is a linear combination of $\{I, A, A^2, \ldots, A^{n-1}\}$. Thus it is true that

$$\begin{aligned} e^{At} &= I + At + A^2\frac{t^2}{2!} + A^3\frac{t^3}{3!} + \cdots \\ &= \alpha_0(t)I + \alpha_1(t)A + \alpha_2(t)A^2 + \cdots + \alpha_{n-1}(t)A^{n-1} \end{aligned}$$

or

$$\boxed{e^{At} = \sum_{i=0}^{n-1} A^{i-1}\alpha_i(t)} \tag{10-45}$$

The problem is to find the proper scalar functions $\alpha_i(t)$ of this linear combination.

This would seem untenable; however, it is approached like so many previous problems of this type—form n linear equations for the n unknowns $\{\alpha_0(t), \alpha_1(t), \ldots, \alpha_{n-1}(t)\}$. To do this we go back to the fact that the eigenvalues also satisfy the characteristic polynomial. Thus for each λ_i

$$\lambda_i^n = -a_{n-1}\lambda_i^{n-1} - a_{n-2}\lambda_i^{n-2} - \cdots - a_1\lambda_i - a_0 \tag{10-46}$$

$$\lambda_i^{n+1} = (a_{n-1}^2 - a_{n-2})\lambda_i^{n-1} + (a_{n-1}a_{n-2} - a_{n-3})\lambda_i^{n-2} + \cdots + (a_{n-1}a_1 - a_0)\lambda_i + (a_{n-1}a_0) \tag{10-47}$$

and so on for λ_i^{n+2}, etc. But comparing Eqs. (10-43) and (10-44) with Eqs. (10-46) and (10-47), it is seen that the linear combinations for powers of A are the same as the linear combination relations for each λ_i. Therefore, for each λ_i and each t it is true that

$$e^{\lambda_i t} = \alpha_0(t) + \alpha_1(t)\lambda_i + \alpha_2(t)\lambda_i^2 + \cdots + \alpha_{n-1}(t)\lambda_i^{n-1} \tag{10-48}$$

If there are n distinct eigenvalues, Eq. (10-48) provides n equations in n unknowns:

$$\begin{bmatrix} 1 & \lambda_1 & \lambda_1^2 & \cdots & \lambda_1^{n-1} \\ 1 & \lambda_2 & \lambda_2^2 & \cdots & \lambda_2^{n-1} \\ \cdots & \cdots & \cdots & \cdots & \cdots \\ 1 & \lambda_n & \lambda_n^2 & \cdots & \lambda_n^{n-1} \end{bmatrix} \begin{bmatrix} \alpha_0(t) \\ \alpha_1(t) \\ \cdots \\ \alpha_{n-1}(t) \end{bmatrix} = \begin{bmatrix} e^{\lambda_1 t} \\ e^{\lambda_2 t} \\ \cdots \\ e^{\lambda t} \end{bmatrix} \tag{10-49}$$

This coefficient matrix is a special form termed a Vandermonde matrix. Special techniques may be used in solving a linear equation problem involving the Vandermonde matrix (e.g., Ref. 10).

If there are complex-conjugate-pair eigenvalues, then it is convenient to formulate the linear equation problem Eq. (10-49) in terms of the real and imaginary parts, thus giving a real-valued coefficient matrix. For example, if $\lambda_{1,2} = \sigma_1 \pm \omega_1 j$ are complex conjugate pairs, then the first two rows would be reformed

$$\begin{bmatrix} \mathrm{Re}\,\{1 & \lambda_1 & \lambda_1^2 & \cdots & \lambda_1^{n-1}\} \\ \mathrm{Im}\,\{1 & \lambda_1 & \lambda_1^2 & \cdots & \lambda_1^{n-1}\} \end{bmatrix} \begin{bmatrix} \alpha_0(t) \\ \alpha_1(t) \\ \vdots \\ \alpha_{n-1}(t) \end{bmatrix} = \begin{bmatrix} \mathrm{Re}\,\{e^{\lambda_1 t}\} \\ \mathrm{Im}\,\{e^{\lambda_1 t}\} \end{bmatrix} = \begin{bmatrix} e^{\sigma_1 t}\cos(\omega_1 t) \\ e^{\sigma_1 t}\sin(\omega_1 t) \end{bmatrix} \tag{10-50}$$

This modification makes the problem real-valued. Notice that the real and imaginary parts of the first two rows are taken *after* the complex powers λ_i^m, $m = 0, 1, 2, \ldots, m - 1$ are computed.

If a λ_i has a multiplicity greater than one, then the same basic procedure is used, only derivatives of the rows are taken a sufficient number of times to yield n independent equations. For example, if λ_1 had multiplicity 3, two derivatives with respect to λ_1 could be taken to yield the first three linear equation problems

$$\begin{bmatrix} 1 & \lambda_1 & \lambda_1^2 & \cdots & \lambda_1^{n-1} \\ 0 & 1 & 2\lambda_1 & \cdots & (n-1)\lambda^{n-2} \\ 0 & 0 & 2 & \cdots & (n-1)(n-2)\lambda^{n-3} \end{bmatrix} \begin{bmatrix} \alpha_0(t) \\ \alpha_1(t) \\ \alpha_2(t) \\ \vdots \\ \alpha_{n-1}(t) \end{bmatrix} = \begin{bmatrix} e^{\lambda_1 t} \\ te^{\lambda_1 t} \\ t^2 e^{\lambda_1 t} \end{bmatrix} \tag{10-51}$$

Note that both sides of the linear equation problem Eq. (10-51) are differentiated with respect to the repeated λ_i, producing terms of the form $e^{\lambda_i t}$, $te^{\lambda_i t}$, $t^2 e^{\lambda_i t}$, and so forth up to the multiplicity of λ_i.

Example 10-17 Let

$$A = \begin{bmatrix} 0 & 1 & 0 \\ 3 & 0 & 1 \\ -13 & -17 & -5 \end{bmatrix}$$

This is the same A as Examples 7-3, 10-12, 10-14, and 10-16. The eigenvalues are $\lambda_1 = -1$, $\lambda_{2,3} = -2 \pm 3j$. Hence the three linear equations for $\alpha_0(t)$, $\alpha_1(t)$, and $\alpha_2(t)$ are given by

$$\begin{bmatrix} 1 & -1 & 1 \\ \text{Re}\{1 & -2+3j & -5-12j\} \\ \text{Im}\{1 & -2+3j & -5-12j\} \end{bmatrix} \begin{bmatrix} \alpha_0(t) \\ \alpha_1(t) \\ \alpha_2(t) \end{bmatrix} = \begin{bmatrix} e^{-t} \\ \text{Re}\{e^{(-2+3j)t}\} \\ \text{Im}\{e^{(-2+3j)t}\} \end{bmatrix}$$

or

$$\begin{bmatrix} 1 & -1 & 1 \\ 1 & -2 & -5 \\ 0 & 3 & -12 \end{bmatrix} \begin{bmatrix} \alpha_0(t) \\ \alpha_1(t) \\ \alpha_2(t) \end{bmatrix} = \begin{bmatrix} e^{-t} \\ e^{-2t}\cos 3t \\ e^{-2t}\sin 3t \end{bmatrix}$$

Inverting the coefficient Vandermonde matrix gives

$$\begin{bmatrix} \alpha_0(t) \\ \alpha_1(t) \\ \alpha_2(t) \end{bmatrix} = \begin{bmatrix} 13/10 & -3/10 & 7/30 \\ 2/5 & -2/5 & 1/5 \\ 1/10 & -1/10 & -1/30 \end{bmatrix} \begin{bmatrix} e^{-t} \\ e^{-2t}\cos 3t \\ e^{-2t}\sin 3t \end{bmatrix}$$

Using this result gives

$$e^{At} = \alpha_0(t)I + \alpha_1(t)A + \alpha_2(t)A^2$$

where $\alpha_0(t)$, $\alpha_1(t)$, and $\alpha_2(t)$ are now solved from above. However, it is usually desirable to group terms of e^{At} into common factors of the basic time functions, or modes. This is easily done from the columns of the coefficient matrix inverse. Thus

$$\begin{aligned} e^{At} &= \left(\frac{13}{10}I + \frac{2}{5}A + \frac{1}{10}A^2\right)e^{-t} \\ &+ \left(-\frac{3}{10}I - \frac{2}{5}A - \frac{1}{10}A^2\right)e^{-2t}\cos 3t \\ &+ \left(\frac{7}{30}I + \frac{1}{5}A - \frac{1}{30}A^2\right)e^{-2t}\sin 2t \end{aligned}$$

Written in this form, the expression for e^{At} is seen to be equal to that obtained in Example 10-16 using the block-diagonal canonical form to compute e^{At}. By hand, this Cayley-Hamilton theorem method is the authors preference for finding e^{At}. On the computer it is roughly comparable to the block-diagonal eigenvalue and eigenvector method. Again, though, there is really no good method to determine e^{At} for continuous t (see Ref. 9).

10-12 SUMMARY

Quite a bit of linear algebra and matrix theory has been covered in this chapter. Of necessity, though, many topics have been omitted. However, the material that has been covered provides the essential background to understand the time-response characteristics of the linear dynamic system and, in particular, prepares us to now reexamine the issues of controllability and observability.

PROBLEMS

10-1 Find the eigenvalues for the following matrices:

(*a*) $A = \begin{bmatrix} 0 & 1 \\ -26 & -2 \end{bmatrix}$ (*b*) $A = \begin{bmatrix} 0 & -26 \\ 1 & -2 \end{bmatrix}$

(*c*) $A = \begin{bmatrix} -2 & 1 \\ -26 & 0 \end{bmatrix}$ (*d*) $A = \begin{bmatrix} 3 & 1 & 2 \\ 0 & 3 & 4 \\ 0 & 0 & 3 \end{bmatrix}$

(*e*) $A = \begin{bmatrix} 0 & 1 & 0 \\ 0 & 0 & 1 \\ 0 & 0 & 0 \end{bmatrix}$ (*f*) $A = \begin{bmatrix} 0 & 0 & 0 \\ 0 & 0 & 1 \\ 0 & -1 & 0 \end{bmatrix}$

10-2 Find the eigenvalues for the following matrices:

(*a*) $A = \begin{bmatrix} 10 & 0 & 0 \\ 0 & 5 & 4 \\ 0 & -4 & 5 \end{bmatrix}$ (*b*) $A = \begin{bmatrix} 2 & 3 \\ 3 & 2 \end{bmatrix}$

(c) $A = \begin{bmatrix} 2 & 3 \\ -3 & 2 \end{bmatrix}$ (d) $A = \begin{bmatrix} 4 & 0 & 0 \\ -1 & 5 & 0 \\ 2 & 6 & 7 \end{bmatrix}$

(e) $A = \begin{bmatrix} 1 & 0 & 0 \\ 1 & 1 & 0 \\ -2 & 3 & 1 \end{bmatrix}$ (f) $A = \begin{bmatrix} 1 & 0 & 0 \\ 5 & 1 & 0 \\ 0 & 7 & 1 \end{bmatrix}$

10-3 The two matrices A and $\tilde{A} = P^{-1}AP$ are related by what is called a similarity transformation. It is true that A and $\tilde{A}$ have the same eigenvalues. Prove this fact by showing that

$$\det[\lambda I - A] = \det[\lambda I - P^{-1}AP]$$

Hint: Start with $\det[\lambda I - P^{-1}AP]$. Then use $I = P^{-1}P$ to write this expression as $\det\{P^{-1}[\lambda I - A]P\}$. Then use the determinant identities (see Appendix B)

$$\det(AB) = \det A \cdot \det B \qquad \text{square } A \text{ and } B$$

$$\det(A^{-1}) = 1/\det A$$

to prove the desired result.

10-4 Let A and $\tilde{A}$ be related by a similarity transformation such that $\tilde{A} = P^{-1}AP$. Let λ_i and ξ_i be an eigenvalue and eigenvector pair of A. Using the results of Prob. 10-3, show that $\tilde{\xi}_i = P^{-1}\xi_i$ is an eigenvector for $\tilde{A}$. *Hint:* Start with the fact that $\tilde{A}\tilde{\xi}_i = \lambda_i\tilde{\xi}_i$ by definition of the eigenvalue and eigenvector linear equation problem. Then make substitutions for $\tilde{A}$.

10-5 Let A have eigenvalue λ_i with corresponding eigenvector ξ_i. Prove that λ_i^N and ξ_i are the eigenvalue and eigenvector of A^N for any positive scalar N. *Hint:* Start with the eigenvalue and eigenvector linear equation $A\xi_i = \lambda_i\xi_i$. Multiply both sides by A to show $A^2\xi_i = \lambda_i^2\xi_i$. Continue proof by induction.

10-6 Same conditions as Prob. 10-5. Assume A is nonsingular so that A^{-1} exists. Prove that the eigenvalues and eigenvectors of A^{-N} are $(\lambda_i)^{-N}$ and ξ_i for any positive scalar N. *Hint:* Start with the eigenvalue and eigenvector linear equation $A\xi_i = \lambda_i\xi_i$. Multiply both sides by A^{-1} to show that $A^{-1}\xi_i = (1/\lambda_i)\xi_i$. Then continue proof by induction.

10-7 Use elementary row operations to find the complete solution to the following linear equation problems. If there is no solution, state this fact.

(a) $\begin{bmatrix} 1 & 2 & 0 \\ 2 & 4 & -3 \\ 0 & 0 & 1 \end{bmatrix}\begin{bmatrix} x_1 \\ x_2 \\ x_3 \end{bmatrix} = \begin{bmatrix} -1 \\ 2 \\ 3 \end{bmatrix}$ (b) $\begin{bmatrix} 1 & 2 & -1 \\ 0 & 1 & 3 \\ 0 & 0 & 1 \end{bmatrix}\begin{bmatrix} x_1 \\ x_2 \\ x_3 \end{bmatrix} = \begin{bmatrix} 4 \\ 1 \\ 2 \end{bmatrix}$

(c) $\begin{bmatrix} 0 & 2 & 3 \\ 1 & 0 & 3 \\ 1 & 2 & 6 \end{bmatrix}\begin{bmatrix} x_1 \\ x_2 \\ x_3 \end{bmatrix} = \begin{bmatrix} 1 \\ 2 \\ 3 \end{bmatrix}$ (d) $\begin{bmatrix} 2 & -3 \\ 1 & 2 \\ 4 & 0 \end{bmatrix}\begin{bmatrix} x_1 \\ x_2 \end{bmatrix} = \begin{bmatrix} 1 \\ 4 \\ 8 \end{bmatrix}$

10-8 Repeat Prob. 10-7 for the following linear equation problems:

(a) $\begin{bmatrix} 2 & 0 & 6 \\ 1 & 1 & 7 \end{bmatrix}\begin{bmatrix} x_1 \\ x_2 \\ x_3 \end{bmatrix} = \begin{bmatrix} 6 \\ 7 \end{bmatrix}$ (b) $\begin{bmatrix} 0 & -1 & -3 & 4 \\ 0 & 1 & 3 & 0 \\ 0 & 0 & 0 & 2 \end{bmatrix}\begin{bmatrix} x_1 \\ x_2 \\ x_3 \\ x_4 \end{bmatrix} = \begin{bmatrix} 5 \\ -4 \\ 2 \end{bmatrix}$

(c) $\begin{bmatrix} 1 & 2 & 0 \\ -1 & 0 & 1 \\ 0 & 2 & 1 \\ 3 & 8 & 1 \end{bmatrix}\begin{bmatrix} x_1 \\ x_2 \\ x_3 \end{bmatrix} = \begin{bmatrix} -1 \\ 1 \\ 0 \\ -3 \end{bmatrix}$ (d) $\begin{bmatrix} 2 & 4 & 6 & 4 \\ 0 & 1 & 4 & 5 \end{bmatrix}\begin{bmatrix} x_1 \\ x_2 \\ x_3 \\ x_4 \end{bmatrix} = \begin{bmatrix} 6 \\ 7 \end{bmatrix}$

10-9 Find the rank, nullity, range space, and null space for the coefficient matrices of Prob. 10-7. Demonstrate that rank + nullity = column dimension.

10-10 Repeat Prob. 10-9 for the coefficient matrices of Prob. 10-8.

10-11 Demonstrate the validity of Proposition 10-3 for the linear equation problems of Prob. 10-7.

10-12 Repeat Prob. 10-11 for the linear equation problems of Prob. 10-8.

10-13 Using the method of elementary row operations, find the eigenvectors for the matrices of Prob. 10-1.

10-14 Repeat Prob. 10-13 for the matrices of Prob. 10-2.

10-15 Repeat Prob. 10-13 for the following matrices:

$$(a)\ A = \begin{bmatrix} 0 & 1 \\ -4 & -5 \end{bmatrix} \quad (b)\ A = \begin{bmatrix} 0 & 1 \\ 4 & 0 \end{bmatrix} \quad (c)\ A = \begin{bmatrix} 0 & 1 \\ 0 & -4 \end{bmatrix}$$

10-16 Repeat Prob. 10-13 for the following matrices:

$$(a)\ \begin{bmatrix} 0 & 1 \\ 6 & -1 \end{bmatrix} \quad (b)\ A = \begin{bmatrix} 0 & 1 \\ 6 & 0 \end{bmatrix} \quad (c)\ A = \begin{bmatrix} 0 & 1 \\ 0 & 6 \end{bmatrix}$$

10-17 Find the eigenvectors via use of Eq. (10-16) for the matrices of Prob. 10-15. Note that each of these matrices is in phase-variable form.

10-18 Find the eigenvectors via use of Eq. (10-16) for the matrices of Prob. 10-16. Note that each of these matrices is in phase-variable form.

10-19 What is the modal matrix, P, such that $P^{-1}AP$ is the diagonal matrix of eigenvalues for the matrices of Prob. 10-15?

10-20 Repeat Prob. 10-19 for the matrices of Prob. 10-16.

10-21 Transform the system

$$\dot{\mathbf{x}} = \begin{bmatrix} 0 & 1 \\ -10 & -2 \end{bmatrix}\mathbf{x} + \begin{bmatrix} 0 \\ K \end{bmatrix}u$$

$$y = [3 \quad 4]\mathbf{x}$$

to the diagonal and block-diagonal canonical forms by using the modal matrix P and P_b as applicable.

10-22 Repeat Prob. 10-21 for the system

$$\dot{\mathbf{x}} = \begin{bmatrix} 0 & 1 \\ -29 & -4 \end{bmatrix}\mathbf{x} + \begin{bmatrix} 0 \\ K \end{bmatrix}u$$

$$y = [-5 \quad 1]\mathbf{x}$$

10-23 Use the fact that

$$P^{-1}P = \begin{bmatrix} \boldsymbol{\eta}_1^T \\ \vdots \\ \boldsymbol{\eta}_n^T \end{bmatrix}_{n\times n} [\boldsymbol{\xi}_1, \quad \ldots, \quad \boldsymbol{\xi}_n]_{n\times n}$$

$$= [\boldsymbol{\eta}_i^T\boldsymbol{\xi}_j]_{n\times n}$$

to show that $\boldsymbol{\eta}_i^T\boldsymbol{\xi}_j = \begin{cases} 1 & \text{for } i = j \\ 0 & \text{for } i \neq j \end{cases}$

10-24 Let A have n independent eigenvectors so that $P^{-1}AP = \Lambda$ where $P = [\boldsymbol{\xi}_1, \boldsymbol{\xi}_1, \ldots, \boldsymbol{\xi}_n]$ and $\Lambda = \text{diagonal}\,[\lambda_1, \lambda_2, \ldots, \lambda_n]$.

(*a*) Show that $AP = P\Lambda$ where

$$AP = [A\boldsymbol{\xi}_1, \quad A\boldsymbol{\xi}_2, \quad \ldots, \quad A\boldsymbol{\xi}_n]$$

$$P\Lambda = [\lambda_1\boldsymbol{\xi}_1, \quad \lambda_2\boldsymbol{\xi}_2, \quad \ldots, \quad \lambda_n\boldsymbol{\xi}_n]$$

(*b*) Show that $P^{-1}A = \Lambda P^{-1}$ by using the results of (*a*).

(*c*) Use the matrix transpose property that $(AB)^T = B^T A^T$ (see Appendix B) to show that $A^T P^{-T} = P^{-T}\Lambda$ where P^{-T} is the transpose of P^{-1}.

(*d*) Letting $\boldsymbol{\eta}_i^T$ be the rows of P^{-1} ($\boldsymbol{\eta}_i$ are the columns of P^{-T}), use *a* and *c* to show that

$$A^T \boldsymbol{\eta}_i = \lambda_i \boldsymbol{\eta}_i \qquad i = 1, 2, \ldots, n$$

10-25 Every matrix is similar to its Jordan canonical form, $P^{-1}AP = J$, where J has the n eigenvalues of A down its diagonal and, perhaps, unity elements on its superdiagonal above blocks of repeated eigenvalues. Using the result of Prob. 10-3 that similar matrices have the same eigenvalues, show that $\det(A)$ is equal to the product of its n eigenvalues, including any repeated eigenvalues. *Hint:* What is the determinant of J?

10-26 Using the fact that

$$\det[\lambda I - A] = \prod_{i=1}^{n} (\lambda - \lambda_i)$$

where λ_i are the eigenvalues of A (including any repeated eigenvalues), show that $\det(-A) = (-1)^n \det(A)$. *Hint:* To what is $[\lambda I - A]$ at $\lambda = 0$ equal? Then use the result of Prob. 10-25.

10-27 The matrix

$$A = \begin{bmatrix} 0 & 1 & 0 \\ 0 & 0 & 1 \\ -1 & -3 & -3 \end{bmatrix}$$

has Jordan canonical form

$$J = \begin{bmatrix} -1 & 1 & 0 \\ 0 & -1 & 1 \\ 0 & 0 & -1 \end{bmatrix}$$

Find the matrix P such that $P^{-1}AP = J$.

10-28 Find linear equation problems for the generalized eigenvectors for

$$A = \begin{bmatrix} 0 & 1 & 0 & 0 & 0 \\ 0 & 0 & 1 & 0 & 0 \\ 0 & 0 & 0 & 1 & 0 \\ 0 & 0 & 0 & 0 & 1 \\ -9 & -33 & -46 & -30 & -9 \end{bmatrix}$$

with Jordan canonical form

$$J = \begin{bmatrix} -1 & 1 & 0 & 0 & 0 \\ 0 & -1 & 1 & 0 & 0 \\ 0 & 0 & -1 & 0 & 0 \\ 0 & 0 & 0 & -3 & 1 \\ 0 & 0 & 0 & 0 & -3 \end{bmatrix}$$

10-29 Find the orthogonal Q (i.e., $Q^T = Q^{-1}$) such that $Q^T A Q = \text{diagonal}\,[\lambda_1, \ldots, \lambda_n]$ for the following symmetric matrix:

$$A = \begin{bmatrix} -1 & 2 \\ 2 & -3 \end{bmatrix}$$

10-30 Repeat Prob. 10-29 for the following symmetric matrix: $A = \begin{bmatrix} -4 & 1 \\ 1 & 4 \end{bmatrix}$

10-31 Suppose that Q has the property that each of its columns, $\mathbf{q}_i$, $i = 1, 2, \ldots, n$, is orthogonal (i.e., $\mathbf{q}_i^T\mathbf{q}_j = 0$ for $i \neq j$) and has unit norm ($\|\mathbf{q}_i\|^2 \equiv \mathbf{q}_i^T\mathbf{q}_i = 1$). Show that $Q^T = Q^{-1}$; i.e., show that $Q^TQ = QQ^T = I$. *Hint:* Use matrix partitioning to show that $Q^TQ = I$. Then use the result that Q has orthonormal rows if it has orthonormal columns.

10-32 Use the power-series definition of e^{At} to show that $e^{\Lambda t} = \text{diagonal}\,[e^{\lambda_1 t}, e^{\lambda_2 t}, \ldots, e^{\lambda_n t}]$ where $\Lambda = \text{diagonal}\,[\lambda_1, \lambda_2, \ldots, \lambda_n]$.

10-33 What is e^{At} for $A = \text{diagonal}\,[-1, -2 + 3j, -2 - 3j]$?

10-34 What is e^{At} for $A = \text{diagonal}\,[-4 + 5j, -4 - 5j, 1]$?

10-35 Find e^{At} in the form of Eq. (10-28*b*) for the matrices of Prob. 10-15.

10-36 Find e^{At} in the form of Eq. (10-28*b*) for the matrices of Prob. 10-16.

10-37 A matrix has eigenvalues $\lambda_{1,2} = -1 \pm 2j$, $\lambda_{3,4} = -5 \pm 4j$, $\lambda_5 = -3$.
(*a*) What is the block-diagonal-form A matrix, Λ_b, for this system?
(*b*) What is $e^{\Lambda_b t}$?

10-38 Repeat Prob. 10-37 for a system with eigenvalues $\lambda_{1,2} = -0.1 \pm 10j$, $\lambda_3 = -1$, $\lambda_{4,5} = -2 \pm 7j$.

10-39 What is e^{At} for the matrix

$$A = \begin{bmatrix} -2 & 0 & 0 & 0 \\ 0 & -3 & 1 & 0 \\ 0 & 0 & -3 & 1 \\ 0 & 0 & 0 & -3 \end{bmatrix}$$

10-40 Repeat Prob. 10-39 for the matrix

$$A = \begin{bmatrix} -2 & 1 & 0 & 0 \\ 0 & -2 & 0 & 0 \\ 0 & 0 & -3 & 1 \\ 0 & 0 & 0 & -3 \end{bmatrix}$$

10-41 Suppose that A has n independent eigenvectors. Then it is true that $A = P\Lambda P^{-1}$ where $\Lambda = \text{diagonal}\,[\lambda_1, \lambda_2, \ldots, \lambda_n]$. Use this fact to show that A satisfies its own characteristic polynomial; i.e.,

$$\Delta(A) = A^n + a_{n-1}A^{n-1} + \cdots + a_0 I = 0$$

where

$$\Delta(\lambda) = \det\,[\lambda I - A] = \lambda^n + a_{n-1}\lambda^{n-1} + \cdots + a_0$$

Hint: Recall that $\Delta(\lambda_i) = 0$ for all λ_i, $i = 1, 2, \ldots, n$. Also, from Prob. 10-3 it is true that $\det\,[\lambda I - A] = \det\,[\lambda I - P\Lambda P^{-1}] = \det\,[\lambda I - \Lambda]$.

10-42 Using the Cayley-Hamilton theorem technique, find e^{At} in the form

$$e^{At} = \begin{bmatrix} \quad\quad \end{bmatrix} e^{\sigma t} \cos \omega t + \begin{bmatrix} \quad\quad \end{bmatrix} e^{\sigma t} \sin \omega t$$

for the matrix

$$A = \begin{bmatrix} -1 & 3 \\ -3 & -1 \end{bmatrix}$$

10-43 Repeat Prob. 10-42 for the matrix:

$$A = \begin{bmatrix} 0 & 1 \\ -10 & -2 \end{bmatrix}$$

10-44 Using the Cayley-Hamilton theorem technique, find e^{At} in the form

$$e^{At} = \left[\qquad \right] e^{\lambda_1 t} + \left[\qquad \right] e^{\sigma_2 t} \cos \omega_2 t$$

$$+ \left[\qquad \right] e^{\sigma_2 t} \sin \omega_2 t$$

for the matrix

$$A = \begin{bmatrix} 0 & 1 & 0 \\ 0 & 0 & 1 \\ -2 & -1 & -2 \end{bmatrix}$$

Note: A has eigenvalues $\lambda_1 = -2$, $\lambda_{2,3} = \pm j$.

10-45 Repeat Prob. 10-48 for the matrix

$$A = \begin{bmatrix} 0 & 1 & 0 \\ 0 & 0 & 1 \\ -5 & -7 & -3 \end{bmatrix}$$

Note: A has eigenvalues $\lambda_1 = -1$, $\lambda_{2,3} = -1 \pm 2j$.

10-46 Set up the linear equation problem that you would use to find $\alpha_0(t), \alpha_1(t), \ldots, \alpha_{n-1}(t)$, in the Cayley-Hamilton technique for computing e^{At} with A having the following eigenvalues. Set up the linear equation problem so that there are all real calculations required (see Eq. (10-50)).

(*a*) $\lambda_1 = -2, \quad \lambda_2 = -2, \quad \lambda_{3,4} = -1 \pm 3j.$

(*b*) $\lambda_{1,2} = -2 \pm j, \quad \lambda_{3,4} = -1 \pm 2j.$

10-47 Repeat Prob. 10-46 for the A matrices with eigenvalues

(*a*) $\lambda_{1,2} = 1 \pm 2j, \quad \lambda_3 = 0, \quad \lambda_4 = 0$

(*b*) $\lambda_{1,2} = \pm j, \quad \lambda_{3,4} = 1 \pm j$

10-48 Let $\mathbf{d}$ be an arbitrary n-dimensional vector such that

$$M \equiv [\mathbf{d}, \quad A\,\mathbf{d}, \quad \ldots, \quad A^{n-1}\,\mathbf{d}]_{n \times n}$$

is nonsingular. Use the Cayley-Hamilton theorem to show that solution to the linear equation problem

$$M\mathbf{x} = (A^n\,\mathbf{d})$$

is

$$\mathbf{x} = [-a_0, \quad -a_1, \quad \ldots, \quad -a_{n-1}]^T$$

where $a_0, a_1, \ldots, a_{n-1}$ are coefficients of the characteristic polynomial

$$\det [\lambda I - A] = \lambda^n + a_{n-1}\lambda^{n-1} + \cdots + a_0$$

Hint: Use the Cayley-Hamilton theorem to show that

$$A^n\,\mathbf{d} = -a_0\,\mathbf{d} - a_1 A\,\mathbf{d} - \cdots - a_{n-1}A^{n-1}\,\mathbf{d}$$

Then show that

$$M\mathbf{x} = x_1\,\mathbf{d} + x_2 A\,\mathbf{d} + \cdots + x_n A^{n-1}\,\mathbf{d}$$

where $x_1, x_2, \ldots, x_n$ are the n unknown elements of the vector $\mathbf{x}$. Equate terms in $M\mathbf{x} = A^n\,\mathbf{d}$ and you are done.

10-49 Use the results of Prob. 10-48 to find the characteristic polynomial coefficients a_0, a_1 for the matrix

$$A = \begin{bmatrix} -1 & 2 \\ -2 & -1 \end{bmatrix}, \qquad \text{pick } \mathbf{d} = \begin{bmatrix} 1 \\ 1 \end{bmatrix}$$

10-50 Repeat Prob. 10-49 for the matrix

$$A = \begin{bmatrix} -3/2 & 1/2 \\ -3 & 1 \end{bmatrix}, \qquad \text{pick } \mathbf{d} = \begin{bmatrix} 1 \\ 1 \end{bmatrix}$$

10-51 Using the Cayley-Hamilton theorem, we may write the matrix function sin (At) as

$$\sin(At) = \sum_{i=0}^{n-1} A^i \beta_i(t)$$

Assume that A has n independent eigenvalues. Find a linear equation problem for finding the n scalar functions $\beta_i(t)$, $i = 0, 1, \ldots, n-1$.

10-52 Suppose that A has n independent eigenvectors so that $P^{-1}AP = \Lambda$, where Λ is the diagonal matrix of eigenvalues. Use the power-series definition of sin (λt) to show that

$$\sin(At) = P \cdot \text{diagonal}\,[\sin(\lambda_1 t), \quad \sin(\lambda_2 t), \quad \ldots, \quad \sin(\lambda_n t)]P^{-1}$$

NOTES AND REFERENCES

The hand-analysis method to use the modified Hermite normal form to solve the general linear-equation problem and find the null space by inspection as presented in Sec. 10-2 is original jointly to the author and Capt. John Wheeler. However, the reader must be cautioned that this is not a practical digital computation method for solving the linear equation problem or the eigenvalue and eigenvector problem. For theory, the reader should consult Ref. 1. For solving the linear equation problem on the digital computer the reader should utilize expertly written software packages such as Ref. 2 or Ref. 3. Software for the general eigenvalue and eigenvector problem can be found in packages such as Refs. 3, 5, or 6. For practical computer methods to determine rank, range space, and null space, see Appendix E.

For greater theory on the Jordan canonical form, the reader should consult Ref. 1 or Ref. 7. However, the reader should be cautioned of computational pitfalls that can arise with the Jordan canonical form, and Ref. 8 provides an in-depth analysis of these problems. There are equally as many computational pitfalls in finding the continuous e^{At}, and Ref. 9 provides an excellent analysis of this problem.

1. Stewart, G. W.: *Introduction to Matrix Computations,* Academic, New York, 1974.
2. Dongarra, J. J., et al.: *LINPACK User's Guide,* SIAM, Philadelphia, 1979.
3. *IMSL Reference Manual,* IMSL Inc., Houston 1979.
4. Wilkinson, J. H.: The Algebraic *Eigenvalue Problem,* Oxford University Press, New York, 1965.
5. Smith, et al.: *Matrix Eigensystem Routine: EISPACK Guide,* Springer-Verlag, Berlin, 1976.
6. Garbow, et al.: *Matrix Eigensystem Routine: EISPACK Guide Extension,* Springer-Verlag, Berlin, 1977.
7. Nering, E. D.: *Linear Algebra and Matrix Theory,* 2d ed., Wiley, New York, 1970.
8. Golub, G. H., and Wilkinson: "Ill Conditioned Eigensystems and the Computation of the Jordan Canonical Form", *SIAM Review,* vol. 18 no. 4, October 1976, pp. 578–619.
9. Moler, C., and C. Van Loan: "Nineteen Dubious Ways to Compute the Exponential of a Matrix," *SIAM Review,* vol. 20 no. 4, Oct. 1978, pp. 801–836.
10. Csaki, F. G.: "Some Notes on the Inversion of Confluent Vandermonde Matrices" *IEEE Trans. Auto. Control,* AC-20, February 1975, pp. 154–157.
11. Laub, A. J.: "Efficient Multivariable Frequency Response Computations," *Proc. 1980 IEEE Conf. Dec. Cntrl., San Diego,* December 1980, pp. 39–41.

CHAPTER

ELEVEN

CONTROLLABILITY AND OBSERVABILITY

Armed now with the tools of linear algebra and matrix theory developed in Chaps. 9 and 10, we may once again consider the state-space system properties of controllability and observability. Controllability is introduced in Chap. 8 in conjunction with solving the discrete time-control problem for a set of piecewise constant, sample-and-hold inputs. Observability is introduced in Chap. 8 in conjunction with solving the initial-state estimation problem from a set of sampled output measurements. Using the results of Chap. 10 these concepts have direct extension to the continuous-time state-space model.

The topics of controllability and observability are at the heart of modern state-space estimation and control theory. State estimation can be an important adjunct to modern state-variable feedback control methods; if the complete state vector is not directly measured, which generally it would not be, the current state of a system must be estimated from available output measurements to provide the necessary state feedback control. Such estimation and control-theory techniques are beyond our scope, but this chapter provides a brief introduction to the controllability and observability properties that are so important in these techniques. The interested reader is referred to Refs. 1 through 4, for example, for treatment of the more advanced control-theory topics.

11-1 CONTROLLABILITY ANALYZED VIA APPLYING THE CAYLEY-HAMILTON THEOREM

Consider the state model

$$\dot{\mathbf{x}}(t) = A\mathbf{x}(t) + B\mathbf{u}(t) \tag{11-1}$$

$$\mathbf{y}(t) = C\mathbf{x}(t) + D\mathbf{u}(t) \tag{11-2}$$

The controllability question regards the following: Starting from the zero initial state, what parts of state space can be reached by using some control input function $\mathbf{u}(t)$, $t \in (0, t_f]$, finite t_f? Those states which can be reached are called *controllable states*. Those states that cannot be reached (in the vernacular, you just can't get there) are called *uncontrollable states*. If it is possible for the system to reach any state in state space, $\mathbf{x} \in R^n$, then the system is *completely controllable*. If it is not possible to do so, the system is said to be *uncontrollable;* starting from the zero initial state there are some regions of state space that cannot be reached no matter what the control input.

The Cayley-Hamilton theorem of Sec. 10-11 may be used to analyze this controllability issue for the continuous-time system. It is convenient to first consider the single-input case. Then denote $B = \mathbf{b}$, a single n-dimensional column vector.

For the single input system the zero-state response is given by

$$\mathbf{x}_{zs}(t) = \int_0^t e^{A(t-\tau)}\mathbf{b}u(\tau)\, d\tau \tag{11-3a}$$

$$= \int_0^t e^{-A\xi}\mathbf{b}u(t - \xi)\, d\xi \tag{11-3b}$$

Using the Cayley-Hamilton theorem we find it true that

$$e^{At} = \sum_{i=0}^{n-1} A^i \alpha_i(t) \tag{11-4}$$

where the $\alpha_i(t)$, $i = 0, 1, \ldots, n - 1$, are linearly independent scalar functions. Their determination is made from a Vandermonde linear equation problem as described in Sec. 10-11. However, it is not really of concern here just what values the $\alpha_i(t)$ functions actually have. The main concern is that the form Eq. (11-4) is *true* regardless of eigenvalue multiplicities or anything else. Furthermore, the $\alpha_i(t)$ functions are linearly independent; that is, the only linear combination of the $\alpha_i(t)$ functions that is identically zero for all t is the zero linear combination.

Using Eq. (11-4) in Eq. (11-3b) gives

$$\mathbf{x}_{zs}(t) = \sum_{i=0}^{n-1} (A_i\mathbf{b}) \int_0^t \alpha_i(\xi)u(t - \xi)\, d\xi \tag{11-5}$$

But for each t the multiplier

$$\beta_i(t) \equiv \int_0^t \alpha_i(\xi)u(t - \xi)\, d\xi \tag{11-6}$$

is just a *scalar* dependent on the input $u(\tau)$, $0 < \tau \leq t$. Also for each $i = 0, 1, \ldots, n - 1$, the factor $(A^i\mathbf{b})$ is an n-dimensional vector in state space. Therefore, a way to interpret Eq. (11-5) is to say that $\mathbf{x}_{zs}(t)$ is a linear combination of the n vectors

$(A^i\mathbf{b})$, $i = 0, 1, \ldots, n$. *All* linear combinations of these n vectors is the range space of the matrix formed from the $(A^i\mathbf{b})$ column vectors

$$M_c = [\mathbf{b}, \quad A\mathbf{b}, \quad \ldots, \quad A^{n-1}\mathbf{b}]_{n\times n} \tag{11-7}$$

Thus Eq. (11-4) is equivalent to saying

$$\mathbf{x}_{zs}(t) \in \mathcal{R}\{M_c\} \tag{11-8}$$

The range space of M_c is, in fact, the controllable subspace of the system Eq. (11-1). If a state $\mathbf{x}$ is not in the range space of M_c, it is not a linear combination of the $(A^i\mathbf{b})$ vectors. By Eq. (11-5) it is impossible for $\mathbf{x}_{zs}(t)$ to ever equal $\mathbf{x}$ regardless of the system input function $u(\tau)$, $0 < \tau \leq t$. The $\mathbf{x}$ state is, therefore, an uncontrollable state. From an initial rest condition the system can never reach states that are outside the range space of M_c, the controllable subspace.

The controllable subspace is the range space of M_c. A basis for the controllable subspace may be found by performing elementary row operations on M_c^T; then $\mathcal{R}\{M_c\}$, the controllable subspace, is the span of the nonzero rows of the resulting Hermite normal form of M_c^T. The uncontrollable subspace is the null space of M_c^T. This null space may also be found by performing elementary row operations on M_c^T as discussed in Sec. 10-3.* These results are illustrated in the example to follow.

The system is completely controllable if and only if the rank of M_c is n. If the rank of M_c is less than n, the system is uncontrollable.

Example 11-1 In Example 6-7, Sec. 6-4, a three-state physical-variable model is derived for an *RLC* network:

$$\dot{\mathbf{x}} = \begin{bmatrix} -5/2 & 0 & 1/2 \\ 0 & -5/2 & -1/2 \\ 0 & 2 & 0 \end{bmatrix} \mathbf{x} + \begin{bmatrix} 1/4 \\ -1/4 \\ 1 \end{bmatrix} u$$

$$y = [0 \quad 0 \quad 1]\mathbf{x}$$

However, analysis of the transfer function shows that this system has only a single pole

$$H(s) = C[sI - A]^{-1}B = \frac{1}{s + 1/2}$$

and thus could be represented by a single state model.

Consider now the controllability matrix for this system:

$$M_c = [\mathbf{b}, \quad A\mathbf{b}, \quad A^2\mathbf{b}]$$

$$= \begin{bmatrix} 1/4 & -1/8 & 1/16 \\ -1/4 & 1/8 & -1/16 \\ 1 & -1/2 & 1/4 \end{bmatrix}$$

*On the computer it is best to determine the rank, range space of M_c, and null space of M_c^T by employing singular value analysis as described in Appendix E.

The rank of M_c is seen to be one; the second row of M_c is -1 times the first, and the third row is 4 times the first. There is one controllable direction in state space, and the uncontrollable subspace is two-dimensional.

The controllable and uncontrollable subspaces are found by performing elementary row operations on M_c^T. Namely,

$$\begin{bmatrix} 1/4 & -1/4 & 1 \\ -1/8 & 1/8 & -1/2 \\ 1/16 & -1/16 & 1/4 \end{bmatrix} \sim \begin{bmatrix} 1 & -1 & 4 \\ 0 & 0 & 0 \\ 0 & 0 & 0 \end{bmatrix}$$

The null space of M_c^T is the uncontrollable subspace

$$\text{Uncontrollable subspace} = \mathfrak{N}\{M_c^T\} = \text{span}\left\{\begin{bmatrix} -1 \\ -1 \\ 0 \end{bmatrix} \begin{bmatrix} 4 \\ 0 \\ -1 \end{bmatrix}\right\}$$

The controllable subspace is the span of the nonzero rows of M_c^T in its Hermite normal form

$$\text{Controllable subspace} \equiv \mathcal{R}\{M_c\} = \text{span}\left\{\begin{bmatrix} 1 \\ -1 \\ 4 \end{bmatrix}\right\}$$

Controllability analysis in this example reveals that there are two redundant states and thus that the three-state model could be reduced to a single-state model. We would not know from this analysis just how to reduce the original state model—we only know that it can be reduced. Later sections introduce methods to accomplish this reduction. It should also be remarked that the methods of Appendix E provide a much more reliable method, in the general case, for the digital computation of the rank, range space, and null space of a given matrix.

Generalization to the Multi-Input System

The previous results may be generalized to the multi-input state model

$$\dot{\mathbf{x}}(t) = A\mathbf{x}(t) + B\mathbf{u}(t) \tag{11-1}$$

where $\mathbf{u}(t)$ is an ℓ-dimensional input vector. Since

$$B = [\mathbf{b}_1, \quad \mathbf{b}_2, \quad \ldots, \quad \mathbf{b}_\ell]_{n\times\ell}$$

the previous analysis with the Cayley-Hamilton theorem may be applied by considering one column of B at a time:

$$\mathbf{x}_{zs}(t) = \text{linear combination } \{\mathbf{b}_1, \quad A\mathbf{b}_1, \quad \ldots, \quad A^{n-1}\mathbf{b}_1, \quad \ldots, \quad \mathbf{b}_\ell, \quad A\mathbf{b}_\ell, \quad \ldots, \quad A^{n-1}\mathbf{b}_\ell\}$$

By regrouping it is seen that

$$\mathbf{x}_{zs}(t) \in \mathcal{R}\{M_c\} \tag{11-9}$$

where $$M_c \equiv [B, \quad AB, \quad \ldots, \quad A^{n-1}B]_{n\times n\ell} \tag{11-10}$$

The condition for the complete controllability of the system Eq. (11-1) is that the rank of M_c be n. This is the same condition as for the discrete time system as studied in Chap. 8.

Example 11-2 Consider

$$H(s) = \left[\frac{6s+5}{s^2+3s+2}, \quad \frac{2K}{s+1}\right]$$

a two-input, single-output system. Using a phase-variable state-space model for each subtransfer function gives the three-state model

$$\dot{\mathbf{x}} = \begin{bmatrix} 0 & 1 & 0 \\ -2 & -3 & 0 \\ 0 & 0 & -1 \end{bmatrix} \mathbf{x} + \begin{bmatrix} 0 & 0 \\ 1 & 0 \\ 0 & K \end{bmatrix} \begin{bmatrix} u_1 \\ u_2 \end{bmatrix}$$

$$y = [5 \quad 6 \quad 2]\mathbf{x}$$

The controllability matrix is

$$M_c = \begin{bmatrix} 0 & 0 & 1 & 0 & -3 & 0 \\ 1 & 0 & -3 & 0 & 7 & 0 \\ 0 & K & 0 & -K & 0 & K \end{bmatrix}$$

The first three columns of M_c are linearly independent. Thus the rank of M_c is 3, and the state model is completely controllable.

From this example it is noticed that the last two columns—the (A^2B) columns—are not necessary for the determination of controllability. It is true that

$$\text{rank}\,[B, \quad AB, \quad A^2B] = \text{rank}\,[B, \quad AB]$$

For the general nth-order, ℓ-input system with rank $(B) = \ell$

$$\text{rank}\,[B, \quad AB, \quad \ldots, \quad A^{n-1}B] = \text{rank}\,[B, \quad AB, \quad \ldots, \quad A^{n-\ell}B] \tag{11-11}$$

Therefore, instead of Eq. (11-10) we may use as the definition of M_c

$$\boxed{M_c \equiv [B, \quad AB, \quad \ldots, \quad A^{n-\ell}B]_{n\times(n\ell-\ell^2+\ell)}} \tag{11-12}$$

assuming that the rank of B is ℓ. Note that Eq. (11-12) also applies to the single-input system with $\ell = 1$.

The three-state, two-input model in Example 11-2 may be modeled by a two-state, two-input model

$$\begin{bmatrix} \dot{x}_1(t) \\ \dot{x}_2(t) \end{bmatrix} = \begin{bmatrix} -1 & 0 \\ 0 & -2 \end{bmatrix} \begin{bmatrix} x_1(t) \\ x_2(t) \end{bmatrix} + \begin{bmatrix} -1 & 2K \\ 7 & 0 \end{bmatrix} \begin{bmatrix} u_1(t) \\ u_2(t) \end{bmatrix}$$

$$y(t) = [1 \quad 1] \begin{bmatrix} x_1(t) \\ x_2(t) \end{bmatrix}$$

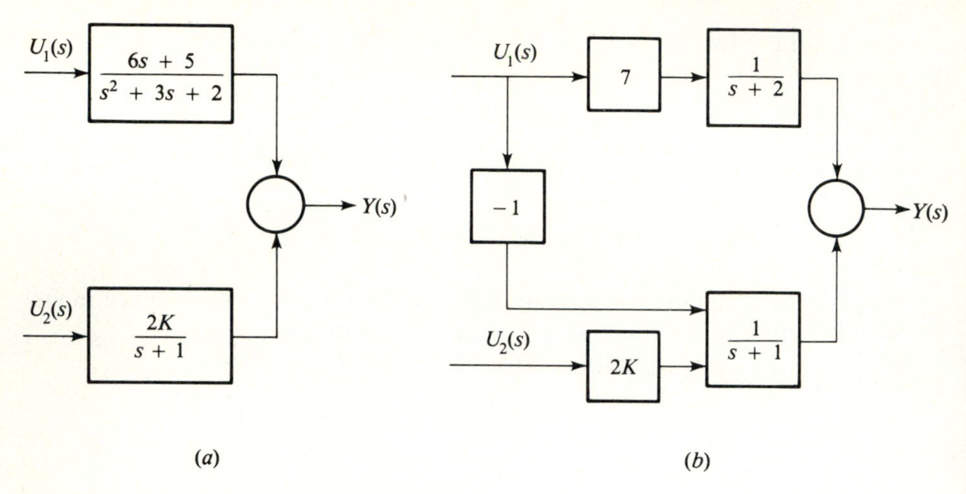

Figure 11-1 Simulation diagram, Example 11-2.

This state model may be derived by first expanding $H(s)$ in partial fraction expansions:

$$H(s) = \left[\frac{-1}{s+1} + \frac{7}{s+2} \quad \frac{2K}{s+1}\right]$$

Then a simulation diagram is used to derive the two-state model. The simulation diagram for this is shown in Fig. 11-1. But the three-state model is completely controllable. Controllability alone does not give a clue as to the reducibility of this particular model.

We next consider state observability, the *dual* property to controllability. Perhaps observability will indicate whether or not the Example 11-2 state model is reducible.

11-2 OBSERVABILITY ANALYZED VIA THE CAYLEY-HAMILTON THEOREM

To examine the issue of observability of the state-space model, Eqs. (11-1) and (11-2), we analyze the system zero-input response:

$$\mathbf{y}_{\text{zi}}(t) = Ce^{At}\mathbf{x}(0) \qquad t \geq 0 \tag{11-13}$$

The observability issue deals with the following question: Are there any states, $\mathbf{x}_{uo} \in R^n$, $\mathbf{x}_{uo} \neq \mathbf{0}$, such that the zero-input response starting from these initial states is identically zero? If so, then these initial states never contribute anything to the system response. They can never be identified or estimated because they never influence the output. They are said to be *unobservable* states. If the system has no unobservable states, the system is said to be *completely observable*. If there are unobservable states, the system is said to be *unobservable*.

The unobservable states, $\mathbf{x}_{uo}$, are in the null space of the linear mapping (Ce^{At}) since

$$\mathbf{y}_{zi}(t) = Ce^{At}\mathbf{x}_{uo} \equiv \mathbf{0} \qquad t \geq 0 \tag{11-14a}$$

The Cayley-Hamilton theorem may be used to help determine this null space. By the Cayley-Hamilton theorem

$$e^{At} = \sum_{i=0}^{n-1} A^i \alpha_i(t) \tag{11-4}$$

where the $\alpha_i(t)$ are linearly independent over any finite interval. This may be used in Eq. (11-14*a*) to write

$$\mathbf{y}_{zi}(t) = \sum_{i=0}^{n-1} CA^i \mathbf{x}_{uo}\alpha_i(t) = \mathbf{0} \qquad t \geq 0 \tag{11-14b}$$

Since the $\alpha_i(t)$ are linearly independent scalar functions, Eq. (11-14*b*) is zero if and only if

$$CA^i\mathbf{x}_{uo} = \mathbf{0} \qquad i = 0, 1, \ldots, n-1 \tag{11-15}$$

The n conditions represented by Eq. (11-15) may be written as a partitioned matrix equation:

$$\begin{bmatrix} C\mathbf{x}_{uo} \\ CA\mathbf{x}_{uo} \\ \vdots \\ CA^{n-1}\mathbf{x}_{uo} \end{bmatrix}_{mn\times 1} = \begin{bmatrix} \mathbf{0} \\ \mathbf{0} \\ \vdots \\ \mathbf{0} \end{bmatrix}_{mn\times 1} \tag{11-16a}$$

The unobservable initial state, x_{uo}, may be factored from the expression on the left to give

$$\begin{bmatrix} C \\ CA \\ \vdots \\ CA^{n-1} \end{bmatrix}_{mn\times n} \mathbf{x}_{uo} = \begin{bmatrix} \mathbf{0} \\ \mathbf{0} \\ \vdots \\ \mathbf{0} \end{bmatrix}_{mn\times 1} \tag{11-16b}$$

This expression is interpreted as saying that the unobservable state $\mathbf{x}_{uo}$ is a member of null space of the $mn \times n$ *observability matrix*

$$M_o \equiv \begin{bmatrix} C \\ CA \\ \vdots \\ CA^{n-1} \end{bmatrix}_{mn\times n} \tag{11-17}$$

If rank M_o is n, the observability matrix M_o has no null space and the system is completely observable. If not, the system is unobservable. The unobservable subspace is, in fact, the null space of M_o. Any initial state in the unobservable subspace would produce an identically zero, zero-input response.

Like the controllability matrix, the observability matrix may be constructed with only $(n-m)$ powers of A provided that the rank of C is m; therefore, because

$$\text{rank}\begin{bmatrix} C \\ CA \\ \vdots \\ CA^{n-m} \end{bmatrix}_{m(n-m+1)\times n} = \text{rank}\begin{bmatrix} C \\ CA \\ \vdots \\ CA^{n-1} \end{bmatrix}_{mn\times n} \tag{11-18}$$

it is convenient to use

$$\boxed{M_o \equiv \begin{bmatrix} C \\ CA \\ \vdots \\ CA^{n-m} \end{bmatrix}} \tag{11-19}$$

as the definition of M_o rather than Eq. (11-17) when the rank of C is m.

Example 11-3 Consider the same three-state system as Example 11-1:

$$\dot{\mathbf{x}} = \begin{bmatrix} -5/2 & 0 & 1/2 \\ 0 & -5/2 & -1/2 \\ 0 & 2 & 0 \end{bmatrix}\mathbf{x} + \begin{bmatrix} 1/4 \\ -1/4 \\ 1 \end{bmatrix}u$$

$$y = [0 \quad 0 \quad 1]\mathbf{x}$$

The observability matrix is

$$M_o = \begin{bmatrix} 0 & 0 & 1 \\ 0 & 2 & 0 \\ 0 & -5 & -1 \end{bmatrix}$$

The rank of M_o is 2, since M_o has only two linearly independent columns. Elementary row operations allow determination of the null space of M_o (the unobservable subspace):

$$\mathfrak{N}\{M_o\} = \text{span}\left\{\begin{bmatrix} -1 \\ 0 \\ 0 \end{bmatrix}\right\} = \text{span}\left\{\begin{bmatrix} 1 \\ 0 \\ 0 \end{bmatrix}\right\}$$

The x_1 state element is unobservable. From Example 6-7, $x_1 = i_1 - i_{in}/2$ and the output is $x_3 = v_3$ (see Fig. 11-2). The current i_1 does not influence the voltage drop across the capacitor, C. Thus, by measuring the voltage across C, we can never infer the current i_1.

Example 11-4 Consider the same system as Example 11-2:

$$\dot{\mathbf{x}} = \begin{bmatrix} 0 & 1 & 0 \\ -2 & -3 & 0 \\ 0 & 0 & -1 \end{bmatrix}\mathbf{x} + \begin{bmatrix} 0 & 0 \\ 1 & 0 \\ 0 & K \end{bmatrix}\begin{bmatrix} u_1 \\ u_2 \end{bmatrix}$$

$$y = [5 \quad 6 \quad 2]\mathbf{x}$$

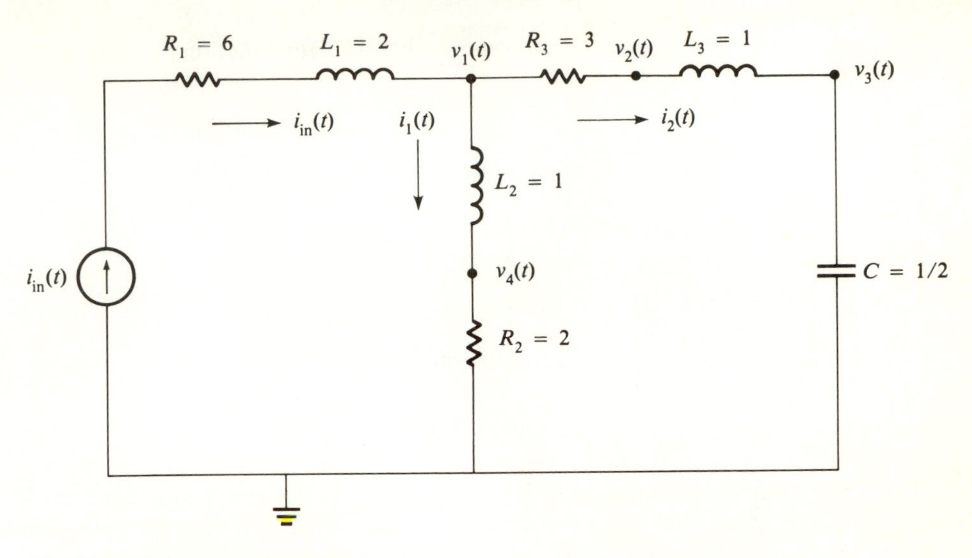

Figure 11-2 Circuit of Example 6-7.

The observability matrix is

$$M_o = \begin{bmatrix} C \\ CA \\ CA^2 \end{bmatrix} = \begin{bmatrix} 5 & 6 & 2 \\ -12 & -13 & -2 \\ 26 & 27 & 2 \end{bmatrix}$$

The rank of M_o is 2; M_o is singular. The system is unobservable.

This three-state, two-input system may be represented by a two-state model

$$\begin{bmatrix} \dot{x}_1 \\ \dot{x}_2 \end{bmatrix} = \begin{bmatrix} -1 & 0 \\ 0 & -2 \end{bmatrix} \begin{bmatrix} x_1 \\ x_2 \end{bmatrix} + \begin{bmatrix} -1 & 2K \\ 7 & 0 \end{bmatrix} \begin{bmatrix} u_1 \\ u_2 \end{bmatrix}$$

$$y = [1 \quad 1] \begin{bmatrix} x_1 \\ x_2 \end{bmatrix}$$

as derived previously by partial fraction expansion analysis. Controllability and observability analysis reveals this, as it did in the SISO case.

Controllability, observability, and model reduction are complex topics for the MIMO situation. See, for example, Ref. 8 for further discussion.

11-3 CONTROLLABILITY, OBSERVABILITY, AND THE TRANSFER FUNCTION FOR THE SISO SYSTEM

Controllability and observability for a system are not intrinsic properties, as they are dependent on the state realization chosen. A state model might be completely controllable and unobservable in one coordinate system, but uncontrollable and completely observable in another.

Example 11-5 Consider the system with transfer function

$$H(s) = \frac{s+2}{s^2+3s+2} = \frac{s+2}{(s+1)(s+2)}$$

The phase-variable (controllability) canonical form is

$$\dot{\mathbf{x}} = \begin{bmatrix} 0 & 1 \\ -2 & -3 \end{bmatrix}\mathbf{x} + \begin{bmatrix} 0 \\ 1 \end{bmatrix}u$$

$$y = [2 \quad 1]\mathbf{x}$$

This state model is completely controllable, since

$$M_c = \begin{bmatrix} 0 & 1 \\ 1 & -3 \end{bmatrix}$$

has rank 2. In phase-variable form for the SISO system, the controllability matrix always has this lower-triangular form. Thus it always has full rank and is completely controllable.

This model is unobservable, though, since

$$M_o = \begin{bmatrix} 2 & 1 \\ -2 & -1 \end{bmatrix}$$

has rank one; M_o is singular. This corresponds to the fact that $H(s)$ has a pole-zero cancellation and is thus reducible.

The dual-state coordinate system*, the observability canonical form or states derived by the simulation diagram, has just the opposite properties. This state model is

$$\dot{\mathbf{x}}' = \begin{bmatrix} 0 & -2 \\ 1 & -3 \end{bmatrix}\mathbf{x}' + \begin{bmatrix} 2 \\ 1 \end{bmatrix}u$$

$$y = [0 \quad 1]\mathbf{x}'$$

The controllability and observability matrices are

$$M_c' = \begin{bmatrix} 2 & -2 \\ 1 & -1 \end{bmatrix} \qquad M_o' = \begin{bmatrix} 0 & 1 \\ 1 & -3 \end{bmatrix}$$

with ranks 1 and 2, respectively. Hence the observability canonical form is uncontrollable and completely observable. For the SISO system the observability canonical form is always completely observable.

*These state models are said to be *dual* because

$$A' = A^T \qquad B' = C^T \qquad C' = B^T$$

Any two state models with such a property are termed *duals* to each other. As seen, $M_o' = M_c^T$ and $M_c' = M_o^T$ for dual-state models. Dual-state models have other interesting estimation and control properties, the discussion of which is beyond our scope. See, for example, Ref. 2.

For the SISO system, uncontrollability and unobservability taken together indicate a reducibility of the transfer function (a pole-zero cancellation) if there is one. This is regardless of the state coordinate system; but the properties of observability and controllability must both be considered, and not just one or the other.

Theorem 11-1 The n-state model for an nth-order SISO transfer function, $H(s)$, is completely controllable and completely observable if and only if $H(s)$ is irreducible; i.e., it has no pole-zero cancellations.

Because a state model has uncontrollable and/or unobservable states does not necessarily mean that these uncontrollable and unobservable states should not be retained in the state model. When deriving the physical-variable state model, uncontrollable or unobservable states might represent important internal dynamics. Some of these internal dynamics may be uncontrollable or unobservable simply because of the output variable chosen or because of parameter values which cause a pole-zero cancellation. These internal dynamics are really there. A change in the input variable, the output variable, or the model parameter values might provide the controllability and/or observability of these states; the corresponding poles would then appear in the transfer function without pole-zero cancellation.

Example 11-6 The three-state model used in Examples 11-1 and 11-3 is a physical-variable state model of an RLC circuit which was derived in Example 6-7. See Fig. 11-2. The three-state model could be modeled with a single-state variable and still retain the same input-output properties as the three-state model. This fact was revealed by the controllability analysis of Example 11-1.

One state variable was intentionally made redundant by the method of derivation; the two inductors L_2 and L_3 are not independent, and so they yield only a single state. The other state (pole) is eliminated because of pole-zero cancellation. Appendix A shows that the general transfer function of the circuit is

$$H(s) = \frac{V_3(s)}{I_{\text{in}}(s)} = \frac{L_2 s + R_2}{C(L_2 + L_3)s^2 + C(R_2 + R_3)s + 1}$$

By the choice of parameter values $L_2 = L_3 = 1$, $C = 1/2$, $R_2 = 2$, $R_3 = 3$ this transfer function is

$$H(s) = \frac{s + 2}{s^2 + 5/2s + 1} = \frac{s + 2}{(s + 2)(s + 1/2)}$$

and there is pole-zero cancellation of the pole at -2.

If the parameter values of the circuit are changed, such a pole-zero cancellation may not occur. The system would then require a two-state model.

For example, suppose $R_3 = 4$ instead of 3. The transfer function is

$$H(s) = \frac{s+2}{s^2+3s+1} = \frac{s+2}{\left(s + \frac{3+\sqrt{5}}{2}\right)\left(\frac{3-\sqrt{5}}{2}\right)}$$

with no pole-zero cancellation.

By methods of Example 6-7 the corresponding physical-variable state model is

$$\dot{\mathbf{x}} = \begin{bmatrix} -3 & 0 & 1/2 \\ 0 & -3 & -1/2 \\ 0 & 2 & 0 \end{bmatrix} \mathbf{x} + \begin{bmatrix} 1/2 \\ -1/2 \\ 1 \end{bmatrix} u$$

$$y = [0 \quad 0 \quad 1]\mathbf{x}$$

Forming the controllability and observability matrices:

$$M_c = \begin{bmatrix} 1/2 & -1 & 5/2 \\ -1/2 & 1 & -5/2 \\ 1 & -1 & 2 \end{bmatrix} \qquad M_o = \begin{bmatrix} 0 & 0 & 1 \\ 0 & 2 & 0 \\ 0 & -6 & -1 \end{bmatrix}$$

They both have rank 2; the three-state model may be reduced by one state agreeing with the transfer-function analysis.

11-4 KALMAN'S CANONICAL DECOMPOSITION OF STATE SPACE

Theorem 11-1 says that uncontrollability and/or unobservability of the state model indicates that the state model is reducible without sacrificing any of the input-output transfer-function properties of the system.

But can the uncontrollable and/or unobservable states be separated so that the model can be reduced by elimination of these states? Kalman [7] has shown that it is always possible to transform to a state coordinate system, $\tilde{\mathbf{x}}$, such that the state $\tilde{\mathbf{x}}$ is separated as

$$\tilde{\mathbf{x}} = \begin{bmatrix} \tilde{\mathbf{x}}_{cc/co} \\ \tilde{\mathbf{x}}_{uc/co} \\ \tilde{\mathbf{x}}_{cc/uo} \\ \tilde{\mathbf{x}}_{uc/uo} \end{bmatrix}_{n\times 1} \tag{11-25}$$

where $\tilde{\mathbf{x}}_{cc/co} \equiv$ the completely controllable and completely observable states

$\tilde{\mathbf{x}}_{uc/co} \equiv$ the uncontrollable, but completely observable states

$\tilde{\mathbf{x}}_{cc/uo} \equiv$ the unobservable, but completely controllable states

$\tilde{\mathbf{x}}_{uc/uo} \equiv$ the uncontrollable and unobservable states

Then $\tilde{\mathbf{x}}_{cc/co}$ is the only portion of state space that affects the input-output transfer-function properties of the system. Letting $\tilde{A}_{cc/co}$, $\tilde{B}_{cc/co}$, $\tilde{C}_{cc/co}$ be the corresponding submatrices of $\tilde{A}$, $\tilde{B}$, and $\tilde{C}$ it is thus true that

$$H(s) = \tilde{C}[sI - \tilde{A}]^{-1}\tilde{B} = \tilde{C}_{cc/co}[sI - \tilde{A}_{cc/co}]^{-1}\tilde{B}_{cc/co} \tag{11-26}$$

Example 11-7 Consider a hypothetical four-state SISO system

$$\begin{bmatrix} \dot{x}_1 \\ \dot{x}_2 \\ \dot{x}_3 \\ \dot{x}_4 \end{bmatrix} = \begin{bmatrix} -1 & * & 0 & 0 \\ 0 & -2 & 0 & 0 \\ * & * & -3 & * \\ 0 & * & 0 & -4 \end{bmatrix} \begin{bmatrix} x_1 \\ x_2 \\ x_3 \\ x_4 \end{bmatrix} + \begin{bmatrix} 5 \\ 0 \\ 6 \\ 0 \end{bmatrix} u$$

$$y = [7 \quad 8 \quad 0 \quad 0] \begin{bmatrix} x_1 \\ x_2 \\ x_3 \\ x_4 \end{bmatrix}$$

where the * indicates a possible nonzero entry in the A matrix. Regardless of the * value it would be true that x_1 is cc/co; x_2 is uc/co; x_3 is cc/uo; and x_4 is uc/uo. This may be seen by tracing how the states ultimately are influenced by the input and how they influence the output. See Fig. 11-3.

Now, taking each * to be zero, the transfer function is computed to be

$$\begin{aligned} H(s) &= [7 \quad 8 \quad 0 \quad 0] \begin{bmatrix} \dfrac{1}{s+1} & 0 & 0 & 0 \\ 0 & \dfrac{1}{s+2} & 0 & 0 \\ 0 & 0 & \dfrac{1}{s+3} & 0 \\ 0 & 0 & 0 & \dfrac{1}{s+4} \end{bmatrix} \begin{bmatrix} 5 \\ 0 \\ 6 \\ 0 \end{bmatrix} \\ &= \frac{35}{s+1} \\ &= C_{cc/co}[sI - A_{cc/co}]^{-1}B_{cc/co} \end{aligned}$$

where $A_{cc/co} = -1$, $B_{cc/co} = 5$, and $C_{cc/co} = 7$.

It is beyond our scope to consider how to transform to the state-space coordinate system that separates the states in the manner of Eq. (11-25). This is an advanced topic area. The interested reader is referred to Moore [8].

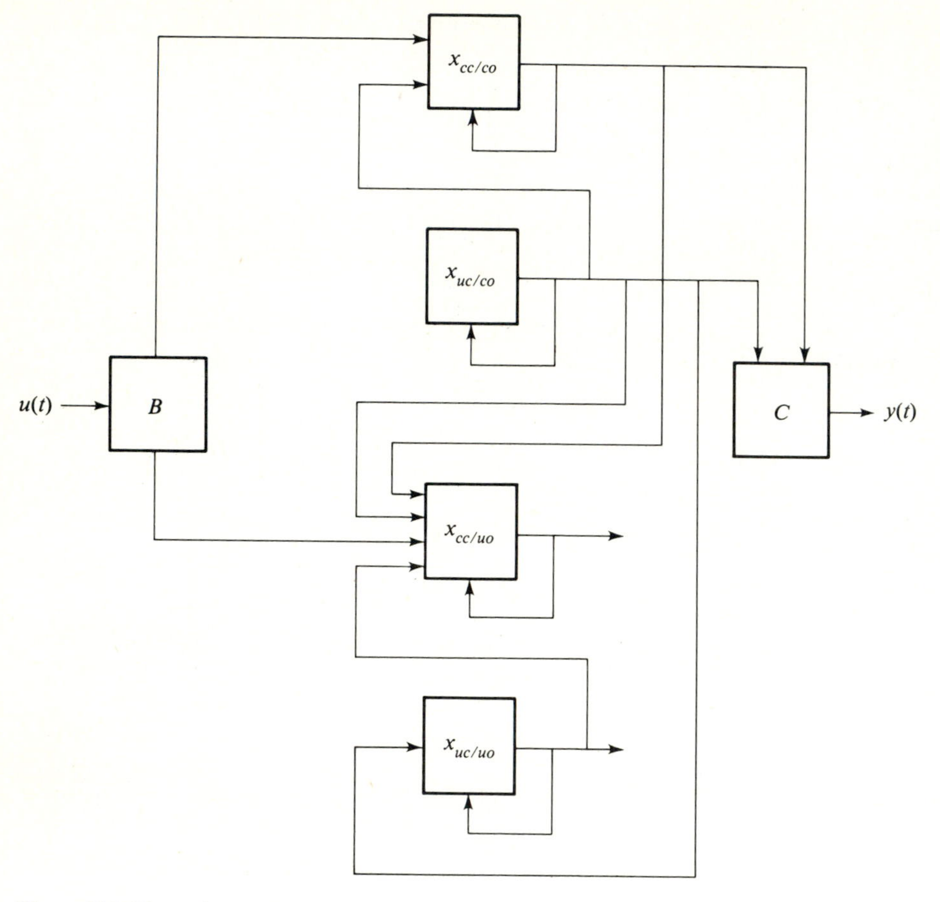

Figure 11-3 Illustration of Kalman's canonical decomposition of state space.

However, there is one state coordinate system with which we have frequently dealt which provides much the same type of decomposition. This is the diagonal (or block-diagonal) coordinate frame. (If a complex pole is uncontrollable and/or unobservable, then so too must be its complex conjugate.)

11-5 CONTROLLABILITY AND OBSERVABILITY ANALYZED IN DIAGONAL STATE COORDINATES

One of the methods introduced in Chap. 6 for deriving the diagonal state coordinate model for a given system $H(s)$ is to expand $H(s)$ in partial fraction form:

$$H(s) = \sum_{i=1}^{n} \frac{r_i}{(s - \lambda_i)} \tag{11-27}$$

where λ_i are the system poles or eigenvalues and r_i are the residues associated with poles, λ_i. (See Appendix C.) For simplicity of discussion it is assumed that the λ_i are distinct; however, the results may be extended to the repeated root situation by similar analysis, but with further complications (see Ref. 7).

The MIMO situation may be handled by performing the partial fraction expansion of each sub-block of the transfer function. States having the same poles or eigenvalues between blocks may then be combined by use of the simulation diagram. The procedure for doing this is described in Sec. 6-5. Regardless, though, the procedure for partial fraction expansion to find the residues is the same as with the SISO transfer function.

Now consider the equation for calculating the residue r_i (again assuming a distinct pole, λ_i):

$$r_i = (s - \lambda_i)H(s)|_{s=\lambda_i} \tag{11-28}$$

But if the system has a zero at λ_i, then r_i will be identically zero. Indeed, for distinct λ_i it may be shown that r_i is zero if and only if $H(s)$ has a zero at λ_i. But if $H(s)$ has a zero at λ_i it implies that $H(s)$ has a pole-zero cancellation. The corresponding pole λ_i is then uncontrollable and/or unobservable. If λ_i is complex and has a pole-zero cancellation, then the complex conjugate of λ_i will also be cancelled.

Example 11-8 Consider the same $H(s)$ as in Example 11-5:

$$H(s) = \frac{s+2}{s^2+3s+2} = \frac{(s+2)}{(s+1)(s+2)}$$

In diagonal form the state-space model is

$$\begin{bmatrix} \dot{x}_1 \\ \dot{x}_2 \end{bmatrix} = \begin{bmatrix} -1 & 0 \\ 0 & -2 \end{bmatrix} \begin{bmatrix} x_1 \\ x_2 \end{bmatrix} + \begin{bmatrix} 1 \\ 1 \end{bmatrix} u$$

$$y = [r_1 \quad r_2] \begin{bmatrix} x_1 \\ x_2 \end{bmatrix}$$

where the residues r_1 and r_2 are computed from

$$r_1 = (s+1)H(s)\Big|_{s=-1} = \frac{s+2}{s+2}\Big|_{s=-1} = 1$$

$$r_2 = (s+2)H(s)\Big|_{s=-2} = \frac{s+2}{s+1}\Big|_{s=-2} = 0$$

Since r_2 is zero, the pole $\lambda_2 = -2$ is unobservable; $H(s)$ has a pole-zero cancellation of this pole. Note that it is a moot point whether one puts the $r_2 = 0$ in the B matrix, making λ_2 uncontrollable, or whether one puts the $r_2 = 0$ in both the B and C matrices, giving the state model

$$\begin{bmatrix} \dot{x}'_1 \\ \dot{x}'_2 \end{bmatrix} = \begin{bmatrix} -1 & 0 \\ 0 & -2 \end{bmatrix} \begin{bmatrix} x'_1 \\ x'_2 \end{bmatrix} + \begin{bmatrix} 1 \\ 0 \end{bmatrix} u$$

$$y = [1 \quad 0] \begin{bmatrix} x'_1 \\ x'_2 \end{bmatrix}$$

This makes the pole $\lambda_2 = -2$ both unobservable and uncontrollable. The important point is that the factor $(s + 2)$ is cancelled from the transfer function, making the pole $\lambda_2 = -2$ uncontrollable and/or unobservable.

One may also start with a given state-space model and transform to the diagonal-form states via use of the modal matrix, the matrix of eigenvectors. The results are the same.

Example 11-9 Consider the state model from Examples 11-1 and 11-3:

$$\dot{\mathbf{x}} = \begin{bmatrix} -5/2 & 0 & 1/2 \\ 0 & -5/2 & -1/2 \\ 0 & 2 & 0 \end{bmatrix} \mathbf{x} + \begin{bmatrix} 1/4 \\ -1/4 \\ 1 \end{bmatrix} u$$

$$y = [0 \quad 0 \quad 1]\mathbf{x}$$

The eigenvalues of A are $\lambda_1 = -1/2$, $\lambda_2 = -2$, $\lambda_3 = -5/2$. The modal P matrix is

$$P = [\xi_1, \quad \xi_2, \quad \xi_3] = \begin{bmatrix} 1 & 1 & 1 \\ -1 & -1 & 0 \\ 4 & 1 & 0 \end{bmatrix}$$

Transforming to diagonal states via

$$\dot{\mathbf{x}}' = P^{-1}AP\mathbf{x}' + P^{-1}Bu$$

$$y = CPx'$$

gives

$$\dot{\mathbf{x}}' = \begin{bmatrix} -1/2 & 0 & 0 \\ 0 & -2 & 0 \\ 0 & 0 & -5/2 \end{bmatrix} \mathbf{x}' + \begin{bmatrix} 1/4 \\ 0 \\ 0 \end{bmatrix} u$$

$$y = [4 \quad 1 \quad 0]\mathbf{x}'$$

From the diagonal form it is seen that two states are uncontrollable and one state is unobservable. This agrees with the results from Examples 11-1 and 11-3, except now we also know which eigenvalues or modes are uncontrollable and unobservable; $\lambda_2 = -2$ is observable but uncontrollable, while $\lambda_3 = -5/2$ is both unobservable and uncontrollable.

From the foregoing analysis and examples it is seen that uncontrollability and unobservability are interchangeable properties; what represents uncontrollability in one state coordinate system manifests itself as unobservability in another state coordinate system, and vice versa. The important point is that only the completely controllable and completely observable dynamics have an influence on the input-output transfer function. This key fact is now used to tie up one loose end from previous discussions—the issue of how the input-output system Hankel matrix can provide useful information regarding system model order.

11-6 THE HANKEL MATRIX AND DETERMINATION OF SYSTEM MODEL ORDER

The results of Theorem 11-1 are also applicable to the discrete state model as studied in Chap. 8:

$$\mathbf{x}_T(k) = A_T\mathbf{x}_T(k-1) + B_T u_T(k-1) \tag{11-29}$$

$$y_T(k) = C_T\mathbf{x}_T(k) \tag{11-30}$$

For simplicity it is assumed that the system has no feedforward. The discrete state model, Eqs. (11-29) and (11-30), is irreducible if and only if the system is completely controllable and completely observable. Then

$$\operatorname{rank} M_{T_c} = n \qquad \operatorname{rank} M_{T_o} = n$$

where M_{T_c} and M_{T_o} are the discrete controllability and observability matrices

$$M_{T_c} \equiv [B_T,\quad A_TB_T,\quad \ldots,\quad A_T^{n-1}B_T] \qquad M_{T_o} \equiv \begin{bmatrix} C_T \\ C_TA_T \\ \vdots \\ C_TA_T^{n-1} \end{bmatrix} \tag{11-31}$$

But in Sec. 8-9 it is shown that the $n \times n$ Hankel matrix

$$\mathcal{H}_T \equiv \begin{bmatrix} h_T(1) & h_T(2) & \cdots & h_T(n) \\ h_T(2) & h_T(3) & \cdots & h_T(n+1) \\ \cdots & \cdots & \cdots & \cdots \\ h_T(n) & h_T(n+1) & \cdots & h_T(2n-1) \end{bmatrix} \tag{11-32}$$

is equal to the product of the observability and controllability matrices

$$\mathcal{H}_T = M_{T_o}M_{T_c} \tag{11-33}$$

This is derived from the fact that the discrete pulse-response model $h_T(k)$ is related to the discrete state model by

$$h_T(k) = C_TA_T^{k-1}B_T \qquad k = 1, 2, \ldots \tag{11-34}$$

In consideration of the system model order, the properties of rank (see Proposition 10-2) give the result that the rank of the Hankel matrix, $\mathcal{H}_T = M_{T_o}M_{T_c}$, is always less than or equal to the lesser of the rank M_{T_o} or rank M_{T_c}; however, the rank of $\mathcal{H}_T$ equals n if and only if rank M_{T_o} = rank $M_{T_c} = n$. Thus the Hankel matrix has rank n if and only if the state model is completely controllable and completely observable; the state model is irreducible.

This is an interesting result, because the pulse-response model, $h_T(k)$, may be found on an experimental basis without knowing anything about the underlying system model, including its order. Techniques for finding $h_T(k)$ are introduced in Chap. 1, but much more sophisticated methods may be applied to help reduce the effect of noise in the measurements (see Notes and References, Chap. 1).

Regardless, though, the Hankel matrix of an arbitrarily high-order $N \times N$ may be formed using the $h_T(k)$. Then when the rank of this large Hankel matrix is determined, it will have (in theory) rank n where n is the proper-order, completely controllable and completely observable state dimension. Determination of rank is most reliably found on the computer via singular value analysis as discussed in Appendix E.

Example 11-10 Using a $T = 1$ s, the 4×4 Hankel matrix for the rocket-sled system introduced in Chap. 1 is

$$\mathcal{H}_T = \begin{bmatrix} 0.14287 & 0.73603 & 1.1658 & 0.96007 \\ 0.73603 & 1.1658 & 0.96007 & 0.32998 \\ 1.1658 & 0.96007 & 0.32998 & -0.14931 \\ 0.96007 & 0.32998 & -0.14931 & -0.13050 \end{bmatrix}$$

Picking a 4×4-dimension Hankel matrix is arbitrary. The four singular values (see Appendix E) of $\mathcal{H}_T$ are found to be

$$\sigma_1 = 2.6404 \qquad \sigma_3 = 0.34049$$

$$\sigma_2 = 1.4727 \qquad \sigma_4 = 1.24 \times 10^{-8}$$

Since σ_4 is very close to zero, this indicates that the rocket-sled system is of rank 3; the 4×4 Hankel matrix for the rocket-sled system at $T = 1$ is found to have three nonzero singular values, and so it may be concluded that the model order is 3. Indeed, it is third-order, since the rocket-sled system has three system poles, as found in Chap. 2.

Example 11-11 The system model

$$H(s) = \frac{5000(s^2 + 1.11s + 0.101)}{(s^4 + 15s^3 + 83s^2 + 359s + 290)}$$

$$= \frac{5000(s + 1.01)(s + 0.1)}{(s + 1)(s + 10)(s + 2 + 5j)(s + 2 - 5j)}$$

was analyzed in Sec. 4-4. It was found that this fourth-order system could be well approximated by the third-order system

$$H_2(s) = \frac{5050(s + 0.1)}{(s + 10)(s + 2 + 5j)(s + 2 - 5j)}$$

because of the near pole-zero cancellation of the pole at -1 and the zero at -1.01.

Analysis of the Hankel matrix also give us this information regarding the underlying model order. In particular, if we discretize the fourth-order $H(s)$ and form the 5×5 discrete Hankel matrix using a $T = 0.1$, it is equal to

$$\mathcal{H}_T = \begin{bmatrix} 15.649 & 22.671 & 12.357 & -1.0218 & -11.071 \\ 22.671 & 12.357 & -1.0218 & -11.071 & -15.110 \\ 12.357 & -1.0218 & -11.071 & -15.110 & -15.336 \\ -1.0218 & -11.071 & -15.110 & -15.336 & -11.575 \\ -11.071 & -15.110 & -15.336 & -11.575 & -6.3788 \end{bmatrix}$$

The singular values of this $\mathcal{H}_T$ are

$$\sigma_1 = 47.669 \qquad \sigma_4 = 2.25 \times 10^{-4}$$

$$\sigma_2 = 44.929 \qquad \sigma_5 = 9.23 \times 10^{-9}$$

$$\sigma_3 = 7.4929$$

Only the first three singular values are large. The rank of $\mathcal{H}_T$ is effectively three. This indicates that the *important* state dimension is three, not four. This agrees with the Chap. 4 analysis.

The previous two examples would indicate that the Hankel matrix can give a fairly reliable indication of the important model order for the SISO system. Indeed this is true, for the Hankel matrix combines both the properties of controllability and observability. However, because this analysis is based on the discrete system, some care must be exercised in picking the discretization interval, T; the singular values and hence the rank of $\mathcal{H}_T$ can be very sensitive to the choice of T.

Example 11-12 Again consider the rocket-sled system, but now we pick $T = \pi/1.044 = 3.009$. Recall that the three continuous poles of the rocket-sled system are $p_1 = -0.2$, $p_{2,3} = -0.1 \pm 1.044$. Therefore, this $T = \pi/1.044$ is exactly twice the Nyquist sampling interval.

The three discrete system poles at this T are found to be

$$q_1 = 0.5478 \qquad q_2 = q_3 = -0.900865$$

The two complex conjugate poles q_2 and q_3 collapse to a repeated real pole at -0.900865. The singular values of the 4×4 Hankel matrix are

$$\sigma_1 = 2.9129 \qquad \sigma_3 = 6.39 \times 10^{-5}$$

$$\sigma_2 = 6.59 \times 10^{-4} \qquad \sigma_4 = 4.53 \times 10^{-9}$$

indicating a *practical* rank of one (there is only one large singular value). At this T the three-pole system looks like a single-pole system. Accidentally picking this T for formation of $h_T(k)$ and the Hankel matrix would thus give a completely erroneous result regarding the rank of the underlying system.

Actually any T in a neighborhood of $T = \pi/1.044 = 3.009$ would give similar poor results. Figure 11-4 shows a plot of the singular values, normalized by taking the ratio to σ_1, of the Hankel matrix for the rocket-sled system for varying T. The best value of T to pick for discretizing the system occurs at

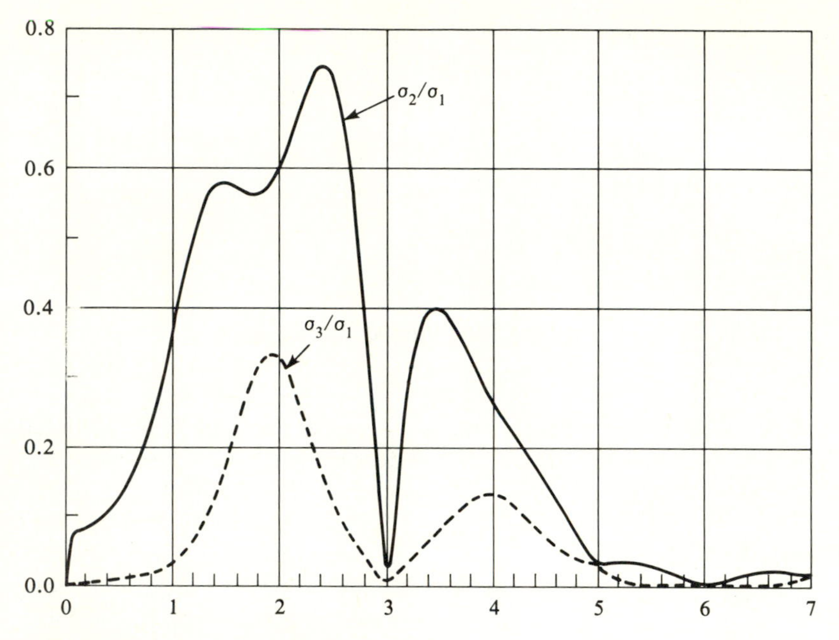

Figure 11-4 Plot of singular value ratios for 3×3 Hankel matrix, rocket-sled system, $0 < T < 7$.

the peak of σ_3/σ_1; this is the value of T that occurs at approximately $T = 2$. Picking $T = 2$ would most clearly reveal that the Hankel matrix is rank 3 and hence that the underlying system is third-order.

Moore [8] shows that the rank properties of the Hankel matrix are directly related to model order determination of the continuous-time linear state-space model. See also Ref. 9.

11-7 ALTERNATIVE METHODS FOR CALCULATION OF CONTROLLABILITY (OBSERVABILITY)

Formation of the controllability matrix M_c to determine controllability properties of a given system provides a useful way to introduce the concept of system controllability for both the discrete and continuous-time system. However, for large-dimension systems the formation of M_c can lead to numerical inaccuracies related to the repeated multiplication by A in the formation of the columns of M_c. In such cases the matrix M_c can be a poor indicator of system controllability properties. The same remarks apply to observability and the M_o matrix.

An alternative way to determine controllability of separate eigenvalues or modes is via the fact that [10]:

$$\text{rank}\,[\lambda_i I - A, B] = n \qquad \text{if and only if } \lambda_i \text{ is a completely controllable eigenvalue}$$

This technique for determination of controllability can also lead to numerical problems if the subject A matrix is *poorly conditioned* relative to finding the system eigenvalues. The interested reader is referred to Ref. 11 for further discussion and presentation of a numerically stable technique for determination of system controllability. But none of these techniques of controllability analysis gives a good indication of the *quality* of controllability. This concept is addressed by the methods of the next section.

11-8 MINIMUM EFFORT CONTROL AND CONTROLLABILITY

Section 11-1 demonstrates that $\mathbf{x}_{zs}$ could never be brought to a state outside the controllable subspace, the range space of M_c. But what is not clear from this development is that a control input can be found, $\mathbf{u}^*(t)$, $0 < t \leq t_f$, such that $\mathbf{x}_{zs}(t_f) = \mathbf{x}$ for a given $\mathbf{x} \in \mathcal{R}\{M_c\}$. There are actually an infinite number of inputs that do this, but this section demonstrates just one—a particular solution, if you will.

To demonstrate such a control input, $\mathbf{u}^*(t)$, the notation of the state transition matrix is adopted:

$$\Phi(t) \equiv e^{At} \tag{11-35}$$

In terms of the state transition matrix, the zero-state response may be written as

$$\mathbf{x}_{zs}(t) = \Phi(t) \int_0^t \Phi(-\tau) B \mathbf{u}(\tau)\, d\tau \tag{11-36}$$

Now suppose it is desired to find the $\mathbf{u}^*(t)$, $0 < t \leq t_f$, such that $\mathbf{x}_{zs}(t_f) = \mathbf{x}$, where it is assumed that $\mathbf{x}$ is a controllable state. The following theorem demonstrates such a control input. For simplicity it is assumed that the system is completely controllable (rank $M_c = n$). This theorem can, however, be extended to an uncontrollable system so long as the desired final state, $\mathbf{x}$, is in the controllable subspace (see Kalman [7]).

Theorem 11-2 Assume a system state model $\{A, B\}$ is completely controllable. Then an input $u^*(t)$, $0 < t \leq t_f$, such that $\mathbf{x}_{zs}(t_f) = \mathbf{x}$ is given by

$$\mathbf{u}^*(t) = B^T \Phi^T(t_f - t) W_c^{-1}(t_f)\mathbf{x} \qquad 0 < t \leq t_f \tag{11-37}$$

where

$$W_c(t_f) \equiv \Phi(t_f)\left(\int_0^{t_f} \Phi(-\tau) B B^T \Phi^T(-\tau)\, d\tau\right)\Phi^T(t_f) \tag{11-38}$$

PROOF Substitution of the input $\mathbf{u}^*(t)$, Eq. (11-37), into the zero-state solution as given by Eq. (11-36) gives

$$\mathbf{x}_{zs}(t_f) = \Phi(t_f) \int_0^{t_f} \Phi(-\tau) B B^T \Phi^T(t_f - \tau) W_c^{-1}(t_f)\mathbf{x}\, d\tau \tag{11-39a}$$

But using $\Phi^T(t_f - \tau) = \Phi^T(-\tau)\Phi^T(t_f)$ this may be written as

$$\mathbf{x}_{zs}(t_f) = \left[\Phi(t_f)\left(\int_0^{t_f} \Phi(-\tau)BB^T\Phi^T(-\tau)\,d\tau\right)\Phi^T(t_f)\right] W_c^{-1}(t_f)\mathbf{x} \tag{11-39b}$$

The term in brackets is $W_c(t_f)$, and so the result that $\mathbf{x}_{zs}(t_f) = \mathbf{x}$ has been proven.

This theorem is due to Kalman [5]. The control input $u^*(t)$ is a special input in that its norm

$$\|\mathbf{u}^*\| \equiv \left(\int_0^{t_f} \mathbf{u}^{*T}(\tau)\mathbf{u}^*(\tau)\,d\tau\right)^{1/2} \tag{11-40}$$

is smaller than the norm of any other $\mathbf{u}(t)$, $0 < t \leq t_f$, which yields the solution $\mathbf{x}_{zs}(t_f) = \mathbf{x}$. The control input, $u^*(t)$, might thus be called the *minimum energy* control solution because it has the least integral squared value.

The $n \times n$ symmetric matrix $W_c(t_f)$, Eq. (11-38), is also a special matrix. It is called the *controllability Gram matrix,* or *controllability Grammian.* Some of the important properties of the controllability Gram matrix are given in the following theorem:

Theorem 11-3 Let $W_c(t)$ be defined by

$$W_c(t) \equiv \Phi(t)\left(\int_0^{t} \Phi(-\tau)BB^T\Phi^T(-\tau)\,d\tau\right)\Phi^T(t) \tag{11-41}$$

Then:

1. $W_c(t)$ is symmetric for all $t \geq 0$.
2. $W_c(t)$ has eigenvalues that are either positive or zero.
3. The state model $\{A, B\}$ is completely controllable if and only if $W_c(t)$, $t > 0$, has rank n or, equivalently, it has all strictly positive eigenvalues.
4. $W_c(t) = \int_0^t \Phi(\xi)BB^T\Phi^T(\xi)\,d\xi$
5. $W_c(t)$ is solution of the $n \times n$ matrix *Lyapunov* differential equation

$$\frac{d}{dt} W_c(t) = AW_c(t) + W_c(t)A^T + BB^T \tag{11-42}$$

with zero boundary condition

$$W_c(0) = 0 \tag{11-43}$$

6. If $\{A, B\}$ represents a stable system (that is, A has all negative real-part eigenvalues) then

$$\lim_{t\to\infty} W_c(t) = W_{c_{ss}} \tag{11-44}$$

where $W_{c_{ss}}$ is the solution of the matrix Lyapunov equation

$$0 = AW_{c_{ss}} + W_{c_{ss}}A^T + BB^T \tag{11-45}$$

7. For achieving the terminal state $\mathbf{x}_{zs}(t_f) = \mathbf{x}$, the norm of the minimum energy control, $u^*(t)$, Eq. (11-37), satisfies the inequality

$$\|\mathbf{u}^*\|^2 \equiv \left(\int_0^{t_f} \mathbf{u}^{*T}(\tau)\mathbf{u}^*(\tau)\, d\tau\right)^{1/2} \leq (1/\lambda_{\min})\|\mathbf{x}\|^2 \qquad (11\text{-}46)$$

where $\lambda_{\min}$ is the smallest eigenvalue of $W_c(t_f)$ and $\|\mathbf{x}\|^2 \equiv (\mathbf{x}^T\mathbf{x})^{1/2}$.

8. For $\mathbf{x} = \alpha\boldsymbol{\xi}_i$ where $\boldsymbol{\xi}_i$ is the unit norm eigenvector of $W_c(t_f)$ corresponding to the eigenvalue λ_i, then

$$\|\mathbf{u}^*\|^2 \equiv \int_0^{t_f} \mathbf{u}^{*T}(\tau)\mathbf{u}^*(\tau)\, d\tau = (1/\lambda_i)\|x\|^2 \qquad (11\text{-}47)$$

where $\|\mathbf{x}\|^2 = \alpha^2$.

PROOF Proofs of properties 1 through 4 are left for the Problems section. To derive property 5, differentiate the expression for $W_c(t)$ giving

$$\begin{aligned}\frac{d}{dt}[W_c(t)] &= \frac{d}{dt}[\Phi(t)] \int_0^t \Phi(-\tau)BB^T\Phi^T(-\tau)\, d\tau\Phi^T(t) \\ &+ \Phi(t)\frac{d}{dt}\left[\int_0^t \Phi(-\tau)BB^T\Phi^T(-\tau)\, d\tau\right]\Phi^T(t) \\ &+ \Phi(t)\int_0^t \Phi(-\tau)BB^T\Phi^T(-\tau)\, d\tau\frac{d}{dt}[\Phi^T(t)] \qquad (11\text{-}48)\end{aligned}$$

It is true that

$$\frac{d}{dt}[\Phi(t)] = A\Phi(t) \qquad \frac{d}{dt}[\Phi^T(t)] = \Phi^T(t)A^T \qquad (11\text{-}49)$$

and, from the Leibnitz rule.

$$\frac{d}{dt}\left[\int_0^t \Phi(-\tau)BB^T\Phi^T(-\tau)\, d\tau\right] = \Phi(-t)BB^T\Phi^T(-t) \qquad (11\text{-}50)$$

These results may be used in Eq. (11-48) to give the desired result, property 5.

To show property 6 is then a simple matter. The differential equation for $W_c(t)$ reaches steady state when the derivative is zero. Letting

$$\frac{d}{dt}W_c(t) = 0 \qquad (11\text{-}51)$$

then yields expression 6 for $W_{c_{ss}}$.

See Probs. 11-18 through 11-21 for proof of the remaining properties.

Computational methods are available to find $W_{c_{ss}}$ when A is stable (i.e., it has eigenvalues with strictly negative real parts). See, for example, Ref. 12. The calculation of $W_{c_{ss}}$ provides a reliable method for determination of state controllability—actually a more reliable method than determination of controllability from M_c. The eigenvalues of $W_{c_{ss}}$ give a qualitative measure of the quality of controllability in certain directions of state space. If $\lambda_{\min}$ is the smallest eigenvalue of $W_{c_{ss}}$, then the control energy it takes to reach $\mathbf{x} = \alpha \xi_{\min}$ is $\alpha^2/\lambda_{\min}$. The eigenvector $\xi_{\min}$ is the unit norm eigenvector corresponding to $\lambda_{\min}$. This result follows from property 7. If $\lambda_{\min}$ is very small, this result may be interpreted as saying that it is very difficult (it takes a great deal of control energy) to reach any state in the direction $\xi_{\min}$. This is a useful interpretation of the property of controllability. By *duality,* similar results apply to observability and a suitably defined *observability Gram matrix*.

The following example shows how the controllability Gram matrix may be computed in closed form for the diagonal state coordinates.

Example 11-13 Consider the diagonal state model

$$\dot{\mathbf{x}} = \begin{bmatrix} -1 & 0 \\ 0 & -2 \end{bmatrix} \mathbf{x} + \begin{bmatrix} 1 \\ 1 \end{bmatrix} u$$

By Theorem 11-3, property 4, the differential equation for $W_c(t)$ is

$$\frac{d}{dt} W_c(t) = \begin{bmatrix} -1 & 0 \\ 0 & -2 \end{bmatrix} W_c(t) + W_c(t) \begin{bmatrix} -1 & 0 \\ 0 & -2 \end{bmatrix} + \begin{bmatrix} 1 & 1 \\ 1 & 1 \end{bmatrix}$$

with zero boundary condition at $t = 0$. Denote

$$W_c(t) = \begin{bmatrix} W_1(t) & W_2(t) \\ W_2(t) & W_3(t) \end{bmatrix}$$

Note that the (2, 1) element of $W_c(t)$ is the same as the (1, 2) element. This is because $W_c(t)$ is always symmetric. The differential equation for $W_c(t)$ is thus equivalent to the three scalar differential equations

$$\dot{W}_1 = -2W_1 + 1 \qquad W_1(0) = 0$$
$$\dot{W}_2 = -3W_2 + 1 \qquad W_2(0) = 0$$
$$\dot{W}_3 = -4W_3 + 1 \qquad W_3(0) = 0$$

These differential equations have solution

$$W_1(t) = e^{-2t} \int_0^t e^{2\tau}\, d\tau = \frac{1 - e^{-2t}}{2}$$

$$W_2(t) = e^{-3t} \int_0^t e^{3\tau}\, d\tau = \frac{1 - e^{-3t}}{3}$$

$$W_3(t) = e^{-4t} \int_0^t e^{4\tau}\, d\tau = \frac{1 - e^{-4t}}{4}$$

The steady-state controllability Gram matrix is found by taking the limit of $W_c(t)$ as $t \to \infty$. This gives

$$W_{c_{ss}} = \lim_{t \to \infty} W_c(t) = \begin{bmatrix} 1/2 & 1/3 \\ 1/3 & 1/4 \end{bmatrix}$$

The eigenvalues of $W_{c_{ss}}$ are

$$\lambda_1 = 0.731 \qquad \lambda_2 = 0.0189$$

The eigenvalues are both nonzero, and so the system is completely controllable. However, a new piece of information is now available. It takes much more "energy" to reach a state in the $\boldsymbol{\xi}_2$ direction than to reach a state in the $\boldsymbol{\xi}_1$ direction. The unit norm $\boldsymbol{\xi}_1$ and $\boldsymbol{\xi}_2$ eigenvectors of $W_{c_{ss}}$ are found to be

$$\boldsymbol{\xi}_1 = \begin{bmatrix} 0.8220 \\ 0.5695 \end{bmatrix} \qquad \boldsymbol{\xi}_2 = \begin{bmatrix} 0.5695 \\ -0.8220 \end{bmatrix}$$

Note that these two eigenvectors are orthogonal. From Theorem 10-2, Sec. 10-7, this is always true of the eigenvectors of a real symmetric matrix.

By property 8 of Theorem 11-3 it is true that the minimum energy control input to reach the terminal state

$$\mathbf{x} = \begin{bmatrix} 0.8220 \\ 0.5695 \end{bmatrix} = \boldsymbol{\xi}_1$$

has the norm squared value

$$\|u^*\|^2 = \left(\int_0^\infty u^2(\tau)\, d\tau\right)^{1/2} = 1/\lambda_1 = 1.368$$

However, for a state in the $\boldsymbol{\xi}_2$ direction

$$\mathbf{x} = \begin{bmatrix} 0.5695 \\ -0.8220 \end{bmatrix} = \boldsymbol{\xi}_2$$

the norm square of the minimum effort control to reach this terminal state is

$$\|u^*\|^2 = \left(\int_0^\infty u^2(\tau)\, d\tau\right)^{1/2} = 1/\lambda_2 = 52.91$$

It thus takes some 38 times as much control energy to reach a state in the $\boldsymbol{\xi}_2$ direction as it does to reach a state in the $\boldsymbol{\xi}_1$ direction. Such a directional sensitivity would probably not have been suspected from the original state model or from the M_c controllability matrix. This information is available via analysis of the controllability Gram matrix.

The computational technique of this example may be generalized to the nth-order multi-input system, provided A has distinct eigenvalues. If A has distinct eigenvalues, then it is possible to diagonalize A by transformation with the modal P matrix as discussed in Sec. 10-5. Closed-form expressions may then be found for $\widehat{W}_c(t)$ with $\widehat{A} = \Lambda$, the diagonal form. Then

$$W_c(t) = P\widehat{W}_c(t)P^T \tag{11-52}$$

is the controllability Gram matrix in the original state coordinates (see Prob. 11-23).

11-9 SUMMARY

Utilizing the tools of linear algebra and matrix theory presented in Chaps. 9 and 10, the important system properties of state controllability and observability are examined for the general nth-order MIMO state-space model. Use of the Cayley-Hamilton theorem is fundamental to Secs. 11-1 and 11-2. In Sec. 11-3 it is seen that only completely controllable and completely observable states have any influence on the system input-output transfer function. This is reconfirmed by consideration of the diagonal-form states in Sec. 11-5.

Section 11-6 takes a brief hiatus to reexamine the system Hankel matrix, first introduced in Sec. 5-1 and then reencountered in Sec. 8-9. In Sec. 8-9 it is shown that the Hankel matrix is the product of the discrete observability and discrete controllability matrices. This is seen to offer a technique for the experimental determination of the model order of an unknown system; however, there are many subtle points left unexplored in the procedure for doing this.

Finally, Sec. 11-8 introduces some of the original results of Kalman [5] on the subject of controllability and minimum-effort control laws. The steady-state controllability Gram matrix provides useful information regarding the quality of controllability in various state directions.

The controllability Gram matrix and the dual observability Gram matrix are useful for several reasons. One of these reasons is that they readily generalize to study controllability and observability for the time-varying linear system.

PROBLEMS

11-1 Determine whether or not the following systems are completely controllable (i.e., rank $M_c = n$):

$$(a)\ \dot{\mathbf{x}} = \begin{bmatrix} 0 & 1 & 0 \\ 0 & 0 & 1 \\ 0 & -2 & -3 \end{bmatrix}\mathbf{x} + \begin{bmatrix} 0 \\ 0 \\ 1 \end{bmatrix}u$$

$$y = [0 \quad 0 \quad 1]\mathbf{x}$$

$$(b)\ \dot{\mathbf{x}} = \begin{bmatrix} 0 & 1 & 0 \\ 0 & 0 & 1 \\ 0 & -2 & -3 \end{bmatrix}\mathbf{x} + \begin{bmatrix} 0 \\ 1 \\ 0 \end{bmatrix}u$$

$$y = [0 \quad 1 \quad 0]\mathbf{x}$$

$$(c)\ \dot{\mathbf{x}} = \begin{bmatrix} 0 & 1 & 0 \\ 0 & 0 & 1 \\ 0 & -2 & -3 \end{bmatrix}\mathbf{x} + \begin{bmatrix} 1 \\ 0 \\ 0 \end{bmatrix}u$$

$$y = [1 \quad 0 \quad 0]\mathbf{x}$$

$$(d)\ \dot{\mathbf{x}} = \begin{bmatrix} 0 & 1 & 0 \\ 0 & 0 & 1 \\ 0 & -2 & -3 \end{bmatrix}\mathbf{x} + \begin{bmatrix} 1 & 0 \\ 0 & 1 \\ 0 & 0 \end{bmatrix}\begin{bmatrix} u_1 \\ u_2 \end{bmatrix}$$

$$\begin{bmatrix} y_1 \\ y_2 \end{bmatrix} = \begin{bmatrix} 0 & 1 & 0 \\ 0 & 0 & 1 \end{bmatrix}\mathbf{x}$$

11-2 Determine whether or not the following systems are completely controllable (that is, rank $M_c = n$):

$$(a)\ \dot{\mathbf{x}} = \begin{bmatrix} 0 & 0 & 0 \\ 1 & 0 & -5 \\ 0 & 1 & -4 \end{bmatrix} \mathbf{x} + \begin{bmatrix} 0 \\ 0 \\ 1 \end{bmatrix} u$$

$$y = [0 \quad 0 \quad 1]\mathbf{x}$$

$$(b)\ \dot{\mathbf{x}} = \begin{bmatrix} 0 & 0 & 0 \\ 1 & 0 & -5 \\ 0 & 1 & -4 \end{bmatrix} \mathbf{x} + \begin{bmatrix} 0 \\ 1 \\ 0 \end{bmatrix} u$$

$$y = [0 \quad 1 \quad 0]\mathbf{x}$$

$$(c)\ \dot{\mathbf{x}} = \begin{bmatrix} 0 & 0 & 0 \\ 0 & -2 & 1 \\ 0 & -1 & -2 \end{bmatrix} \mathbf{x} + \begin{bmatrix} 1 \\ 0 \\ 0 \end{bmatrix} u$$

$$y = [1 \quad 0 \quad 0]\mathbf{x}$$

$$(d)\ \dot{\mathbf{x}} = \begin{bmatrix} 0 & 0 & 0 \\ 1 & 0 & -5 \\ 0 & 1 & -4 \end{bmatrix} \mathbf{x} + \begin{bmatrix} 0 & 0 \\ 0 & 1 \\ 1 & 0 \end{bmatrix} u$$

$$\begin{bmatrix} y_1 \\ y_2 \end{bmatrix} = \begin{bmatrix} 1 & 0 & 0 \\ 0 & 1 & 0 \end{bmatrix} \mathbf{x}$$

11-3 For the systems in Prob. 11-1 determine a basis for the controllable subspace and a basis for the uncontrollable subspace. *Note:* These bases are both found by performing elementary row operations on M_c^T. The uncontrollable subspace is the null space on M_c^T. Basis vectors for the controllable subspace are the nonzero rows of M_c^T after elementary row operations. If M_c has full rank n, the null space of M_c^T is the zero vector.

11-4 Repeat Prob. 11-3 for the systems of Prob. 11-2.

11-5 Determine whether or not the systems in Prob. 11-1 are completely observable (that is, rank $M_o = n$).

11-6 Determine whether or not the systems in Prob. 11-2 are completely observable (that is, rank $M_o = n$).

11-7 For each of the systems in Prob. 11-1 find the observable subspace and the unobservable subspace. Both of these subspaces may be determined by performing elementary row operations on M_o. The unobservable subspace is the null space of M_o. Basis vectors for the observable subspace are the nonzero rows of M_o after elementary row operations.

11-8 Repeat Prob. 11-7 for the systems of Prob. 11-2.

11-9 (*a*) Find $H(s) = C[sI - A]^{-1}B$ for the systems in Prob. 11-1. If there are any pole-zero cancellations, then make these reductions. *Note:* the system eigenvalues are $\lambda_1 = 0$, $\lambda_2 = -1$, $\lambda_3 = -2$.

(*b*) Do the results of *a* agree with controllability and observability analysis of Probs. 11-1, 11-3, 11-5, and 11-7? Explain.

11-10 (*a*) Find $H(s) = C[sI - A]^{-1}B$ for the systems in Prob. 11-2. If there are any pole-zero cancellations, then make these reductions. *Note:* the system eigenvalues are $\lambda_1 = 0$, $\lambda_{2,3} = -1 \pm 2j$.

(*b*) Do the results of *a* agree with controllability and observability analysis of Probs. 11-2, 11-4, 11-6, and 11-8? Explain.

11-11 (*a*) Transform the systems in Prob. 11-1 to diagonal form.

(*b*) Analyze controllability and observability of these models.

(*c*) Do the results correlate with the results from Probs. 11-1, 11-3, 11-5, 11-7 and 11-9? Discuss.

11-12 (*a*) Transform the systems of Prob. 11-2 to block-diagonal form.

(*b*) Analyze controllability and observability of these models.

(*c*) Do the results correlate with the results from Probs. 11-2, 11-4, 11-6, and 11-10? Discuss.

11-13 Consider the following state-space model:

$$\dot{\mathbf{x}} = \begin{bmatrix} -1 & 4 & 0 & 2 \\ 0 & -2 & 0 & 1 \\ 0 & 0 & -3 & 3 \\ 0 & 0 & 0 & -4 \end{bmatrix} \mathbf{x} + \begin{bmatrix} 1 \\ 1 \\ 0 \\ 0 \end{bmatrix} u$$

$$y = [0 \quad 1 \quad 0 \quad 2]\mathbf{x}$$

(*a*) Draw a simulation diagram of the system.

(*b*) Identify the states as being *cc/co, cc/uo, uc/co, uc/uo.*

11-14 Repeat Prob. 11-13 for the system

$$\dot{\mathbf{x}} = \begin{bmatrix} -1 & 0 & 0 & 0 \\ 1 & -2 & 0 & 0 \\ 2 & 1 & -3 & 1 \\ 1 & 0 & 0 & -4 \end{bmatrix} \mathbf{x} + \begin{bmatrix} 0 \\ 0 \\ 1 \\ 1 \end{bmatrix}$$

$$y = [-1 \quad 0 \quad 0 \quad 1]\mathbf{x}$$

11-15 Values of the discrete pulse-response model for an unknown system using a sample spacing $T = 0.5$ are found to be $h_T(1) = 1$, $h_T(2) = 3$, $h_T(3) = 2$, $h_T(4) = -1$, $h_T(5) = -3$.

(*a*) What is the state model order?

(*b*) Is it possible that the state model order could be of a higher order than that determined in part *a*? Explain.

11-16 Repeat Prob. 11-15 for $h_T(1) = 1$, $h_T(2) = 4$, $h_T(3) = 9$, $h_T(4) = 14$, $h_T(5) = 14$. Assume $T = 0.2$.

11-17 Use the definition Eq. (11-41) to show that $W_c(t)$ is symmetric for all $t \geq 0$; that is, $W_c^T(t) = W_c(t)$. *Hint:* Use the properties of the matrix transpose that $(A^T)^T = A$ and $(ABC)^T = C^T B^T A^T$.

11-18 Using the input

$$\mathbf{u}^*(t) = B^T \Phi^T(t_f - t) W_c^{-1}(t_f)\mathbf{x}$$

show that

$$\|\mathbf{u}^*\|^2 \equiv \int_0^{t_f} \mathbf{u}^{*T}(\tau)\mathbf{u}^*(\tau)\, d\tau$$

$$= \mathbf{x}^T W_c^{-1}(t_f)\mathbf{x}$$

where $W_c(t_f)$ is given by Eq. (11-41).

11-19 Assume $W_c(t_f)$ has rank n. From Prob. 10-6 it is true that the eigenvalues and eigenvectors of $W_c^{-1}(t_f)$ are $(1/\lambda_i)$ and $\boldsymbol{\xi}_i$, $i = 1, 2, \ldots, n$, where λ_i and $\boldsymbol{\xi}_i$ are eigenvalues and eigenvectors of $W_c(t_f)$. Let $\mathbf{x} = \alpha\boldsymbol{\xi}_i$ where $\boldsymbol{\xi}_i$ is picked to have unit norm, $\|\boldsymbol{\xi}_i\| = (\boldsymbol{\xi}_i^T\boldsymbol{\xi}_i)^{1/2} = 1$. Use the result of Prob. 11-18 to show that the norm square of $\mathbf{u}^*(t)$ to take the system from the zero initial state to the terminal state $\mathbf{x}(t_f) = \mathbf{x}$ is given by $\|\mathbf{u}^*\|^2 = (1/\lambda_i)\alpha^2$.

11-20 Assume $W_c(t_f)$ has rank n. Since $W_c(t_f)$ is symmetric, its n eigenvectors are orthogonal, and they form an orthogonal basis for R^n (see Theorem 10-2, Sec. 10-7). Therefore, any state $\mathbf{x} \in R^n$ in state space may be uniquely represented in terms of the n orthogonal eigenvectors by

$$\mathbf{x} = \sum_{i=1}^{n} \alpha_i \boldsymbol{\xi}_i$$

where $\alpha_i = \mathbf{x}^T\xi_i$. Show that the norm square of the minimum-energy control to reach $\mathbf{x}(t_f) = \mathbf{x}$ starting from the zero initial state is given by

$$\|\mathbf{u}^*\|^2 = \sum_{i=1}^{n} \frac{\alpha_i^2}{\lambda_i}$$

Use the results of Prob. 11-19.

11-21 Use the results of Prob. 11-20 to show that the norm square of the minimum-energy control to reach any state in state space, $\mathbf{x} \in R^n$, by time t_f starting from the zero initial state satisfies the upper bound

$$\|\mathbf{u}^*\|^2 \le \frac{\|\mathbf{x}\|^2}{\lambda_{\min}}$$

11-22 Suppose that the initial state is $\mathbf{x}(0)$ at $t_0 = 0$. Show that the control input

$$\mathbf{u}^*(t) = B^T\Phi^T(t_f - t)W_c^{-1}(t_f)(\mathbf{x} - \Phi(t_f)\mathbf{x}(0))$$

takes the system from $\mathbf{x}(0)$ to $\mathbf{x}(t_f) = \mathbf{x}$.

11-23 (*a*) Suppose the state coordinates $\mathbf{x}(t)$ and $\hat{\mathbf{x}}(t)$ are related by $x(t) = P\hat{\mathbf{x}}(t)$. Show that their respective controllability Gram matrices are related by

$$W_c(t) = P\hat{W}_c(t)P^T$$

Hint: Use the facts that $A = P\hat{A}P^{-1}$ and $B = P\hat{B}$. How are $\Phi(t) = e^{At}$ and $\hat{\Phi}(t) = e^{\hat{A}t}$ related? How are $\Phi^T(t) = e^{A^Tt}$ and $\hat{\Phi}^T(t) = e^{\hat{A}^Tt}$ related? Use these facts in Eq. (11-41).

(*b*) Suppose that the system is stable so that $W_{c_{ss}} = \lim_{t\to\infty} W_c(t)$ exists. Use the results of *a* to show that

$$W_{c_{ss}} = P\hat{W}_{c_{ss}}P^T$$

11-24 Write the three scalar differential equations, with boundary conditions, that may be used to compute $W_c(t)$ for the system

$$\dot{\mathbf{x}} = \begin{bmatrix} 0 & 1 \\ -10 & -11 \end{bmatrix}\mathbf{x} + \begin{bmatrix} 0 \\ 20 \end{bmatrix}u$$

$$y = [0.1 \quad 1]x$$

11-25 Repeat Prob. 11-24 for the system

$$\dot{\mathbf{x}}' = \begin{bmatrix} 0 & -10 \\ 1 & -11 \end{bmatrix}\mathbf{x}' + \begin{bmatrix} 0.1 \\ 1 \end{bmatrix}u$$

$$y = [0 \quad 20]\mathbf{x}'$$

11-26 (*a*) Use the controllability matrices, M_C and $\hat{M}_c$, to find the matrix of transition, P, such that $\hat{\mathbf{x}} = P\mathbf{x}$ where

$$\dot{\hat{\mathbf{x}}} = \begin{bmatrix} -1 & 0 \\ 0 & -10 \end{bmatrix}\hat{\mathbf{x}} + \begin{bmatrix} 1 \\ 1 \end{bmatrix}u$$

for the system of Prob. 11-24. *Hint:* How are M_c and $\hat{M}_c$ related? See Sec. 9-7.

(*b*) Find closed-form expressions for $\hat{W}_c(t)$. See Example 11-10.

(*c*) Take the limit of $\hat{W}_c(t)$ as $t \to \infty$ to find $\hat{W}_{c_{ss}}$.

(*d*) What are the eigenvalues and unit-norm eigenvectors of $\hat{W}_{c_{ss}}$?

(*e*) Use the result of Prob. 11-23 that $W_{c_{ss}} = P\hat{W}_{c_{ss}}P^T$ to find $W_{c_{ss}}$.

(*f*) What are the eigenvalues and unit-norm eigenvectors of $W_{c_{ss}}$?

(*g*) Is the system completely controllable?

(*h*) Qualitatively compare the amount of energy it takes to reach a state in the ξ_1 direction versus a state in the ξ_2 direction.

11-27 Repeat Prob. 11-26 for the system in Prob. 11-25 finding the matrix of transition P' such that $\mathbf{x}'(t) = P'\hat{\mathbf{x}}(t)$ and $\hat{\mathbf{x}}(t)$ is the same as in Prob. 11-26.

11-28 Based on the results of Probs. 11-26 and 11-27, comment on the validity of the statement: "The quality of controllability in different state coordinate systems is invariant."

11-29 Derive property 4 of Theorem 11-3. *Hint:* Let $\xi = t - \tau$ and substitute into Eq. (11-41).

NOTES AND REFERENCES

The concepts of controllability and observability were introduced by Kalman [5] in relation to solving the problem of optimal control (see also Ref. 6 and Sec. 11-8). Then in Ref. 7 Kalman presents a comprehensive overview of the general topics of controllability, observability, and minimal realizations for a state-space model. Many of the results of this chapter parallel those of this early paper. The interested reader is referred to Ref. 6 for deeper discussions and more comprehensive results, particularly in regard to the MIMO case. Recently Moore [8] has extended the early work of Kalman by deriving computationally stable means to transform to a balanced state coordinate system in which states having the least influence on the system input-output properties may be identified and thus eliminated from the state model. The concepts in Ref 8 are also associated with those of the Hankel matrix as discussed in Sec. 11-6. These are advanced topic areas, but the interested reader is also referred to Refs. 8, 9, and 13 for deeper discussions.

1. D'Azzo, J., and C. H. Houpis: *Linear Control Systems: Conventional and Modern,* 2d ed., McGraw-Hill, New York, 1981.
2. Fortmann, T. E., and K. L. Hitz: *An Introduction to Linear Control Systems,* Marcel Dekker, New York, 1977.
3. Maybeck, P. S.: *Stochastic Models, Estimation, and Control,* vol. 1, Academic Press, New York, 1979.
4. Meditch, J. S.: *Stochastic Optimal Linear Estimation and Control,* McGraw-Hill, New York, 1969.
5. Kalman, R. E.: "On the General Theory of Control Systems," *Proceedings, 1st International Congress on Automatic Control,* 1960. Butterworths, London, vol. 1, 1961, pp. 481–492.
6. Kalman, R. E., P. L. Falb, and M. Arbib: *Topics in Mathematical Systems Theory,* McGraw-Hill, New York, 1969.
7. Kalman, R. E.: "Mathematical Description of Linear Dynamical Systems," *Journal* S.I.A.M. *Control,* vol. 1, no. 2, 1963, pp. 152–192.
8. Moore, B. C.: "Principle Component Analysis in Linear Systems: Controllability, Observability, and Model Reduction," *IEEE Trans. Auto. Control,* vol. AC-26, no. 1, Feb. 1981, pp. 17–31.
9. Kung, S. Y., and D. W. Lin: "Optimal Hankel-Norm Model Reductions: Multivariable Systems," *IEEE Trans. Auto. Control,* vol. AC-26, no. 4, Aug. 1981, pp. 832–852.
10. Rosenbrock, H. H.: *State Space and Multivariable Theory,* Nelson, London, 1970.
11. Paige, C. C.: "Properties of Numerical Algorithms Related to Computing Controllability," *IEEE Trans. Auto. Control,* vol. AC-26, no. 1, Feb. 1981, pp. 130–138.
12. Bartels, R. H., and G. W. Stewart: "Solutions of the Matrix Equation $AX + XB = C$," *Commun. Assoc. Comput. Mach.,* vol. 15, 1972, pp. 820–826.
13. Laub, A. J.: "Computation of 'Balancing' Transformations," *Proceedings, 1980 IEEE Conference on Decision and Control,* San Diego, CA, December 1980, pp. FA8-E.

CHAPTER

TWELVE

LINEAR TIME-VARYING STATE MODELS

In Sec. 1-9 the time-varying linear system is briefly introduced. Little analysis could be performed on an input-output basis with the time-varying linear system. However, with state variable methods there is a great deal more that can be done to analyze the time-varying linear system. This chapter is an introduction to some of the important topics regarding state variable methods for linear time-varying systems.

12-1 THE GENERAL LINEAR TIME-VARYING STATE-MODEL SOLUTION

The general linear time-varying state-space model has the form

$$\dot{\mathbf{x}}(t) = A(t)\mathbf{x}(t) + B(t)\mathbf{u}(t) \tag{12-1}$$

$$\mathbf{y}(t) = C(t)\mathbf{x}(t) + D(t)\mathbf{u}(t) \tag{12-2}$$

where at least one of the matrices A, B, C, and D varies in time t; if they are all constant this is the time-invariant state model. We assume that all four are functions of t. For the general case it is assumed that the state $\mathbf{x}(t)$ is n-dimensional; the output $\mathbf{y}(t)$ is m-dimensional; and the input $\mathbf{u}(t)$ is ℓ-dimensional. The matrices $A(t)$, $B(t)$, $C(t)$, and $D(t)$ are then dimensioned accordingly.

The time-varying linear system model of the form of Eqs. (12-1) and (12-2) is frequently encountered in the process of linearizing a nonlinear system model about some nominal solution path. This topic is developed in Sec. 12-6. This section considers the analytic solution of the state model [Eqs. (12-1) and (12-2)] in terms of system state transition matrix, denoted $\Phi(t, t_0)$.

The Zero-Input Solution

Assume the system in Eqs. (12-1) and (12-2) starts at the initial state $\mathbf{x}(t_0)$ at initial time t_0 and has zero input for $t > t_0$. Solution of the state model then yields the zero-input response.

The form of this zero-input response may be developed purely on the basis of the linearity of the model, Eqs. (12-1) and (12-2). The initial state vector may be written as

$$\mathbf{x}(t_0) = \begin{bmatrix} x_1(t_0) \\ x_2(t_0) \\ \vdots \\ x_n(t_0) \end{bmatrix}$$

$$= x_1(t_0)\begin{bmatrix} 1 \\ 0 \\ \vdots \\ 0 \end{bmatrix} + x_2(t_0)\begin{bmatrix} 0 \\ 1 \\ \vdots \\ 0 \end{bmatrix} + \cdots + x_n(t_0)\begin{bmatrix} 0 \\ 0 \\ \vdots \\ 1 \end{bmatrix} \qquad (12\text{-}3)$$

Define the n-dimensional unit vectors $\phi_i(t_0, t_0)$, $i = 1, 2, \ldots, n$, by

$$\phi_i(t_0, t_0) \equiv \begin{bmatrix} 0 \\ 0 \\ \vdots \\ 1 \\ \vdots \\ 0 \end{bmatrix}_{n\times 1} \quad i\text{th row element} \qquad (12\text{-}4)$$

Then

$$\Phi(t_0, t_0) \equiv [\phi_1(t_0, t_0), \quad \phi_2(t_0, t_0), \quad \ldots, \quad \phi_n(t_0, t_0)]_{n\times n}$$
$$= I \qquad \text{the identity matrix} \qquad (12\text{-}5)$$

Equation (12-3) may be written in terms of the $\phi_i(t_0, t_0)$ unit vectors as

$$\mathbf{x}(t_0) = \sum_{i=1}^{n} x_i(t_0)\boldsymbol{\phi}_i(t_0, t_0) \qquad (12\text{-}6a)$$

or equivalently as

$$\mathbf{x}(t_0) = \Phi(t_0, t_0)\mathbf{x}(t_0) \qquad (12\text{-}6b)$$

With zero input the state differential equation, Eq. (12-1), may be written as the homogeneous time-varying differential equation

$$\dot{\mathbf{x}}_{zi}(t) = A(t)\mathbf{x}_{zi}(t) \qquad (12\text{-}7)$$

Then define $\phi_i(t, t_0)$, $i = 1, 2, \ldots, n$, $t \geq t_0$, as the *unit vector solutions* to the n time-varying homogeneous differential equation problems:

$$\frac{d}{dt}\boldsymbol{\phi}_i(t, t_0) = A(t)\boldsymbol{\phi}_i(t, t_0) \qquad i = 1, 2, \ldots, n \qquad (12\text{-}8)$$

Boundary conditions are Eq. (12-4). It is generally a nontrivial matter to find these n unit solutions, $\boldsymbol{\phi}_i(t, t_0)$, $i = 1, 2, \ldots, n$. Usually they must be found by numerical integration on the digital computer. Numerical integration is introduced in Sec. 12-5. However, for now suppose that the $\boldsymbol{\phi}_i(t, t_0)$, $i = 1, 2, \ldots, n$, $t \geq t_0$, solutions to Eq. (12-8) have been found.

Because of the relation between $\mathbf{x}(t_0)$ and the $\boldsymbol{\phi}_i(t_0, t_0)$, Eq. (12-6*b*), we may go all the way back to the Chap. 1 principle of linearity of the zero-input response to write the solution to Eq. (12-7) as the linear combination of the n unit vector solutions:

$$\mathbf{x}_{zi}(t) = \sum_{i=1}^{n} x_i(t_0)\boldsymbol{\phi}_i(t, t_0) \qquad t \geq t_0 \tag{12-9}$$

Define the $n \times n$ *state transition matrix,* $\Phi(t, t_0)$, from the n unit vector solutions $\boldsymbol{\phi}_i(t, t_0)$ as

$$\Phi(t, t_0) \equiv [\boldsymbol{\phi}_1(t, t_0), \quad \boldsymbol{\phi}_2(t, t_0), \quad \ldots, \quad \boldsymbol{\phi}_n(t, t_0)]_{n\times n} \tag{12-10}$$

Note that this definition is compatible with Eq. (12-5). Then Eq. (12-9) written as the linear combination of unit solutions is equivalent to

$$\boxed{\mathbf{x}_{zi}(t) = \Phi(t, t_0)\mathbf{x}(t_0)} \tag{12-11}$$

The corresponding output response to the initial state $\mathbf{x}(t_0)$ is therefore

$$\boxed{\mathbf{y}_{zi}(t) = C(t)\Phi(t, t_0)\mathbf{x}(t_0)} \tag{12-12}$$

Equations (12-11) and (12-12) are equivalent to the zero-input solutions for the linear time-invariant system

$$\mathbf{x}_{zi}(t) = e^{A(t-t_0)}\mathbf{x}(t_0) = \Phi(t - t_0)\mathbf{x}(t_0) \tag{12-13}$$

$$\mathbf{y}_{zi}(t) = Ce^{A(t-t_0)}\mathbf{x}(t_0) = C\Phi(t - t_0)\mathbf{x}(t_0) \tag{12-14}$$

However, unlike the $\Phi(t) = e^{At}$ for the constant A matrix, there is no general "closed-form" solution for $\Phi(t, t_0)$ with the time-varying $A(t)$ matrix. There are some special exceptions; the interested reader is referred to Refs. 1 to 3.

In general one must solve $\Phi(t, t_0)$ from the n unit vector time-varying differential equations, Eq. (12-8). Each of these n unit vector differential equations, in turn, represents n-scalar, -coupled, linear time-varying differential equations. Hence Eq. (12-8) really represents n^2-coupled linear time-varying differential equations. These may be written compactly as

$$\boxed{\frac{d}{dt}\Phi(t, t_0) = A(t)\Phi(t, t_0)} \tag{12-15}$$

with boundary condition

$$\boxed{\Phi(t_0, t_0) = I} \tag{12-16}$$

Equations (12-15) and (12-16) are equivalent to

$$\dot{\Phi}(t) = A\Phi(t) \qquad \Phi(0) = I \tag{12-17}$$

for the constant A. However, unlike the constant A matrix case, numerical solution of Eqs. (12-15) and (12-16) is about the only good way, in general, to find $\Phi(t, t_0)$. Furthermore, $\Phi(t, t_0)$ varies not only with changing t but also with changing t_0.

Example 12-1 Consider the linear time-varying system model

$$\dot{\mathbf{x}}(t) = \begin{bmatrix} -1 & t \\ 0 & -2 \end{bmatrix}\mathbf{x}(t) + \begin{bmatrix} 0 \\ 1 \end{bmatrix}u(t)$$

$$y(t) = [\sin t \quad \cos t]\mathbf{x}(t)$$

The state transition matrix differential equation is

$$\frac{d}{dt}\Phi(t, t_0) = \begin{bmatrix} -1 & t \\ 0 & -2 \end{bmatrix}\Phi(t, t_0)$$

with boundary condition

$$\Phi(t, t_0) = \begin{bmatrix} 1 & 0 \\ 0 & 1 \end{bmatrix}$$

For notational convenience denote

$$\Phi(t, t_0) = \begin{bmatrix} \phi_{1,1}(t, t_0) & \phi_{1,2}(t, t_0) \\ \phi_{2,1}(t, t_0) & \phi_{2,2}(t, t_0) \end{bmatrix}$$

Making this substitution into the state transition matrix differential equation gives the four coupled scalar differential equations:

$$\begin{aligned} \dot{\phi}_{1,1} &= -\phi_{1,1} + t\phi_{2,1} & \phi_{1,1}(t_0, t_0) &= 1 \\ \dot{\phi}_{2,1} &= -2\phi_{2,1} & \phi_{2,1}(t_0, t_0) &= 0 \\ \dot{\phi}_{1,2} &= -\phi_{1,2} + t\phi_{2,2} & \phi_{1,2}(t_0, t_0) &= 0 \\ \dot{\phi}_{2,2} &= -2\phi_{2,2} & \phi_{2,2}(t_0, t_0) &= 1 \end{aligned}$$

Note that the differential equations for each column of $\Phi(t, t_0)$ are inter-coupled with elements in that column but not between columns. Solving for the first column elements gives

$$\phi_{2,1}(t, t_0) = 0 \qquad \text{for all } t$$

$$\phi_{1,1}(t, t_0) = e^{-(t-t_0)}$$

For the second column it is seen that

$$\phi_{2,2}(t, t_0) = e^{-2(t-t_0)}$$

$$\phi_{1,2}(t, t_0) = \int_{t_0}^{t} e^{-(t-\tau)}[\tau\phi_{2,2}(\tau, t_0)]\, d\tau$$

$$= \int_{t_0}^{t} e^{-t}e^{2t_0}\tau e^{-\tau}\, d\tau$$

From a table of integrals and a bit of algebra the expression for $\phi_{1,2}(t, t_0)$ becomes

$$\phi_{1,2}(t, t_0) = e^{-(t-t_0)}(t_0 + 1) - e^{-2(t-t_0)}(t + 1)$$

All elements of the 2×2 $\Phi(t, t_0)$ state transition matrix have been found. The state transition matrix may then be used to find either $\mathbf{x}_{zi}(t)$ or $y_{zi}(t)$ from Eqs. (12-11) and (12-12) for a given $\mathbf{x}(t_0)$ at a specified initial time t_0. Note that $\Phi(t, t_0)$ is a function of both t and t_0 [the $\phi_{1,2}(t, t_0)$ element in particular] and not just on the time difference $t - t_0$. This is a basic characteristic of the time-varying system which distinguishes it from the time-invariant system. This is the same property as that discussed in Sec. 1-9.

The Zero-State Solution

Quite analogous to the time-invariant zero-state solution, the zero-state solution to the time-varying system of Eqs. (12-1) and (12-2) is

$$\boxed{\mathbf{x}_{zs}(t) = \int_{t_0}^{t} \Phi(t, \tau)B(\tau)\mathbf{u}(\tau)\, d\tau} \tag{12-18}$$

$$\boxed{\mathbf{y}_{zs}(t) = \int_{t_0}^{t} C(t)\Phi(t, \tau)B(\tau)\mathbf{u}(\tau)\, d\tau + D(t)\mathbf{u}(t)} \tag{12-19}$$

To demonstrate that Eqs. (12-18) and (12-19) are correct, it is necessary to show that $\mathbf{x}_{zs}(t_0) = \mathbf{0}$ and that $\mathbf{x}_{zs}(t)$, Eq. (12-18), satisfies the differential equation, Eq. (12-1). The first condition is true since the integral from t_0 to $t = t_0$ is zero for a nonimpulsive input.

For the second condition, simply substitute Eq. (12-18) into Eq. (12-1) and see if there is equality:

$$\frac{d}{dt}\left(\int_{t_0}^{t} \Phi(t, \tau)B(\tau)\mathbf{u}(\tau)\, d\tau\right) \stackrel{?}{=} A(t)\int_{t_0}^{t} \Phi(t, \tau)B(\tau)\mathbf{u}(\tau)\, d\tau + B(t)\mathbf{u}(t) \tag{12-20}$$

Leibnitz rule (e.g., [4]) must be used to differentiate the integral:

$$\frac{d}{dt}\left(\qquad \right) = \Phi(t, t)B(t)\mathbf{u}(t) + \int_{t_0}^{t} \frac{d}{dt}[\Phi(t, \tau)]B(\tau)\mathbf{u}(\tau)\, d\tau \tag{12-21}$$

Using

$$\Phi(t, t) = I \qquad \text{and} \qquad \frac{d}{dt}\Phi(t, \tau) = A(t)\Phi(t, \tau)$$

in Eq. (12-21) shows that Eq. (12-20) is correct. Thus Eqs. (12-18) and (12-19) are the correct zero-state solution for the linear time-varying system.

The Superposition Integral

The zero-state solution of Eqs. (12-18) and (12-19) is even less appealing than the zero-state solution for the linear time-invariant system. At least in the time-invariant system the undesirable convolution integral could be eliminated:

$$\int_{t_0}^{t} e^{A(t-\tau)}B\mathbf{u}(\tau)\, d\tau = e^{At}\int_{t_0}^{t} e^{-A\tau}B\mathbf{u}(\tau)\, d\tau \tag{12-22}$$

The time-varying state transition matrix does not separate this way, so the integrals in Eqs. (12-18) and (12-19) do not simplify.

The integrals in Eqs. (12-18) and (12-19) are called *superposition integrals.* They are difficult to compute. Note that within the integral, the integral involves $\Phi(t, \tau)$ with variable of integration τ. This makes the integral especially difficult because $\Phi(t, \tau)$ changes with *both* t and τ independently and is not simply a function of the time difference $(t - \tau)$. Equations (12-18) and (12-19) are useful for analysis purposes but are seldom used for computational purposes.

The Time-Varying Impulse Response Function

In Sec. 1-9 the zero-state response is written as the superposition integral

$$y_{zs}(t) = \int_{t_0}^{t} h(t, \tau)u(\tau)\, d\tau \tag{12-23}$$

where $h(t, \tau)$ is the time-varying linear system impulse response function. Comparing Eq. (12-23) with Eq. (12-19) then provides an expression for $h(t, \tau)$ in terms of the time-varying state-space model. Namely,

$$\boxed{h(t, \tau) = C(t)\Phi(t, \tau)B(\tau) + D(t)\delta(t - \tau)} \tag{12-24}$$

This is equivalent to

$$h(t) = Ce^{At}B + D\delta(t) \tag{12-25}$$

for the linear time-invariant system.

Example 12-2 For the system of Example 12-1

$$h(t, \tau) = \sin(t)\phi_{1,2}(t, \tau) + \cos(t)\phi_{2,2}(t, \tau)$$

The Complete Solution

The time-varying system of Eqs. (12-1) and (12-2) is linear, so it satisfies

$$\mathbf{x}(t) = \mathbf{x}_{zi}(t) + \mathbf{x}_{zs}(t) \tag{12-26}$$

$$\mathbf{y}(t) = \mathbf{y}_{zi}(t) + \mathbf{y}_{zs}(t) \tag{12-27}$$

Substitution of previous results then gives

$$\boxed{\mathbf{x}(t) = \Phi(t, t_0)\mathbf{x}(t_0) + \int_{t_0}^{t} \Phi(t, \tau)B(\tau)\mathbf{u}(\tau)\, d\tau} \tag{12-28}$$

$$\boxed{\mathbf{y}(t) = C(t)\Phi(t, t_0)\mathbf{x}(t_0) + \int_{t_0}^{t} C(t)\Phi(t, \tau)B(\tau)\mathbf{u}(\tau)\, d\tau + D(t)\mathbf{u}(t)} \tag{12-29}$$

Equations (12-28) and (12-29) again emphasize the fundamental property of the state-space model: the state $\mathbf{x}(t_0)$, t_0 arbitrary, is a complete summary of past information. The only other thing that one needs to know to predict the future response is the state model and the future input. This fact plays an important role in the numerical integration solution of the linear time-varying state model. However, before going into this topic it is valuable to consider some basic properties of the state transition matrix and analysis insights which the solution to Eqs. (12-28) and (12-29) provides.

12-2 PROPERTIES OF THE STATE TRANSITION MATRIX

Important properties of the state transition matrix for a given real, square, piecewise continuous $A(t)$ are summarized in the following theorem:

Theorem 12-1 *Properties of* $\Phi(t, t_0)$. Let $\Phi(t, t_0)$ be the unique solution of

$$\frac{d}{dt}\Phi(t, t_0) = A(t)\Phi(t, t_0) \qquad \Phi(t_0, t_0) = I \tag{12-30}$$

Then,

1. $\Phi(t, t) = I$ for all t.
2. For any t_0, t_1, and t_2 (semigroup property)

$$\Phi(t_2, t_0) = \Phi(t_2, t_1)\Phi(t_1, t_0) \tag{12-31}$$

3. For any t and t_0, $\Phi(t, t_0)$ is nonsingular and

$$[\Phi(t, t_0)]^{-1} = \Phi(t_0, t) \tag{12-32}$$

4.
$$\frac{d}{d\tau}\Phi(t, \tau) = -\Phi(t, \tau)A(\tau) \tag{12-33}$$

with boundary condition $\Phi(t, t) = I$.

5. The *row* vectors of $\Phi(t, \tau)$, $\boldsymbol{\beta}_i(t, \tau)$, $i = 1, 2, \ldots, n$, satisfy the adjoint differential equations

$$\frac{d}{d\tau}\boldsymbol{\beta}_i(t, \tau) = -A^T(\tau)\boldsymbol{\beta}_i(t, \tau) \tag{12-34}$$

with boundary condition $\boldsymbol{\beta}_i(t, t) = \boldsymbol{\delta}_i$, where $\boldsymbol{\delta}_i$ is the n-dimensional vector with unity ith element and otherwise zero.

PROOF Property 1 stems immediately from the definition of $\Phi(t, t_0)$, since t_0 is an arbitrary initial time.

Property 2, the semigroup property, follows from the uniqueness of solution to (see, e.g., [12])

$$\dot{\mathbf{x}}(t) = A(t)\mathbf{x}(t) \tag{12-35}$$

With boundary condition $\mathbf{x}(t_0)$, it is true that $\mathbf{x}(t_1) = \Phi(t_1, t_0)\mathbf{x}(t_0)$. Starting from $\mathbf{x}(t_1)$, the solution at t_2 is $\mathbf{x}(t_2) = \Phi(t_2, t_1)\mathbf{x}(t_1)$. Substitution for $\mathbf{x}(t_1)$ then yields

$$\begin{aligned}\mathbf{x}(t_2) &= \Phi(t_2, t_1)\Phi(t_1, t_0)\mathbf{x}(t_0)\\ &= \Phi(t_2, t_0)\mathbf{x}(t_0)\end{aligned}$$

Property 3 is derived from 1 and 2 since

$$\Phi(t, t_0)\Phi(t_0, t) = \Phi(t, t) = I$$

$$\Phi(t_0, t)\Phi(t, t_0) = \Phi(t_0, t_0) = I$$

which together form the definition of a matrix inverse.

Property 4 then uses property 3 via

$$\frac{d}{d\tau}\Phi(t, \tau) = \frac{d}{d\tau}[\Phi(\tau, t)]^{-1}$$

But the derivative of a matrix inverse satisfies (see Prob. 12.9)

$$\frac{d}{d\tau}[\Phi(\tau, t)]^{-1} = -[\Phi(\tau, t)]^{-1}\left[\frac{d}{d\tau}\Phi(\tau, t)\right][\Phi(\tau, t)]^{-1}$$

Manipulating this expression gives

$$\begin{aligned}\frac{d}{d\tau}\Phi(t, \tau) &= -[\Phi(t, \tau)][A(\tau)\Phi(\tau, t)][\Phi(t, \tau)]\\ &= -\Phi(t, \tau)A(\tau)\end{aligned}$$

as claimed.

Equation (12-33) may be transposed to yield

$$\frac{d}{d\tau}\Phi^T(t, \tau) = -A^T(\tau)\Phi^T(t, \tau)$$

But the rows of $\Phi(t, \tau)$, $\boldsymbol{\beta}_i(t, \tau)$, $i = 1, 2, \ldots, n$, are the columns of $\Phi^T(t, \tau)$. Thus the $\boldsymbol{\beta}_i(t, \tau)$ are the unit solutions of the system Eq. (12-34) so Property 5 readily follows.

The properties of the time-varying state transition matrix are useful to the time-varying linear system in many of the same ways that properties of $\Phi(t) = e^{At}$ are useful to analysis of the linear time-invariant system. One of these applications is to analyze controllability and observability.

*12-3 CONTROLLABILITY OF THE LINEAR TIME-VARYING SYSTEM

The controllable subspace of the linear time-varying system is the range space of the zero-state linear mapping

$$\mathbf{x}_{zs}(t_f) = \int_{t_0}^{t_f} \Phi(t_f, \tau)B(\tau)\mathbf{u}(\tau)\, d\tau \qquad (12\text{-}18)$$

for all possible inputs, $\mathbf{u}(\tau)$, $t_0 < \tau < t_f$. But this is difficult to analyze, because no closed-form expression exists for $\Phi(t_f, \tau)$ as it does for $\Phi(\tau) = e^{A\tau}$.

An alternative approach is to define a controllability Gram matrix for the time-varying linear system as done for the time-invariant system in Sec. 11.8. Defining

$$\boxed{W_c(t_f, t_0) \equiv \int_{t_0}^{t_f} \Phi(t_f, \tau)B(\tau)B^T(\tau)\Phi^T(t_f, \tau)\, d\tau} \qquad (12\text{-}36)$$

then the linear time-varying system, Eq. (12-1), is completely state controllable over the interval (t_0, t_f) if and only if the rank of $W_c(t_f, t_0)$ is n. The zero-state solution, $\mathbf{x}_{zs}(t_f)$, may be driven to anywhere in state space with a suitably defined input function.

Indeed

$$\boxed{u^*(t) = B^T(t)\Phi^T(t_f, t)W_c^{-1}(t_f, t_0)\mathbf{x}} \qquad (12\text{-}37)$$

is the minimum energy-control solution such that $\mathbf{x}_{zs}(t_f) = \mathbf{x}$ starting from $\mathbf{x}_{zs}(t_0) = \mathbf{0}$. Verification of this solution is made by substitution of Eq. (12-37) into Eq. (12-18). (See Ref. 5.)

If $W_c(t_f, t_0)$ does not have rank n, the system is uncontrollable on the interval (t_0, t_f). The controllable subspace is the range space of $W_c(t_f, t_0)$ and the uncontrollable subspace is the null space of $W_c(t_f, t_0)$. Again $\mathbf{x}_{zs}(t_f)$ can never be forced to equal a state in the uncontrollable subspace.

The controllability Gram matrix for this time-varying linear system shares many properties with controllability Gram matrix of the time-invariant system. It is always symmetric; its eigenvalues are always zero or positive; eigenvectors corresponding to zero eigenvalues correspond to uncontrollable state directions; the size of each eigenvalue is inversely proportional to the amount of energy it takes to reach states in the corresponding eigenvector directions; and $W_c(t, t_0)$ may be found from the Lyapunov differential equation

$$\boxed{\frac{d}{dt} W_c(t, t_0) = A(t)W_c(t, t_0) + W_c(t, t_0)A^T(t) + B(t)B^T(t)} \tag{12-38}$$

$$\boxed{W_c(t_0, t_0) = 0} \tag{12-39}$$

In general, $W_c(t, t_0)$ has no steady-state value as $t \to \infty$.

*12-4 OBSERVABILITY OF THE LINEAR TIME-VARYING SYSTEM

The unobservable subspace over the interval (t_0, t_f) is defined as the space of $\mathbf{x}(t_0)$ such that

$$\mathbf{y}_{zi}(t) = C(t)\Phi(t, t_0)\mathbf{x}(t_0) \tag{12-12}$$

is "zero" for all t over the interval $t_0 < t < t_f$. If there is no such subspace in state space, the system of Eqs. (12-1) and (12-2) is said to be completely observable over that interval. It is important to specify the interval; like the property of controllability, the property of observability may change for different intervals with the time-varying system.

Also like controllability, the null space of the linear mapping $C(t)\Phi(t, t_0)$ is not easy to determine by a direct analysis. An *observability Gram matrix* may instead be defined as

$$\boxed{W_o(t_f, t_0) \equiv \int_{t_0}^{t_f} \Phi^T(\tau, t_0)C^T(\tau)C(\tau)\Phi(\tau, t_0)\, d\tau} \tag{12-40}$$

Then the system of Eqs. (12-1) and (12-2) is completely observable if and only if the rank of $W_o(t_f, t_0)$ is n. If not, the system is unobservable on the interval (t_0, t_f). The unobservable subspace is then the null space of $W_o(t_f, t_0)$. The observable subspace is the range space of $W_o(t_f, t_0)$.

The observability Gram matrix is applicable to this situation because $\mathbf{y}_{zi}(t)$ is defined to be zero if and only if

$$\int_{t_0}^{t_f} \mathbf{y}_{zi}^T(\tau)\mathbf{y}_{zi}(\tau)\, d\tau = 0 \tag{12-41}$$

This is a norm squared of the $\mathbf{z}_{zi}(t)$ function. Thus $\mathbf{y}_{zi}(t)$ is defined to be zero if it is zero in norm.

Substitution of Eq. (12-12) into Eq. (12-41) gives

$$\int_{t_0}^{t_f} \mathbf{y}_{zi}^T(\tau)\mathbf{y}_{zi}(\tau)\, d\tau = \mathbf{x}^T(t_0)\left(\int_{t_0}^{t_f} \Phi^T(\tau, t_0)C^T(\tau)C(\tau)\Phi(\tau, t_0)\, d\tau\right)\mathbf{x}(t_0) \tag{12-42}$$

The term inside parentheses is the observability Gram matrix.

The observability Gram matrix shares many properties that are similar to the controllability Gram matrix. For instance, it is symmetric and has eigenvalues which are always zero or positive. Its eigenvectors are always orthogonal and bear similar information regarding the "quality" of observability in certain directions of state space. Finally, it also may be calculated as solution to a time-varying Lyapunov matrix differential equation

$$\boxed{\frac{d}{dt} W_o(t_f, t) = -W_o(t_f, t)A(t) - A^T(t)W_o(t_f, t) - C^T(t)C(t)} \tag{12-43}$$

backward in time to t_0 starting from the terminal boundary condition

$$\boxed{W_o(t_f, t_f) = 0} \tag{12-44}$$

Derivation of Eq. (12-43) is made by differentiation of Eq. (12-40) using the Leibnitz rule for differentiation of an integral and the fact that

$$\frac{d}{dt}\Phi(\tau, t) = -\Phi(\tau, t)A(t) \tag{12-45}$$

(See Property 4 of Theorem 12-1 and Prob. 12.14.)

12-5 DISCRETE-TIME SIMULATION OF THE LINEAR TIME-VARYING SYSTEM

Generally it is not possible to find a discrete-time representation of the linear time-varying state solution, Eq. (12-28), analogous to the discrete-time-invariant state model

$$\mathbf{x}_T(k) = A_T\mathbf{x}_T(k-1) + B_T\mathbf{u}_T(k-1) \tag{8-19}$$

Even if the sample interval T is constant and $\mathbf{u}(t)$ is piecewise constant, the coefficient matrices in the discrete update

$$\mathbf{x}_T(k) = \Phi(kT, kT - T)\mathbf{x}_T(k-1) + \left(\int_{(k-1)T}^{kT} \Phi(kT, \tau)B(\tau)\,d\tau\right)\mathbf{u}_T(k-1) \tag{12-46}$$

are varying in time; the matrices $\Phi(kT, kT - T)$ and

$$\left(\int_{(k-1)T}^{kT} \Phi(kT, \tau)B(\tau)\,d\tau\right) \tag{12-47}$$

would have to be recomputed for each k. If this is the case, it would generally be impractical to use Eq. (12-46) for discrete simulation.

An alternative approach for discrete simulation of linear time-varying system is via numerical integration. Indeed, numerical integration may be used to simulate any state differential equation model, linear or nonlinear. It may be used also to simulate the linear time-invariant system, but Eq. (8-19) and the methods discussed in Chap. 8 would usually be more efficient for the linear time-invariant system.

The literature in the area of numerical integration is vast. (See, e.g., Refs. 6 to 9.) This section attempts to give only a brief introduction to one common technique—fourth-order Runge-Kutta numerical integration.

Runge-Kutta numerical integration is based on the concept of the Taylor's series approximation. Namely, the state at time kT may be approximated by the first-order Taylor's series

$$\mathbf{x}(kT) \simeq \mathbf{x}(kT - T) + \dot{\mathbf{x}}|_{(k-1)T}T \tag{12-48}$$

But the state derivative at time $t = (k-1)T$ is known from the state differential equation

$$\dot{\mathbf{x}}(t) = A(t)\mathbf{x}(t) + B(t)\mathbf{u}(t) \tag{12-1}$$

Namely, at the instant $t = (k-1)T$, the state derivative is

$$\dot{\mathbf{x}}(kT - T) = A(kT - T)\mathbf{x}(kT - T) + B(kT - T)\mathbf{u}(kT - T) \tag{12-49}$$

Making this substitution into Eq. (12-48) and rearranging terms gives the first-order Taylor's series discrete-state update equation

$$\mathbf{x}_T(k) = [I + A(kT - T)T]\mathbf{x}_T(k-1) + B(kT - T)\mathbf{u}_T(k-1)T \tag{12-50}$$

Equation (12-50) is a finite-difference equation. If A is constant, this finite-difference equation will actually be unstable if any of the eigenvalues of

$$[I + AT] \tag{12-51}$$

have magnitude greater than 1. This sets an upper bound on the step size T to maintain stability, but usually one would have to pick T much shorter to maintain accuracy of integration. Generally speaking, the problems of stability are obviated by using higher-order integration formulas.

The first-order Taylor's series, Eq. (12-50), yields rectangular integration and generally rather poor simulation accuracy. But it is known that the higher-order Taylor's series can improve this accuracy. For instance, in integrating a given scalar differential equation

$$\dot{x}(t) = f(x, t) \tag{12-52}$$

the second-order Taylor series approximation update is

$$x_T(k) \simeq x_T(k-1) + \dot{x}\Big|_{(k-1)T} T + \ddot{x}\Big|_{(k-1)T} \frac{T^2}{2!} \tag{12-53}$$

It has an error bound for each iteration step proportional to $T^2/2$. The error bound for rectangular integration is proportional to the step size itself, T. Suppose a fourth-order Taylor series is used; the update equation is

$$x_T(k) \simeq x_T(k-1) + x^{(1)}T + x^{(2)}\frac{T^2}{2!} + x^{(3)}\frac{T^3}{3!} + x^{(4)}\frac{T^4}{4!} \tag{12-54}$$

where all derivatives are evaluated at $t = (k-1)T$. Then the error between the true $x(kT)$ and the computed $x_T(k)$ is bounded by a factor proportional to $(T^4/4!)$. The approximation gets better and better the higher the order of Taylor series approximation one uses.

The problem with the high-order Taylor series approximation is that one usually does not know an expression for $x^{(2)}$ and higher-order derivatives. The differential equation provides an expression for the first derivative

$$\dot{x}(kT) \simeq f(x_T(k), kT) \tag{12-55}$$

at any time $t = kT$; generally, though, there is no convenient expression to use for the higher derivatives.

The strategy in Runge-Kutta integration is to numerically approximate the higher derivatives by using first-derivative values computed at points intermediate to the integration interval. A given-order Runge-Kutta numerical integration formula is fully equivalent to the same-order Taylor series approximation. The fourth-order Runge-Kutta is frequently used, and it has an error bound equal to a fourth-order Taylor series.

There are several different forms of the fourth-order Runge-Kutta algorithm. A frequently used form is given in Table 12-1.

This algorithm is expressed as the integration of the general nonlinear differential equation

$$\dot{\mathbf{x}}(t) = \mathbf{f}(\mathbf{x}, t) \tag{12-56}$$

from the state condition $\mathbf{x}(kT - T) = \mathbf{x}_T(k-1)$ to find $\mathbf{x}_T(k) \simeq \mathbf{x}(kT)$. The step interval is T. Note that the algorithm calls the derivative subroutine $\mathbf{f}(\mathbf{x}, t)$ four times in total to compute the state update $\mathbf{x}_T(k)$. This algorithm may be used to numerically integrate any set of n first-order differential equations. The key is to

Table 12-1 Fourth-order Runge-Kutta integration algorithm

1. Given $\mathbf{x}_T(k-1)$, $t=(k-1)T$, and a derivative subroutine $\mathbf{f}(\mathbf{x}, t)$ that computes $\dot{\mathbf{x}}(t) = \mathbf{f}(\mathbf{x}, t)$ for any $\mathbf{x}$ and t.
2. Compute, in turn, the four n-dimensional vectors:

$$\mathbf{g}_0 \equiv \mathbf{f}[\mathbf{x}_T(k-1), (k-1)T]$$

$$\mathbf{g}_1 \equiv \mathbf{f}\left[\mathbf{x}_T(k-1) + \mathbf{g}_0\left(\frac{T}{2}\right), \left(\frac{k-1}{2}\right)T\right]$$

$$\mathbf{g}_2 \equiv \mathbf{f}\left[\mathbf{x}_T(k-1) + \mathbf{g}_1\left(\frac{T}{2}\right), \left(\frac{k-1}{2}\right)T\right]$$

$$\mathbf{g}_3 \equiv \mathbf{f}[\mathbf{x}_T(k-1) + \mathbf{g}_2 T, kT]$$

3. Compute the n-dimensional vector

$$\mathbf{g}_4 \equiv \frac{\mathbf{g}_0 + 2\mathbf{g}_1 + 2\mathbf{g}_2 + \mathbf{g}_3}{6}$$

4. Update the state and time

$$\mathbf{x}_T(k) = \mathbf{x}_T(k-1) + \mathbf{g}_4 T$$

$$t = kT$$

write the appropriate derivative subroutine for $\dot{\mathbf{x}}(t)$. By proper choice of derivative subroutine, this same numerical integration routine may be applied to integrate $\Phi(t, t_0)$, $W_c(t, t_0)$, $W_o(t, t_0)$, or anything else that is desired.

Example 12-3 Set up the derivative subroutine $\mathbf{f}(\mathbf{x}, t)$ to simultaneously integrate

$$\begin{bmatrix} \dot{x}_1(t) \\ \dot{x}_2(t) \end{bmatrix} = \begin{bmatrix} 0 & 1 \\ -t & -2 \end{bmatrix} \begin{bmatrix} x_1(t) \\ x_2(t) \end{bmatrix} + \begin{bmatrix} 0 \\ e^{-t} \end{bmatrix} u(t)$$

and its state transition matrix $\Phi(t, t_0)$. Take $t_0 = 0$, assume initial state

$$\begin{bmatrix} x_1(0) \\ x_2(0) \end{bmatrix} = \begin{bmatrix} 2 \\ -3 \end{bmatrix}$$

and input $u(t) = \cos(5t)$.

The first task is to set up an augmented "state" vector differential equation. The differential equation for $\Phi(t, 0)$ is

$$\frac{d}{dt}\begin{bmatrix} \phi_{1,1}(t) & \phi_{1,2}(t) \\ \phi_{2,1}(t) & \phi_{2,2}(t) \end{bmatrix} = \begin{bmatrix} 0 & 1 \\ -t & -2 \end{bmatrix} \begin{bmatrix} \phi_{1,1}(t) & \phi_{1,2}(t) \\ \phi_{2,1}(t) & \phi_{2,2}(t) \end{bmatrix}$$

with boundary conditions

$$\begin{bmatrix} \phi_{1,1}(0) & \phi_{1,2}(0) \\ \phi_{2,1}(0) & \phi_{2,2}(0) \end{bmatrix} = \begin{bmatrix} 1 & 0 \\ 0 & 1 \end{bmatrix}$$

Define the "augmented" state vector X by

$$[X(1),\ X(2),\ X(3),\ X(4),\ X(5),\ X(6)]^T \equiv [x_1,\ x_2,\ \phi_{1,1},\ \phi_{2,1},\ \phi_{1,2},\ \phi_{2,2}]^T$$

Then in terms of the X state vector, the derivative relations DX are

$$DX(1) = X(2)$$

$$DX(2) = -tX(1) - 2X(2) + \exp(-t)u(t)$$

$$DX(3) = X(4)$$

$$DX(4) = -tX(3) - 2X(4)$$

$$DX(5) = X(6)$$

$$DX(6) = -tX(5) - 2X(6)$$

The input $u(t)$ is set at $u(t) = \cos(5t)$, but it could be made to be anything that is desired. These derivative relations would be set up in subroutine form for call by the Runge-Kutta algorithm subroutine. Usually this derivative subroutine would be set up in the form:

```
SUBROUTINE F(TIME, X, DX)
DIMENSION X(n), DX(n)
```

The current time t and the current state value X are input variables to the derivative subroutine.

Before starting the numerical integration one must initialize the time variable t, the initial state X, and the discrete step size T. These are variables necessary to initiate the Runge-Kutta numerical integration algorithm.

Important to the Runge-Kutta numerical integration algorithm is the step length T. If T is picked too large, the accuracy is poor. If T is picked too small, computation time is increased. Round-off errors can also become a problem if T is too small because of the large number of cycles required to cover a desired absolute time interval, (t_0, t_f). Numerous methods have been devised for step-size selection and control but a usually effective and simple method is to select an initial trial T; integrate forward using T; then halve T and again integrate forward; if there is no appreciable change in the solution at the corresponding kT solution points, the original T is probably small enough; if there is a significant change in the solution, set $T = T/2$ and repeat the test process.

The subject of numerical integration is a broad and important topic area. We have barely scratched the surface of the topic here. The interested reader is referred to the previously cited reference sources. However, like the various linear algebra and eigenvalue and eigenvector routines previously mentioned, there are high-quality numerical integration routines available at nearly all computational centers (e.g., IMSL [10]). Therefore, one does not have to be a numerical analyst specialist to make effective use of the technique of numerical integration. Numerical integration is an effective method to provide the discrete time simulation of the state differential equation, whether linear time-invariant, linear time-varying, or nonlinear. However, if the system is linear and time-invariant the methods presented in Chap. 8 would be a viable alternative to numerical integration.

12-6 LINEARIZING A NONLINEAR STATE-DIFFERENTIAL EQUATION

This last section of the text deals with the problem of linearizing a nonlinear state-differential equation for the purpose of deriving a linear state-differential equation model. Chapter 1 emphasizes that no physical system is ever linear by the strict definition of system linearity. However, it is also emphasized that almost all physical systems obey linear system relations if the operating region is sufficiently restricted. Therefore, it is only fitting that the text end with the mathematical procedure for actually accomplishing this.

We consider the problem of linearizing some given nonlinear state-differential equation model

$$\dot{\mathbf{x}}(t) = \mathbf{f}(\mathbf{x}, \mathbf{u}, t) \tag{12-57}$$

$$\mathbf{y}(t) = \mathbf{h}(\mathbf{x}, \mathbf{u}, t) \tag{12-58}$$

where $\mathbf{x}(t)$ is the n-dimensional state vector, $\mathbf{y}(t)$ is the m-dimensional output variable, and $\mathbf{u}(t)$ is the ℓ-dimensional input vector. The $\mathbf{f}(\cdot)$ and $\mathbf{h}(\cdot)$ are n- and m-dimensional nonlinear functions of the state $\mathbf{x}(t)$, the input $\mathbf{u}(t)$, and perhaps time itself, t.

Example 12-4

$$\begin{bmatrix} \dot{x}_1(t) \\ \dot{x}_2(t) \end{bmatrix} = \begin{bmatrix} -x_2(t) \\ x_1^2(t) - u(t)x_2(t) - u^2(t) \end{bmatrix}$$

$$y(t) = [x_1(t) + u(t)x_2(t)]$$

The nonlinear $\mathbf{f}(\cdot)$ vector is

$$\mathbf{f}(\mathbf{x}, \mathbf{u}, t) = \begin{bmatrix} f_1(\mathbf{x}, \mathbf{u}, t) \\ f_2(\mathbf{x}, \mathbf{u}, t) \end{bmatrix} = \begin{bmatrix} -x_2(t) \\ x_1^2(t) - u(t)x_2(t) - u^2(t) \end{bmatrix}$$

There is no explicit dependence on t. The nonlinear $h(\cdot)$ is

$$h(\mathbf{x}, \mathbf{u}, t) = [x_1(t) + u(t)x_2(t)]$$

The Nominal State Trajectory

Fundamental to linearizing the nonlinear system, Eqs. (12-57) and (12-58), is the concept of a nominal state solution or nominal state trajectory, $\mathbf{x}_{\text{nom}}(t)$, $t_0 \le t \le t_f$. The nominal state trajectory initiates from some given initial state, $\mathbf{x}_{\text{nom}}(t_0)$. Then the nominal state trajectory propagates forward in time being driven by a nominal forcing input, $\mathbf{u}_{\text{nom}}(t)$, $t_0 \le t \le t_f$.

Usually $\mathbf{x}_{nom}(t)$ would be found by numerical integration of the nonlinear state-differential equation, Eq. (12-57), using the $\mathbf{u}_{nom}(t)$ input. However, there are cases in which $\mathbf{u}_{nom}$ is a *constant* input. Then if $\mathbf{x}_{nom}$ happens to be just the right value it is true that

$$\dot{\mathbf{x}}_{nom} = \mathbf{0} = \mathbf{f}(\mathbf{x}_{nom}, \mathbf{u}_{nom}, t) \tag{12-59}$$

Such a condition is termed an equilibrium condition of the nonlinear system, and $\mathbf{x}_{nom} = \mathbf{x}_{eq}$ is constant; the system has reached a steady-state equilibrium value. The state dynamics do not change unless the input changes to perturb the state solution away from the equilibrium value.

A physical illustration of this situation is that of dropping a marble into a cup. The marble oscillates around the bottom of the cup until finally coming to a rest or equilibrium state at the bottom of the cup. If one jostles the cup—applies a perturbing input—the state dynamics of the marble are once again set in motion. To model the time evolution of these state dynamics, one would have to numerically integrate the nonlinear state-differential equations of motion.

Example 12-5 Again consider the nonlinear system of Example 12-4

$$\begin{bmatrix} \dot{x}_1(t) \\ \dot{x}_2(t) \end{bmatrix} = \begin{bmatrix} -x_2(t) \\ x_1^2(t) - u(t)x_2(t) - u^2(t) \end{bmatrix}$$

$$y(t) = [x_1(t) + u(t)x_2(t)]$$

An example equilibrium condition is

$$x_{1_{eq}} = 2 \qquad x_{2_{eq}} = 0 \qquad u_{eq} = 2$$

From the nonlinear differential equation it is seen that

$$\dot{\mathbf{x}}_{eq} = \mathbf{f}(\mathbf{x}_{eq}, u_{eq}) = 0$$

for these "equilibrium" values. The state trajectory would remain constant. The nominal output variable would also be constant at $y_{nom} = y_{eq} = 2$.

If the nominal state trajectory is started at some other initial state at $t_0 = 0$, say

$$\mathbf{x}_{1_{nom}}(0) = 1 \qquad \mathbf{x}_{2_{nom}}(0) = 1$$

and using the same constant input $u_{nom}(t) = 2u_{-1}(t)$, the system would not be in equilibrium. The nominal state trajectory would be changing in time. The nominal output response and nominal state trajectories are shown in Fig. 12-1.

Note that the nominal state trajectory eventually comes to the constant steady-state equilibrium value of $x_{1_{eq}} = 2$ and $x_{2_{eq}} = 0$, because it is a stable equilibrium point. State perturbations in a vicinity of this equilibrium condition come back to this steady-state equilibrium value.

Nonlinear systems can have multiple equilibrium conditions and not all of them need to be stable. With the same $u_{nom}(t) = 2u_{-1}(t)$ constant input, it

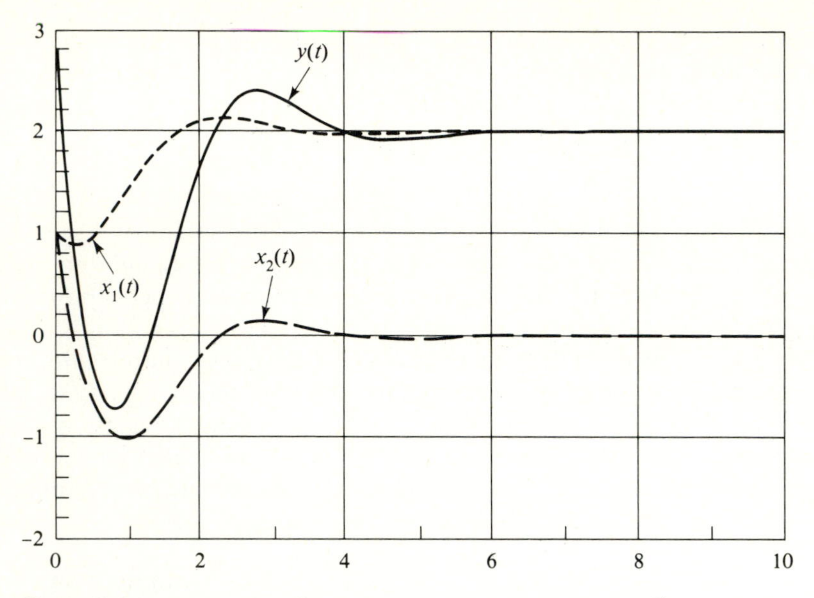

Figure 12-1 Response of nonlinear system to $u(t) = 2$, $\mathbf{x}(0) = [1, 1]^T$.

is seen that

$$x'_{1_{\text{eq}}} = -2 \qquad x'_{2_{\text{eq}}} = 0$$

is also an equilibrium point of the system. Once again

$$\dot{\mathbf{x}}'_{\text{eq}} = \mathbf{f}(\mathbf{x}'_{\text{eq}}, u_{\text{eq}}) = 0$$

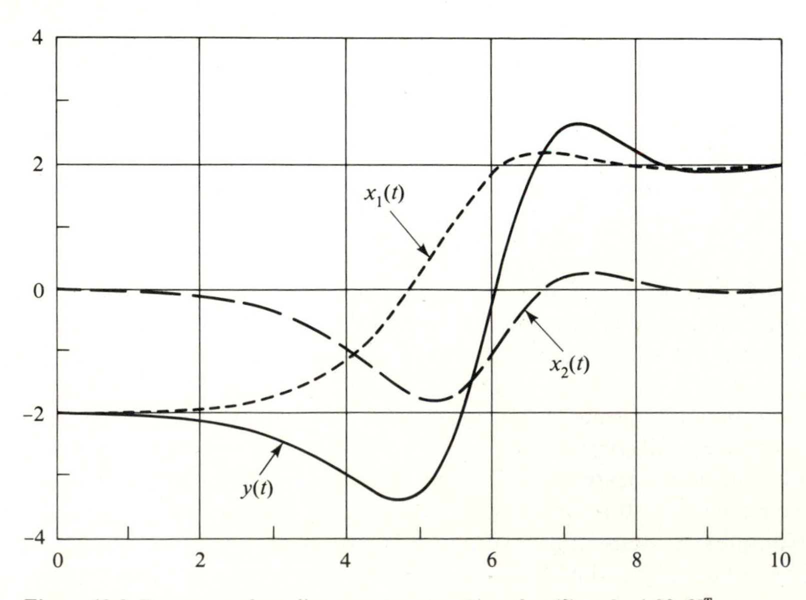

Figure 12-2 Response of nonlinear system to $u(t) = 2$, $\mathbf{x}(0) = [-1.99, 0]^T$.

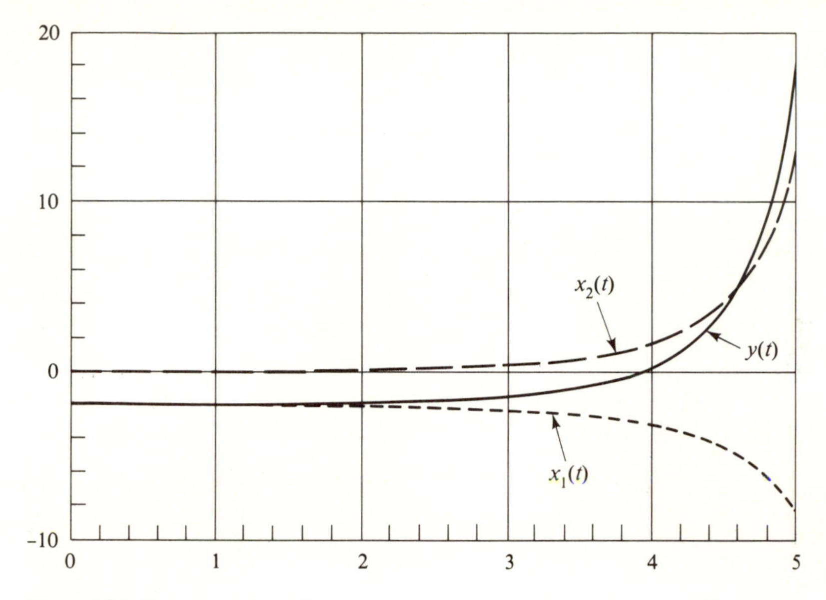

Figure 12-3 Response of nonlinear system to $u(t) = 2$, $\mathbf{x}(0) = [-2.01, 0]^T$.

for these state and input values. With no perturbations the system would remain at this constant equilibrium condition. However, just perturb the system ever so slightly and the system does not return to this equilibrium point. For instance, consider the initial state

$$x'_{1_{\text{nom}}}(0) = -1.99 \qquad x'_{2_{\text{nom}}}(0) = 0$$

with the same $u_{\text{nom}}(t) = 2u_{-1}(t)$ constant input. The state and output trajectory response is shown in Fig. 12-2. Note that the system eventually settles at the *stable* equilibrium state, $x_{1_{\text{eq}}} = 2$, $x_{2_{\text{eq}}} = 0$.

Next consider the initial conditions for the nominal trajectory

$$x''_{1_{\text{nom}}}(0) = -2.01 \qquad x''_{2_{\text{nom}}}(0) = 0$$

The resulting state and output trajectory is shown in Fig. 12-3. This time the response goes off to ∞.

The equilibrium state $x'_{1_{\text{eq}}} = -2$, $x'_{2_{\text{eq}}} = 0$ is said to be unstable. Small perturbations in the conditions tend to cause the system to depart dramatically from the equilibrium point without returning. This is illustrated graphically in Fig. 12-4. The stable equilibrium point is at the bottom of the cup at $x = 2$. With perturbations from this point the ball still tends to settle in the bottom of the cup. The unstable equilibrium point is at the top of the cup at $x = -2$. Perturbations to the right cause the ball to roll down into the cup. Perturbations to the left cause the ball to go off to $-\infty$. This is basically what is happening in this example.
Shortly it is seen that the linearized model gives us the means to analyze the stability of such an equilibrium condition. However, this is but a small glimpse

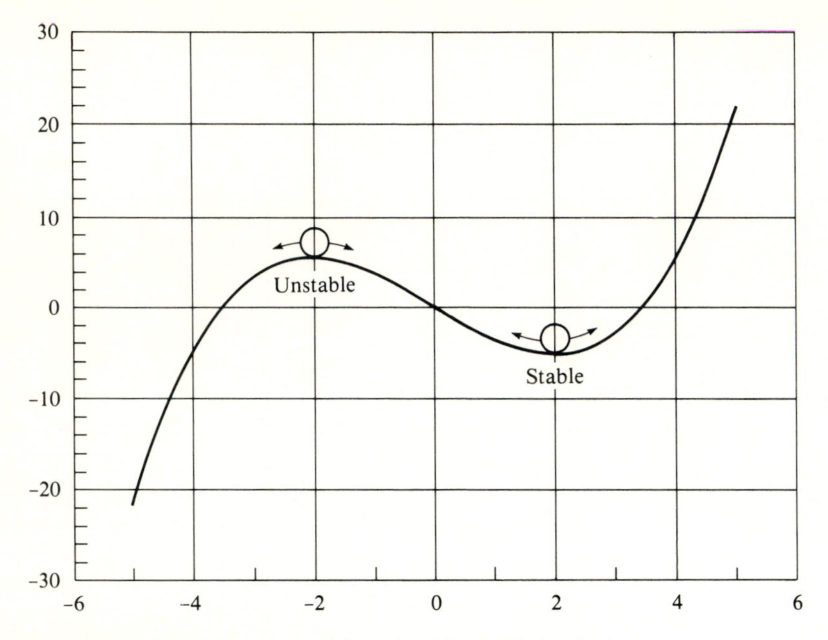

Figure 12-4 Illustration of unstable and stable equilibrium values.

into the fascinating topic of nonlinear systems theory. Some unusual occurrences can happen in nonlinear systems theory which do not have any counterpart in the linear system theory. The interested reader is referred to, for example, Ref. 11.

The Perturbed-State Trajectory

Consider next the process of finding the linearized dynamics of the system. These linearized dynamics are found referenced to some nominal state trajectory, $\mathbf{x}_{\text{nom}}(t)$, with a nominal input function, $\mathbf{u}_{\text{nom}}(t)$, producing the nominal output $\mathbf{y}_{\text{nom}}(t)$. What the linearized dynamics describe is how small perturbations from the nominal conditions will influence the computed state and output. The perturbations might take the form of perturbed initial state value

$$\mathbf{x}_{\text{pert}}(t_0) = \mathbf{x}_{\text{nom}}(t_0) + \Delta\mathbf{x}(t_0) \tag{12-60}$$

or a perturbed input value

$$\mathbf{u}_{\text{pert}}(t) = \mathbf{u}_{\text{nom}}(t) + \Delta\mathbf{u}(t) \qquad t \geq t_0 \tag{12-61}$$

In either event the perturbed state trajectory will differ from the nominal state trajectory

$$\mathbf{x}_{\text{pert}}(t) = \mathbf{x}_{\text{nom}}(t) + \Delta\mathbf{x}(t) \tag{12-62}$$

Likewise, the perturbed output will differ from the nominal output

$$\mathbf{y}_{\text{pert}}(t) = \mathbf{h}(\mathbf{x}_{\text{pert}}, \mathbf{u}_{\text{pert}}, t)$$
$$= \mathbf{y}_{\text{nom}}(t) + \Delta\mathbf{y}(t) \tag{12-63}$$

These concepts are illustrated in Fig. 12-5.

The Linear Perturbation-State Model

The concept behind the linearized trajectory is to find a first-order linear approximation to the true $\Delta\mathbf{x}(t)$ and the true $\Delta\mathbf{y}(t)$. We denote these first-order linear approximations by

$$\delta\mathbf{x}(t) \simeq \Delta\mathbf{x}(t) \qquad \delta\mathbf{y}(t) \simeq \Delta\mathbf{y}(t) \tag{12-64}$$

The linear perturbation state $\delta\mathbf{x}(t)$ satisfies the linear differential equation

$$\frac{d}{dt}\delta\mathbf{x}(t) = \left[\frac{\partial\mathbf{f}}{\partial\mathbf{x}}\right]_{n\times n}\delta\mathbf{x}(t) + \left[\frac{\partial\mathbf{f}}{\partial\mathbf{u}}\right]_{n\times \ell}\delta\mathbf{u}(t) \tag{12-65}$$

while $\delta\mathbf{y}(t)$ satisfies the linear output relation

$$\delta\mathbf{y}(t) = \left[\frac{\partial\mathbf{h}}{\partial\mathbf{x}}\right]_{m\times n}\delta\mathbf{x}(t) + \left[\frac{\partial\mathbf{h}}{\partial\mathbf{u}}\right]_{m\times \ell}\delta\mathbf{u}(t) \tag{12-66}$$

The initial condition on $\delta\mathbf{x}(t_0)$ is set at $\delta\mathbf{x}(t_0) = \Delta\mathbf{x}(t_0)$, the true state perturbation at t_0; while $\delta\mathbf{u}(t)$ is set at $\delta\mathbf{u}(t) = \Delta\mathbf{u}(t)$, the true perturbation in the input.

The $n \times n$, $n \times \ell$, $m \times n$, and $m \times \ell$ Jacobian matrices

$$\left[\frac{\partial\mathbf{f}}{\partial\mathbf{x}}\right] \quad \left[\frac{\partial\mathbf{f}}{\partial\mathbf{u}}\right] \quad \left[\frac{\partial\mathbf{h}}{\partial\mathbf{x}}\right] \quad \left[\frac{\partial\mathbf{h}}{\partial\mathbf{u}}\right] \tag{12-67}$$

are each evaluated along $\mathbf{x}_{\text{nom}}(t)$ and $\mathbf{u}_{\text{nom}}(t)$. These four matrices correspond to the A, B, C, and D matrices of the general linear state variable model. The perturbation variables $\partial\mathbf{x}(t)$, $\partial\mathbf{u}(t)$, and $\partial\mathbf{y}(t)$ correspond to the state, input, and output variables, respectively.

If the nominal state trajectory $\mathbf{x}_{\text{nom}}(t)$ and nominal input $\mathbf{u}_{\text{nom}}(t)$ are varying in time, the four matrices, Eq. (12-67), are, in general, time-varying. The linearized system, Eq. (12-65) and (12-66), is then a time-varying linear system model. If

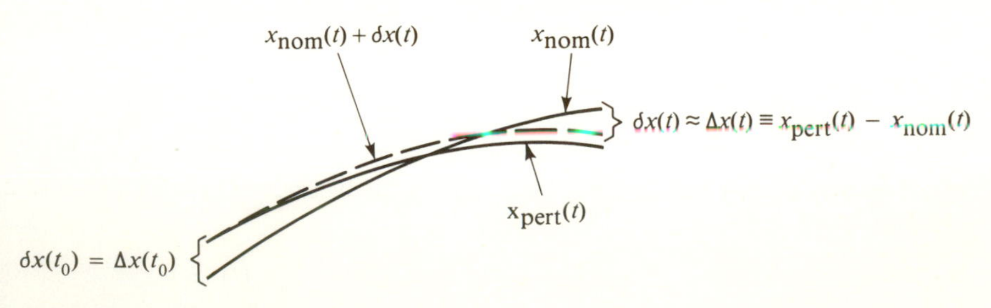

Figure 12-5 Illustration of relations between $x_{\text{nom}}(t)$, $x_{\text{pert}}(t)$, $x(t)$, and $x(t)$.

$\mathbf{x}_{nom}$ and $\mathbf{u}_{nom}$ are each at a constant or equilibrium condition the four matrices, Eq. (12-67), are constant. The linearized system, Eqs. (12-65) and (12-66), is then a linear time-invariant state-space model with constant A, B, C, and D matrices. These concepts are illustrated in Example 12-6.

Example 12-6 Again consider the nonlinear state dynamics of Example 12-4:

$$\begin{bmatrix} \dot{x}_1(t) \\ \dot{x}_2(t) \end{bmatrix} = \begin{bmatrix} -x_2(t) \\ x_1^2(t) - u(t)x_2(t) - u^2(t) \end{bmatrix}$$

$$y(t) = [x_1(t) + u(t)x_2(t)]$$

The linearized state-space model corresponding to Eqs. (12-65) and (12-66) is

$$\begin{bmatrix} \delta\dot{x}_1(t) \\ \delta\dot{x}_2(t) \end{bmatrix} = \begin{bmatrix} 0 & -1 \\ 2x_{1_{nom}}(t) & -u_{nom}(t) \end{bmatrix} \begin{bmatrix} \delta x_1(t) \\ \delta x_2(t) \end{bmatrix} + \begin{bmatrix} 0 \\ -x_{2_{nom}}(t) - 2u_{nom}(t) \end{bmatrix} \delta u(t)$$

$$\delta y(t) = [1, u_{nom}(t)] \begin{bmatrix} \delta x_1(t) \\ \delta x_2(t) \end{bmatrix}$$

where

$$\delta u(t) \equiv u_{pert}(t) - u_{nom}(t)$$

$$\begin{bmatrix} \delta x_1(t_0) \\ \delta x_2(t_0) \end{bmatrix} = \begin{bmatrix} x_{1_{pert}}(t_0) - x_{1_{nom}}(t_0) \\ x_{2_{pert}}(t_0) - x_{2_{nom}}(t_0) \end{bmatrix}$$

From the linear perturbation state model it is seen that this is a time-invariant linear system model or a time-varying linear system depending upon whether $\mathbf{x}_{nom}(t)$ and $\mathbf{u}_{nom}(t)$ are constant (they are at an equilibrium condition) or whether they are varying in time.

If the nominal trajectory values are varying in time, this represents a time-varying linear perturbation state model. This is one of the frequent ways in which a time-varying linear system model comes about in practice.

In either event, whether a constant $\mathbf{x}_{nom}$ and $\mathbf{u}_{nom}$ or a time-varying $\mathbf{x}_{nom}(t)$ and $\mathbf{u}_{nom}(t)$, the perturbation variables $\delta\mathbf{x}(t)$ and $\delta\mathbf{y}(t)$ represent the same thing. They are first-order linear approximations to the true

$$\Delta\mathbf{x}(t) \equiv \mathbf{x}_{pert}(t) - \mathbf{x}_{nom}(t) \tag{12-68}$$

$$\Delta\mathbf{y}(t) \equiv \mathbf{y}_{pert}(t) - \mathbf{y}_{nom}(t) \tag{12-69}$$

Example 12-7 Using the same nonlinear state model and resulting linear perturbation model as Example 12-6, we choose to linearize about the stable equilibrium condition

$$x_{1_{nom}}(t) = x_{1_{eq}} = 2 \qquad x_{2_{nom}}(t) = x_{2_{eq}} = 0 \qquad u_{nom}(t) = u_{eq} = 2$$

Then the linear perturbation state model is time-invariant

$$\begin{bmatrix} \delta\dot{x}_1(t) \\ \delta\dot{x}_2(t) \end{bmatrix} = \begin{bmatrix} 0 & -1 \\ 4 & -2 \end{bmatrix} \begin{bmatrix} \delta x_1(t) \\ \delta x_2(t) \end{bmatrix} + \begin{bmatrix} 0 \\ -4 \end{bmatrix} \delta u(t)$$

$$\delta y(t) = [1 \quad 2]\, \delta\mathbf{x}(t)$$

For simplicity we pick $u_{\text{pert}}(t) = 2u_{-1}(t)$. Then

$$\delta u(t) = u_{\text{pert}}(t) - u_{\text{nom}}(t) = 0$$

However, we pick boundary conditions on the perturbation initial state of $x_{1_{\text{pert}}}(0) = 1$, $x_{2_{\text{pert}}}(0) = 1$. Then boundary conditions on $\delta\mathbf{x}(0)$ are

$$\delta x_1(0) = x_{1_{\text{pert}}}(0) - x_{1_{\text{nom}}}(0) = -1$$

$$\delta x_2(0) = x_{2_{\text{pert}}}(0) - x_{2_{\text{nom}}}(0) = 1$$

Using these initial conditions we find the state solution for $\delta\mathbf{x}(t)$ to be

$$\delta\mathbf{x}(t) = \exp\left\{\begin{bmatrix} 0 & -1 \\ 4 & -2 \end{bmatrix} t\right\} \begin{bmatrix} -1 \\ 1 \end{bmatrix}$$

and $\delta y(t)$ is

$$\delta y(t) = [1, \quad 2] \exp\left\{\begin{bmatrix} 0 & -1 \\ 4 & -2 \end{bmatrix} t\right\} \begin{bmatrix} -1 \\ 1 \end{bmatrix}$$

The solution for $\delta\mathbf{x}(t)$ versus $\Delta\mathbf{x}(t) \equiv \mathbf{x}_{\text{pert}}(t) - \mathbf{x}_{\text{nom}}(t)$ and $\delta y(t)$ versus $\Delta y(t) \equiv y_{\text{pert}}(t) - y_{\text{nom}}(t)$ is shown in Fig. 12-6. The first-order linear approx-

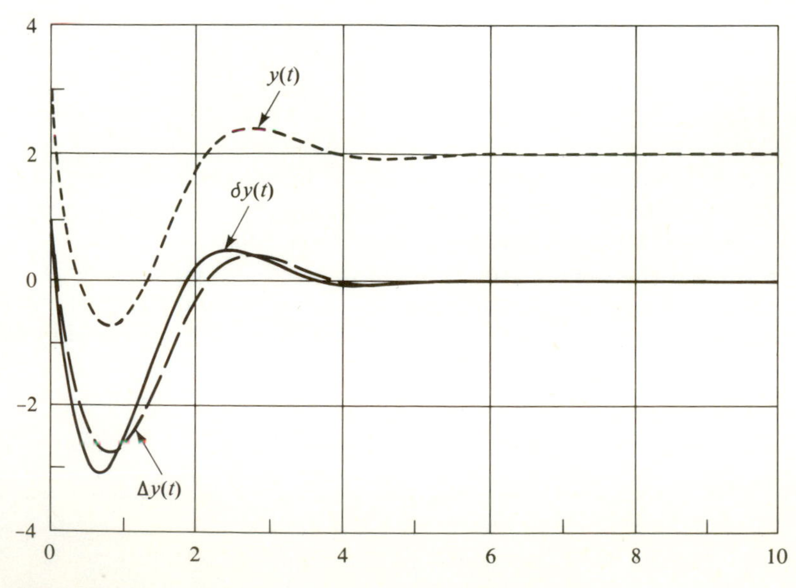

Figure 12-6 Plot of $\delta y(t)$ vs. $\Delta y(t)$ at stable equilibrium point.

imations $\delta\mathbf{x}(t)$ and $\delta y(t)$ are seen to be good approximations to $\Delta\mathbf{x}(t)$ and $\Delta y(t)$.

Stability of the Equilibrium Condition

In the last example, the linear perturbation state model is seen to give a good approximation to the true perturbations, because the system is linearized about a stable equilibrium point. If the same experiment is tried with linearization about the unstable equilibrium point, $x'_{1_{\text{eq}}} = -2$, $x'_{2_{\text{eq}}} = 0$, $u_{\text{eq}} = 2$, the results are very poor.

Example 12-8 With $x'_{1_{\text{nom}}}(t) = x'_{1_{\text{eq}}} = -2$, $x'_{2_{\text{nom}}}(t) = x'_{2_{\text{eq}}} = 0$, $u_{\text{nom}}(t) = u_{\text{eq}} = 2$, the linearized system is still time-invariant with the model

$$\begin{bmatrix} \delta\dot{x}_1(t) \\ \delta\dot{x}_2(t) \end{bmatrix} = \begin{bmatrix} 0 & -1 \\ -4 & -2 \end{bmatrix} \begin{bmatrix} \delta x_1(t) \\ \delta x_2(t) \end{bmatrix} + \begin{bmatrix} 0 \\ -4 \end{bmatrix} \delta u(t)$$

$$\delta y(t) = [1 \quad 2]\, \delta\mathbf{x}(t)$$

Again pick $\delta u(t) = 0$ and pick $\delta\mathbf{x}(0)$ very close to zero:

$$\begin{bmatrix} \delta x_1(0) \\ \delta x_2(0) \end{bmatrix} = \begin{bmatrix} -0.01 \\ 0 \end{bmatrix} \quad \text{or} \quad \begin{bmatrix} x_{1_{\text{pert}}}(0) \\ x_{2_{\text{pert}}}(0) \end{bmatrix} = \begin{bmatrix} -1.99 \\ 0 \end{bmatrix}$$

The linearized $\delta\mathbf{x}(t)$ and $\delta y(t)$ are given by

$$\delta\mathbf{x}(t) = \exp\left\{\begin{bmatrix} 0 & -1 \\ -4 & -2 \end{bmatrix} t\right\} \begin{bmatrix} -0.01 \\ 0 \end{bmatrix}$$

$$\delta y(t) = [1, \quad 2] \exp\left\{\begin{bmatrix} 0 & -1 \\ -4 & -2 \end{bmatrix} t\right\} \begin{bmatrix} -0.01 \\ 0 \end{bmatrix}$$

But now the linearized dynamics give a very poor approximation to the true $\Delta\mathbf{x}(t)$ and $\Delta y(t)$. This is seen in Fig. 12-7.

The difference between Examples 12-7 and 12-8 is fundamental. In Example 12-7 linearization is made about a stable equilibrium point. In Example 12-8 linearization is made about an unstable equilibrium point.

How in general can one determine whether a selected equilibrium point is stable or unstable? The answer is fairly direct. The linearized model is stable at a stable equilibrium point; it is unstable at an unstable equilibrium point. In Example 12-7 the linearized plant matrix

$$\left[\frac{\partial \mathbf{f}}{\partial \mathbf{x}}\right] = A = \begin{bmatrix} 0 & -1 \\ 4 & -2 \end{bmatrix}$$

has eigenvalues $\lambda_1 = -1 + \sqrt{3}j$ and $\lambda_2 = -1 - \sqrt{3}j$. Both eigenvalues are stable. But in Example 12-8 the linearized plant matrix

$$\left[\frac{\partial \mathbf{f}}{\partial \mathbf{x}}\right] = A = \begin{bmatrix} 0 & -1 \\ -4 & -2 \end{bmatrix}$$

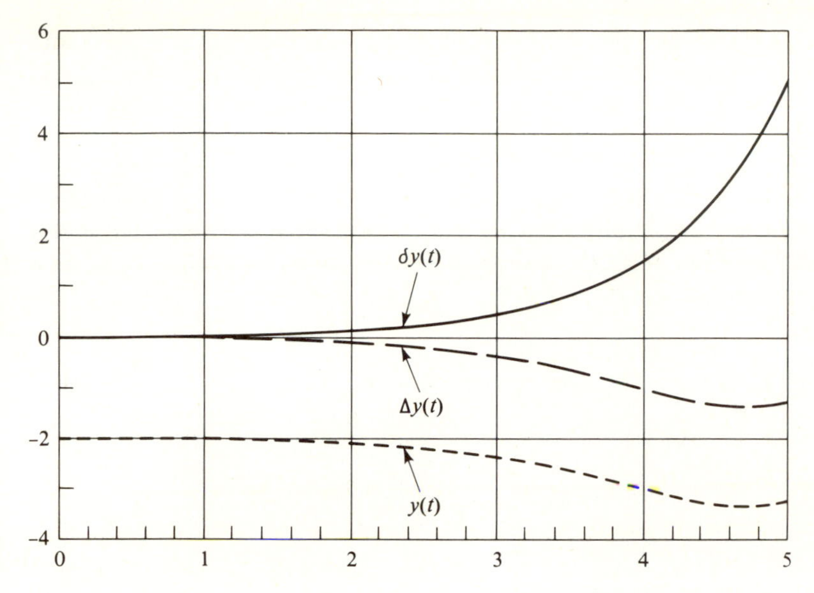

Figure 12-7 Plot of $\delta y(t)$ vs. $\Delta y(t)$ at unstable equilibrium point.

has eigenvalues $\lambda_1 = -1 - \sqrt{5}$, $\lambda_2 = -1 + \sqrt{5}$, one stable and one unstable. Since the linearized perturbation state model is unstable, it indicates that the underlying nonlinear equilibrium point is unstable.

Proof of these facts is beyond our scope, but they generalize to the nth-order case. This is an interesting connection between nonlinear systems theory and linear systems theory. It is a fitting note on which to end.

12-7 SUMMARY

This chapter has introduced analysis and simulation methods applicable to the linear time-varying state space model. The system state transition matrix is used to write an analytical solution to the general nth-order state-space model. This solution may again be decomposed into a sum of the zero-input and zero-state solutions. The zero-state solution takes the form of a superposition integral. A relation between the system state transition matrix and the system impulse response function is thereby derived.

The state transition matrix solution forms the basis of useful analysis of the system properties of controllability and observability. To conduct this analysis, one must form the controllability and observability Gram matrices. These are, in turn, most conveniently determined by solving a linear time-varying matrix differential equation.

Numerical integration is seen to be an effective method for digital simulation of the linear time-varying state differential equation. Numerical integration methods are quite general, and may be applied to linear time-varying, linear time-invariant, and nonlinear state models alike. However, the methods of Chap. 8 would generally lead to a "less-expensive" discrete time simulation of the linear time-invariant system.

The last section deals with the problem of linearizing a nonlinear state model about a nominal state solution trajectory. This can lead to either a linear time-invariant or linear time-varying "perturbation state" model, depending on whether or not the nominal state trajectory is at a constant equilibrium condition.

PROBLEMS

12-1 Find $\Phi(t, t_0)$ for the linear time-varying system

$$\dot{\mathbf{x}}(t) = \begin{bmatrix} -1 & 0 \\ e^{2t} & -20 \end{bmatrix}\mathbf{x}(t) + \begin{bmatrix} 1 \\ 0 \end{bmatrix}u(t)$$

$$y(t) = [t \quad 0]\mathbf{x}(t)$$

12-2 Find $\Phi(t, t_0)$ for the linear time-varying system

$$\dot{\mathbf{x}}(t) = \begin{bmatrix} -10 & 0 \\ t & -1 \end{bmatrix}\mathbf{x}(t) + \begin{bmatrix} 1 \\ 2 \end{bmatrix}u(t)$$

$$y(t) = [e^{-t} \quad 0]\mathbf{x}(t)$$

12-3 What is $h(t, \tau)$ for the system in Prob. 12-1?

12-4 What is $h(t, \tau)$ for the system in Prob. 12-2?

12-5 The system in Prob. 12-1 starts from the initial state $\mathbf{x}(0) = [1 \quad -1]^T$, and the unit step input is applied.

(*a*) What is $y_{zi}(t)$, $t \geq 0$?

(*b*) What is $y_{zs}(t)$, $t \geq 0$? Use the result from Prob. 12-3.

(*c*) What is $y(t)$?

12-6 Repeat Prob. 12-5 for the system of Prob. 12-2. Assume the same initial state and input.

12-7 For the state transition matrix of Prob. 12-1 demonstrate the first two properties of Theorem 12-1.

12-8 For the state transition matrix of Prob. 12-2 demonstrate the third property of Theorem 12-1.

12-9 Suppose that $A(t)$ is nonsingular and differentiable. Show that

$$\frac{d}{dt}[A^{-1}(t)] = -A^{-1}(t)\frac{d}{dt}[A(t)]A^{-1}(t)$$

Hint: $A^{-1}(t)A(t) = I$ and $(d/dt)I = 0$. Then take the derivative of both sides of $A^{-1}(t)A(t) = I$ and solve for the derivative of $A^{-1}(t)$.

12-10 Show that Eq. (12-37) for $\mathbf{u}^*(t)$ is the control which takes the linear time-varying system from $\mathbf{x}(t_0) = 0$ to $\mathbf{x}(t_f) = \mathbf{x}$. *Hint:* Substitute Eq. (12-37) into Eq. (12-18) using the definition Eq. (12-36) for the controllability Gram matrix.

12-11 Find the three scalar differential equations and boundary conditions that may be used to calculate $W_c(t, t_0)$ for the system of Prob. 12-1. Note: There are only three scalar differential equations required because $W_c(t, t_0)$ is symmetric.

12-12 Repeat Prob. 12-11 for the system of Prob. 12-12.

12-13 Show that the observability Gram matrix is symmetric.

12-14 Derive the differential equation for the observability Gram matrix, Eq. (12-43) with boundary condition Eq. (12-44), by differentiation of

$$W_o(t_f, t) = \int_t^{t_f} \Phi^T(\tau, t)C^T(\tau)C(\tau)\Phi(\tau, t)\, d\tau$$

with respect to t. Use the Leibnitz rule for differentiation of the integral

$$\frac{d}{dt}\left(\int_a^b f\right) = f(b)\frac{db}{dt} - f(a)\frac{da}{dt} + \int_a^b \frac{d}{dt}f$$

and use Property 4 of Theorem 12-1

$$\frac{d}{dt}\Phi(\tau, t) = -\Phi(\tau, t)A(t)$$

12-15 Find the three scalar differential equations and boundary conditions that may be used to compute $W_o(t_f, t)$ for the system of Prob. 12-1. There are only three scalar differential equations because $W_o(t_f, t)$ is symmetric.

12-16 Repeat Prob. 12-15 for the system of Prob. 12-2.

12-17 (*a*) Is the system of Prob. 12-1 stable or unstable? *Hint:* What happens to $A(t)$ as $t \to \infty$?

(*b*) Is the completely controllable/completely observable part of the system of Prob. 12-1 stable or unstable? *Hint:* Does $h(t, t_0)$ decay exponentially to zero as $t \to \infty$?

(*c*) Comment on the statement: "A time-varying linear system model may have strictly negative eigenvalues for all t and still be unstable."

12-18 Repeat Prob. 2-17 for the system of Prob. 2-2.

12-19 What is the state finite-difference equation for rectangular integration of the state differential equation of the system of Prob. 12-1?

12-20 Repeat Prob. 12-19 for the system of Prob. 12-2.

12-21 (*a*) For the system of Prob. 12-1 find a six-dimensional state vector, X, and corresponding differential equations that you could use to numerically integrate both $\mathbf{x}(t)$ and $\Phi(t, t_0)$ for $t \geq t_0$ with a given $u(t)$ input function. Set these up in "subroutine" format.

(*b*) What boundary conditions would you use on X at t_0?

12-22 Repeat Prob. 12-21 for the system of Prob. 12-2.

12-23 Repeat Prob. 12-21 to find the nine-dimensional state vector to simultaneously integrate $\mathbf{x}(t)$, $\Phi(t, t_0)$, and $W_c(t, t_0)$ for $t \geq t_0$ for the system of Prob. 12-1.

12-24 Repeat Prob. 12-23 for the system of Prob. 12-2.

12-25 (*a*) Suppose t_f is specified. Find a three-dimensional state vector $X(t)$ and corresponding differential equations to numerically integrate the three elements of $W_o(t_f, t)$. Set these up in "subroutine" form.

(*b*) What boundary condition on $X(t)$ would you use?

(*c*) Could you combine these three "states" with the nine of Prob. 12-21 to form a twelve-dimensional state vector to simultaneously numerically integrate $\mathbf{x}(t)$, $\Phi(t, t_0)$, $W_c(t, t_0)$, $W_o(t_f, t)$? Explain.

12-26 Repeat Prob. 12-5 for the system of Prob. 12-2.

12-27 Find the linear time-varying perturbation system model of the nonlinear system

$$\dot{x}_1(t) = -x_1(t) \sin u(t) + u(t)$$

$$\dot{x}_2(t) = x_1^2(t) - x_2^2(t) \sin x_1(t)$$

$$y(t) = x_1(t)x_2(t)$$

linearized about the nominal state solution $x_{1_{\text{nom}}}(t)$, $x_{2_{\text{nom}}}(t)$, with nominal input $u_{\text{nom}}(t)$.

12-28 Repeat Prob. 12-27 for the nonlinear system

$$\dot{x}_1(t) = -\sin x_1(t) + \cos u(t)$$

$$\dot{x}_2(t) = -x_2^2(t) + 1 + x_1(t)$$

$$y(t) = x_1^2(t) + x_1(t)x_2(t)$$

12-29 (*a*) Show that $x_{1_{nom}} = \pi/2$, $x_{2_{nom}} = \pi/2$, $u_{nom} = \pi/2$ is an equilibrium solution of the nonlinear system of Prob. 10-27.

(*b*) Find the linear time-invariant perturbation model for the equilibrium condition of part *a*.

(*c*) Determine whether the equilibrium condition is stable or unstable.

12-30 Repeat Prob. 12-29 for the conditions $x_{1_{nom}} = 0$, $x_{2_{nom}} = 1$, $u_{nom} = \pi/2$ with the system of Prob. 12-28.

12-31 Repeat Prob. 12-29 for $x_{1_{nom}} = \pi/2$, $x_{2_{nom}} = \pi/2$, and $u_{nom} = -\pi/2$.

12-32 Repeat Prob. 12-30 for $x_{1_{nom}} = \pi$, $x_{2_{nom}} = (1 + \pi)^{1/2}$, $u_{nom} = -\pi/2$.

NOTES AND REFERENCES

References 1 to 3 provide additional background on the linear-time-varying state space model. The results on controllability given in Ref. 5 are general to the linear time-varying system by using the system state transition matrix. References 6 to 9 are general references on numerical analysis for which numerical integration is a very important topic area. References 11 and 12 are just two of many good sources dealing with nonlinear systems. The material of this text, and this chapter in particular, is important background before embarking on a serious study into nonlinear systems theory.

1. DeRusso, P., R. Roy, and C. Close, *State Variables for Engineers,* Wiley, New York, 1965.
2. D'Angelo, H., *Linear Time-Varying Systems: Analysis and Synthesis,* Allyn and Bacon, Boston, 1970.
3. Chen, C. T., *Introduction to Linear System Theory,* Holt, New York, 1970.
4. Apostle, T. M., *Mathematical Analysis,* Addison-Wesley, Reading, MA, 1977, p. 99.
5. Kalman R. E., Y. C. Ho, and K. S. Narendra, "Controllability of Linear Dynamic Systems," *Contributions to Differential Equations,* vol. 1, no. 1, Wiley, New York, 1961, pp. 128–151.
6. Hildebrand, F. B., *Introduction to Numerical Analysis,* 2d ed., McGraw-Hill, New York, 1973.
7. Ralston, A., *A First Course in Numerical Analysis,* 2d ed., McGraw-Hill, New York, 1978.
8. Sheid, F. J., *Schaum's Outline of Theory and Problems of Numerical Analysis,* McGraw-Hill, New York, 1968.
9. Fox, L., *Numerical Solution of Ordinary and Partial Differential Equations,* Addison-Wesley, Reading, MA, 1962.
10. *IMSL Reference Manual,* IMSL Inc., Houston, TX, 1979.
11. Vidyasagar, M., *Nonlinear Systems Analysis,* Prentice-Hall, Englewood Cliffs, NJ, 1978.
12. Coddington, E. A., and N. Levinson, *Theory of Ordinary Differential Equations,* McGraw-Hill, New York, 1955.

APPENDIX

A

FINDING THE TRANSFER FUNCTION FOR A LINEAR CIRCUIT VIA LINEAR GRAPHS

Linear graphs provide a systematic method by which to analyze and simplify a given linear circuit, be it electrical, mechanical, thermal, fluid, or whatever. In particular, the system transfer function may be derived directly from the linear graph. This appendix develops the technique for passive electrical circuits and simple translational mechanical systems only. For more detail and for application to other types of systems the interested reader is referenced to Shearer, Murphy, and Richardson [A-1].

A-1 GENERALIZED IMPEDANCE RELATIONS

As developed in Chaps. 1–4, the transfer function $H(s)$ is defined as the ratio of the output-variable transform, $Y(s)$, to the input-variable transform, $U(s)$,

$$H(s) = \frac{Y(s)}{U(s)} \tag{A-1}$$

The same concept may be applied to individual circuit elements. The *transfer function* of the circuit element is defined as the *impedance* of that circuit element. The impedance of an individual circuit element is given the symbol $Z(s)$, and it is defined as the transform ratio of the across-variable change [given the general symbol $V(s)$] to the through input variable [given the general symbol $I(s)$]. Thus,

$$Z(s) \equiv \frac{V(s)}{I(s)} \tag{A-2}$$

for a given circuit element.

Electric Systems

For electrical systems the appropriate across variable is the voltage drop across individual elements such as resistors, capacitors, and inductors. The appropriate through variable is the current through these elements. Hence the applicable impedance for each of the three basic circuit elements may be defined as:

Resistor (R) From Ohm's law the voltage drop across a resistor is given by $v(t) = i(t)R$. Taking the Laplace transform and solving for $Z_R(s) = V(s)/I(s)$ gives

$$\boxed{Z_R(s) = R} \tag{A-3}$$

Capacitor (C) The voltage drop across a capacitor is related to the current by

$$\frac{d}{dt}v(t) = \frac{1}{C}i(t) \tag{A-4}$$

Taking the Laplace transform and solving for $Z_C(s) = V(s)/I(s)$ gives

$$\boxed{Z_C(s) = \frac{1}{Cs}} \tag{A-5}$$

Inductor (L) The voltage drop across an inductor is related to the current by

$$v(t) = L\frac{d}{dt}i(t) \tag{A-6}$$

Taking the Laplace transform and solving for $Z_L(s) = V(s)/I(s)$ gives

$$\boxed{Z_L(s) = Ls} \tag{A-7}$$

Mechanical Systems

For translational mechanical systems the appropriate across variable is the velocity of a node (a node is a point connecting an element to one or more elements). The appropriate through variable is the force transmitted through an element of the system. The impedance of a translational mechanical element is then the ratio of the velocity output to the force input.

Mechanical Resistor or Damper (b) The force exerted by a friction or damper element is proportional to the velocity of the moving surfaces or piston

$$F_b(t) = bv(t) \tag{A-8}$$

This $v(t)$ velocity might represent the velocity difference across two ends of a moving piston, or it might represent the velocity of a node across a stationary surface. Taking the Laplace transform and solving for the impedance $Z_b(s) = V(s)/F_b(s)$ gives

$$Z_b(s) = \frac{1}{b} \tag{A-9}$$

Mass (m) By Newton's law the reaction force of a mass to a force input is given by

$$F_m(t) = m\frac{d}{dt}v(t) \tag{A-10}$$

The acceleration force of the mass is always referenced to ground or to a stationary inertial reference frame. Taking the Laplace transform and solving for the impedance $Z_m(s) = V(s)/F_m(s)$ gives

$$Z_m(s) = \frac{1}{ms} \tag{A-11}$$

Spring (k) By Hooke's law the force exerted by a linear spring is proportional to the displacement of the two ends of the spring from their rest position:

$$F_k(t) = k[x_2(t) - x_1(t)] \tag{A-12}$$

Differentiating gives

$$\frac{d}{dt}F_k(t) = k[v_2(t) - v_1(t)] \tag{A-13}$$

or the force derivative is proportional to the velocity change across the spring. We represent this velocity change simply by $v(t)$. A generalized expression for the spring force is

$$\frac{d}{dt}F_k(t) = kv(t) \tag{A-14}$$

where $v(t)$ denotes the velocity change across the spring. Taking the Laplace transform and solving for the impedance $Z_k(s) = V(s)/F_k(s)$ gives

$$Z_k(s) = \frac{s}{k} \tag{A-15}$$

A summary of these generalized impedance relations along with their circuit diagram depiction is shown in Table A-1.

A-2 DERIVING THE LINEAR GRAPH

The first step in using the linear graph to find the transfer function of a given linear network—be it electrical, mechanical, or whatever—is to draw the linear

Table A-1 Summary of passive electrical and mechanical elements

Electrical element	Differential equation	Impedance
(*a*) Resistor 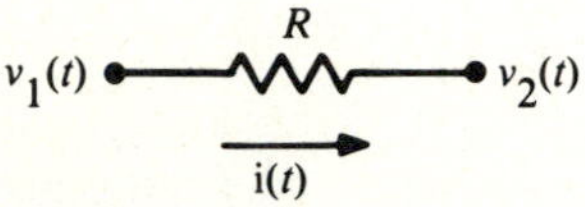 R $v_1(t)$ —— $v_2(t)$ $i(t)$	$v(t) = Ri(t)$	$Z_R(s) = R$
(*b*) Capacitor 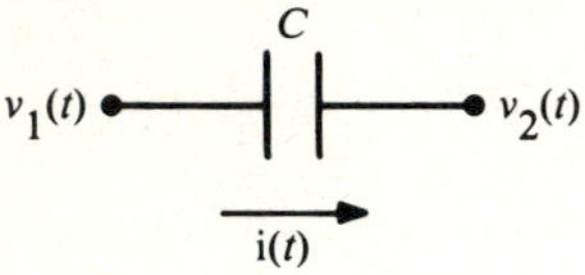 C $v_1(t)$ —— $v_2(t)$ $i(t)$	$C\frac{d}{dt}v(t) = i(t)$	$Z_C(s) = \frac{1}{Cs}$
(*c*) Inductor 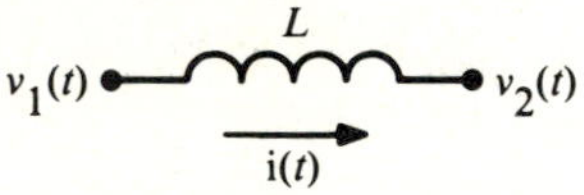 L $v_1(t)$ —— $v_2(t)$ $i(t)$	$v(t) = L\frac{d}{dt}i(t)$	$Z_L(s) = Ls$

Mechanical element	Differential equation	Impedance
(*a*) Damper or friction 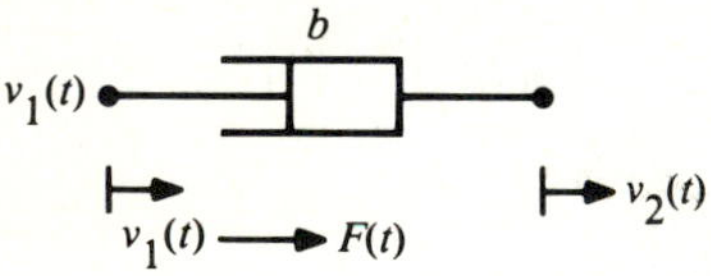 b $v_1(t)$ $v_2(t)$ $v_1(t)$ → $F(t)$	$v(t) = \frac{1}{b}F(t)$	$Z_b(s) = \frac{1}{b}$
(*b*) Mass 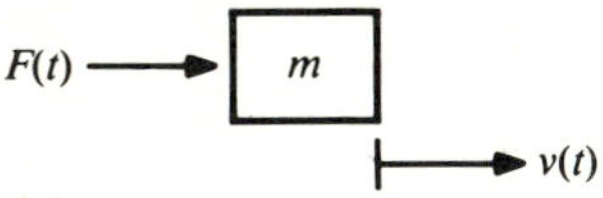 $F(t)$ → m $v(t)$	$m\frac{d}{dt}v(t) = F(t)$	$Z_m(s) = \frac{1}{ms}$
(*c*) Spring 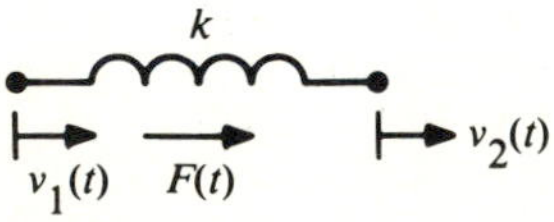 k $v_1(t)$ $F(t)$ $v_2(t)$	$v(t) = \frac{1}{k}\frac{d}{dt}F(t)$	$Z_k(s) = \frac{s}{k}$

graph showing the generalized impedance relations between all nodes. The node points in the network are points in the network having a distinct value of the across variable. For electrical networks this means finding and labeling all distinct voltage nodes. For mechanical translational systems this means finding and labeling all separate velocity nodes. Also to be defined and labeled is the *ground node* to be used as a reference for the other nodes. When one measures the voltage at a node in an electrical circuit, it is the voltage relative to some ground level. When one measures the velocity in a mechanical system, it is the velocity relative to ground or an inertial reference. For instance, it is common to use the earth as the reference for velocity measurements, but the earth is actually moving itself.

Once defining the ground level and all of the separate nodes, one draws all of the system connections between the nodes, including all input sources. These input sources might be through-variable input sources, such as current input sources in electrical systems or force input sources in mechanical systems (e.g., a rocket thrust force input). Or else the input source might be an across-variable input source such as a voltage source (e.g., a battery) in an electrical system or a velocity field (e.g., the current stream in a river) in a mechanical system. Through-variable input sources are labeled and shown by a circle containing an arrow in the direction of the input. Across-variable input sources are labeled and shown by a circle with a + and − showing the polarity relative to ground of the across-variable source.

The last step in the linear graph is to label all connections between nodes by the applicable impedance existing between the nodes. This depends on the type of circuit element existing between the nodes. Note that two nodes might have multiple or parallel paths between them. The next section has a number of examples illustrating these points.

A-3 USING THE LINEAR GRAPH TO FIND THE INPUT-OUTPUT TRANSFER FUNCTION, *H*(*s*)

One begins simplification of the linear graph by combining, in succession, impedances in series and impedances in parallel. Impedances connected in series have the equivalent impedance which is the sum of the individual impedances. Thus for $Z_1(s)$ and $Z_2(s)$ connected in series, the equivalent impedance is

$$Z_{\text{eq}}(s) = Z_1(s) + Z_2(s) \tag{A-16}$$

This also applies to n impedances connected in series

$$Z_{\text{eq}}(s) = \sum_{i=1}^{n} Z_i(s) \tag{A-17}$$

Two impedances connected in parallel have the equivalent impedance given by

$$Z_{\text{eq}}(s) = \left(\frac{1}{Z_1(s)} + \frac{1}{Z_2(s)}\right)^{-1} \tag{A-18}$$

This also applies to n impedances connected in parallel

$$Z_{\text{eq}}(s) = \left(\sum_{i=1}^{n} \frac{1}{Z_i(s)}\right)^{-1} \tag{A-19}$$

While performing the network simplifications one keeps track of intermediate transfer functions between the defined output variable, intermediate variables, and finally to the desired input variable. Three basic rules are used in defining the intermediate transfer functions:

1. The impedance definition itself

$$Z(s) \equiv \frac{V(s)}{I(s)} \tag{A-2}$$

2. The voltage divider rule (or across-variable divider rule) for impedances connected in series. This rule states that the ratio of the voltage drop across impedance $Z_i(s)$ to the voltage drop across all n impedances connected in series with equivalent impedance $Z_{eq}(s)$ is given by

$$\frac{V_i(s)}{V_{eq}(s)} = \frac{Z_i(s)}{Z_{eq}(s)} = \frac{Z_i(s)}{Z_1(s) + \cdots + Z_n(s)} \tag{A-20}$$

For a translational mechanical system, *velocity drop* would replace *voltage drop*.

3. The current divider rule (or through-variable divider rule) for impedances connected in parallel. This rule states that the ratio of current going through branch j to the total current entering the set of parallel branches is given by

$$\frac{I_j(s)}{I_{in}(s)} = \frac{1/Z_j(s)}{1/Z_{eq}(s)} = \frac{1/Z_j(s)}{\sum_{i=1}^{n} 1/Z_i(s)} \tag{A-21}$$

For a translational mechanical system, the term *force* would replace *current,* because force is the applicable through variable.

Example A-1 Consider the series electrical circuit of Fig. A-1, with the linear graph shown in Fig. A-2. The voltage divider rule gives the input-output transfer function of the output capacitor voltage

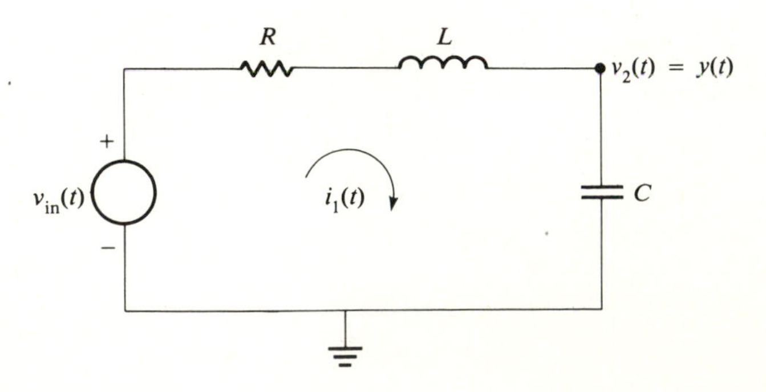

Figure A-1 Series *RLC* circuit.

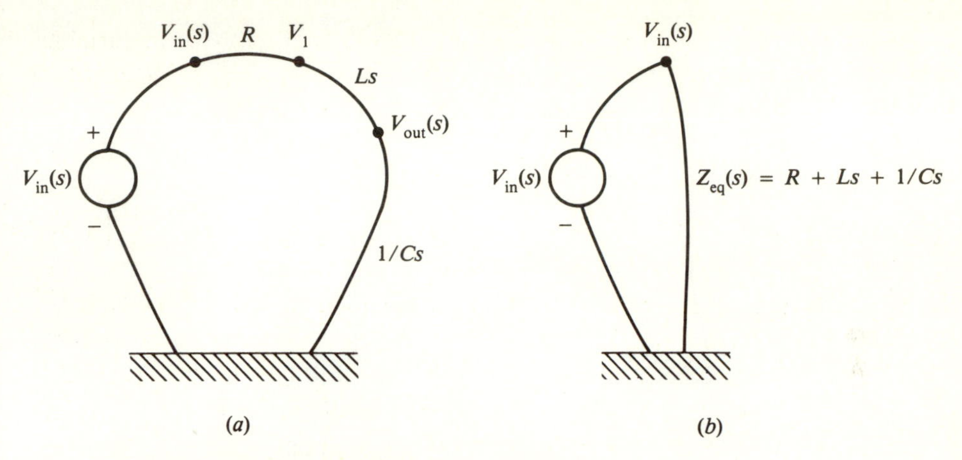

Figure A-2 Linear graph of series electric circuit of Fig. A-1.

$$\frac{V_{out}(s)}{V_{in}(s)} = \frac{1/Cs}{R + Ls + 1/Cs} = \frac{1}{LCs^2 + RCs + 1}$$

Example A-2 Consider the parallel electric circuit of Fig. A-3, with the linear graph shown in Fig. A-4. The current divider rule gives the input-output transfer function of the output current through inductor L as

$$\frac{I_{out}(s)}{I_{in}(s)} = \frac{1/Ls}{1/R + 1/Ls + Cs} = \frac{1}{LCs^2 + (L/R)s + 1}$$

If $V_C(s)$, the voltage drop across the capacitor, is chosen as the output, then

$$\frac{V_C(s)}{I_{in}(s)} \equiv Z_{eq}(s) = \left(\frac{1}{R} + \frac{1}{Ls} + Cs\right)^{-1}$$

$$= \frac{Ls}{LCs^2 + (L/R)s + 1}$$

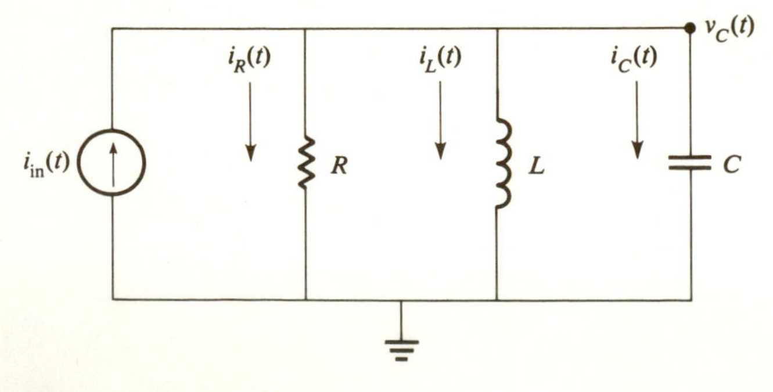

Figure A-3 Parallel RLC network.

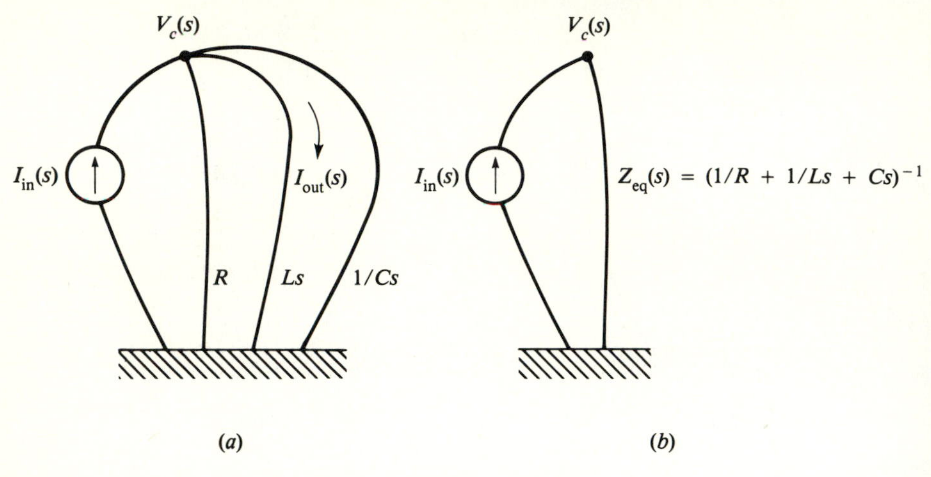

Figure A-4 Linear graph of parallel *RLC* circuit of Fig. A-3.

Notice that the characteristic polynomial does not change when selecting a different output; only the zero-numerator polynomial changes.

Example A-3 Consider the electric circuit of Fig. A-5, with the linear graph shown in Fig. A-6. This is the same electrical circuit as considered in Example 6-7, Sec. 6-4. The output is the voltage drop across the capacitor, $V_3(s)$. By definition of the capacitor impedance,

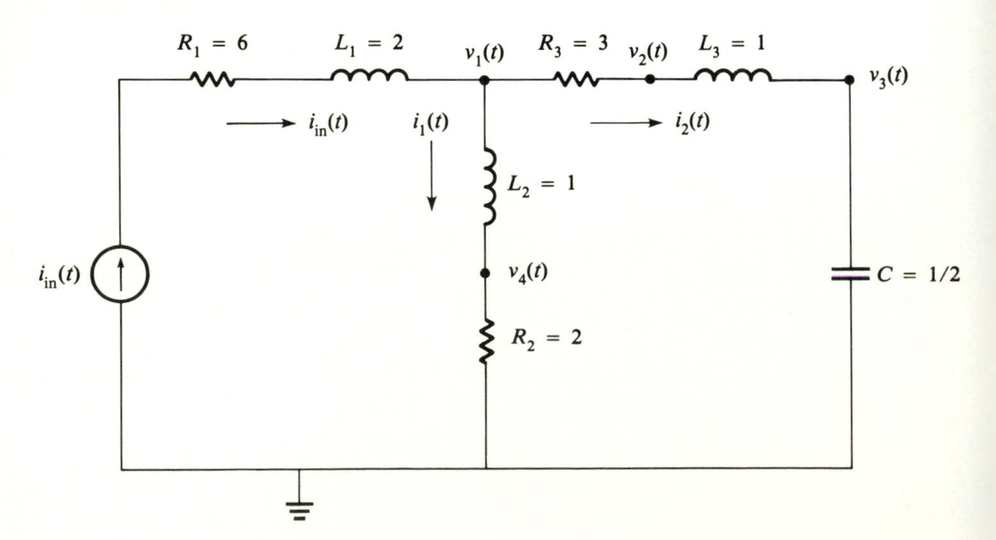

Figure A-5 Electric circuit of Example 6-7.

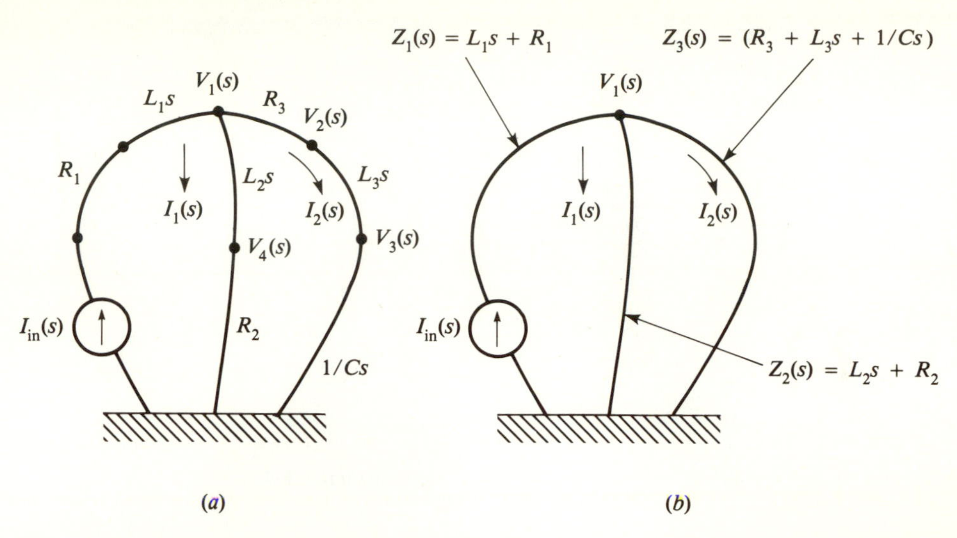

Figure A-6 Linear graph of electric circuit of Fig. A-5.

$$Z_C(s) = \frac{1}{Cs} \equiv \frac{V_3(s)}{I_2(s)}$$

But a current divider relation gives a relationship between $I_2(s)$ and $I_{in}(s)$:

$$\frac{I_2(s)}{I_{in}(s)} = \frac{1/Z_3(s)}{1/Z_2(s) + 1/Z_3(s)}$$

$$= \frac{Cs(L_2s + R_2)}{C(L_2 + L_3)s^2 + C(R_2 + R_3)s + 1}$$

Therefore, the transfer function between $V_3(s)$ and $I_{in}(s)$ may be found as

$$\frac{V_3(s)}{I_{in}(s)} = \frac{V_3(s)}{I_2(s)} \frac{I_2(s)}{I_{in}(s)}$$

$$= \frac{L_2s + R_2}{C(L_2 + L_3)s^2 + C(R_2 + R_3)s + 1}$$

Substitution of $R_2 = 2$, $R_3 = 3$, $L_2 = 1$, $L_3 = 1$, $C = 1/2$ gives

$$\frac{V_3(s)}{I_{in}(s)} = \frac{s + 2}{s^2 + \frac{5}{2}s + 1} = \frac{s + 2}{(s + 2)(s + 1/2)} = \frac{1}{s + 1/2}$$

Hence the system has a pole-zero cancellation and it may be modeled by a first-order system.

Example A-4 Consider the parallel mass, spring, and damper system of Fig.

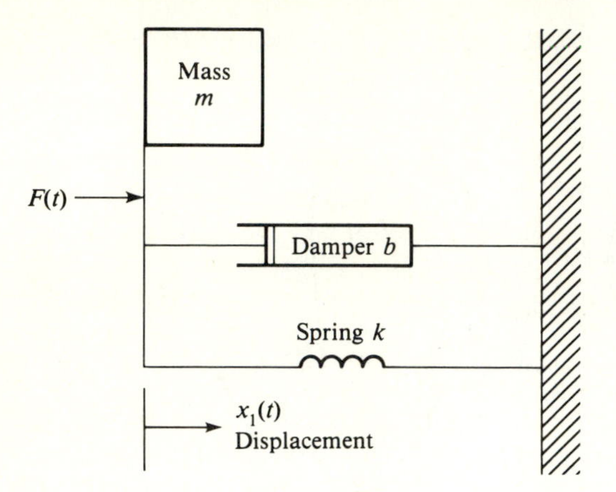

Figure A-7 Parallel mass, damper, and spring system, Example 6-4.

A-7, with the linear graph shown in Fig. A-8. Notice that the mass is connected to ground by a broken line; the acceleration force of a mass is always with respect to an inertial reference. Choosing the velocity $V_1(s)$ as the output gives

$$\frac{V_1(s)}{F_{in}(s)} \equiv Z_{eq}(s) = \frac{s}{m_1 s^2 + bs + k}$$

If the displacement $X_1(s)$ is chosen as the output, then

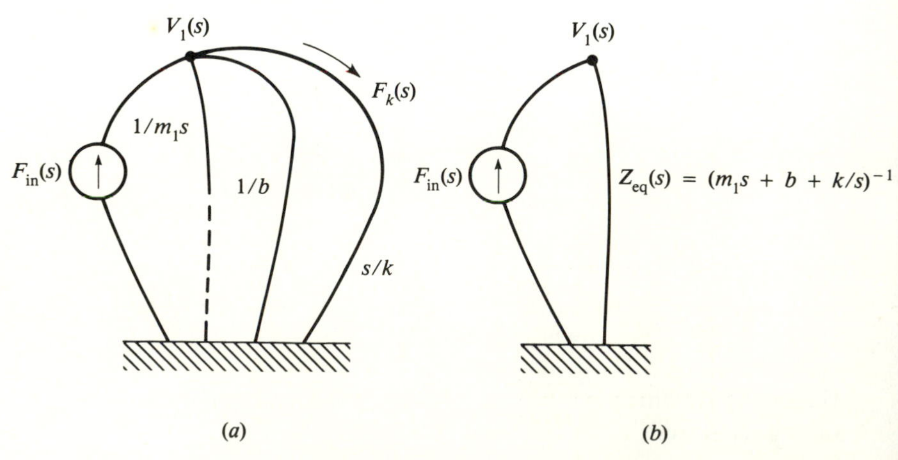

Figure A-8 Linear graph of parallel mass, spring, and damper system of Fig. A-7. (*a*) Original. (*b*) Simplified.

$$\frac{X_1(s)}{V_1(s)} = \frac{1}{s}$$

and
$$\frac{X_1(s)}{F_{in}(s)} = \frac{X_1(s)}{V_1(s)}\frac{V_1(s)}{F_{in}(s)} = \frac{1}{m_1 s^2 + bs + k}$$

If the force through the spring is selected as the output, a force divider relation gives

$$\frac{F_k(s)}{F_{in}(s)} = \frac{k/s}{1/Z_{eq}(s)} = \frac{k}{m_1 s^2 + bs + k}$$

Example A-5 Consider the rocket-sled linear system in Fig. A-9. Its linear graph is shown in Fig. A-10. The output is the velocity of mass m_2. Using intermediate relations of the linear graph and the velocity divider rule it is seen that

$$\frac{V_{out}(s)}{V_1(s)} = \frac{Z_2(s)}{Z_2(s) + s/k}$$

where
$$Z_2(s) = \frac{1}{m_2 s + b_2}$$

But from Fig. A-10*d* the velocity $V_1(s)$ is related to $F_{in}(s)$ by the impedance definition

$$\frac{V_1(s)}{F_{in}(s)} \equiv Z_{eq}(s) = \left(\frac{1}{Z_1(s)} + \frac{1}{Z_3(s)}\right)^{-1}$$

where
$$Z_1(s) = \frac{1}{m_1 s + b_1}$$

$$Z_3(s) = \frac{s}{k} + \frac{1}{m_2 s + b_2}$$

The final input-output transfer function is given as the product

$$\frac{V_{out}(s)}{F_{in}(s)} = \frac{V_{out}(s)}{V_1(s)}\frac{V_1(s)}{F_{in}(s)}$$

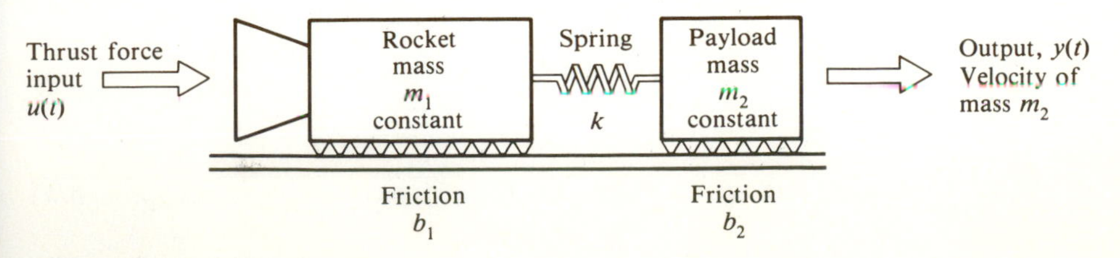

Figure A-9 Rocket-sled system.

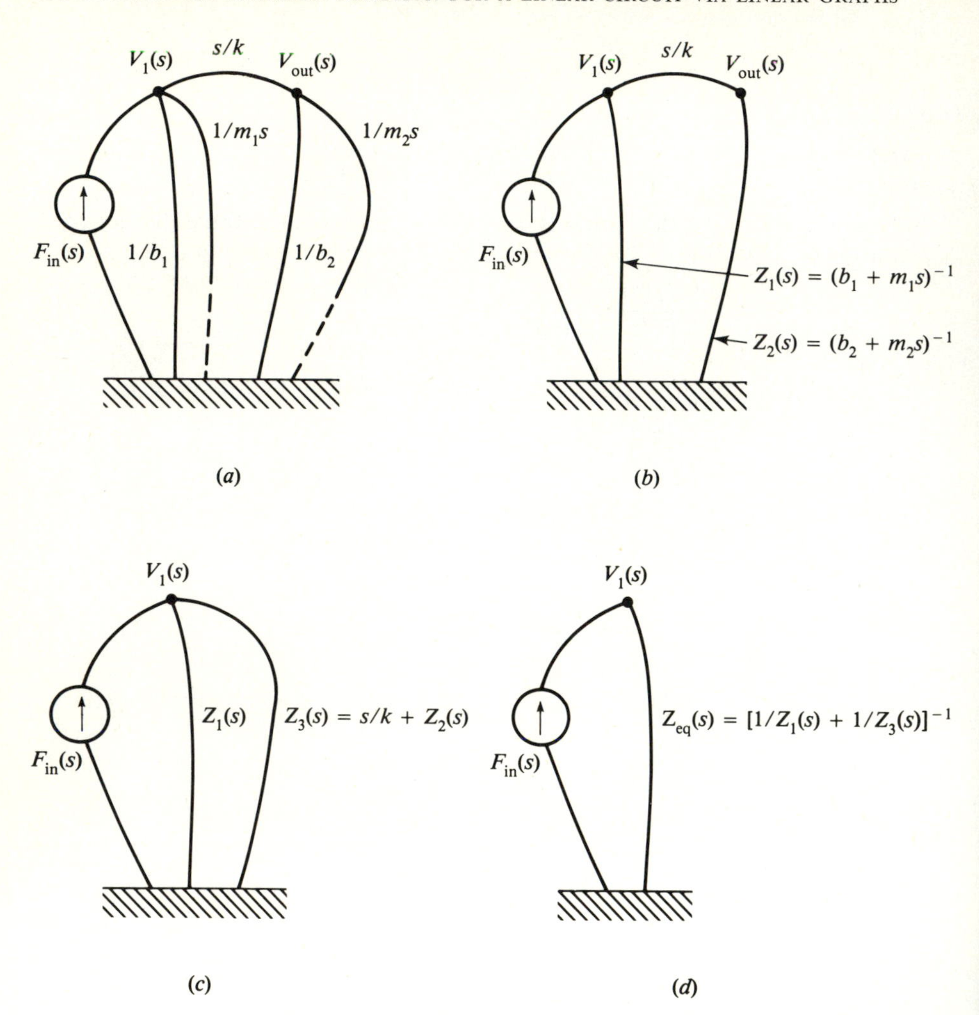

Figure A-10 Linear graph of rocket-sled linear system, Fig. A-9. (*a*) Original. (*b*) Simplified once. (*c*) Simplified twice. (*d*) Final simplification.

After a bit of algebra this is found to equal

$$\frac{V_{out}(s)}{F_{in}(s)} = \frac{k}{m_1 m_2 s^3 + (m_1 b_2 + m_2 b_1)s^2 + (m_1 k + m_2 k + b_1 b_2)s + (b_1 k + b_2 k)}$$

This agrees with Eq. (2-11) in Sec. 2-1.

A-4 FURTHER NOTES

The techniques to simplify the linear graph and find the input-output transfer function may be applied to multi-input and multi-output systems. The previous examples show how multiple outputs could be considered for a single-input system. For multiple inputs, one simply considers one input at a time. One removes all but one input from the network. One does this for each input in succession. When through-variable input sources (e.g., a current source) are removed from a circuit, it leaves an open circuit where they are removed. When across-variable input sources (e.g., a voltage source) are removed from a circuit, it leaves a short circuit where they are removed. See Ref. A-1 for further details.

Also it is mentioned that the previous techniques for linear graph simplification and deriving the transfer function apply only to networks that are simply connected. They do not apply, for example, to the bridge network as shown in Fig. A-11. More sophisticated circuit-analysis methods must be applied to analyze and find the transfer function of such bridge networks.

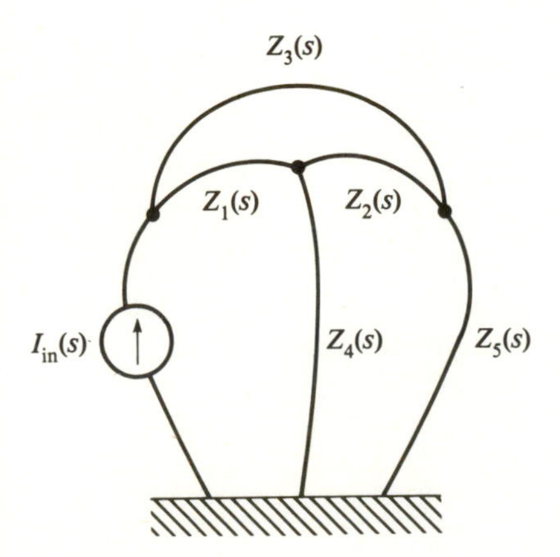

Figure A-11 Linear graph for which simplification techniques here do not apply.

REFERENCES

A-1. Shearer, J. L., A. T. Murphy, and H. H. Richardson; *Introduction to System Dynamics*, Addison-Wesley, Reading, Mass., 1971.

APPENDIX

B

INTRODUCTION TO MATRICES

This text uses matrices and matrix methods extensively. Most of the required results are developed within the body of the text. However, some basic skills and definitions associated with matrices are presumed. This appendix is intended to give a brief review of these basic skills and definitions.

To begin, a matrix is defined as a rectangular array of elements whose values might be real or complex. Here we assume that the elements are all real-valued. In the text matrices are always assigned a capital letter (e.g., A) and the elements of a matrix are given the corresponding lower-case symbol with subscripts denoting the row and column index (e.g., a_{ij} is the ith row and jth column element of the matrix A). Thus we write, for example,

$$A = \begin{bmatrix} a_{11} & a_{12} & \cdots & a_{1n} \\ a_{21} & a_{22} & \cdots & a_{2n} \\ \cdots & \cdots & \cdots & \cdots \\ a_{m1} & a_{m2} & \cdots & a_{mn} \end{bmatrix} = [a_{ij}]_{m \times n} \tag{B-1a}$$

to denote the general $m \times n$ matrix, A. Often it is convenient to partition A into its column vectors. This would be denoted by

$$A = [\mathbf{a}_1, \quad \mathbf{a}_2, \quad \ldots, \quad \mathbf{a}_n]_{m \times n} \tag{B-1b}$$

where

$$\mathbf{a}_j = \begin{bmatrix} a_{1j} \\ a_{2j} \\ \vdots \\ a_{mj} \end{bmatrix} \tag{B-2}$$

A basic operation with a matrix is to take the transpose. The transpose operation interchanges the columns and the rows of the matrix. Thus for the general

$m \times n$ A its transpose is

$$A^T = [a_{ij}]^T_{m\times n} = [a_{ji}]_{n\times m}$$

$$= [\mathbf{a}_1, \quad \mathbf{a}_2, \quad \ldots, \quad \mathbf{a}_n]^T_{m\times n} = \begin{bmatrix} \mathbf{a}_1^T \\ \mathbf{a}_2^T \\ \vdots \\ \mathbf{a}_n^T \end{bmatrix}_{n\times m} \tag{B-3}$$

To add two matrices (say A and B), the matrices must be of the same dimension. Then one simply adds each individual element as in

$$A + B = [a_{ij}]_{m\times n} + [b_{ij}]_{m\times n} = [a_{ij} + b_{ij}]_{m\times n} \tag{B-4}$$

To multiply a scalar (say k) by a matrix (say A) one multiplies every element of A by the scalar k.

$$kA = [ka_{ij}]_{m\times n} = [k\mathbf{a}_1, \quad k\mathbf{a}_2, \quad \ldots, \quad k\mathbf{a}_n]_{m\times n} \tag{B-5}$$

To multiply two matrices (say C and A), they must have compatible dimensions; the column dimension of C must equal the row dimension of A. Then it is true that

$$C_{p\times m}A_{m\times n} = \left[\sum_{k=1}^{n} c_{ik}a_{kj}\right]_{p\times n} \tag{B-6a}$$

Additionally, if one partitions C into its p rows as

$$C = \begin{bmatrix} \mathbf{c}_1^T \\ \mathbf{c}_2^T \\ \vdots \\ \mathbf{c}_p^T \end{bmatrix}_{p\times m} \tag{B-7}$$

where
$$\mathbf{c}_i^T \equiv [c_{i1}, \quad c_{i2}, \quad \ldots, \quad c_{im}]_{1\times m} \tag{B-8}$$

one may write the product CA as the partitioned product

$$C_{p\times m}A_{m\times n} = \begin{bmatrix} \mathbf{c}_1^T \\ \mathbf{c}_2^T \\ \vdots \\ \mathbf{c}_p^T \end{bmatrix} [\mathbf{a}_1, \quad \mathbf{a}_2, \quad \ldots, \quad \mathbf{a}_n]$$

$$= \begin{bmatrix} \mathbf{c}_1^T\mathbf{a}_1 & \mathbf{c}_1^T\mathbf{a}_2 & \cdots & \mathbf{c}_1^T\mathbf{a}_n \\ \mathbf{c}_2^T\mathbf{a}_1 & \mathbf{c}_2^T\mathbf{a}_2 & \cdots & \mathbf{c}_2^T\mathbf{a}_n \\ \cdots & \cdots & \cdots & \cdots \\ \mathbf{c}_p^T\mathbf{a}_1 & \mathbf{c}_p^T\mathbf{a}_2 & \cdots & \mathbf{c}_p^T\mathbf{a}_n \end{bmatrix}_{p\times n} \tag{B-6b}$$

The algebra of matrices enjoys basic properties such as being associative

$$(A + B) + C = A + (B + C) \tag{B-9}$$

$$(AB)C = A(BC) \tag{B-10}$$

provided the dimensions are compatible for the above operations. Matrix operations are also distributive

$$C(A + B) = CA + CB \tag{B-11}$$

$$k(A + B) = kA + kB \tag{B-12}$$

$$A(k_1 + k_2) = Ak_1 + Ak_2 \tag{B-13}$$

and commutative with respect to scalars

$$kA = Ak \tag{B-14}$$

However, it must be emphasized that matrix multiplication is *generally* not commutative

$$CA \neq AC \text{ in general} \tag{B-15}$$

even if C and A both have compatible dimensions (i.e., A and C are both square $n \times n$).

The transpose operation mixed with the product and sums of matrices also has some important results:

$$(A + B)^T = A^T + B^T \tag{B-16}$$

$$(CA)^T = A^T C^T \tag{B-17}$$

$$(ABC)^T = C^T B^T A^T \tag{B-18}$$

These results are easily proven.

When A is square $n \times n$, there are a number of other important definitions and operations. The trace of the square A is defined by the sum of its diagonal elements

$$\text{trace}(A) = \sum_{i=1}^{n} a_{ii} \tag{B-19}$$

Then it is true that

$$\text{trace}(A + B) = \text{trace } A + \text{trace } B \tag{B-20}$$

$$\text{trace}(kA) = k \text{ trace}(A) \tag{B-21}$$

$$\text{trace}(AB) = \text{trace}(BA) \qquad \text{for } A_{m\times n} \text{ and } B_{n\times m} \tag{B-22}$$

$$\text{trace}(A^T) = \text{trace}(A) \tag{B-23}$$

For a square A the determinant of A is also a very important function. It is defined by the following general expression:

$$\det \begin{bmatrix} a_{11} & a_{12} & \cdots & a_{1n} \\ a_{21} & a_{22} & \cdots & a_{2n} \\ \cdots & \cdots & \cdots & \cdots \\ a_{n1} & a_{n2} & \cdots & a_{nn} \end{bmatrix} = \sum (-1)^k a_{1j_1} a_{2j_2} \cdots a_{nj_n} \tag{B-24}$$

where the summation is over all permutations of values of $j_1, j_2, \ldots, j_n$, such that $j_1 \neq j_2 \neq \cdots \neq j_n$, and k equals the number of inversions it requires to put $j_1, j_2, \ldots, j_n$ in the order 1, 2, 3, ..., n. Thus, for example,

$$\det \begin{bmatrix} a_{11} & a_{12} \\ a_{21} & a_{22} \end{bmatrix} = (-1)^0 a_{11}a_{22} + (-1)^1 a_{12}a_{21} \tag{B-25}$$

and

$$\begin{aligned} \det \begin{bmatrix} a_{11} & a_{12} & a_{13} \\ a_{21} & a_{22} & a_{23} \\ a_{31} & a_{32} & a_{33} \end{bmatrix} &= (-1)^0 a_{11}a_{22}a_{33} + (-1)^1 a_{11}a_{23}a_{32} \\ &\quad + (-1)^1 a_{12}a_{21}a_{33} + (-1)^2 a_{12}a_{23}a_{31} \\ &\quad + (-1)^2 a_{13}a_{21}a_{32} + (-1)^1 a_{13}a_{22}a_{31} \end{aligned} \tag{B-26}$$

The calculation of the determinant of A is thus seen to take $n!$ multiplications and a like number of additions. Thus it is an expensive calculation for large A. The following properties of the determinant may be derived from its definition Eq. (B-24) and are often helpful in analysis and computation:

1. If two rows (columns) in a matrix are interchanged, the determinant will change its sign.
2. The value of the determinant is unchanged if any scalar multiple of any row (or column) is added to any other row (or column).
3. If any row (or column) of a matrix is identically zero, the value of the determinant is zero.
4. If any row (or column) of a matrix is linearly dependent on the other rows (or columns), the determinant is zero.
5. The determinant of an upper-triangular (or lower-triangular) matrix is the product of all elements on the diagonal. *Note:* an upper-triangular matrix (or lower-triangular matrix) is one having all zero elements below (or above) the main diagonal.
6. Multiplication of any row (or column) of a matrix by a nonzero scalar multiplies the determinant by k.
7. $\det(A) = det(A^T)$
8. $\det(AB) = \det(A)\det(B) = \det(BA)$ for square A and B.

Using properties 1 through 6 and the elementary row operations introduced in Sec. 2-7, the computational cost of calculating the determinant can be considerably reduced. Namely, elementary row operations may be used to transform the matrix to upper-triangular form while keeping track of all row interchanges that would change the sign of the determinant (property 1). Then property 5 may be applied.

The concept of cofactors and the adjoint matrix are also important quantities related to the determinant of a square matrix. The i, j cofactor of A, denoted $\text{cof}_{i,j}(A)$, is the determinant of the $(n-1) \times (n-1)$ matrix obtained by deleting the ith row and jth column of A and choosing the weighting $(-1)^{i+j}$. Thus for the general 3×3 A matrix

$$A = \begin{bmatrix} a_{11} & a_{12} & a_{13} \\ a_{21} & a_{22} & a_{23} \\ a_{31} & a_{32} & a_{33} \end{bmatrix} \tag{B-27}$$

$$\text{cof}_{1,1}(A) = (-1)^2 \det \begin{bmatrix} a_{22} & a_{23} \\ a_{32} & a_{33} \end{bmatrix} = a_{22}a_{33} - a_{23}a_{32} \tag{B-28}$$

$$\text{cof}_{1,2}(A) = (-1)^3 \det \begin{bmatrix} a_{21} & a_{23} \\ a_{31} & a_{33} \end{bmatrix} = -a_{21}a_{33} + a_{23}a_{31} \tag{B-29}$$

etc. The adjoint matrix [denoted adj (A)] is then defined as the transpose of the matrix of cofactors:

$$\text{adj}\,(A) \equiv [\text{cof}_{i,j}(A)]^T \tag{B-30}$$

Cramer's rule for finding the inverse, A^{-1}, of a square matrix A is given by

$$A^{-1} = \frac{\text{adj}\,(A)}{\det\,(A)} \tag{B-31}$$

provided $\det\,(A) \neq 0$. When A^{-1} exists, the matrix is said to be *nonsingular,* and so a necessary and sufficient condition for the square A to be nonsingular is for its determinant to be nonzero. The matrix inverse, A^{-1}, has the property that

$$A^{-1}A = AA^{-1} = I \tag{B-32}$$

where I is the matrix with ones on the diagonal, and otherwise it is zero:

$$I \equiv \text{diagonal}\,[1, \quad 1, \quad \ldots, \quad 1]_{n \times n} \tag{B-33}$$

The cofactors may also be used to define an alternative way to calculate the determinant of a matrix. The method is called Laplace's method, and the method states that

$$\det\,(A) = \sum_{i=1}^{n} a_{ij}\,\text{cof}_{i,j}(A) \qquad \text{for any column } j \tag{B-34}$$

or

$$\det\,(A) = \sum_{j=1}^{n} a_{ij}\,\text{cof}_{i,j}(A) \qquad \text{for any row } i \tag{B-35}$$

Usually one selects the row or column having the greatest number of zero elements, for this simplifies computations.

Example B-1

$$A = \begin{bmatrix} 1 & -2 & -1 \\ 3 & -1 & 0 \\ 0 & 1 & 2 \end{bmatrix}$$

Then by Eq. (B-26)

$$\begin{aligned} \det(A) &= (1)(-1)(2) - (1)(0)(1) - (-2)(3)(2) \\ &\quad + (-2)(0)(0) + (-1)(3)(1) - (-1)(-1)(0) \\ &= 7 \end{aligned}$$

Alternatively, by Laplace expansion, choosing the first column, since it has a zero entry,

$$\det(A) = (1)(-2) - (3)(-3) = 7$$

The adjoint matrix is found via Eq. (B-30) to be

$$\operatorname{adj}(A) = \begin{bmatrix} -2 & -6 & 3 \\ 3 & 2 & -1 \\ -1 & -3 & 5 \end{bmatrix}^T = \begin{bmatrix} -2 & 3 & -1 \\ -6 & 2 & -3 \\ 3 & -1 & 5 \end{bmatrix}$$

Finally, by Cramer's rule

$$A^{-1} = \frac{\operatorname{adj}(A)}{\det(A)} = \begin{bmatrix} -2/7 & 3/7 & -1/7 \\ -6/7 & 2/7 & -3/7 \\ 3/7 & -1/7 & 5/7 \end{bmatrix}$$

and $AA^{-1} = A^{-1}A = I$.

APPENDIX

C

PARTIAL FRACTION EXPANSIONS

An important skill is the ability to go from a transfer function in the form

$$H(s) = \frac{b_m s^m + b_{m-1} s^{m-1} + \cdots + b_0}{(s - p_1)(s - p_2) \cdots (s - p_n)} \equiv \frac{N(s)}{\Delta(s)} \tag{C-1}$$

with $m < n$, to the equivalent partial fraction expansion of the general (or closely related) form

$$H(s) = \sum_{i=1}^{n} \frac{r_i}{s - p_i} \tag{C-2}$$

The r_i, $i = 1, 2, \ldots, n$, are termed the residues associated with the roots p_i, $i = 1, 2, \ldots, n$.

C-1 DISTINCT ROOTS

If p_i is real and distinct, the formula for r_i is quite straightforward:

$$r_i = (s - p_i)H(s)|_{s=p_i} \tag{C-3}$$

This also applies to the complex distinct pole, p_i, as well. Then, in general, r_i is complex. Furthermore, if p_i and p_{i+1} are complex conjugate pairs, then r_{i+1} is the complex conjugate of r_i. Therefore, if one finds r_i, one also knows r_{i+1}.

Example C-1

$$H(s) = \frac{s^3 + 5s - 1}{s(s+1)(s+2-3j)(s+2+3j)}$$

$$= \frac{r_1}{s} + \frac{r_2}{s+1} + \frac{r_3}{s+2-3j} + \frac{r_4}{s+2+3j}$$

where
$$r_1 = sH(s)|_{s=0} = -1/13$$
$$r_2 = (s+1)H(s)|_{s=-1} = 7/10$$
$$r_3 = (s+2-3j)H(s)|_{s=-2+3j} = -0.9115 + 0.2910j$$
$$r_4 = \bar{r}_3 = -0.9115 - 0.2910j$$

C-2 COMPLEX DISTINCT ROOTS

The previous method applies to finding the complex residues for complex distinct roots. However, calculations with complex numbers can be quite tedious. A method by which to eliminate complex number calculations would be desirable.

To this end, suppose the denominator of $H(s)$ has m real roots, $p_1, p_2, \ldots, p_m$, and k complex-conjugate-pair roots, $c_1 \pm d_1 j$, $c_2 \pm d_2 j$, $\ldots$, $c_k \pm d_k j$. The partial fraction expansion form sought is

$$H(s) = \frac{N(s)}{(s-p_1)\cdots(s-p_m)(s-c_1 \pm d_1 j)\cdots(s-c_k \pm d_k j)}$$

$$= \sum_{i=1}^{m} \frac{r_i}{s-p_i} + \sum_{i=1}^{k} \frac{f_i s + g_i}{(s-c_i)^2 + d_i^2} \qquad \text{(C-4)}$$

Note that

$$(s - c_i + d_i j)(s - c_i - d_i j) = (s - c_i)^2 + d_i^2$$

Also note that Eq. (C-4) is a useful form because, for example, the inverse Laplace transform of terms

$$\frac{f_i s + g_i}{(s-c_i)^2 + d_i^2}$$

may be found directly (see Table 4-1, Chap. 4).

To find the form Eq. (C-4), the procedure is fairly direct. The r_i are first calculated as described in the previous section:

$$r_i = (s - p_i)H(s)|_{s=p_i} \qquad i = 1, 2, \ldots, m \qquad \text{(C-3)}$$

Once finding the r_i, the unknowns f_i and g_i, $i = 1, 2, \ldots, k$, may be determined. There are $2k$ unknowns f_i and g_i altogether. These are found by establishing $2k$

linear equations in the $2k$ unknowns, f_i and g_i, $i = 1, 2, \ldots, k$. The $2k$ linear equations are found by evaluating both sides of Eq. (C-4) at $2k$ separate values of s. Usually it is convenient to pick $s = c_i$, $i = 1, 2, \ldots, k$, provided that $H(c_i)$ exists. The other k values of s may be picked arbitrarily. In all $2k$ values of s are selected leading to $2k$ equations in the $2k$ unknowns, f_i and g_i, $i = 1, 2, \ldots, k$.

Example C-2 Again consider $H(s)$ from Example C-1:

$$H(s) = \frac{s^3 + 5s - 1}{s(s + 1)(s + 2 - 3j)(s + 2 + 3j)}$$

$$= \frac{s^3 + 5s - 1}{s(s + 2)[(s + 2)^2 + 3^2]}$$

The partial fraction expansion form sought is

$$H(s) = \frac{r_1}{s} + \frac{r_2}{s + 1} + \frac{fs + g}{(s + 2)^2 + 3^2}$$

The residues r_1 and r_2 are found the same as in Example C-1. The problem remaining is to find f and g.

The solution to f and g is set up as a linear equation problem. We have two unknowns, and so we select two convenient values of s to establish two equations in two unknowns. In particular, suppose we first pick $s = -2$. Then evaluating $H(s)$ at $s = -2$ gives

$$H(s)\bigg|_{s=-2} = \frac{-8 - 10 - 1}{(-2)(-1)(3^2)} = \frac{-19}{18}$$

$$= \frac{-1/13}{-2} + \frac{7/10}{-1} + \frac{-2f + g}{9}$$

Simplifying gives

$$-2f + g = -3.5462$$

This is the first equation for the two unknowns, f and g.

To get the second equation we need to evaluate $H(s)$ at a second value of s. Picking $s = 0$ is usually convenient, but it cannot be used here because the denominator of $H(s)$ has a root at $s = 0$. We arbitrarily pick $s = 1$. Then

$$H(s)\bigg|_{s=1} = \frac{1 + 5 - 1}{(1)(2)(18)} = \frac{5}{36}$$

$$= -\frac{1}{13} + \frac{7}{20} + \frac{f + g}{18}$$

Simplifying gives the second equation in two unknowns:

$$f + g = -2.4154$$

These two equations in two unknowns may be set up in standard linear equation problem format:

$$\begin{bmatrix} -2 & 1 \\ 1 & 1 \end{bmatrix} \begin{bmatrix} f \\ g \end{bmatrix} = \begin{bmatrix} -3.5462 \\ -2.4154 \end{bmatrix}$$

Solving gives $f = 0.3769$ and $g = -2.7923$. The final partial fraction expansion form of $H(s)$ is

$$H(s) = \frac{-1/13}{s} + \frac{7/10}{s+1} + \frac{0.3769s - 2.7923}{(s+2)^2 + (3)^2}$$

C-3 REPEATED ROOTS

Partial fraction expansion of the $H(s)$ having repeated roots in the denominator polynomial is slightly more complicated. First the form of the partial fraction expansion must be determined. Suppose $H(s)$ has the form

$$H(s) = \frac{N(s)}{(s - p_1)^q (s - p_{q+1}) \cdots (s - p_n)} \tag{C-5}$$

where the root p_1 is repeated q times. The partial fraction expansion form of $H(s)$ then would be

$$H(s) = \frac{r_1}{(s - p_1)^q} + \frac{r_2}{(s - p_1)^{q-1}} + \cdots + \frac{r_q}{(s - p_1)} + \sum_{i=q+1}^{n} \frac{r_i}{(s - p_i)} \tag{C-6}$$

The r_i, $i = q + 1, \ldots, n$, are determined as previously:

$$r_i = (s - p_i)H(s)|_{s=p_i} \tag{C-3}$$

The r_1 coefficient is also determined this way.

$$r_1 = (s - p_1)^q H(s)|_{s=p_1} \tag{C-7}$$

However, to determine $r_2, r_3, \ldots, r_q$ differentiation may be used as

$$r_2 = \frac{d}{ds}\{(s - p_1)^q H(s)\}\bigg|_{s=p_1} \tag{C-8}$$

$$r_3 = \frac{1}{2}\frac{d^2}{ds^2}\{(s - p_1)^q H(s)\}\bigg|_{s=p_1} \tag{C-9}$$

. . . .

$$r_q = \frac{1}{(q-1)!}\frac{d^{q-1}}{ds^{q-1}}\{(s - p_1)^q H(s)\}\bigg|_{s=p_1} \tag{C-10}$$

Example C-3 Consider

$$H(s) = \frac{1}{(s+1)^3(s+3)}$$

Then, the partial fraction expression is

$$H(s) = \frac{r_1}{(s+1)^3} + \frac{r_2}{(s+1)^2} + \frac{r_3}{(s+1)} + \frac{r_4}{(s+3)}$$

where $r_1 = (s+1)^3 H(s)|_{s=-1} = 1/2$

$$r_2 = \frac{d}{ds}\{(s+1)^3 H(s)\}|_{s=-1} = -\left.\frac{1}{(s+3)^2}\right|_{s=-1} = -1/4$$

$$r_3 = \frac{1}{2}\frac{d^2}{ds^2}\{(s+1)^3 H(s)\}|_{s=-1} = \left.\frac{1}{(s+3)^3}\right|_{s=-1} = 1/8$$

$$r_4 = (s+3)H(s)|_{s=-3} = \left.\frac{1}{(s+1)^3}\right|_{s=-3} = -1/8$$

APPENDIX

D

RECURSIVE CALCULATION OF THE RESOLVENT MATRIX

This appendix presents a recursive algorithm for computing the $n \times n$ resolvent matrix in the form

$$[SI - A]^{-1} = \frac{Q(s)}{\Delta(s)} = \frac{Q_{n-1}s^{n-1} + \cdots + Q_1 s + Q_0}{s^n + a_{n-1}s^{n-1} + \cdots + a_1 s + a_0} \tag{D-1}$$

where $$\Delta(s) \equiv \det[sI - A] = s^n + a_{n-1}s^{n-1} + \cdots + a_1 s + a_0 \tag{D-2}$$

and Q_i, $i = 1, 2, \ldots, n$, are each $n \times n$ matrices. This algorithm is generally attributed to Leverrier or Faddeev. The recursive algorithm for computing both the Q_i making up $Q(s)$ and the a_i making up $\Delta(s)$ is given in the following theorem:

Theorem D-1 The Q_i matrices and a_i scalars may be computed recursively via

$$Q_{n-1} = I \tag{D-3}$$

$$a_{n-1} = -\text{trace}\,(Q_{n-1}A) \tag{D-4}$$

$$Q_{n-2} = Q_{n-1}A + a_{n-1}I \tag{D-5}$$

$$a_{n-2} = -\frac{1}{2}\,\text{trace}\,(Q_{n-2}A) \tag{D-6}$$

. . . .

$$Q_1 = Q_2 A + a_2 I \tag{D-7}$$

$$a_1 = -\frac{1}{n-1}\,\text{trace}\,(QA) \tag{D-8}$$

$$Q_0 = Q_1 A + a_1 I \tag{D-9}$$

$$a_0 = \frac{1}{n}\,\text{trace}\,(Q_0 A) \tag{D-10}$$

$$0 = Q_0 A + a_0 I \tag{D-11}$$

PROOF Multiply both sides of Eq. (D-1) by $\Delta(s)[sI - A]$ to give

$$\Delta(s)I = Q(s)[sI - A] \tag{D-12}$$

Substitution of expressions for $\Delta(s)$ and $Q(s)$ into Eq. (D-12) and multiplying the terms of $Q(s)$ by $[sI - A]$ gives

$$\begin{aligned}(s^n + a_{n-1}s^{n-1} + \cdots + a_1 s + a_0)I \\ = Q_{n-1}s^n + (Q_{n-2} - Q_{n-1}A)s^{n-1} + \cdots \\ + (Q_1 - Q_2 A)s^2 + (Q_0 - Q_1 A)s - Q_0 A\end{aligned} \tag{D-13}$$

Equating like powers of s gives

$$I = Q_{n-1} \tag{D-14}$$

$$a_{n-1}I = Q_{n-2} - Q_{n-1}A \tag{D-15}$$

$$\cdots\cdots\cdots\cdots\cdots\cdots$$

$$a_2 I = Q_1 - Q_2 A \tag{D-16}$$

$$a_1 I = Q_0 - Q_1 A \tag{D-17}$$

$$a_0 I = Q_0 A \tag{D-18}$$

Equations (D-14) through (D-18) may be rearranged to yield Eqs. (D-3), (D-5), (D-7), (D-9), and (D-11).

Proof of the recursion for the a_i coefficients is rather involved, and the interested reader is referred to Ref. D-1 or Ref. D-2 for proof.

We remark that this algorithm is known to be numerically sensitive and prone to wordlength-related computational errors. It is wise to use double precision in applying this algorithm. Equation (D-11) provides a convenient check, though, for computational accuracy.

REFERENCES

D-1. Faddeev, D. K., and Faddeeva, V. N.: *Computational Methods of Linear Algebra,* Freeman, San Francisco, 1963.

D-2. Gantmacher, F. R.: *The Theory of Matrices, Vol. I,* Chelsea, New York, 1959.

APPENDIX

E

INTRODUCTION TO THE SINGULAR-VALUE DECOMPOSITION

The singular-value decomposition for real square matrices was established by Sylvester in 1889 and extended to general matrices by Eckart and Young in 1939.* Through the extensive research of numerical analysis specialists since the 1960s, the singular-value decomposition now stands as one of the cornerstones of modern numerical analysis. Furthermore, high-quality, expertly written software to perform the singular-value decomposition is readily available at nearly all major computational centers. Its use may be made by the numerical analysis layperson with nothing more than the appropriate subroutine call to an existing library package (e.g., EISPACK [E-2], LINPACK [E-3], and IMSL [E-4]). For further discussion of the theory and applications of the singular-value decomposition than that which is presented in this brief appendix the interested reader is referred to Ref. E-5.

E-1 FUNDAMENTAL THEOREM

The fundamental theorem of the singular-value decomposition states:

Theorem E-1: Singular-Value Decomposition Let M be an arbitrary real† $m \times n$-dimension matrix. Then orthogonal U and V matrices (that is,

*See Stewart [E-1] for discussion.

†The singular-value decomposition results also apply to the complex-matrix case. See Stewart [E-1] for details.

$U^T = U^{-1}$ and $V^T = V^{-1}$) may be found such that

$$M = U \begin{bmatrix} \Sigma & 0 \\ 0 & 0 \end{bmatrix}_{m \times n} V^T \tag{E-1a}$$

where

$$\Sigma = \text{diagonal}\,[\sigma_1, \quad \sigma_2, \quad \ldots, \quad \sigma_r] \tag{E-2}$$

is a diagonal matrix of the strictly positive singular values of M, σ_i, $i = 1, 2, \ldots, r$. The orthogonal column vectors of U,

$$U = [\mathbf{u}_1, \quad \mathbf{u}_2, \quad \ldots, \quad \mathbf{u}_m]_{m \times m} \tag{E-3}$$

are called the *left singular vectors* of M. The orthogonal column vectors of V,

$$V = [\mathbf{v}_1, \quad \mathbf{v}_2, \quad \ldots, \quad \mathbf{v}_n]_{n \times n} \tag{E-4}$$

are called the *right singular vectors* of M.

PROOF See Stewart [E-1] for proof and discussion.

It is beyond our scope to prove this fundamental theorem. The primary interest to us lies in its applications.

E-2 CALCULATION RANK AND SUBSPACES OF A MATRIX

Corollary E-1 Let M have singular-value decomposition (SVD)

$$M = [U_1 \quad U_2] \begin{bmatrix} \Sigma & 0 \\ 0 & 0 \end{bmatrix}_{m \times n} \begin{bmatrix} V_1^T \\ V_2^T \end{bmatrix} \tag{E-1b}$$

where the orthogonal $U_{m \times m}$ has been partitioned such that U_1 has r orthogonal columns $\mathbf{u}_1, \mathbf{u}_2, \ldots, \mathbf{u}_r$, and U_2 has $(m - r)$ orthogonal columns $\mathbf{u}_{r+1}, \ldots, \mathbf{u}_m$. Likewise, the orthogonal V has been partitioned such that V_1 has r orthogonal columns $\mathbf{v}_1, \mathbf{v}_2, \ldots, \mathbf{v}_r$, and V_2 has $(n - r)$ orthogonal columns $\mathbf{v}_{r+1}, \ldots, \mathbf{v}_n$. Then

1. The rank of M equals r, the number of nonzero singular values.
2. The range space of M is the span $\{\mathbf{u}_1, \ldots, \mathbf{u}_r\}$.
3. The null space of M is the span $\{\mathbf{v}_{r+1}, \ldots, \mathbf{v}_n\}$.
4. The range space of M^T is the span $\{\mathbf{v}_1, \ldots, \mathbf{v}_r\}$.
5. The null space of M^T is the span $\{\mathbf{u}_{r+1}, \ldots, \mathbf{u}_m\}$.

PROOF For number 1, premultiply both sides of Eq. (E-1a) by U^T and postmultiply both sides by V, giving

$$U^T M V = \begin{bmatrix} \Sigma & 0 \\ 0 & 0 \end{bmatrix} \tag{E-5}$$

(recall $U^TU = I$ and $V^TV = I$, since U and V are orthogonal matrices). But pre- and postmultiplication of M by nonsingular matrices does not change its rank. Therefore, the rank of M equals the rank of

$$\begin{bmatrix} \Sigma & 0 \\ 0 & 0 \end{bmatrix}_{m\times n} \tag{E-6}$$

which obviously has rank r.

For number 2, postmultiply both sides of Eq. (E-1b) by V to yield

$$MV = [U_1 \quad U_2]\begin{bmatrix} \Sigma & 0 \\ 0 & 0 \end{bmatrix} = [U_1\Sigma \quad 0]_{n\times n} \tag{E-7a}$$

But the range space of MV equals the range space of M since V is nonsingular. From Eq. (E-7a), though, the span of the columns of MV equals the span of the columns of U_1.

For number 3, partition $V = [V_1 \quad V_2]$ in Eq. (E-7a) to yield

$$M[V_1 \quad V_2] = [MV_1 \quad MV_2]_{m\times n} = [U_1\Sigma \quad 0]_{m\times n} \tag{E-7b}$$

Thus $MV_2 = 0$ and the $(n - r)$ orthogonal columns of V_2 form an orthogonal basis for the $(n - r)$-dimension null space of M.

Finally, results numbers 4 and 5 follow by taking the transpose of M in Eq. (E-1b) to give

$$M^T = [V_1 \quad V_2]\begin{bmatrix} \Sigma & 0 \\ 0 & 0 \end{bmatrix}\begin{bmatrix} U_1^T \\ U_2^T \end{bmatrix}$$

Results numbers 4 and 5 then follow analogous to numbers 2 and 3.

It is generally acknowledged by the numerical analysis specialist that the singular-value decomposition provides the most suitable and computationally stable means for finding the rank and the four fundamental subspaces of a matrix. Furthermore, the orthogonal columns of U and V provide orthogonal bases for these subspaces; use of an orthogonal basis to represent a subspace can be valuable for many reasons. These are particularly important results for us since it is so very easy to find the singular-value decomposition of a given matrix by using readily available software.

Example E-1

$$M = \begin{bmatrix} 1 & 0 & 1 \\ 2 & 1 & 1 \\ 0 & -2 & 2 \\ 3 & 2 & 1 \end{bmatrix}$$

This is the same matrix M as considered in Example 10-6, Sec. 10-3. Using elementary row operations it is there shown that the rank of M is two and its null space is

$$\mathfrak{N}(M) = \text{span}\left\{\begin{bmatrix} 1 \\ -1 \\ -1 \end{bmatrix}\right\}$$

The singular-value decomposition of M is (shown to three significant figures)

$$M = \begin{bmatrix} -0.242 & -0.295 & 0.847 & -0.371 \\ -0.526 & -0.116 & -0.490 & -0.685 \\ 0.0859 & -0.946 & -0.201 & 0.238 \\ -0.811 & 0.0627 & 0.0444 & 0.580 \end{bmatrix} \begin{bmatrix} 4.61 & 0 & 0 \\ 0 & 2.96 & 0 \\ 0 & 0 & 1.01 \times 10^{-14} \\ 0 & 0 & 0 \end{bmatrix}$$

$$\times \begin{bmatrix} -0.808 & -0.503 & -0.305 \\ -0.114 & 0.643 & -0.757 \\ -0.577 & 0.577 & 0.577 \end{bmatrix}$$

There are only two "nonzero" singular values, and so the rank of M is two. The null space of M is given as the span of the third right singular vector:

$$\mathfrak{N}(M) = \text{span}\left\{\begin{bmatrix} -0.577 \\ 0.577 \\ 0.577 \end{bmatrix}\right\}$$

This agrees with the result found by elementary row operations. An orthogonal basis for the range space of M is given by the first two left singular vectors:

$$\mathfrak{R}(M) \equiv \text{span}\left\{\begin{bmatrix} 1 \\ 2 \\ 0 \\ 3 \end{bmatrix} \begin{bmatrix} 0 \\ 1 \\ -2 \\ 2 \end{bmatrix} \begin{bmatrix} 1 \\ 1 \\ 2 \\ 1 \end{bmatrix}\right\}$$

$$= \text{span}\left\{\begin{bmatrix} -0.242 \\ -0.526 \\ 0.0859 \\ -0.811 \end{bmatrix} \begin{bmatrix} -0.295 \\ -0.116 \\ -0.946 \\ 0.0627 \end{bmatrix}\right\}$$

E-3 SOLUTION OF THE GENERAL LINEAR-EQUATION PROBLEM

In addition to being useful for finding the rank and subspaces of a matrix, the SVD can be particularly valuable in solution of the basic linear equation problem—particularly the overdetermined linear-equation problem where there are more equations than unknowns. Such a situation arises, for example, in the process of linear regression analysis.

Theorem E-2: Generalized Inverse Let M have SVD as in Theorem E-1. Then the best approximate solution to the linear equation problem

$$M\mathbf{x} = \mathbf{d}$$

is given by

$$\hat{\mathbf{x}} = V_1\Sigma^{-1}U_1^T\mathbf{d} + \text{span}\,\{V_2\}$$

That is, the $\hat{\mathbf{x}}$ is the solution that minimizes the norm-squared error

$$\|M\hat{\mathbf{x}} - \mathbf{d}\|^2$$

The minimum possible for this norm-squared error is zero when $M\hat{\mathbf{x}} = \mathbf{d}$. Furthermore,

$$\mathbf{x}^* = V_1\Sigma^{-1}U_1^T\mathbf{d}$$

is the $\hat{\mathbf{x}}$ solution of minimum norm itself; that is, $\|\mathbf{x}^*\| < \|\hat{\mathbf{x}}\|$ for all $\hat{\mathbf{x}} \neq \mathbf{x}^*$.

PROOF See Stewart [E-1].

Example E-2 Consider

$$\begin{bmatrix} 1 & 0 & 1 \\ 2 & 1 & 1 \\ 0 & -2 & 2 \\ 3 & 2 & 1 \end{bmatrix} \begin{bmatrix} x_1 \\ x_2 \\ x_3 \end{bmatrix} = \begin{bmatrix} 1 \\ 0 \\ 0 \\ 0 \end{bmatrix}$$

This is the same M coefficient matrix as considered in Example E-1. This problem has no exact solution because $\mathbf{d}$ is not contained in the range space of M.

However, by Theorem E-2 the SVD of M gives us the best approximate solution. In particular,

$$\hat{\mathbf{x}} = V_1\Sigma^{-1}U_1^T\mathbf{d} + \text{span}\,\{V_2\}$$

$$= \begin{bmatrix} 0.0497 \\ -0.0147 \\ 0.0645 \end{bmatrix} + \text{span}\left\{\begin{bmatrix} -0.577 \\ 0.577 \\ 0.577 \end{bmatrix}\right\}$$

and $$\mathbf{x}^* = \begin{bmatrix} 0.0497 \\ -0.0147 \\ 0.0645 \end{bmatrix}$$

is the best approximate solution of least norm. Note that

$$M\mathbf{x}^* = \begin{bmatrix} 0.114 \\ 0.149 \\ 0.158 \\ 0.184 \end{bmatrix}$$

is not too close to the given **d**, but it is the best **x**, in a norm-squared error sense, that can be made to the desired **d**.

The solution $\mathbf{x}^*$ from Theorem E-2 is also called the generalized or pseudoinverse solution of the linear equation problem $M\mathbf{x} = \mathbf{d}$, and

$$M^\dagger \equiv V_1 \Sigma^{-1} U_1^T$$

is called the generalized or pseudoinverse solution of the matrix M (see Ref. E-1).

E-4 CONDITION NUMBER OF A GIVEN MATRIX

A number of times in the text it is mentioned that a certain linear equation problem might be highly sensitive to small numerical errors. Such a situation arises, for example, in using the Hankel matrix or the controllability matrix, M_c (or observability matrix, M_o). When such a situation occurs the coefficient matrix, M, is said to be *ill-conditioned.* The singular values of M give a definitive means to determine the *condition number* of a given matrix.

We consider only square $n \times n$ matrices, but the concepts are extendable to the nonsquare case also (see Stewart [E-6]). A square $n \times n$ matrix, M, is nonsingular (invertible or has rank n) if and only if it has n strictly positive singular values. However, the separation between the largest and smallest singular value gives a strong indication of how sensitive the linear equation problem $M\mathbf{x} = \mathbf{d}$ will be to small errors. Indeed, the solution of the error linear equation problem

$$(M + E)\mathbf{x}_\varepsilon = (\mathbf{d} + \boldsymbol{\varepsilon})$$

is related to the true solution, **x**, to the true linear equation problem $M\mathbf{x} = \mathbf{d}$ by the normalized bound

$$\frac{\|\mathbf{x} - \mathbf{x}_\varepsilon\|}{\|\mathbf{x}\|} \leq (\sigma_1/\sigma_n)\left\{\frac{\|E\|}{\|M\|} + \frac{\|\boldsymbol{\varepsilon}\|}{\|d\|}\right\}$$

where σ_1 is the largest singular value of M and σ_n is the smallest (see Ref. E-6). The $n \times n$ matrix E is an error matrix to the true coefficient matrix M, and $\boldsymbol{\varepsilon}$ is an $n \times 1$ error vector to the data vector **d**. These errors may be introduced for many reasons such as computer rounding, computational errors, or simply noise. Regardless of how they are introduced, the ratio of the largest to smallest singular value of M gives a good indication of how sensitive the linear equation problem will be to small errors. If the ratio is small (close to unity), the linear equation problem is relatively insensitive to errors. If the ratio is large (as a worst case it can become infinite), the linear equation problem is said to be ill-conditioned, and the solution is highly sensitive to small errors.

The ratio of the largest to smallest singular value is called the *condition number* of a matrix. It is a fascinating topic, and the interested reader is referred to, for example, Stewart [E-1 and E-6].

REFERENCES

E-1. Stewart, G. W.: *Introduction to Matrix Computations,* Academic, New York, 1973.

E-2. Garbow, et al.: *Matrix Eigensystem Routines: EISPACK Guide Extension,* Springer-Verlag, Berlin, 1976.

E-3. Dongarra, J. J., et al.: *LINPACK User's Guide,* SIAM, Philadelphia, 1979.

E-4. *IMSL Reference Manual,* IMSL Inc., Houston, 1979.

E-5. Klema, V. C., and A. J. Laub: "The Singular Value Decomposition: Its Computation and Some Applications," *IEEE Trans. Auto. Control,* vol. AC-25, April 1980, pp. 164–176.

E-6. Stewart, G. W.: "On the Perturbation of Pseudo-Inverses, Projections, and Linear Least Squares Problems," *SIAM Review,* vol. 4, 1977, pp. 634–662.

APPENDIX

F

APPLICABLE SOFTWARE

This appendix presents a brief summary of applicable software routines that are readily available either commercially or essentially at no cost. These routines generally fit the category of being expertly written and highly transportable to various computer systems; they generally support topics discussed in the text (as well as many more functions not discussed). The discussion is split into two basic categories of software: subroutine libraries and interactive packages.

F-1 SUBROUTINE LIBRARIES

Two subroutine libraries have been mentioned a number of times throughout the text: EISPACK (Refs. F-1 and F-2) and LINPACK (Ref. F-3). The former includes extensive subroutines for solving the eigenvalue-eigenvector problem, while the latter includes a variety of routines for solving and analyzing the basic linear equation problem. Both are highly portable and available at essentially no cost.

Commercially available subroutines which use many basic subroutines from EISPACK and LINPACK are:

IMSL. International Mathematical and Statistical Libraries Inc., Sixth Floor, NBC Building, 7500 Bellair Boulevard, Houston, Texas 77036; telephone: (713) 772-1927

NAG. Numerical Algorithms Group (USA) Inc., 1250 Grace Court, Downers Grove, Illinois 60516; telephone: (312) 971-2337

HARWELL. The Harwell Subroutine Library is issued by the Computer Science and Systems Division of the United Kingdom Atomic Energy Authority, Harwell, England.

These packages include a variety of capabilities extending well beyond the basics covered in this text. Reference F-4 presents an excellent review of these packages plus several more packages not mentioned above.

The aforementioned packages are quite general. A more specialized package of subroutines, tailored especially for estimation and control theory applications, is available from Dr. David Kleinman of the University of Connecticut, Storrs, Conn. (Ref. F-5). It also contains many general purpose matrix operation subroutines and is especially portable.

F-2 INTERACTIVE PACKAGES

As a labor-saving mechanism the subroutine packages described in the previous section are highly recommended; these packages contain time-tested, expertly written subroutines that could never be duplicated by the novice. But as a learning mechanism there is no substitute for interactive software whereby the user can carry out desired operations in a terminal session. This section describes a small sample of such interactive packages with which the author is personally familiar.

A package which carries out nearly all the operations discussed in the text, plus many more which are more advanced, is the MATLAB package. MATLAB employs EISPACK and LINPACK subroutines as its foundation. It is also very flexible, allowing the user to easily construct matrix manipulation subroutines which are not contained in MATLAB. It is available at essentially no cost from its architect:

Professor Cleve Moler
Department of Computer Science
University of New Mexico
Albuquerque, New Mexico 87131

If I had to recommend only one library to supplement the material of this text, I would recommend MATLAB.

A derivate of MATLAB is the MATRIX_X package. It carries out all the control-related operations of Chap. 11 plus many more not covered in the text. It is commercially available from

Integrated Systems, Inc.
151 University Avenue, Suite 400
Palo Alto, California 94301

A final interactive package with which the author is personally familiar is the program TOTAL. TOTAL was developed in 1977 by Stanley Larimer as a masters thesis project at the Air Force Institute of Technology, Wright Patterson AFB, Ohio. Since then it has been extensively used, class-tested, and augmented. Its capabilities are mostly related to controls, but it is also flexible for general-analysis purposes. It is available at essentially no cost from its current custodians:

AFWAL/FIGC
Wright Patterson AFB, Ohio 45433

REFERENCES

F-1. Smith et al.: *Matrix Eigensystem Routines: EISPACK Guide,* Springer-Verlag, Berlin, 1976.

F-2. Garbow et al.: *Matrix Eigensystem Routines: EISPACK Guide Extension,* Springer-Verlag, Berlin, 1977.

F-3. Dongarra, J. J. et al.: *LINPACK Users' Guide*, SIAM, Philadelphia, 1979.

F-4. Gaffnew, P. W.: "Information and Advice on Numerical Software," Rep. ORNL/CSD/TM-147, Oak Ridge National Laboratory, Oak Ridge, Tenn., May 1981.

F-5. Kleinman, D. L.: "A Description of Computer Programs for Use in Linear Systems Studies," University of Connecticut Technical Report, EECS-TR-77-2, July, 1977.

APPENDIX

G

LIST OF SYMBOLS

SYMBOL	DEFINITION	CHAPTER IN WHICH INTRODUCED
$a_0, a_1, \ldots, a_{n-1}$	Coefficients characteristic polynomial	2
$b_0, b_1, \ldots. b_m$	Coefficients zero polynomial	2
$a_{T_0}, \ldots, a_{T_{n-1}}$	Coefficients discrete characteristic polynomial	5
$b_{T_0}, \ldots, b_{T_m}$	Coefficients discrete zero polynomial	5
a_{ij}; b_{ij}; c_{ij}	The ith row and jth column element of A, B, C, respectively	6
b	Damping or friction coefficient	2
$\mathbf{d}$	Data vector in general linear equation problem, $M\mathbf{x} = \mathbf{d}$	2
e	Exponential	1
$f(\cdot)$, $g(\cdot)$	Arbitrary nonlinear functions	1
$h(t)$	Continuous impulse response function	1
$[h(t)]$	Impulse response matrix for MIMO system	7
$h_T(k)$	Discrete pulse response sequence	1
i	Electric current	2
j	Complex number $\sqrt{-1}$	2
k	Scalar multiplier	1
k	Spring constant	2
ℓ	Number of inputs in MIMO system	6
m	System mass	1
m	Order of zero polynomial	2
m	Number of outputs in a MIMO system	6
n	Order of characteristic polynomial	2
n	State dimension	6
p_i	System poles	2
p_{T_i}	Discrete system poles	5
q	General complex-valued exponent	2
$r(t)$	Unit step response	1
$[r(t)]$	Step response matrix for MIMO system	7

SYMBOL	DEFINITION	CHAPTER IN WHICH INTRODUCED
$r(kT)$	Unit step response sampled at $t = kT$	1
$r_T(k)$	Discrete step response model sequence	1
r_i	Residue in partial fraction expansion	2
s	Laplace variable	2
t	Time variable	1
t_0	Initial time	1
$u(t)$	Input function	1
$u_{-1}(\cdot)$	Unit step function	1
$u_T(k)$	Piecewise constant input	1
$\mathbf{v}$	General vector	9
$\mathbf{x}(t)$	State vector	1
$\mathbf{x}_T(k)$	Discrete model state vector	8
$y(t)$	System output	1
$y_T(k)$	Discrete system model output	5
z	z transform variable	5
z_i	System zeros	2
A, B, C	Plant, input, and output matrices	6
A_T, B_T, C_T	Discrete system plant, input, and output matrices	8
C	Electrical capacitance	2
D	Feedforward matrix	6
E, F	Intermediate matrices used in computing A_T and B_T	8
$F(j\omega)$ or $F(s)$	Fourier or Laplace transform of $f(s)$ or $f(t)$	3 or 4
$G(s)$	Feedback transfer function	4
$H(s)$	System transfer function	2
$H_T(z)$	Discrete system transfer function	5
I	Identify matrix	2
J	Matrix in its Jordan canonical form	10
K	System gain factor	2
L	Electrical inductance	2
$L\{\cdot\}$	General linear operation	2
M	General matrix	2
M_c, M_o	Controllability and observability matrices	9
M_{T_c}, M_{T_o}	Discrete system controllability and observability matrices	8
N	Upper integer index	1
$N(s)$	Zero (numerator) polynomial	2
P	Coordinate transformation matrix	9
Q	Orthogonal transformation matrix	10
R	Electric resistance	2
R_i	Matrix partial fraction residue	7
$U(j\omega)$ or $U(s)$	Fourier or Laplace transform of $u(t)$	3 or 4
U, V	Orthogonal matrices of left and right singular vectors	11, E
V	Vector subspace	9
W_c, W_o	Controllability and observability Gram matrices	11
$X(s)$	Laplace transform of $x(t)$	7
$Y(j\omega)$ or $Y(s)$	Fourier or Laplace transform of $y(t)$	3 or 4
$\mathfrak{f}$	Frequency in hertz	3
$\mathcal{A}$, $\mathcal{B}$	Vector space bases	9
$\mathcal{E}$	Energy of signal	3

SYMBOL	DEFINITION	CHAPTER IN WHICH INTRODUCED
$\mathcal{F}\{\cdot\}$	Fourier transform	3
$\tilde{\mathcal{F}}\{\cdot\}$	Fourier transform, cycles per time unit	3
$\mathcal{H}_T$	System Hankel matrix	5
$\mathcal{L}\{\cdot\}$	Laplace transform	4
$\mathcal{N}\{\cdot\}$	Null space	10
$\mathcal{R}\{\cdot\}$	Range space	10
$\mathcal{Z}\{\cdot\}$	z transform	5
α	General transform variable	3
$\alpha, \beta, \zeta, \xi, \tau$	Variables of integration	1
$\gamma_T(\cdot)$	Unit comb function	3
$\delta(\cdot)$	Unit Dirac delta function	1
$\delta_T(\cdot)$	Unit pulse input	1
$\boldsymbol{\xi}_i$	Right eigenvector	10
$\boldsymbol{\eta}_i$	Left (reciprical) eigenvector	10
λ_i	Eigenvalue (pole)	7
σ_i	Real part of complex variable	2
σ_i	Singular value	11, E
ν	Nullity	10
ρ	Rank	10
ω	Frequency in radians per time unit	2
ω_i	Imaginary part of complex variable	2
$\Delta(s)$	Characteristic polynomial	2
$\Gamma_T(\cdot)$	Transform of $\gamma_T(\cdot)$	3
Λ	Diagonal matrix of eigenvalues	7
Λ_T	Diagonal matrix of $e^{\lambda_i T}$	8

Subscripts

h	Homogeneous	2
p	Particular	2
zi	Zero input	1
zs	Zero state	1
ss	Steady state	1
T	Discrete system	1

Superscripts

T	Transpose	2
(1), (2), (3)	First, second, third derivative	2
$\dot{}, \ddot{}, \dddot{}$	First, second, third derivative	1

APPENDIX

H

PROBLEM SOLUTIONS

Chapter 1

1-1 (*a*) Linearity of the zero-input and zero-state response.
(*b*) Additive decomposition, linearity of zero-state response. (*c*) All three properties.

1-3 $10(e^{-2}t + 3\sin 2t + \cos 2t)u_{-1}(t)$.

1-5 $2[e^{-(t-5)} - 1 + (t-5)]u_{-1}(t-5)$.

1-7 (*a*) t.i.; (*b*) t.v.; (*c*) t.i.; (*d*) t.i.

1-9 (*a*) 2.2, 3.2, 3.6; (*b*) 2.2, 5.4, 9.48; (*c*) 2.2, −1.2, 1.6; (*d*) −0.4, −0.3, −0.1;
(*e*) 22, 5.6, 13; (*f*) 21.6, 5.3, 12.9.

1-11 (*a*) $y_{zi}(t) = 18,\ 0 < t \le 1;\ y_{zi}(t) = 24 - 6t,\ 1 < t \le 4;\ y_{zi}(t) = 0,\ t > 4;$
(*b*) $y_{zs}(t) = 6t,\ 0 < t \le 3;\ y_{zs}(t) = 18,\ 3 < t \le 4;\ y_{zs}(t) = 42 - 6t,\ 4 < t \le 7;\ y_{zs}(t) = 0,\ t > 7;$
(*c*) $y(t) = 18 + 6t,\ 0 < t \le 1;\ y(t) = 24,\ 1 < t \le 3;\ y(t) = 42 - 6t,\ 3 < t < 7$: $y(t) = 0,\ t > 7$.

1-13 (*a*) $y_{zs}(t) = 2(1 - e^{-t})u_{-1}(t)$; (*b*) $y(t) = 2(1 - e^{-t})u_{-1}(t)$; (*c*) $y(t) = 2e^{-t} + 2t - 2$.

1-15 $\sin 2(t - 3)$.

1-17 0.39347, 0.23865, 0.14475, 0.08779.

1-19 1.0, 0.8, −0.4, 0.8.

1-21 $h(t) = (4e^{-4t} - 2e^{-2t}\sin t + e^{-2t}\cos t)u_{-1}(t)$.

1-23 rise time $\approx$ 0.5, duplication time $\approx$ 0.9
peak time $\approx$ 1.6, settling time $\approx$ 8.2
peak value $\approx$ 0.36, final value $\approx$ 0.25

1-25 (*a*) stable; (*b*) unstable; (*c*) unstable.

Chapter 2

2-1 (*a*) $\ddot{y} + (R/L)\dot{y} + (1/LC)y = (1/L)\dot{u}$; (*b*) $\ddot{y} + (R/L)\dot{y} + (1/LC)y = (1/\mathrm{LC})u$.

2-3 (*a*) $\dot{y} = (1/m)u$; (*b*) $\dot{y} + (k/b)y = (k/b)u$.

2-5 $\ddot{y} + \left(\dfrac{k(m_1 + m_2)}{m_1 m_2}\right)\dot{y} = \left(\dfrac{k}{m_1 m_2}\right)u.$

2-7 (*a*) $3/(s + 2)$; (*b*) $(3s + 2)/(s + 2)$; (*c*) $s/(s^2 + 4)$; (*d*) $(s^2 + 2s + 10)/s^2$.

2-9 (*a*) $c_1e^{-7t} + c_2e^{-8t}$; (*b*) $c_1e^{-7t} + c_2e^{-8t}$; (*c*) $c_1 + c_2e^{(-0.2+2j)t} + c_3e^{(-0.2-2j)t}$, $c_1 + c_2'e^{-0.2t}\cos(2t) + c_3'e^{-0.2t}\sin(2t)$, $c_1 + c_2''e^{-0.2t}\sin(2t + c_3'')$; (*d*) $c_1e^{-t} + c_2te^{-t}$;

2-11 (*a*) stable, $p_1 = -2$; (*b*) stable, $p_1 = -2$, $z_1 = -2/3$; (*c*) marginally stable, $p_{1,2} = \pm 2j$, $z_1 = 0$; (*d*) unstable, $p_{1,2} = 0$, $z_{1.2} = -1 \pm 3j$.

2-13 $y_p(t) = -25 + 3t - \dfrac{1}{6}e^t$.

2-15 (*a*) $y_p(t) = \dfrac{4}{5}\cos 2t - \dfrac{3}{5}\sin 2t = \sin(2t - 53.13°)$;

(*b*) $y_p(t) = \dfrac{4}{5}\cos 2t + \dfrac{3}{5}\sin 2t = \sin(2t + 53.13°)$.

2-17 $y_p(t) = 2e^{-3t} + 4te^{-3t}$.

2-19 $y_p(t) = -12e^{-t} + 12te^{-t}$.

2-21 (*b*) $y_p(t) = \text{Re}\{H(s)e^{st}\}|_{s=\sigma+j\omega}$; (*c*) $y_p(t) = -3/17e^{-2t}\cos 4t + 12/17e^{-2t}\sin 4t$.

2-25 (*a*) $y(t) = -\dfrac{2}{7}e^{-7t} + \dfrac{2}{7}$; (*b*) $y(t) = -\dfrac{153}{29}e^{-2t} + \dfrac{124}{29}\cos 5t + \dfrac{20}{29}\sin 5t$;

(*c*) $y(t) = 7e^{-3t} - 5e^{-4t}$.

2-27 (*a*) $\mathbf{x} = [-6, 7/2, 6]^T$; (*b*) $\mathbf{x} = [7/4, -2, 0, 0]^T$.

2-29 (*a*) $r(t) = \dfrac{1}{5} - \dfrac{1}{5}e^{-2t}\cos t - \dfrac{2}{5}e^{-2t}\sin t$; (*b*) $r(t) = \dfrac{1}{5} - \dfrac{1}{5}e^{-2t}\cos t + \dfrac{3}{5}e^{-2t}\sin t$;

(*c*) $r(t) = -\dfrac{1}{5} + \dfrac{1}{5}e^{-2t}\cos t + \dfrac{7}{5}e^{-2t}\sin t$; (*d*) $r(t) = \dfrac{2}{5} + \dfrac{3}{5}e^{-2t}\cos t + \dfrac{1}{5}e^{-2t}\sin t$;

(*e*) $r(t) = -\dfrac{2}{5} + \dfrac{7}{5}e^{-2t}\cos t - \dfrac{1}{5}e^{-2t}\sin t$.

2-31 (*a*) $\dot{y}(0^+) = 0$, $y_{ss} = 1/5$; (*b*) $\dot{y}(0^+) = 1$, $y_{ss} = 1/5$; (*c*) $\dot{y}(0^+) = 1$, $y_{ss} = -1/5$; (*d*) $\dot{y}(0^+) = -1$, $y_{ss} = 2/5$; (*e*) $\dot{y}(0^+) = -3$, $y_{ss} = -2/5$.

2-33 (*a*) $h(t) = e^{-2t}\sin t$; (*b*) $h(t) = e^{-2t}\cos t - e^{-2t}\sin t$; (*c*) $h(t) = e^{-2t}\cos t - 3e^{-2t}\sin t$; (*d*) $h(t) = -e^{-2t}\cos t - e^{-2t}\sin t + \delta(t)$; (*e*) $h(t) = -3e^{-2t}\cos t - e^{-2t}\sin t + \delta(t)$.

2-35 $h(0^+) = 0$ for a system with no zero. $h(0^+) \neq 0$ for a system with a zero.

Chapter 3

3-1 (*a*) $y_p(t) = 0.467\sin(t + 37.4°)$; (*b*) $y_p(t) = 2.795\sin(2t - 26.6°)$; (*c*) $y_p(t) = 0.134\sin(8t - 94.1°)$.

3-3 (*a*) $y_p(t) = 0.166\sin(2t + 1.9°)$, $z_1 = -0.1$, $p_1 = -1$, $p_2 = -5$;
(*b*) $y_p(t) = 0.166\sin(2t + 7.6°)$, $z_1 = +0.1$, $p_1 = -1$, $p_2 = -5$;
(*c*) $y_p(t) = 0.166\sin(2t - 51.2°)$, $z_1 = -0.1$, $p_1 = +1$, $p_2 = -5$;
(*d*) $y_p(t) = 0.166\sin(2t - 45.5°)$, $z_1 = +0.1$, $p_1 = +1$, $p_2 = -5$;
(*e*) $y_p(t) = 0.166\sin(2t - 181.9°)$, $z_1 = +0.1$, $p_1 = +1$, $p_2 = +5$.

3-5 All five have the same Bode magnitude plot.

3-7 All five have different Bode phase-angle plots.

3-9 Yes, if the phase-angle plots are included.

3-11 (*a*) $2/(j\omega + 3)$; (*b*) $32/[(j\omega + 2)^2 + 64]$; (*c*) $3j\pi[\delta(\omega - 4) - \delta(\omega + 4)]$; (*d*) $2/(j\omega + 3)^2$.

3-13 $f(t) = 5e^{-t}\cos(4t)u_{-1}(t)$.

3-15 (*a*) $80\,\text{sinc}(5\omega)$; (*b*) $160\,\text{sinc}(10\omega)$; (*c*) $e^{2t_0}/(j\omega + 2)$; (*d*) $e^{-2t_0}e^{-j\omega t_0}/(j\omega + 2)$; (*e*) $e^{-j\omega t_0}/(j\omega + 2)$; (*f*) $1/[(j\omega + 2)(j\omega + 5)]$.

3-17 (*b*) $f(t) = e^{-2t}/3 - e^{-5t}/3$; $F(j\omega) = (1/3)/(j\omega + 2) - (1/3)/(j\omega + 5)$; (*c*) Yes (use partial fraction expansion of Prob. 3-15*f*).

3-19 $\sqrt{\pi}/2$.

3-21 $H(j\omega) = (j\omega + 0.5)/[(j\omega)^2 + 2e^{-j\omega}(j\omega) + 3]$.

3-23 5 Hz.

3-25 $T = 1/40$, sampling interval.

Chapter 4

4-1 (*a*) $2/(s + 3)$; (*b*) $32/[(s + 2)^2 + 64]$; (*c*) $12/(s^2 + 16)$; (*d*) $2/[(s + 3)^2]$.

4-3 (*a*) $(-9e^{-t} + 17e^{-2t})u_{-1}(t)$; (*b*) $e^{-t} \sin 4t,\ u_{-1}(t)$; (*c*) $e^{-t} \cos 2t,\ u_{-1}(t)$.

4-5 (*a*) $1/s^2 - 5/s$; (*b*) $(1/s^2)e^{-5s}$; (*c*) $(1/s^2)e^{-5s} + (5/s)e^{-5s}$; (*d*) $1/s^2$.

4-7 (*a*) 0; (*b*) $-2/15$.

4-9 (*a*) $y(t) = y_{zs}(t) = (-2/7)e^{-7t} + 2/7$;
(*b*) $y_{zi}(t) = -5e^{-3t}$, $y_{zs}(t) = 3e^{-3t} + 22 \cos 4t - 4 \sin 4t$;
(*c*) $y_{zi}(t) = 5e^{-3t} - 3e^{-4t}$, $y_{zs}(t) = (4/3)e^{-3t} - (3/2)e^{-4t} + 1/6$.

4-11 (*a*) $h(t) = 5e^{-2t},\ u_{-1}(t)$; (*b*) $h(t) = (-2e^{-t} + 5e^{-2t})u_{-1}(t)$;
(*c*) $h(t) = e^{-0.2t} \sin (2t)u_{-1}(t)$; (*d*) $h(t) = e^{-t}u_{-1}(t) + \delta(t)$.

4-13 (*a*) $r(t) = 1/4 - (1/4)e^{-2t} - (1/2)te^{-2t}$; (*b*) $r(t) = 1/4 - (1/4)e^{-2t} + (1/2)te^{-2t}$;
(*c*) $r(t) = -1/4 + (1/4)e^{-2t} + (3/2)te^{-2t}$.

4-15 (*a*) $r(t) = 10 - 0.909e^{-100t} - 9.09e^{-t}$;
(*b*) $r(t) = 4.95 + 0.05e^{-t}(e^{10jt} + e^{-10jt}) + (-2.53 + 2.47j)e^{-t}e^{jt} + (-2.53 - 2.47j)e^{-t}e^{-jt}$;
(*c*) $r(t) = 220 - 226.6e^{-0.1t} + (3.31 + 4.01j)e^{-t}e^{2jt} + (3.31 - 4.01j)e^{-t}e^{-2jt}$.

4-17 $H(s) = [50(s + 3)(6s + 5)]/[s(s + 5)(s + 2)^2]$.

4-19 $p_1 = -.167$, $p_2 = -5.98$, $z_1 = -.1$, $z_2 = -.2$.

Chapter 5

5-1 $y_T(k) - 8y_T(k - 1) + 3.8y_T(k - 2) = 2u_T(k - 1) - 15u_T(k - 2)$.

5-3 $h_T(1) = 2$, $h_T(2) = 1$, $h_T(3) = 0.4$, $h_T(4) = -0.6$, $h_T(5) = -6.32$, $h_T(6) = -48.28$.

5-5 $r_T(1) = 2$, $r_T(2) = 3$, $r_T(3) = 3.4$, $r_T(4) = 2.8$, $r_T(5) = -3.52$, $r_T(6) = -51.8$.

5-7 $y_{T_h}(k) = c_1(0.507)^k + c_2(7.49)^k$.

5-9. Unstable. Pole at $p_{T_2} = 7.49 > 1$.

5-11 (*a*) Unstable; (*b*) unstable.

5-13 (*a*) $y_T(k) - 0.8187y_T(k - 1) = 0.9063u_T(k - 1)$;
(*b*) $y_T(k) - 1.221y_T(k - 1) = 1.107u_T(k - 1)$;
(*c*) $y_T(k) - 1.489y_T(k - 1) + 0.5488y_T(k - 2) = 0.1977u_T(k - 1) + 0.1081u_T(k - 2)$;
(*d*) $y_T(k) - 1.489y_T(k - 1) + 0.5488y_T(k - 2) = 0.0991u_T(k - 1) - 0.1888u_T(k - 2)$.

5-17 $p_1 = -3.393$, $p_2 = 10.07$.

5-19 (*a*) $y_p(t) = -16.67$, $y_p(t) = 1.471$; (*b*) no, both systems are unstable.

5-21 (*a*) $y_{ss} = 4$; (*b*) $y_{ss} = 5.357$; (*c*) $y_{ss} = 3.371$.

5-23 (*a*) $y_{T_h}(k) = c_1(0.5)^k$, $y_{T_p} = 4$, $y_T(k) = -4(0.5)^k + 4$;
(*b*) $y_{T_h}(k) = c_1(0.2)^k + c_2(0.3)^k$, $y_{T_p} = 5.357$, $y_T(k) = 17.499(0.2)^k - 22.856(0.3)^k + 5.357$;
(*c*) $y_{T_h}(k) = c_1(0.2 + 0.5j)^k + c_2(0.2 - 0.5j)^k =$
$c_1'(0.5385)^k \cos (1.19k) + c_2'(0.5385)^k \sin (1.19k)$; $y_{T_p} = 3.371$;
$y_T(k) = -3.371(0.5385)^k \cos (1.19k) - 3.393(0.5385)^k \sin (1.19k) + 3.371$.

5-25 (*a*) $r_T(1) = 2$, $r_T(2) = 3$, $r_T(3) = 3.5$; (*b*) $r_T(1) = 2$, $r_T(2) = 4$, $r_T(3) = 4.88$;
(*c*) $r_T(1) = 1$, $r_T(2) = 3.4$, $r_T(3) = 4.07$.

5-27 $y_{T_p}(k) = (d/dz)\left\{\frac{2}{z - 0.5}z^k\right\} = -8 - 4k$.

5-29 $y_{T_p}(k) = -\frac{4}{3}(-1)^k$.

5-31 $y_{T_p}(k) = \text{Re}\left\{\frac{2}{q-0.5}(q)^k\right\}$, $q = (1/\sqrt{2} + j/\sqrt{2})$.

5-33 (*a*) $p_1 = -3.466$; (*b*) $p_1 = -8.047$, $p_2 = -6.02$; (*c*) $p_{1,2} = -3.095 \pm 5.95j$.

5-35 $y_T(k) = -4(0.5)^k + 4$.

5-37 Derivation follows along similar lines as Eq. (5-17) itself, only additional equations up to N are used. Yes, this can improve computational accuracy.

Chapter 6

6-1 (*a*) $\dot{x} = -2x + u$, $y = 3x$.
(*b*) $\dot{x} = -2x + u$, $y = 2x + 3u$.

(c) $$\dot{\mathbf{x}} = \begin{bmatrix} 0 & 1 \\ -2 & -3 \end{bmatrix}\mathbf{x} + \begin{bmatrix} 0 \\ 1 \end{bmatrix}u, \; y = [3, 0]\mathbf{x}.$$

(*d*) $$\dot{\mathbf{x}} = \begin{bmatrix} 0 & 1 \\ -2 & -3 \end{bmatrix}\mathbf{x} + \begin{bmatrix} 0 \\ 1 \end{bmatrix}u, \; y = [9, 3]\mathbf{x}.$$

(*e*) $$\dot{\mathbf{x}} = \begin{bmatrix} 0 & 1 \\ -2 & -3 \end{bmatrix}\mathbf{x} + \begin{bmatrix} 0 \\ 1 \end{bmatrix}u, \; y = [-10, -3]\mathbf{x} + 3u.$$

6-3 (*a*) $x(0) = -1/3$; (*b*) $x(0) = 5/2$; (*c*) $\mathbf{x}(0) = [-1/3, 4/3]^T$; (*d*) $\mathbf{x}(0) = [-1, 8/3]^T$; (*e*) $\mathbf{x}(0) = [5/2, -10]^T$.

6-5 (*a*) $\dot{x} = -2x + 3u$, $y = x$.
(*b*) $\dot{x} = -2x + 2u$, $y = x + 3u$.

(*c*) $$\dot{\mathbf{x}} = \begin{bmatrix} 0 & -2 \\ 1 & -3 \end{bmatrix}\mathbf{x} + \begin{bmatrix} 3 \\ 0 \end{bmatrix}u, \; y = [0, 1]\mathbf{x}.$$

(*d*) $$\dot{\mathbf{x}} = \begin{bmatrix} 0 & -2 \\ 1 & -3 \end{bmatrix}\mathbf{x} + \begin{bmatrix} 9 \\ 3 \end{bmatrix}u, \; y = [0, 1]\mathbf{x}.$$

(*e*) $$\dot{\mathbf{x}} = \begin{bmatrix} 0 & -2 \\ 1 & -3 \end{bmatrix}\mathbf{x} + \begin{bmatrix} -10 \\ -3 \end{bmatrix}u, \; y = [0, 1]\mathbf{x} + 3u.$$

6-7 (*a*) $x(0) = -1$; (*b*) $x(0) = 1$; (*c*) $\mathbf{x}(0) = [1, -1]^T$; (*d*) $\mathbf{x}(0) = [7, -1]^T$; (*e*) $\mathbf{x}(0) = [10, 5]^T$.

6-9 (*a*) $\dot{x} = -2x + u$, $y = 3x$.
(*b*) $\dot{x} = -2x + u$, $y = 2x + 3u$.

(*c*) $$\dot{\mathbf{x}} = \begin{bmatrix} -1 & 0 \\ 0 & -2 \end{bmatrix}\mathbf{x} + \begin{bmatrix} 1 \\ 1 \end{bmatrix}u, \; y = [3, -3]\mathbf{x}.$$

(*d*) $$\dot{\mathbf{x}} = \begin{bmatrix} -1 & 0 \\ 0 & -2 \end{bmatrix}\mathbf{x} + \begin{bmatrix} 1 \\ 1 \end{bmatrix}u, \; y = [6, -3]\mathbf{x}.$$

(e) $$\dot{\mathbf{x}} = \begin{bmatrix} -1 & 0 \\ 0 & -2 \end{bmatrix}\mathbf{x} + \begin{bmatrix} 1 \\ 1 \end{bmatrix}u, \; y = [-7, 4]\mathbf{x} + 3u.$$

6-11 (*a*) $x(0) = -1/3$; (*b*) $x(0) = 5/2$; (*c*) $\mathbf{x}(0) = [2/3, 1]^T$; (*d*) $\mathbf{x}(0) = [4/3, 3]^T$; (*e*) $\mathbf{x}(0) = [-5/7, 0]^T$.

6-13 $$\dot{\mathbf{x}} = \begin{bmatrix} 0 & -1/L \\ 1/C & 0 \end{bmatrix}\mathbf{x} + \begin{bmatrix} 1/2 \\ 0 \end{bmatrix}V_{\text{in}}, \; y = [1, 0]\mathbf{x}.$$

6-15 $\dot{x}_1 = \dfrac{1}{m}F_{\text{in}}; \qquad y = V_1 = x_1.$

6-17 (*a*)

$$\dot{\mathbf{x}} = \begin{bmatrix} 0 & 1 & 0 \\ -5 & -6 & 0 \\ 0 & 0 & -5 \end{bmatrix}\mathbf{x} + \begin{bmatrix} 0 & 0 \\ 1 & 0 \\ 0 & 1 \end{bmatrix}\begin{bmatrix} u_1 \\ u_2 \end{bmatrix}$$

$$y = [2 \quad 1 \quad 3]\mathbf{x}$$

(*b*)

$$\dot{\mathbf{x}} = \begin{bmatrix} 0 & 1 & 0 \\ -5 & -6 & 0 \\ 0 & 0 & -5 \end{bmatrix}\mathbf{x} + \begin{bmatrix} 0 \\ 1 \\ 1 \end{bmatrix}u$$

$$\begin{bmatrix} y_1 \\ y_2 \end{bmatrix} = \begin{bmatrix} 2 & 1 & 0 \\ 0 & 0 & 3 \end{bmatrix}\mathbf{x}$$

(*c*)

$$\dot{\mathbf{x}} = \begin{bmatrix} 0 & 1 & 0 & 0 & 0 & 0 & 0 \\ -5 & -6 & 0 & 0 & 0 & 0 & 0 \\ 0 & 0 & 0 & 1 & 0 & 0 & 0 \\ 0 & 0 & -5 & -6 & 0 & 0 & 0 \\ 0 & 0 & 0 & 0 & 0 & 1 & 0 \\ 0 & 0 & 0 & 0 & -5 & -6 & 0 \\ 0 & 0 & 0 & 0 & 0 & 0 & -5 \end{bmatrix}\mathbf{x} + \begin{bmatrix} 0 & 0 \\ 1 & 0 \\ 0 & 0 \\ 0 & 1 \\ 0 & 0 \\ 1 & 0 \\ 0 & 1 \end{bmatrix}\begin{bmatrix} u_1 \\ u_2 \end{bmatrix}$$

$$\begin{bmatrix} y_1 \\ y_2 \end{bmatrix} = \begin{bmatrix} 2 & 1 & 2 & 0 & 0 & 0 & 0 \\ 0 & 0 & 0 & 0 & 5 & 0 & 3 \end{bmatrix}\mathbf{x}$$

6-19 1. (*a*) $H(s) = \left[\dfrac{1/4}{s+1} + \dfrac{3/4}{s+5}, \dfrac{3}{s+5}\right]$

(*b*)

$$H(s) = \begin{bmatrix} \dfrac{1/4}{s+1} + \dfrac{3/4}{s+5} \\ \dfrac{3}{s+5} \end{bmatrix}$$

(*c*)

$$H(s) = \begin{bmatrix} \dfrac{1/4}{s+1} + \dfrac{3/4}{s+5} & \dfrac{1/2}{s+1} + \dfrac{-1/2}{s+5} \\ \dfrac{5/4}{s+1} + \dfrac{-5/4}{s+5} & \dfrac{3}{s+5} \end{bmatrix}$$

3. (*a*)

$$\dot{\mathbf{x}} = \begin{bmatrix} -1 & 0 \\ 0 & -5 \end{bmatrix}\mathbf{x} + \begin{bmatrix} 1 & 0 \\ 3/4 & 3 \end{bmatrix}$$

$$y = [1/4 \quad 1]\mathbf{x}$$

(*b*)

$$\dot{\mathbf{x}} = \begin{bmatrix} -1 & 0 \\ 0 & -5 \end{bmatrix}\mathbf{x} + \begin{bmatrix} 1 \\ 1 \end{bmatrix}u$$

$$\begin{bmatrix} y_1 \\ y_2 \end{bmatrix} = \begin{bmatrix} 1/4 & 3/4 \\ 0 & 3 \end{bmatrix}\mathbf{x}$$

(*c*)

$$\dot{\mathbf{x}} = \begin{bmatrix} -1 & 0 & 0 & 0 \\ 0 & -1 & 0 & 0 \\ 0 & 0 & -5 & 0 \\ 0 & 0 & 0 & -5 \end{bmatrix}\mathbf{x} + \begin{bmatrix} 1 & 0 \\ 0 & 1 \\ 1 & 0 \\ 0 & 1 \end{bmatrix}\begin{bmatrix} u_1 \\ u_2 \end{bmatrix}$$

$$\begin{bmatrix} y_1 \\ y_2 \end{bmatrix} = \begin{bmatrix} 1/4 & 1/2 & 3/4 & -1/2 \\ 5/4 & 0 & -5/4 & 3 \end{bmatrix} \mathbf{x}$$

Chapter 7

7-1 $x(t) = e^{-2(t-1)}7 + e^{-2t} \int_1^t e^{2\tau} 3u(\tau)\, d\tau$; $y(t) = 5e^{-2(t-1)}7 + 5e^{-2t} \int_1^t e^{2\tau} 3u(\tau)\, d\tau + 4u(t)$.

7-5 $y(t) = (55/2)e^{-2(t-1)} + 23/2$.

7-7 Assuming $x(1) = 7$ in both cases, the solution is the same.

7-9 $H(s) = \dfrac{4s + 23}{s + 2}$

7-11 (*a*)

$$[sI - A]^{-1} = \begin{bmatrix} \dfrac{1}{s + 1 - 2j} & 0 \\ 0 & \dfrac{1}{s + 1 + 2j} \end{bmatrix}, \quad e^{At} = \begin{bmatrix} e^{(-1+2j)t} & 0 \\ 0 & e^{(-1-2j)t} \end{bmatrix}$$

(*b*)

$$[sI - A]^{-1} = \begin{bmatrix} \dfrac{s + 1}{(s + 1)^2 + 4} & \dfrac{2}{(s + 1)^2 + 4} \\ \dfrac{-2}{(s + 1)^2 + 4} & \dfrac{s + 1}{(s + 1)^2 + 4} \end{bmatrix}, \quad e^{At} = \begin{bmatrix} \cos 2t & \sin 2t \\ -\sin 2t & \cos 2t \end{bmatrix} e^{-t}$$

(*c*)

$$[sI - A]^{-1} = \begin{bmatrix} \dfrac{s + 2}{(s + 1)^2 + 4} & \dfrac{1}{(s + 1)^2 + 4} \\ \dfrac{-5}{(s + 1)^2 + 4} & \dfrac{s}{(s + 1)^2 + 4} \end{bmatrix}$$

$$e^{At} = \begin{bmatrix} \cos 2t + \dfrac{1}{2} \sin 2t & \dfrac{1}{2} \sin 2t \\ -\dfrac{5}{2} \sin 2t & \cos 2t - \dfrac{1}{2} \sin 2t \end{bmatrix} e^{-t}$$

(*d*) $[sI - A]^{-1}$ and e^{At} are the transpose of (*c*).

7-13 $H(s) = (4s + 16)/(s^2 + 2s + 5)$ for all.

7-15
$$\begin{bmatrix} \dfrac{1}{-1 + 2j} & 0 \\ 0 & \dfrac{1}{-1 - 2j} \end{bmatrix} \begin{bmatrix} e^{(-1+2j)t} - 1 & 0 \\ 0 & e^{(-1-2j)t-} - 1 \end{bmatrix}$$

7-17
$$\begin{aligned} \dot{\Phi}_{1,1}(t) &= (-1 + 2j)\Phi_{1,1}(t), \ \Phi_{1,1}(0) = 1 \\ \dot{\Phi}_{2,1}(t) &= (-1 - 2j)\Phi_{2,1}(t), \ \Phi_{2,1}(0) = 0 \\ \dot{\Phi}_{1,2}(t) &= (-1 + 2j)\Phi_{1,2}(t), \ \Phi_{1,2}(0) = 0 \\ \dot{\Phi}_{2,2}(t) &= (-1 - 2j)\Phi_{2,2}(t), \ \Phi_{2,2}(0) = 1 \end{aligned}$$

7-19
$$\begin{aligned} e^{A(t_1+t_2)} &\equiv I + A(t_1 + t_2) + A^2\frac{(t_1 + t_2)^2}{2!} + A^3\frac{(t_1 + t_2)^2}{3!} + \cdots \\ &= \left(I + At_1 + A^2\frac{t_1^2}{2!} + \cdots\right)\left(I + At_2 + A^2\frac{t_2^2}{2!} + \cdots\right) \\ &= e^{At_1}e^{At_2} \end{aligned}$$

7-21
$$\mathbf{x}(t) = \begin{bmatrix} 20 \cos 2(t + 3) - 10 \sin 2(t + 3) \\ -20 \cos 2(t + 3) - 10 \sin 2(t + 3) \end{bmatrix} e^{-(t+3)} +$$

$$\int_{-3}^{t} \begin{bmatrix} \cos (t - \tau) + \sin (t - \tau) \\ \cos (t - \tau) - \sin (t - \tau) \end{bmatrix} e^{-(t-\tau)}(4 \sin 7\tau)\, d\tau$$

$$y(t) = [120 \cos 2(t+3) - 40 \sin 2(t+3)]e^{-(t+3)} + \int_{-3}^{t} [4 \cos 2(t-\tau) + 6 \sin 2(t-\tau)e^{-(t-\tau)}](4 \sin 7\tau)\, d\tau$$

7-25 (*a*) $h(t) = (4 \cos 2t + 6 \sin 2t)e^{-t}$; (*b*) $H(s) = (4s + 16)/[(s+1)^2 + 4]$.

7-27 $r_{ss} = 16/5$.

7-29 $x(0^+) = \int_{0^-}^{0^+} e^{A\tau}\mathbf{b}\delta(t-\tau)\, d\tau = \mathbf{b}$.

Chapter 8

8-1 $y_T(0.3) = 6.603$.

8-3 $y_T(0.3) = 3.514$.

8-7 $\{[(e^{-0.875})^2]^2\}^2 = e^{-7} = 0.000912$.

8-9 $\|A\|_F = 64.03/2^6 = 0.6006$, $N = 6$.

8-11 $N = 5$.

8-15 (*a*) $u_T(0) = 0$, $u_T(1) = 2$, $u_T(2) = 4$ and $\dot{u}_T(k) = 2$ for all k; (*b*) $u_T(0) = 1$, $u_T(1) = 3$, $u_T(2) = 5$; (*c*) The piecewise linear approximation from (*a*) is exact: $u_T(k) + \dot{u}_T(k)(t - kT) = 2kT + 2(t - kT) = 2t$.

8-17 (*a*) $x_T(k) = 0.1353x_T(k-1) + 4.3233u_T(k-1) + 1.4850\dot{u}_T(k-1)$; $y_T(k) = 5x_T(k)$; (*b*) $y_T(1) = 14.850$, $y_T(2) = 60.093$, $y_T(3) = 109.450$.

8-19 (*a*) $x_T(k) = 0.1353x_T(k-1) + 4.3233u_T(k-1)$; $y_T(k) = 5x_T(k)$; (*b*) $y_T(1) = 21.617$, $y_T(2) = 67.775$, $y_T(3) = 117.255$; (*c*) The result from 8-17(*b*) is exact.

8-23 $h_T(1) = 4.532$, $h_T(2) = 3.710$, $h_T(3) = 3.038$.

8-25
$$\mathbf{x}_T(k) = \begin{bmatrix} 0 & 1 \\ -0.15 & 0.8 \end{bmatrix}\mathbf{x}_T(k-1) + \begin{bmatrix} 0 \\ 1 \end{bmatrix}u_T(k-1)$$
$$y_T(k) = [2, 3]\mathbf{x}_T(k)$$

8-27
$$\mathbf{x}_T(k) = \begin{bmatrix} 0.3 & 0 \\ 0 & 0.5 \end{bmatrix}\mathbf{x}_T(k-1) + \begin{bmatrix} 1 \\ 1 \end{bmatrix}u_T(k-1)$$
$$y_T(k) = [-14.5, 17.5]\mathbf{x}_T(k)$$

8-29 (*a*)
$$A_T = \begin{bmatrix} 0.991 & 0.086 \\ -0.172 & 0.733 \end{bmatrix} \qquad B_T = \begin{bmatrix} 0.0045 \\ 0.0861 \end{bmatrix}$$

(*b*) $h_T(1) = 0.45$, $h_T(2) = 0.119$

(*c*) $y_T(k) - 1.724y_T(k-1) + 0.741y_T(k-2) = 0.045u_T(k-1) + 0.041u_T(k-2)$

(*d*)
$$\mathbf{x}_T(k) = \begin{bmatrix} 0 & 1 \\ -0.741 & 1.724 \end{bmatrix}\mathbf{x}_T(k-1) + \begin{bmatrix} 0 \\ 1 \end{bmatrix}u_T(k-1)$$
$$y_T(k) = [0.045, 0.041]\mathbf{x}_T(k)$$

8-31 (*a*) Uncontrollable; (*b*) completely observable.

8-33 $u_T(0) = 60.82$, $u_T(1) = -39.27$, $u_{T_{ss}} = 3/2$.

8-35 $u_T(0) = -2$, $u_T(1) = 8.2$.

8-37 (*a*)
$$\mathbf{x}(0) = \begin{bmatrix} 3.205 \\ 0.897 \end{bmatrix}$$

(*b*)
$$\mathbf{x}(1) = \begin{bmatrix} 0.897 \\ -1.449 \end{bmatrix}$$

Chapter 9

9-1 (*a*) $\mathbf{v} = [v_1, v_2, v_3, v_4]^T$; (*b*) $\alpha\mathbf{v} = [\alpha v_1, \alpha v_2, \alpha v_3, \alpha v_4]^T$;
(*c*) $\mathbf{v}_a + \mathbf{v}_b = [v_{a_1} + v_{b_1}, v_{a_2} + v_{b_2}, v_{a_3} + v_{b_3}, v_{a_4} + v_{b_4}]^T$; (*d*) $\mathbf{0} = [0, 0, 0, 0]^T$.
9-3 (*a*) $\sqrt{5}$, $\sqrt{10}$, $\sqrt{5/2}$, $\sqrt{45}$; (*b*) $\mathbf{v}_1/\sqrt{5}$, $\mathbf{v}_2/\sqrt{10}$, $\mathbf{v}_3/\sqrt{5/2}$, $\mathbf{v}_4/\sqrt{45}$.
9-5 (*a*) $\mathbf{v}^T\mathbf{v}_1 = 0$, $\mathbf{v}^T\mathbf{v}_2 = -12$, $\mathbf{v}^T\mathbf{v}_3 = 6.5$, $\mathbf{v}^T\mathbf{v}_4 = -26$; (*b*) $\mathbf{v}_1$.
9-9 (*a*) 2; (*b*) 2; (*c*) 2; (*d*) 1; (*e*) 2; (*f*) 2; (g) 2; (*h*) 2; (*i*) 2; (*j*) 2.
9-11 (*a*) $\{\mathbf{v}_1, \mathbf{v}_2\}$, $\{\mathbf{v}_1, \mathbf{v}_3\}$, $\{\mathbf{v}_1, \mathbf{v}_4\}$, $\{\mathbf{v}_2, \mathbf{v}_4\}$, $\{\mathbf{v}_3, \mathbf{v}_4\}$; (*b*) $\{\mathbf{v}_1, \mathbf{v}_2\}$, $\{\mathbf{v}_1, \mathbf{v}_3\}$;
(*c*) $\{\mathbf{v}_1/\sqrt{5}, \mathbf{v}_2/\sqrt{10}\}$, $\{\mathbf{v}_1/\sqrt{5}, \mathbf{v}_3/\sqrt{5/2}\}$.

9-13 (*a*) $P = \begin{bmatrix} 4 & -2 \\ 1 & 0 \end{bmatrix}$

(*b*) $\mathbf{v} = 3\mathbf{v}_1 + \mathbf{v}_2$
(*c*) $\mathbf{v}|_{\mathcal{B}} = [10, 3]^T$
(*d*) $\mathbf{v} = 10\mathbf{v}_3 + 3\mathbf{v}_4$

9-15 $\dot{\tilde{\mathbf{x}}} = \begin{bmatrix} -50 & -28 \\ 84 & 47 \end{bmatrix}\tilde{\mathbf{x}} + \begin{bmatrix} 30 \\ -50 \end{bmatrix}u, \; y = [12, 7]\tilde{\mathbf{x}}$

9-17 $P = \begin{bmatrix} 1 & -1 \\ -1 & 2 \end{bmatrix}$

9-19 $M_c = \begin{bmatrix} 0 & 1 \\ 1 & -3 \end{bmatrix}, \; \widehat{M}_c = \begin{bmatrix} 1 & -1 \\ 1 & -2 \end{bmatrix}, \; P = M_c\widehat{M}_c^{-1} = \begin{bmatrix} 1 & -1 \\ -1 & 2 \end{bmatrix}$

9-21 $M_o = \begin{bmatrix} 5 & 1 \\ -2 & 2 \end{bmatrix}, \; M_o' = \begin{bmatrix} 0 & 1 \\ 1 & -3 \end{bmatrix}, \; P = M_o^{-1}M_o' = \begin{bmatrix} -1/12 & 5/12 \\ 5/12 & -13/12 \end{bmatrix}$

9-23 (*a*) $\dot{\tilde{\mathbf{x}}} = \begin{bmatrix} 0 & 1 \\ -2 & -3 \end{bmatrix}\tilde{\mathbf{x}} + \begin{bmatrix} 1 \\ 2 \end{bmatrix}u, \; y = [1, 0]\tilde{\mathbf{x}}$

(*b*) $M_o = \begin{bmatrix} 5 & 1 \\ -2 & 2 \end{bmatrix}, \; \widetilde{M}_o = \begin{bmatrix} 1 & 0 \\ 0 & 1 \end{bmatrix}, \; P = M_o^{-1}\widetilde{M}_o = \begin{bmatrix} 2/12 & -1/12 \\ 2/12 & 5/12 \end{bmatrix}$

(*c*) $\widetilde{P} = \begin{bmatrix} 1/2 & 1/4 \\ 1/3 & 1/3 \end{bmatrix}$

(*d*) $P' = \begin{bmatrix} 3 & 1 \\ 1 & 0 \end{bmatrix}$

Chapter 10

10-1 (*a*) $-1 \pm 5j$; (*b*) $-1 \pm 5j$; (*c*) $-1 \pm 5j$; (*d*) 3, 3, 3; (*e*) 0, 0, 0; (*f*) 0, $\pm j$;
10-7 (*a*) No solution

(*b*) $\begin{bmatrix} 16 \\ -5 \\ 2 \end{bmatrix}$

(*c*) $\begin{bmatrix} 2 \\ 1/2 \\ 0 \end{bmatrix} + \text{span}\left\{\begin{bmatrix} 3 \\ 3/2 \\ -1 \end{bmatrix}\right\}$

(*d*) $\begin{bmatrix} 2 \\ 1 \end{bmatrix}$

10-9 (*a*) rank = 2, nullity = 1, $\mathcal{R}\{A\} = \text{span}\left\{\begin{bmatrix} 1 \\ 2 \\ 0 \end{bmatrix}\begin{bmatrix} 0 \\ -3 \\ 1 \end{bmatrix}\right\}$ $\quad \mathcal{N}\{A\} = \text{span}\left\{\begin{bmatrix} 2 \\ -1 \\ 0 \end{bmatrix}\right\}$

(*b*) rank = 3, nullity = 0, $\mathcal{R}\{A\} = \text{span}\left\{\begin{bmatrix} 1 \\ 0 \\ 0 \end{bmatrix}\begin{bmatrix} 0 \\ 1 \\ 0 \end{bmatrix}\begin{bmatrix} 0 \\ 0 \\ 1 \end{bmatrix}\right\}$ $\quad \mathcal{N}\{A\} = 0$

(*c*) rank = 2, nullity = 1, $\mathcal{R}\{A\} = \text{span}\left\{\begin{bmatrix} 0 \\ 1 \\ 1 \end{bmatrix}\begin{bmatrix} 2 \\ 0 \\ 2 \end{bmatrix}\right\}$ $\quad \mathcal{N}\{A\} = \text{span}\left\{\begin{bmatrix} 3 \\ 3/2 \\ -1 \end{bmatrix}\right\}$

(*d*) rank = 2, nullity = 0, $\mathcal{R}\{A\} = \text{span}\left\{\begin{bmatrix} 2 \\ 1 \\ 4 \end{bmatrix}\begin{bmatrix} -3 \\ 2 \\ 0 \end{bmatrix}\right\}$ $\quad \mathcal{N}\{A\} = 0$

10-11 (*a*) rank (A) = 2, rank (A, **d**) = 3; (*b*) rank (A) = rank (A, **d**) = 3; (*c*) rank (A) = rank (A, **d**) = 2; (*d*) rank (A) = rank (A, **d**) = 2.

10-13 (*a*)
$$\boldsymbol{\xi}_{1,2} = \begin{bmatrix} \dfrac{1 \pm 5j}{26} \\ -1 \end{bmatrix}$$

(*b*)
$$\boldsymbol{\xi}_{1,2} = \begin{bmatrix} -1 \mp 5j \\ -1 \end{bmatrix}$$

(*c*)
$$\boldsymbol{\xi}_{1,2} = \begin{bmatrix} \dfrac{-1 \pm 5j}{26} \\ -1 \end{bmatrix}$$

(*d*)
$$\boldsymbol{\xi}_1 = \begin{bmatrix} -1 \\ 0 \\ 0 \end{bmatrix}$$

(*e*)
$$\boldsymbol{\xi}_1 = \begin{bmatrix} -1 \\ 0 \\ 0 \end{bmatrix}$$

(*f*)
$$\boldsymbol{\xi}_1 = \begin{bmatrix} -1 \\ 0 \\ 0 \end{bmatrix}, \boldsymbol{\xi}_{2,3} = \begin{bmatrix} 0 \\ \pm j \\ -1 \end{bmatrix}$$

10-15 (*a*)
$$\lambda_1 = -1, \boldsymbol{\xi}_1 = \begin{bmatrix} 1 \\ -1 \end{bmatrix}, \lambda_2 = -4, \boldsymbol{\xi}_2 = \begin{bmatrix} 1/4 \\ -1 \end{bmatrix}$$

(*b*)
$$\lambda_1 = 2, \boldsymbol{\xi}_1 = \begin{bmatrix} -1/2 \\ -1 \end{bmatrix}, \lambda_2 = -2, \boldsymbol{\xi}_2 = \begin{bmatrix} 1/2 \\ -1 \end{bmatrix}$$

(*c*)
$$\lambda_1 = 0, \boldsymbol{\xi}_1 = \begin{bmatrix} -1 \\ 0 \end{bmatrix}, \lambda_2 = -4, \boldsymbol{\xi}_2 = \begin{bmatrix} 1/4 \\ -1 \end{bmatrix}$$

10-17 (*a*)
$$\boldsymbol{\xi}_1 = \begin{bmatrix} 1 \\ -1 \end{bmatrix}, \boldsymbol{\xi}_2 = \begin{bmatrix} 1 \\ -4 \end{bmatrix}$$

(*b*) $\xi_1 = \begin{bmatrix} 1 \\ 2 \end{bmatrix}, \xi_2 = \begin{bmatrix} 1 \\ -2 \end{bmatrix}$

(*c*) $\xi_1 = \begin{bmatrix} 1 \\ 0 \end{bmatrix}, \xi_2 = \begin{bmatrix} 1 \\ -4 \end{bmatrix}$

10-19 (*a*) $P = \begin{bmatrix} 1 & 1 \\ -1 & -4 \end{bmatrix}$

(*b*) $P = \begin{bmatrix} 1 & 1 \\ 2 & -2 \end{bmatrix}$

(*c*) $P = \begin{bmatrix} 1 & 1 \\ 0 & -4 \end{bmatrix}$

10-21 $\dot{\hat{\mathbf{x}}} = \begin{bmatrix} -1 + 3j & 0 \\ 0 & -1 - 3j \end{bmatrix}\hat{\mathbf{x}} + \begin{bmatrix} -jK/6 \\ jK/6 \end{bmatrix}, y = [-1 + 12j, -1 - 12j]\hat{\mathbf{x}}$

$\dot{\mathbf{x}}_b = \begin{bmatrix} -1 & 3 \\ -3 & -1 \end{bmatrix}\mathbf{x}_b + \begin{bmatrix} 0 \\ K/3 \end{bmatrix}u, y = [-1, 12]\mathbf{x}_b$

10-27 $P = \begin{bmatrix} -1 & 2 & 3 \\ 1 & -1 & -1 \\ -1 & 0 & 0 \end{bmatrix}$

10-29 $Q = \begin{bmatrix} 0.851 & 0.526 \\ 0.526 & -0.851 \end{bmatrix}$

10-33 $e^{At} = \text{diagonal}\,[e^{-t}, e^{(-2+3j)t}, e^{(-2-3j)t}]$.

10-35 (*a*) $e^{At} = e^{-t}\begin{bmatrix} 1 \\ -1 \end{bmatrix}[4/3, 1/3] + e^{-4t}\begin{bmatrix} 1/4 \\ -1 \end{bmatrix}[-4/3, -4/3]$

(*b*) $e^{At} = e^{2t}\begin{bmatrix} -1/2 \\ -1 \end{bmatrix}[-1, -1/2] + e^{-2t}\begin{bmatrix} 1/2 \\ -1 \end{bmatrix}[1, -1/2]$

(*c*) $e^{At} = \begin{bmatrix} -1 \\ 0 \end{bmatrix}[-1, -1/4] + e^{-4t}\begin{bmatrix} 1/4 \\ -1 \end{bmatrix}[0, -1]$

10-37 (*a*) $\text{diagonal}\left[\begin{bmatrix} -1 & 2 \\ -2 & -1 \end{bmatrix}, \begin{bmatrix} -5 & 4 \\ -4 & -5 \end{bmatrix}, -3\right]$

(*b*) $\text{diagonal}\left[\begin{bmatrix} \cos 2t & \sin 2t \\ -\sin 2t & \cos 2t \end{bmatrix}e^{-t}, \begin{bmatrix} \cos 4t & \sin 4t \\ -\sin 4t & \cos 4t \end{bmatrix}e^{-5t}, e^{-3t}\right]$

10-39 $e^{At} = \begin{bmatrix} e^{-2t} & 0 & 0 & 0 \\ 0 & e^{-3t} & te^{-3t} & t^2/2\, e^{-3t/2} \\ 0 & 0 & e^{-3t} & te^{-3t} \\ 0 & 0 & 0 & e^{-3t} \end{bmatrix}$

10-43 $e^{At} = \begin{bmatrix} 1 & 0 \\ 0 & 1 \end{bmatrix}e^{-t}\cos 3t + \begin{bmatrix} 1/3 & 1/3 \\ -10/3 & -1/3 \end{bmatrix}e^{-t}\sin 3t$

10-45 $\begin{bmatrix} 5/4 & 1/2 & 1/4 \\ -5/4 & -1/2 & -1/4 \\ 5/4 & 1/2 & 1/4 \end{bmatrix}e^{-t} + \begin{bmatrix} -1/4 & -1/2 & -1/4 \\ 5/4 & 3/2 & 1/4 \\ -5/4 & -1/2 & 3/4 \end{bmatrix}e^{-t}\cos 2t$

$$+\begin{bmatrix} 1/2 & 1/2 & 0 \\ 0 & 1/2 & 1/2 \\ -5/2 & -7/2 & -1 \end{bmatrix} e^{-t}\sin 2t$$

10-47 (*a*)
$$\begin{bmatrix} 1 & 1 & -3 & -11 \\ 0 & 2 & 4 & -2 \\ 1 & 0 & 0 & 0 \\ 0 & 1 & 0 & 0 \end{bmatrix}\begin{bmatrix} \alpha_0(t) \\ \alpha_1(t) \\ \alpha_2(t) \\ \alpha_3(t) \end{bmatrix} = \begin{bmatrix} e^t\cos 2t \\ e^t \sin 2t \\ 1 \\ t \end{bmatrix}$$

(*b*)
$$\begin{bmatrix} 1 & 0 & -1 & 0 \\ 0 & 1 & 0 & -1 \\ 1 & 1 & 0 & -2 \\ 0 & 1 & 2 & 2 \end{bmatrix}\begin{bmatrix} \alpha_0(t) \\ \alpha_1(t) \\ \alpha_2(t) \\ \alpha_3(t) \end{bmatrix} = \begin{bmatrix} \cos t \\ \sin t \\ e^t\cos t \\ e^t \sin t \end{bmatrix}$$

10-49 $a_0 = 5$, $a_1 = 2$.

10-51
$$\begin{bmatrix} 1 & \lambda_1 & \cdots & \lambda_1^n \\ 1 & \lambda_2 & \cdots & \lambda_2^n \\ \vdots & & & \\ 1 & \lambda_n & \cdots & \lambda_n^n \end{bmatrix}\begin{bmatrix} \beta_1(t) \\ \beta_2(t) \\ \vdots \\ \beta_n(t) \end{bmatrix} = \begin{bmatrix} \sin(\lambda_1 t) \\ \sin(\lambda_2 t) \\ \vdots \\ \sin(\lambda_n t) \end{bmatrix}$$

Chapter 11

11-1 (*a*) *cc*; (*b*) *cc*; (*c*) *uc*; (*d*) *cc*.

11-3 (*a*), (*b*), and (*d*) $\mathcal{R}\{M_c\} = R^3$, $\mathcal{N}\{M_c^T\} = \mathbf{0}$.

(*c*)
$$\mathcal{R}\{M_c\} = \text{span}\left\{\begin{bmatrix} 1 \\ 0 \\ 0 \end{bmatrix}\right\}, \quad \mathcal{N}\{M_c^T\} = \text{span}\left\{\begin{bmatrix} 0 \\ 1 \\ 0 \end{bmatrix}\begin{bmatrix} 0 \\ 0 \\ 1 \end{bmatrix}\right\}$$

11-5 (*a*) *uo*; (*b*) *uo*; (*c*) *co*; (*d*) *uo*.

11-7 (*a*), (*b*), and (*d*)
$$\mathcal{R}\{M_o^T\} = \text{span}\left\{\begin{bmatrix} 0 \\ 1 \\ 0 \end{bmatrix}\begin{bmatrix} 0 \\ 0 \\ 1 \end{bmatrix}\right\}, \quad \mathcal{N}\{M_o\} = \text{span}\left\{\begin{bmatrix} 1 \\ 0 \\ 0 \end{bmatrix}\right\}$$

(*c*) $\mathcal{R}\{M_o^T\} = R^3$, $\mathcal{N}\{M_o\} = \mathbf{0}$

11-9 (*a*) $H(s) = \dfrac{s}{(s+1)(s+2)}$

(*b*) $H(s) = \dfrac{s+3}{s^2+3s+2}$

(*c*) $H(s) = 1/s$

(*d*)
$$H(s) = \begin{bmatrix} 0 & \dfrac{s+3}{(s+1)(s+2)} \\ 0 & \dfrac{-2}{(s+1)(s+2)} \end{bmatrix}$$

11-11 For each, $\hat{A} = \text{diagonal}\,[0, -1, -2]$

(*a*) $\hat{B} = [1/2, -1, 1/2]^T$, $\hat{C} = [0, 1, 4]$

(*b*) $\hat{B} = [3/2, -2, 1/2]^T$, $\hat{C} = [0, -1, -2]$

(*c*) $\hat{B} = [1, 0, 0]^T$, $\hat{C} = [1, 1, 1]$

(*d*) $$\widehat{B} = \begin{bmatrix} 1 & 3/2 \\ 0 & -2 \\ 0 & 1/2 \end{bmatrix}, \quad \widehat{C} = \begin{bmatrix} 0 & -1 & -2 \\ 0 & 1/2 & 1/2 \end{bmatrix}$$

11-13 (*a*) x_2 is *cc/co*; x_1 is *cc/uo*; x_4 is *uc/co*; x_3 is *uc/uo*.

11-15 (*a*) 2; (*b*) yes; if there is a poor choice of T.

11-25 $\dot{W}_1 = -20W_2 + 0.01$; $\dot{W}_2 = -10W_2 - 10W_3 + 0.1$; $\dot{W}_3 = 2W_2 - 22W_3 + 1$.

11-27 (*a*) $$P = \begin{bmatrix} 1 & -1.1 \\ -0.1 & 1.1 \end{bmatrix}$$

(*b*) $$\widehat{W}_c(t) = \begin{bmatrix} \dfrac{1 - e^{-2t}}{2} & \dfrac{1 - e^{-11t}}{11} \\ \dfrac{1 - e^{-11t}}{11} & \dfrac{1 - e^{-20t}}{20} \end{bmatrix}$$

(*c*) $$\widehat{W}_{c_{ss}} = \begin{bmatrix} 1/2 & 1/11 \\ 1/11 & 1/20 \end{bmatrix}$$

(*d*) $\widehat{\lambda}_1 = 0.0323$, $\widehat{\boldsymbol{\xi}}_1 = [-0.1908, 0.9816]^T$
$\widehat{\lambda}_2 = 0.5177$, $\widehat{\boldsymbol{\xi}}_2 = [0.9816, 0.1908]^T$

(*e*) $$W'_{c_{ss}} = \begin{bmatrix} 0.3605 & -0.0005 \\ -0.0005 & 0.0455 \end{bmatrix}$$

(*f*) $\lambda'_1 = 0.3605$, $\boldsymbol{\xi}'_1 = [1, 0]^T$
$\lambda'_2 = 0.0455$, $\boldsymbol{\xi}'_2 = [0, 1]^T$

(*g*) Yes

(*h*) $\|u^*\|^2 = 2.774$ for $\mathbf{x} = \boldsymbol{\xi}'_1$

$\|u^*\|^2 = 21.98$ for $\mathbf{x} = \boldsymbol{\xi}'_2$

Chapter 12

12-1 $$\Phi(t, t_0) = \begin{bmatrix} e^{-(t-t_0)} & 0 \\ e^{-20t}e^{t_0}(e^{21t} - e^{21t_0})/21 & e^{-20(t-t_0)} \end{bmatrix}$$

12-3 $h(t, t_0) = te^{-(t-t_0)}$

12-5 (*a*) $y_{zi}(t) = te^{-t}$; (*b*) $y_{zs}(t) = -te^{-t} - e^{-t}$; (*c*) $y(t) = -e^{-t}$.

12-11 $\dot{W}_{c_1}(t, t_0) = -2W_{c_1}(t, t_0) + 1$
$\dot{W}_{c_2}(t, t_0) = e^{2t}W_{c_1}(t, t_0) - 21W_{c_2}(t, t_0)$
$\dot{W}_{c_3}(t, t_0) = 2e^{2t}W_{c_2}(t, t_0) - 40W_{c_3}(t, t_0)$
$W_{c_1}(t_0, t_0) = W_{c_2}(t_0, t_0) = W_{c_3}(t_0, t_0) = 0$

12-15 $\dot{W}_{o_1}(t_f, t) = 2W_{o_1}(t_f, t) - 2e^{2t}W_{o_2}(t_f, t) - t^2$
$\dot{W}_{o_2}(t_f, t) = 21W_{o_2}(t_f, t) - e^{2t}W_{o_3}(t_f, t)$
$\dot{W}_{o_3}(t_f, t) = 40W_{o_3}(t_f, t)$
$W_{o_1}(t_f, t_f) = W_{o_2}(t_f, t_f) = W_{o_3}(t_f, t_f) = 0$

12-17 (*a*) Unstable; (*b*) stable; (*c*) true.

12-19 $$\mathbf{x}_T(k) = \begin{bmatrix} 1 - T & 0 \\ e^{2kT} & 1 - 20T \end{bmatrix}\mathbf{x}_T(k-1) + \begin{bmatrix} T \\ 0 \end{bmatrix}u_T(k-1)$$

12-21 (*a*) $DX(1) = -X(1) + u(t)$ $\quad DX(4) = e^{2t}X(3) - 20X(4)$

$DX(2) = e^{2t}X(1) - 20X(2)$ $\quad DX(5) = -X(5)$

$DX(3) = -X(3)$ $\quad DX(6) = e^{2t}X(5) - 20X(6)$

(*b*) $X(1) = x_1(t_0)$ $\quad X(2) = x_2(t_0)$

$X(3) = X(6) = 1$ $\quad X(4) = X(5) = 0$

12-23 $DX(7) = -2X(7) + 1$ $\quad DX(9) = 2e^{2t}X(8) - 40X(9)$

$DX(8) = e^{2t}X(7) - 21X(8)$

12-25 (*a*) $DX(1) = 2X(1) - 2e^{2t}X(3) - t^2$ $\quad DX(3) = 40X(3)$

$DX(2) = 21X(2) - e^{2t}X(4)$

(*b*) $X(1) = X(2) = X(3) = 0$ at $t = t_f$.

(*c*) No. These integrate backward.

12-27

$$\begin{bmatrix} \delta\dot{x}_1(t) \\ \delta\dot{x}_2(t) \end{bmatrix} =$$

$$\begin{bmatrix} -\sin u_{\text{nom}}(t) & 0 \\ 2x_{1_{\text{nom}}}(t) - x^2_{2_{\text{nom}}}(t)\cos x_{1_{\text{nom}}}(t) & -2x_{2_{\text{nom}}}(t)\sin x_{1_{\text{nom}}}(t) \end{bmatrix}$$

$$\times \begin{bmatrix} \delta x_1(t) \\ \delta x_2(t) \end{bmatrix} + \begin{bmatrix} -x_{1_{\text{nom}}}(t)\cos u_{\text{nom}}(t) + 1 \\ 0 \end{bmatrix} \delta u(t)$$

$$\delta y(t) = [x_{2_{\text{nom}}}(t), x_{1_{\text{nom}}}(t)]\delta\mathbf{x}(t)$$

12-29 (*a*) $\dot{\mathbf{x}}_{\text{nom}} = 0$

(*b*)

$$\delta\dot{\mathbf{x}} = \begin{bmatrix} -1 & 0 \\ \pi & -\pi \end{bmatrix}\delta\mathbf{x} + \begin{bmatrix} 1 \\ 0 \end{bmatrix}\delta u$$

$$\delta y = \left[\frac{\pi}{2}, \frac{\pi}{2}\right]\delta\mathbf{x}$$

(*c*) Stable.

12-31 (*b*)

$$\delta\dot{\mathbf{x}} = \begin{bmatrix} 1 & 0 \\ \pi & -\pi \end{bmatrix}\delta\mathbf{x} + \begin{bmatrix} 1 \\ 0 \end{bmatrix}\delta u$$

$$\delta y = \left[\frac{\pi}{2}, \frac{\pi}{2}\right]\delta\mathbf{x}$$

(*c*) Unstable.

INDEX

INDEX